Harold J. Benson
Pasadena City College

Stanley E. Gunstream
Pasadena City College

Arthur Talaro
Pasadena City College

Kathleen P. Talaro
Pasadena City College

Intermediate Cat Version

ANATOMY &
PHYSIOLOGY
Laboratory Textbook

Fifth Edition

D1359794

Boston   Burr Ridge, IL   Dubuque, IA   Madison, WI   New York   San Francisco   St. Louis
Bangkok   Bogotá   Caracas   Lisbon   London   Madrid   Mexico City   Milan
New Delhi   Seoul   Singapore   Sydney   Taipei   Toronto

McGraw-Hill Higher Education

A Division of The McGraw-Hill Companies

ANATOMY & PHYSIOLOGY LABORATORY TEXTBOOK: INTERMEDIATE
CAT VERSION, FIFTH EDITION

Copyright © 2000, 1996 by The McGraw-Hill Companies, Inc. All rights reserved. Printed in the United States of America. Except as permitted under the United States Copyright Act of 1976, no part of this publication may be reproduced or distributed in any form or by any means, or stored in a data base or retrieval system, without the prior written permission of the publisher.

 This book is printed on recycled, acid-free paper containing 10% postconsumer waste.

1 2 3 4 5 6 7 8 9 0 QPD/QPD 0 9 8 7 6 5 4 3 2 1 0

ISBN 0–697–34231–X

Vice president and editorial director: *Kevin T. Kane*
Publisher: *Colin H. Wheatley*
Sponsoring editor: *Kristine Tibbetts*
Developmental editor: *Patrick F. Anglin*
Marketing manager: *Heather K. Wagner*
Editing associate: *Joyce Watters*
Senior production supervisor: *Mary E. Haas*
Coordinator of freelance design: *Rick Noel*
Supplement coordinator: *Stacy A. Patch*
Compositor: *Carlisle Communications, Ltd.*
Typeface: *11/12 Times Roman*
Printer: *Quebecor Printing Book Group/Dubuque, IA*

Covor photo: © *Dr. Dennis Kunkel/Phototake NYC*

Anatomy and Physiology Laboratory Textbooks

Benson, Gunstream, Talaro, Talaro	Complete Version—Cat, 7th edition Intermediate Version—Cat, 5th edition Intermediate Version—Fetal Pig, 4th edition Short Version, 7th edition
Gunstream, Benson, Talaro, Talaro	Essentials Version, 2nd edition
Benson, Talaro	Human Anatomy, 6th edition

Some of the laboratory experiments included in this text may be hazardous if materials are handled improperly or if procedures are conducted incorrectly. Safety precautions are necessary when you are working with chemicals, glass test tubes, hot water baths, sharp instruments, and the like, or for any procedures that generally require caution. Your school may have set regulations regarding safety procedures that your instructor will explain to you. Should you have any problems with materials or procedures, please ask your instructor for help.

www.mhhe.com

Contents

Contents

In this fifth edition of the intermediate cat version of the *Anatomy and Physiology Laboratory Textbook* we have done the following: (1) expanded the Histology Atlas to incorporate new material, (2) provided four histology self-quizzes, (3) expanded the instructions for using Intelitool equipment, (4) upgraded many of the anatomical illustrations, (5) updated certain concepts that shed new light on previously little understood theories, (6) shortened some exercises and transferred some material to the Instructor's Handbook to make room for new material, and (7) changed many illustrations to improve visualization.

The Histology Atlas has been expanded to forty pages from thirty-six. Although much of the atlas remains essentially the same, additions and alterations were made that relate to the skin, tongue, respiratory passages, lungs, female reproductive organs, and spermatogenesis.

One of the shortcomings of previous editions has been the inability to measure the extent of a student's comprehension of histology. Although some questions in the Laboratory Reports do address this problem, we have felt for some time that more testing was needed.

At many institutions students are subjected to laboratory practical examinations in which microscopes are set up with various types of tissues displayed for identification by the student. This type of exam can be quite traumatic to the student the first time it is experienced. To give the student an opportunity to see how he or she might do on such an exam, we have developed four "self-quizzes." In laboratories where practical exams are used, these self-quizzes can do much to show the student ahead of time what types of questions might be asked. Even if no laboratory practical is used, the self-quizzes are helpful review tests.

The four Self-Quizzes are located at the end of Exercise 69. They encompass eighty-six microscope setups with 261 questions. Answers to the quizzes are located on pages 196 and 298. At appropriate places in the manual students are prompted to take each quiz.

Another shortcoming of the previous edition was that we didn't have complete instructions for performing Intelitool experiments on all computer platforms. At the time of publication of the fourth edition, only instructions for the Apple II were available.

Since then software and instructions for IBM (or compatible PC) and Macintosh have been developed. On one system (Spirocomp) instructions for Windows 95 are also available. In this edition we have included instructions for running all the Intelitool experiments on all platforms.

The expansion of the Intelitool exercises and the addition of the histology self-quizzes has resulted in the addition of thirty-two pages of new material to this edition. To accommodate this information without increasing the size of the book we have eliminated some less important material, compacted some exercises, and transferred some information to the Instructor's Handbook. The bar codes for the *Slice of Life* discs fall in this last category. Instructors who use these bar codes may wish to provide students with photocopies of the codes.

One other item that was transferred from the manual to the Instructor's Handbook was the former appendix B that contains recipes for the various solutions and reagents that are used in the experiments. Since students don't generally need this information, it was felt that placement in the I.H. was appropriate.

In previous editions a set of $2'' \times 2''$ Kodachrome slides pertaining to the Histology Atlas illustrations were made available to users of the manual. They are still available to those who do not have a set. A legend explaining the slides is provided in the back of the Instructor's Handbook, which is helpful when projecting the slides to groups of students. The slides can be had at no cost by simply contacting the Educational Services Department at McGraw-Hill in Dubuque, IA.

The changes in this edition were suggested by the following individuals who have used this book in the past: Michele A. Finn of Jamestown Community College, Olean, NY; Susan K. Gilmore of the University of Pittsburgh, Bradford, PA; Paul Mickelson of Fond Du Lac Community College, Cloquet, MN; Pat Palanker of Middlesex County College, Edison, NJ; Brian W. Hill of Freed-Hardeman University, Henderson, TN; Robert D. Moldenhauer of St. Clair County Community College, Port Huron, MI; Alyce M. St. Pierre of North Shore Community College, Lynn, MA; Donna Sasnow of South Surburban College, South Holland, IL; James Ezell and Barbara L. Stewart of

Sargeant Reynolds Community College, Richmond, VA; Ralph E. Reiner of College of the Redwoods, Eureka, CA; A. Scott Helgeson of Des Moines Area Community College, Des Moines, IA: Juville Dario-Becker of Central Virginia Community College, Lynchburg, VA; Donna Sasnow of South Surburban College and Marian G. Langer of St. Francis College, Loretto, PA. Our sincerest gratitude is extended to these individuals for their valuable assistance. Although we may not have been able to incorporate all the changes requested, most suggestions have been included in this edition.

These laboratory exercises have been developed to provide you with a basic understanding of anatomical and physiological principles that underlie medicine, nursing, dentistry, and other related health professions. Laboratory procedures that reflect actual clinical practices are included wherever feasible. In each exercise you will find essential terminology that will become part of your working vocabulary. Mastery of all concepts, vocabulary, and techniques will provide you with a core of knowledge crucial to success in your chosen profession.

During the first week of this course your instructor will provide you with a schedule of laboratory exercises in the order of their performance. There is an implied expectation that you will have familiarized yourself with the content of each experiment prior to that week's session, thus ensuring that you will be properly prepared so as to minimize disorganization and mistakes.

The *Laboratory Reports* coinciding with each exercise are located at the back of the book. They are perforated for easy removal; be sure to remove each sheet as necessary. This will facilitate data collection, completion of answers, and grading. Your instructor may give further procedural details on the handling of these reports.

The exercises in this laboratory guide consist essentially of four kinds of activities: (1) illustration labeling, (2) anatomical dissections, (3) physiological experiments, and (4) microscopic studies. The following suggestions should be helpful in performing these assignments.

Labeling The activity of labeling illustrations is essentially a determination of your understanding of the written text. Since all labeled structures are explicitly described in the manual, all that is necessary is to read the manual very carefully. Incorrectly labeled illustrations usually indicate a lack of comprehension.

Once the illustrations are labeled they can be useful to you in two other ways. First, the illustrations may be used for reference purposes in dissections or examinations of anatomical specimens. This is particularly true in the skeletal and nervous systems. Second, the illustrations can be used for review purposes. If you cover the number legend of the labels as you mentally attempt to name the structures on the illustration, you can easily determine your level of understanding. Periodic reviews of this type during the semester will be very helpful.

Usually the labeling of illustrations will be performed *prior* to coming to the laboratory. In this way the laboratory time will be used primarily for dissections, experimentation, or microscopic examinations.

Dissections Although cadavers provide the ideal dissection specimens in human anatomy, they are expensive and thus may be impractical for large classes in introductory anatomy and physiology. Since much of cat anatomy is similar to human anatomy, it has been selected as the primary dissection specimen in this manual. The rat will also be used. Occasionally, sheep and beef organs will be studied. Frogs will frequently be used in physiological experiments.

When using live animals in experimental procedures it is imperative that they be handled with great care. Consideration must be exercised to minimize pain in all experiments on vertebrate animals. Inconsiderate or haphazard treatment of any animal will not be tolerated.

Physiological Experiments Before performing any physiological experiments be sure that you understand the overall procedure. Reading the experiment prior to entering the laboratory will help a great deal.

Handle all instruments carefully. Most pieces of equipment are expensive, may be easily damaged, and are often irreplaceable. The best insurance against breakage or damage is to thoroughly understand how the equipment is expected to function.

Maintain astuteness in observations and record keeping. Record data immediately; postponement detracts from precision. Insightful data interpretation will also be expected.

Microscopic Studies Cytological and histological studies will be made to lend meaning to text descriptions. Familiarize yourself with the contents of the *Histology Atlas,* which includes photomicrographs of most of the tissues you will study in this course. Note that the atlas is located in the middle of the book and is readily located due to the color band on the edges of its pages. If drawings are required, execute them with care, and label those structures that are significant.

Histology Self-Quizzes   Periodically you will be prompted to test your comprehension of histological studies by taking *Histology Self-Quizzes,* which are located on pages 407 to 422. Note that there are four of them. The answers to each self-quiz are included on designated pages so that you can determine for yourself your degree of understanding. You should find them quite helpful, particularly if lab practical exams are a part of this course.

Laboratory Efficiency   Success in any science laboratory situation requires a few additional disciplines:

1. Always follow the instructor's verbal comments at the beginning of each laboratory session. It is at this time that difficulties will be pointed out, group assignments will be made, and procedural changes will be announced. Take careful notes on substitutions or changes in methods or materials.
2. In view of the above statement, it is obvious that the beginning of each laboratory period is a critical time. It is for this reason that tardiness is intolerable.
3. Keep your work area tidy. Books, bags, purses, and extraneous supplies should be located away from the work area. Tidiness should also extend to assembly of apparatus.
4. Abstain from eating, drinking, or smoking within the confines of the laboratory.
5. Report immediately to the instructor any injuries that occur.
6. Be serious-minded and methodical. Horseplay, silliness, or flippancy will not be tolerated during experimental procedures.
7. Work independently, but cooperatively, when performing team experiments. Attend to your assigned responsibility, but be willing to lend a hand to others where necessary. Participation and development of laboratory techniques are an integral part of the course.

Laboratory Reports   When seeking answers to the questions and problems on the Laboratory Reports, work independently. The effort you expend to complete these reports is as essential as doing the experiment. The easier route of letting someone else solve the problems for you will handicap you at examination time. You are taking this course to learn anatomy and physiology. No one else can learn it for you.

Some of the laboratory experiments included in this book may be hazardous if materials are handled improperly or if procedures are conducted incorrectly. Safety precautions are necessary when you are working with chemicals, glass test tubes, hot water baths, sharp instruments, and the like, or for any procedures that generally require caution. Your school may have set regulations regarding safety procedures that your instructor will explain to you. Should you have any problems with materials or procedures, please ask your instructor for help.

Anatomical Terminology

Anatomical description would be extremely difficult without specific terminology. A consensus prevails among many students that anatomists synthesize multisyllabic words in a determined conspiracy to harass the beginner's already overburdened mind. Naturally, nothing could be further from the truth.

Scientific terminology is created out of necessity. It functions as a precise tool that allows us to say a great deal with a minimum of words. Conciseness in scientific discussion not only saves time, but also promotes clarity of understanding.

Most of the exercises in this laboratory manual employ the terms defined in this exercise. They are used liberally to help you to locate structures that are to be identified on the illustrations. If you do not know the exact meanings of these words, obviously you will be unable to complete the required assignments. Before you attempt to label any of the illustrations in this exercise, read the text material first.

Relative Positions

Descriptive positioning of one structure with respect to another is accomplished with the following pairs of words. Their Latin or Greek derivations are provided to help you understand their meanings.

Superior and Inferior   These two words denote vertical levels of position. The Latin word *super* means *above;* thus, a structure that is located above another one is said to be superior. Example: The nose is *superior* to the mouth.

The Latin word *inferus* means *below* or *low;* thus, an inferior structure is one that is below or under some other structure. Example: The mouth is *inferior* to the nose.

Anterior and Posterior   Fore and aft positioning of structures is described with these two terms. The word anterior is derived from the Latin *ante,* meaning *before.* A structure that is anterior to another one is in front of it. Example: Bicuspids are *anterior* to molars.

Anterior surfaces are the most forward surfaces of the body. The front portions of the face, chest, and abdomen are anterior surfaces.

Posterior is derived from the Latin *posterus,* which means *following.* The term is the opposite of anterior. Example: The molars are *posterior* to the bicuspids.

When these two terms are applied to the surfaces of the hand and arm, it is assumed that the body is in the **anatomic position,** as shown in figures 1.1 and 1.2. In the anatomic position the individual is standing upright with the face, palms of the hands, and toes facing forward.

Cranial and Caudal   When describing the location of structures of four-legged animals, these terms are often used in place of anterior and posterior. Since the word *cranial* pertains to the skull (Greek: *kranion,* skull), it may be used in place of anterior. The word *caudal* (Latin: *cauda,* tail) may be used in place of posterior.

Dorsal and Ventral   These terms, as used in comparative anatomy of animals, assume all animals, including humans, to be walking on all fours. The dorsal surfaces are thought of as *upper* surfaces and the ventral surfaces as *underneath* surfaces.

The word *dorsal* (Latin: *dorsum,* back) not only applies to the back of the trunk of the body, but may also be used in describing the back of the head and the back of the hand.

Standing in a normal posture, a man's dorsal surfaces become posterior. A four-legged animal's back, on the other hand, occupies a superior position.

The word *ventral* (Latin: *venter,* belly) generally pertains to the abdominal and chest surfaces. However, the underneath surfaces of the head and feet of four-legged animals are also often referred to as ventral surfaces. Likewise, the palm of the hand may also be referred to as being ventral.

Proximal and Distal   These terms are used to describe parts of a structure with respect to its point of attachment to some other structure. In the case of the arm or leg, the point of reference is where the limb is attached to the trunk of the body. In the case of a finger, the point of reference is where it is attached to the palm of the hand.

Proximal (Latin: *proximus,* nearest) refers to that part of the limb nearest to the point of attachment. Example: The upper arm is the *proximal* portion of the arm.

Distal (Latin: *distare,* to stand apart) means just the opposite of proximal. Anatomically, the distal portion of a limb or other part of the body is that portion that is most remote from the point of reference (attachment). Example: The hand is *distal* to the arm.

Medial and Lateral   These two terms are used to describe surface relationships with respect to the median line of the body. The *median line* is an imaginary line on a plane that divides the body into right and left halves.

The term *medial* (Latin: *medius,* middle) is applied to surfaces of structures that are closest to the median line. The medial surface of the arm, for example, is the surface next to the body because it is closest to the median line.

As applied to the appendages, the term *lateral* is the opposite of medial. The Latin derivation of this word is *lateralis,* which pertains to *side.* The lateral surface of the arm is the outer surface, or that surface farthest away from the median line. The sides of the head are said to be lateral surfaces.

Body Sections

To observe the structure and relative positions of internal organs it is necessary to view them in sections that have been cut through the body. Considering the body as a whole, there are only three planes to identify. Figure 1.1 shows these sections.

Sagittal Sections   A section parallel to the long axis of the body (longitudinal section) that divides the body into right and left sides is a *sagittal section.* If such a section divides the body into equal halves, as in figure 1.1, it is said to be a *midsagittal section.*

Frontal Section   A longitudinal section that divides the body into front and back portions is a *frontal* or *coronal* section. The other longitudinal section seen in figure 1.1 is of this type.

Transverse Sections   Any section that cuts through the body in a direction that is perpendicular to the long axis is a *transverse* or *cross section.* This is the third section shown in figure 1.1. In this case it is parallel to the ground.

Although these sections have been described here only in relationship to the body as a whole,

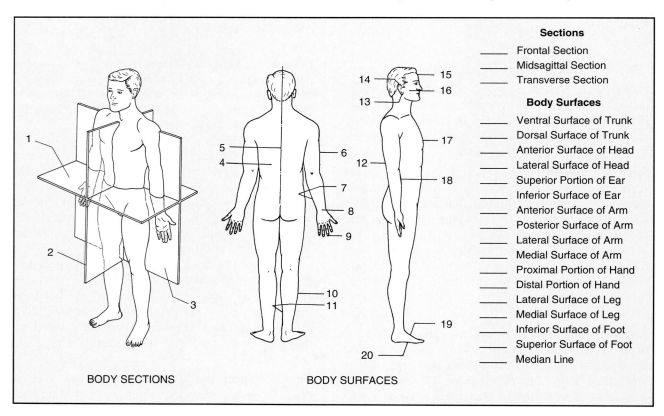

Sections
_____ Frontal Section
_____ Midsagittal Section
_____ Transverse Section

Body Surfaces
_____ Ventral Surface of Trunk
_____ Dorsal Surface of Trunk
_____ Anterior Surface of Head
_____ Lateral Surface of Head
_____ Superior Portion of Ear
_____ Inferior Surface of Ear
_____ Anterior Surface of Arm
_____ Posterior Surface of Arm
_____ Lateral Surface of Arm
_____ Medial Surface of Arm
_____ Proximal Portion of Hand
_____ Distal Portion of Hand
_____ Lateral Surface of Leg
_____ Medial Surface of Leg
_____ Inferior Surface of Foot
_____ Superior Surface of Foot
_____ Median Line

BODY SECTIONS   BODY SURFACES

Figure 1.1   **Body sections and surfaces.**

they can be used on individual organs such as the arm, finger, or tooth.

Assignment:

To test your understanding of the descriptive terminology, identify the labels in figure 1.1 by placing the correct numbers in front of the terms to the right of the illustrations. Also, record these numbers on the Laboratory Report.

Regional Terminology

Various terms such as *flank, groin, brachium,* and *epigastric* have been applied to specific regions of the body to facilitate localization. Figures 1.2 and 1.3 pertain to some of the more predominantly used terminology.

Trunk

The anterior surface of the trunk may be subdivided into two pectoral, two groin, and the abdominal regions. The upper chest region is divided into two **pectoral** or **mammary** regions. Between the two pectoral regions is a long narrow strip called the **sternal** area. The **abdominal** region is the anterior trunk area below the diaphragm, and the area of the abdomen that surrounds the navel is the **umbilical** region. Where the thighs of the legs meet the abdomen are the **groin,** or **inguinal** areas, and the area where the reproductive organs are located is referred to as the **genital** region. A small area between the thighs, which is between the anus and reproductive organs, is called the **perineum.** In females it is equivalent to the pelvic outlet.

The posterior surface or dorsum of the trunk can be differentiated into the costal, lumbar, and buttocks regions. The **costal** (Latin: *costa,* rib) portion is the part of the dorsum that lies over the rib cage. The lower back region between the ribs and hips is the **lumbar** or **loin** region. Since the buttocks are formed by the gluteal muscles, the buttocks or rump area should be called the **gluteal** region. The hip region is the **coxal** area.

The side of the trunk that adjoins the lumbar region is called the **flank.** The armpit region that is between the trunk and upper arm is the **axilla.**

Upper Extremities

Each upper extremity consists of the arm and hand. The arm is defined as that portion from the shoulder to the **carpus** (wrist). To differentiate the parts of

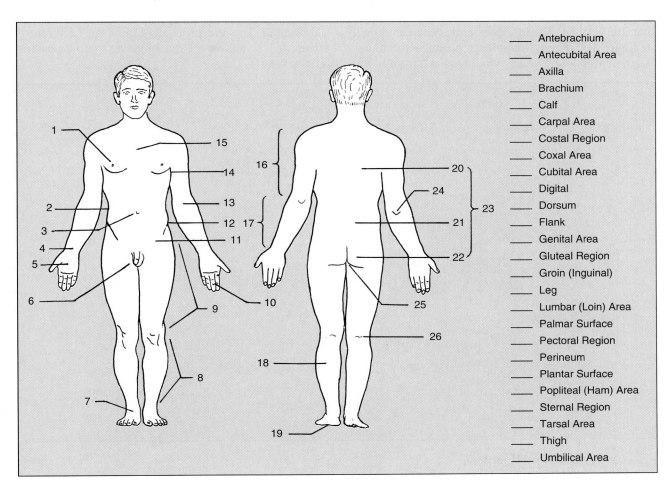

_____ Antebrachium
_____ Antecubital Area
_____ Axilla
_____ Brachium
_____ Calf
_____ Carpal Area
_____ Costal Region
_____ Coxal Area
_____ Cubital Area
_____ Digital
_____ Dorsum
_____ Flank
_____ Genital Area
_____ Gluteal Region
_____ Groin (Inguinal)
_____ Leg
_____ Lumbar (Loin) Area
_____ Palmar Surface
_____ Pectoral Region
_____ Perineum
_____ Plantar Surface
_____ Popliteal (Ham) Area
_____ Sternal Region
_____ Tarsal Area
_____ Thigh
_____ Umbilical Area

Figure 1.2 **Regional terminology.**

the arm, the term **brachium** is used for the upper arm and **antebrachium** for the forearm (between the elbow and carpus). The elbow area on the posterior surface of the arm is the **cubital** area. That area on the opposite side of the elbow is the **antecubital** area. It is also correct to refer to the entire anterior surface of the antebrachium as being antecubital.

The anterior surfaces of the hands are called **palmar** surfaces and the fingers are **digital**.

Lower Extremities

The lower extremity consists of the thigh, leg, tarsus, and foot. The upper portion of this limb is the **thigh,** and the lower portion from the knee to the ankle is the **leg.** The posterior surface of the leg is the **calf.**

Between the thigh and the leg on the posterior surface, opposite to the knee, is a depression called the **ham** or **popliteal** region. The ankle is the **tarsal** area and the sole of the foot is the **plantar** surface.

Abdominal Divisions

The abdominal surface may be divided into quadrants or into nine distinct areas. To divide the abdomen into nine regions one must establish four imaginary planes: two that are horizontal and two that are vertical. These planes and areas are shown in figure 1.3. The **transpyloric plane** is the upper horizontal plane, which would pass through the lower portion of the stomach (pyloric portion). The **transtubercular plane** is the other horizontal plane that touches the top surfaces of the hipbones (iliac crests). The two vertical planes, or **right** and **left lateral planes,** are approximately halfway between the midsagittal plane and the crests of the hips.

The planes describe the umbilical, epigastric, hypogastric, hypochondriac, and lumbar regions. The **umbilical** area lies in the center, includes the navel, and is bordered by the two horizontal and two vertical planes. Immediately above the umbilical area is the **epigastric,** which covers much of the stomach. Below the umbilical zone is the **hypogastric,** or *pubic area.* On each side of the epigastric are right and left **hypochondriac** areas, and beneath the hypochondriac areas are the right and left **lumbar** areas. (Note that although we tend to think of only the lower back as being the lumbar region, we see here that it extends around to the anterior surface as well.)

Assignment:
Label figures 1.2 and 1.3 and transfer these numbers to the Laboratory Report.

Laboratory Report

After transferring all the labels from figures 1.1 through 1.3 to the proper columns on Laboratory Report 1,2, answer the questions that pertain to this exercise.

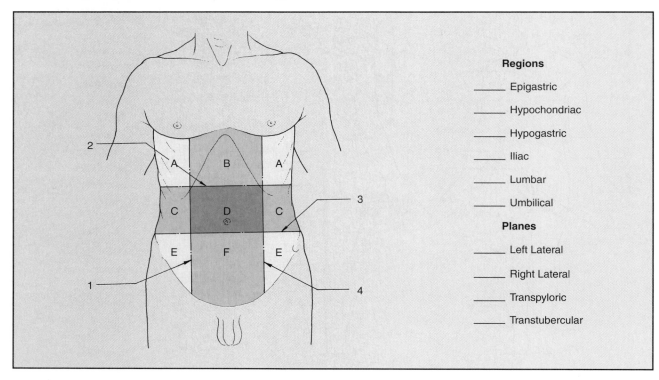

Regions
_____ Epigastric
_____ Hypochondriac
_____ Hypogastric
_____ Iliac
_____ Lumbar
_____ Umbilical

Planes
_____ Left Lateral
_____ Right Lateral
_____ Transpyloric
_____ Transtubercular

Figure 1.3 Abdominal regions.

Body Cavities and Membranes

2

All the internal organs *(viscera)* are contained in body cavities, which are completely or partially lined with smooth membranes. The relationships of these cavities to each other, the organs they contain, and the membranes that line them will be studied in this exercise.

Body Cavities

Figure 2.1 illustrates the seven principal cavities of the body. The two major cavities are the dorsal and ventral cavities. The **dorsal cavity,** which is nearest to the dorsal surface, includes the cranial and spinal cavities. The **cranial cavity** is the hollow portion of the skull that contains the brain. The **spinal cavity** is a long tubular canal within the vertebrae that contains the spinal cord. The **ventral cavity** is the largest cavity and encompasses the chest and abdominal regions.

The superior and inferior portions of the ventral cavity are separated by a dome-shaped thin muscle, the **diaphragm.** The **thoracic cavity,** which is that part of the ventral cavity superior to the diaphragm, is separated into right and left compartments by a membranous partition or septum called the **mediastinum.** The lungs are contained in these right and left compartments. The heart, trachea, esophagus, and thymus gland are enclosed within the mediastinum.

Figure 2.2 reveals the relationship of the lungs to the structures within the mediastinum. Note that within the thoracic cavity there exists a pair of right and left **pleural cavities** that contain the lungs and a **pericardial cavity** that contains the heart.

The **abdominopelvic cavity** is the portion of the ventral cavity that is inferior to the diaphragm. It consists of two portions: the abdominal and pelvic

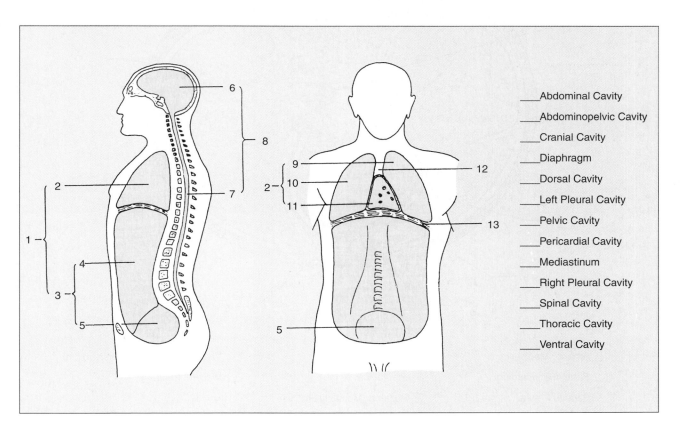

____Abdominal Cavity

____Abdominopelvic Cavity

____Cranial Cavity

____Diaphragm

____Dorsal Cavity

____Left Pleural Cavity

____Pelvic Cavity

____Pericardial Cavity

____Mediastinum

____Right Pleural Cavity

____Spinal Cavity

____Thoracic Cavity

____Ventral Cavity

Figure 2.1   Body cavities.

cavities. The **abdominal cavity** contains the stomach, liver, gallbladder, pancreas, spleen, kidneys, and intestines. The **pelvic cavity** is the most inferior portion of the abdominopelvic cavity and contains the urinary bladder, sigmoid colon, rectum, uterus, and ovaries.

Body Cavity Membranes

The body cavities are lined with serous membranes that provide a smooth surface for the enclosed internal organs. Although these membranes are quite thin, they are strong and elastic. Their surfaces are moistened by a self-secreted *serous fluid* that facilitates ease of movement of the viscera against the cavity walls.

Thoracic Cavity Membranes

The membranes that line the walls of the right and left thoracic compartments are called **parietal pleurae** (*pleura,* singular). The lungs, in turn, are covered with **visceral (pulmonary) pleurae.** Note in figure 2.2 that these pleurae are continuous with each other. The potential cavity between the parietal and visceral pleurae is the **pleural cavity.** Inflammation of the pleural membranes results in a condition called *pleurisy.*

Within the broadest portion of the mediastinum lies the heart. It, like the lungs, is covered by a thin serous membrane, the **visceral pericardium,** or **epicardium.** Surrounding the heart is a double-layered fibroserous sac, the **parietal pericardium.** The inner layer of this sac is a serous membrane that is continuous with the epicardium of the heart. Its outer layer is fibrous, which lends considerable strength to the structure. A small amount of serous fluid produced by the two serous membranes lubricates the surface of the heart to minimize friction as it pulsates within the parietal pericardium. The potential space between the visceral and parietal pericardia is called the **pericardial cavity.**

Abdominal Cavity Membranes

The serous membrane of the abdominal cavity is the peritoneum. It does not extend deep down into the pelvic cavity, however; instead, its most inferior

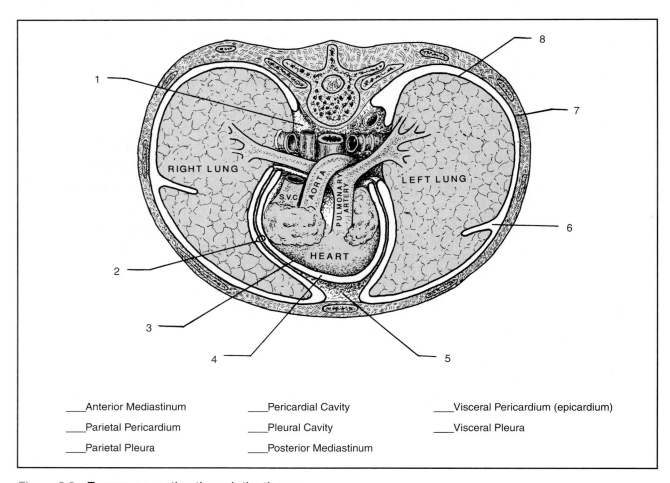

RIGHT LUNG

LEFT LUNG

AORTA

S.V.C.

PULMONARY ARTERY

HEART

1

2

3

4

5

6

7

8

___Anterior Mediastinum   ___Pericardial Cavity   ___Visceral Pericardium (epicardium)

___Parietal Pericardium   ___Pleural Cavity   ___Visceral Pleura

___Parietal Pleura   ___Posterior Mediastinum

Figure 2.2   Transverse section through the thorax.

boundary extends across the abdominal cavity at a level that is just superior to the pelvic cavity. The top portion of the urinary bladder is covered with the peritoneum.

In addition to lining the abdominal cavity, the peritoneum has double-layered folds called **mesenteries,** which extend from the dorsal body wall to the viscera, holding these organs in place. These mesenteries contain blood vessels and nerves that supply the viscera enclosed by the peritoneum.

That part of the peritoneum attached to the body wall is the **parietal peritoneum.** The peritoneum that covers the visceral surfaces is the **visceral peritoneum.** The potential cavity between the parietal and visceral peritoneums is called the **peritoneal cavity.**

Extending downward from the inferior surface of the stomach is a large mesenteric fold called the **greater omentum.** This double-membrane structure passes downward from the stomach in front of the intestines, sometimes to the pelvis, and back up to the transverse colon, where it is attached. Because it is folded upon itself it is essentially a double mesentery consisting of four layers. Protuberance of the abdomen in obese individuals is due to fat accumulation in the greater omentum.

A smaller mesenteric fold, the **lesser omentum,** extends between the liver and the superior surface of the stomach and a short portion of the duodenum. Illustration B of figure 2.3 shows the relationship of these two omenta to the abdominal organs.

Assignment:
Label figures 2.1, 2.2, and 2.3.

Laboratory Report

Complete Laboratory Report 1,2 by answering the questions that pertain to this exercise.

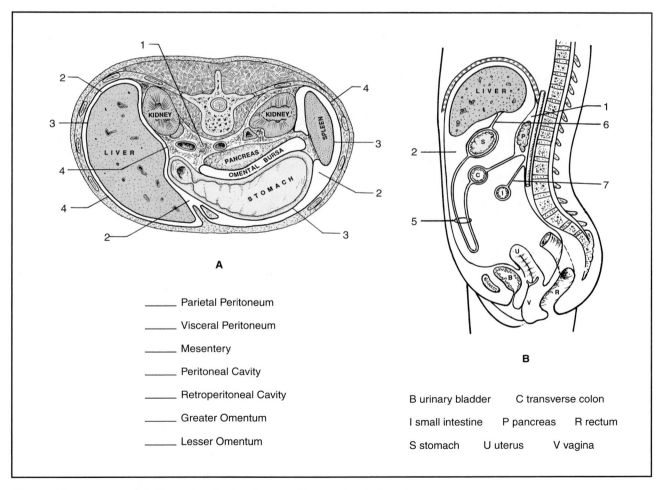

_____ Parietal Peritoneum

_____ Visceral Peritoneum

_____ Mesentery

_____ Peritoneal Cavity

_____ Retroperitoneal Cavity

_____ Greater Omentum

_____ Lesser Omentum

B urinary bladder   C transverse colon

I small intestine   P pancreas   R rectum

S stomach   U uterus   V vagina

Figure 2.3   Transverse and longitudinal sections of the abdominal cavity.

3

During this laboratory period we will dissect a freshly killed rat to perform a cursory study of the majority of the organ systems. Since rats and humans have considerable anatomical and physiological similarities, much will be learned here about human anatomy.

Before beginning the dissection, however, it will be necessary to review the eleven systems of the body. A brief description of each system follows. Keep in mind that an **organ** is defined as a structure composed of two or more tissues that performs one or more physiological functions. A **system,** on the other hand, is a group of organs that directly relate to each other functionally. To make the dissection more meaningful, answer the questions on the Laboratory Report *before* doing the dissection.

The Integumentary System

The Latin word *integumentum* means covering. The body surface covering that makes up this system includes the skin, hair, and nails.

The skin's principal function is to prevent bodily invasion by harmful microorganisms. In addition to being a mechanical barrier, the skin produces sweat and sebum (oil), which contain antimicrobial substances for further protection.

The skin also aids in temperature regulation and excretion. The evaporation of perspiration cools the body. The fact that perspiration contains many of the same excretory products found in urine indicates that the kidneys are aided by the skin in the elimination of water, salts, and some nitrogenous wastes from the blood.

To some extent, the skin also plays a role in nutrition. Evidence of this is seen in the way vitamin D forms in the skin's deeper layers when a vitamin D precursor in those cells is exposed to ultraviolet rays of sunlight.

As long as the skin remains intact, the internal environment is protected. However, serious skin injury, such as deep burns, may result in excessive serous fluid loss and electrolytic imbalances.

The Skeletal System

The skeletal system forms a solid framework around which the body is constructed. It consists of bones, cartilage, and ligaments. This system provides support and protection for the softer parts of the body. Delicate organs such as the lungs, heart, brain, and spinal cord are protected by the bony enclosure of the skeletal system.

In addition to protection, the bones provide points of attachment for muscles, which act as levers when the muscles contract. This arrangement makes movement possible.

Two other important functions of the skeletal system are mineral storage and blood cell production. The mineral component of bones provides a pool of calcium, phosphorus, and other ions that may be utilized to stabilize the mineral content of the blood. With respect to blood cell genesis, both red and white blood cells are formed in the red marrow of certain bones of the body.

The Muscular System

Attached to the skeletal framework of the body are muscles that make up nearly half the weight of the body. The skeletal muscles consist primarily of long multinucleated cells. The ability of these cells to shorten when stimulated by nerve impulses enables the muscles to move parts of the body in walking, eating, breathing, and other activities.

Two other kinds of muscle tissue exist in the body: smooth and cardiac. Smooth muscle tissue is the type found in the walls of internal organs (viscera). Cardiac muscle tissue is found in the walls of the heart. Both types function like skeletal muscle fibers in that they perform work by shortening (contraction).

The Nervous System

The nervous system consists of the brain, spinal cord, nerves, and receptors. While the nervous system has many functions, such as muscular control and regulation of circulation and breathing, its basic function is to facilitate adaptability, that is, the ability to adjust to external and internal environmental changes.

To be able to adapt to environmental changes there is, first of all, a multitude of different kinds of **receptors** throughout the body that are activated by various kinds of stimuli. A receptor may be stimulated by changes in such things as temperature, pressure, chemicals, sound waves, or light. Once activated, nerve impulses pass along **conduction pathways** (nerves and spinal cord) to **interpretation centers** in the brain where recognition and evaluation of stimuli occur. Correct responses, which may be muscular or glandular, are then achieved by the nervous system via outgoing conduction pathways.

The Cardiovascular System

The cardiovascular system consists of the heart, arteries, veins, capillaries, blood, and spleen. The **heart** is a muscular pump that moves blood throughout the body. **Arteries** are thick-walled vessels that carry blood from the heart to the microscopic **capillaries** that permeate all the tissues. **Tissue fluid,** containing nutrients and oxygen, leaves the blood through the capillary walls and passes into the spaces between the cells. **Veins** are large blood vessels that convey blood from the capillaries back to the heart.

Thus, we see that this system provides transportation of various materials from one part of the body to another. In addition to carrying nutrients, oxygen, and carbon dioxide, the cardiovascular system transports hormones from glands, metabolic wastes from cells, and excess heat from muscles to the skin.

Another very important function of the blood is protection against microbial invasion. The presence of phagocytic (cell-eating) white blood cells, antibodies, and special enzymes in the blood prevents invading microorganisms from destroying the body.

The **spleen** is an oval structure on the left side of the abdominal cavity that acts to some extent as a blood reservoir. It also plays an important role in the removal of fragile red blood cells.

Even more significant, this organ is probably the source of B-lymphocytes. Evidence of this is the presence of a large number of these antibody-producing cells in the spleen. This organ is part of the reticuloendothelial system, which is described below under the lymphatic system.

The Lymphatic System

The lymphatic system is a network of lymphatic vessels that returns tissue fluid from the intercellular spaces of tissues to the blood. This system is also responsible for the absorption of fats from the intestines. Although carbohydrates and proteins are absorbed directly into the blood through the intestinal wall, fats must pass first into the lymphatic system and then into the blood.

Once tissue fluid enters the lymphatic vessels it is called **lymph.** As this fluid moves through the lymphatic vessels it passes through nodules of lymphoid tissue called **lymph nodes.** Stationary phagocytic

cells in these nodes remove bacteria and other foreign material, purifying the lymph before it is returned to the blood. Lymphocytes are also produced here.

Lymphoid tissue, as seen in the nodes, is also seen in the thymus gland, liver, spleen, tonsils, adenoids, appendix, Peyer's patches (in the digestive tract), and bone marrow. This diverse collection of lymphoidal tissue is collectively referred to as the **reticuloendothelial system,** an important component of the immune system.

The Respiratory System

The respiratory system consists of two portions: the air passageways and the respiratory portion. The actual exchange of gases between the blood and the air occurs in the respiratory portion. The lungs contain many tiny sacs called **alveoli,** which greatly increase the surface area for the transfer of oxygen and carbon dioxide in breathing. The passageways consist of the **nasal cavity, nasopharynx, larynx, trachea,** and **bronchi.**

The Digestive System

The digestive system includes the **mouth, salivary glands, esophagus, stomach, small intestine, large intestine, rectum, pancreas, liver,** and **gallbladder.** Its function is to convert ingested food to molecules that are small enough to pass through the intestinal lining into the capillaries and lymphatic vessels of the intestinal wall. This digestive process is achieved by hydrolysis of food molecules by **enzymes** produced in the digestive tract.

In addition, this system functions in the elimination of undigested materials and protection against infection.

The Urinary System

Cellular metabolism produces waste materials such as carbon dioxide, excess water, nitrogenous products, and excess metabolites. Although the skin, lungs, and large intestine assist in the removal of some of these wastes from the body, the majority of these products is removed from the blood by the **kidneys.** During this process of waste removal, the kidneys function to regulate the chemical composition of all body fluids to maintain homeostasis.

To assist the kidneys in excretion are the **ureters,** which drain the kidneys, the **urinary bladder,** which stores urine, and the **urethra,** which drains the bladder.

The Endocrine System

This system consists of a number of widely dispersed **endocrine glands** that dispense their secretions directly into the blood. These secretions, which are absorbed directly into capillaries within the glands, are called **hormones.**

Hormones perform such functions as integrating various physiological activities (metabolism and growth), directing the differentiation and maturation of the ovaries and testes, and regulating specific enzymatic reactions.

The endocrine system includes the following glands: **pituitary, thyroid, parathyroid, thymus, adrenal, pancreas,** and **pineal.** In addition, the **ovaries, testes, stomach lining, placenta,** and **hypothalamus** also produce important hormones.

The Reproductive System

Continuity of the species is the function of the reproductive system. Spermatozoa are produced in the testes of the male, and ova are produced by the ovaries of the female. Male reproductive organs include the **testes, penis, scrotum, accessory glands,** and various ducts. The female reproductive organs include the **ovaries, vagina, uterus, uterine** (Fallopian) **tubes,** and **accessory glands.**

Laboratory Report

Complete the Laboratory Report for this exercise.

This Laboratory Report should be completed before starting the rat dissection.

Rat Dissection

You will work with a laboratory partner to perform this part of the exercise. Your principal objective in the dissection is *to expose the organs for study, not to simply cut up the animal.* Most cutting will be performed with scissors. Whereas the scalpel blade will be used only occasionally, the flat blunt end of the handle will be used frequently for separating tissues.

Materials:
    freshly killed rat, dissecting pan with wax bottom, dissecting kit, dissecting pins

Skinning the Ventral Surface

1. Pin the four feet to the bottom of the dissecting pan as illustrated in figure 3.1. Before making any incision examine the oral cavity. Note the large **incisors** in the front of the mouth that are used for biting off food particles. Force the mouth open sufficiently to examine the flattened **molars** at the back of the mouth. These teeth are used for grinding food into small particles.

    Note that the **tongue** is attached at its posterior end. Lightly scrape the surface of

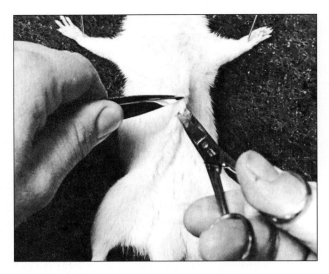

Figure 3.1 Incision is started on the median line with a pair of scissors.

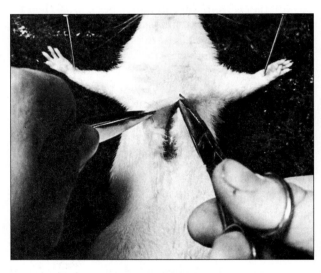

Figure 3.2 First cut is extended from the midtorso up to the lower jaw.

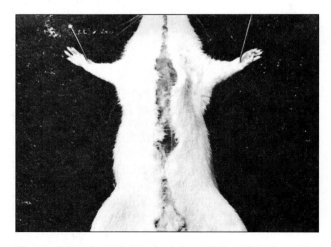

Figure 3.3 Completed incision of the skin from the lower jaw to the anus.

the tongue with a scalpel to determine its texture. The roof of the mouth consists of an anterior **hard palate** and a posterior **soft palate.** The throat is the **pharynx,** which is a component of both the digestive and respiratory systems.

2. Lift the skin along the midventral line with your forceps and make a small incision with scissors as shown in figure 3.2. Cut the skin upward to the lower jaw, turn the pan around, and complete the incision to the anus, cutting around both sides of the genital openings. The completed incision should appear as in figure 3.3.

3. With the handle of the scalpel, separate the skin from the musculature as shown in figure 3.4. The fibrous connective tissue that lies between the skin and musculature is the **superficial fascia** (Latin: *fascia,* band).

4. Skin the legs down to the "knees" or "elbows" and pin the stretched-out skin to the wax. Examine the surfaces of the **muscles** and note that **tendons,** which consist of tough fibrous connective tissue, attach the muscles to the skeleton.

   Covering the surface of each muscle is another thin gray feltlike layer, the **deep fascia.** Fibers of the deep fascia are continuous with fibers of the superficial fascia, so that considerable force with the scalpel handle is necessary to separate the two membranes.

5. At this stage your specimen should appear as in figure 3.5. If your specimen is a female, the

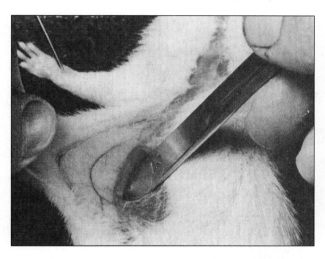

Figure 3.4 The skin is separated from the musculature with scalpel handle.

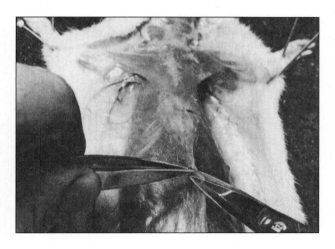

Figure 3.5 Incision of musculature is begun on the median line.

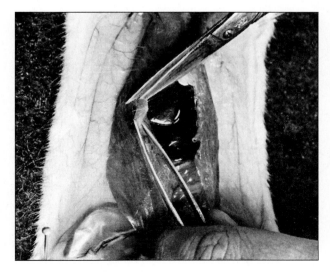

Figure 3.6 Lateral cuts through the musculature are made at base of rib cage in both directions.

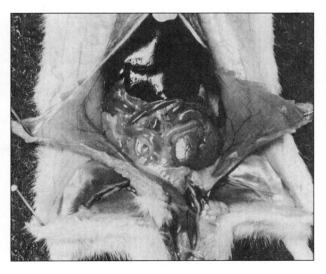

Figure 3.7 Two flaps of the abdominal wall are pinned back to expose the viscera.

mammary glands will probably remain attached to the skin.

Opening the Abdominal Wall

1. As shown in figure 3.5, make an incision through the abdominal wall with a pair of scissors. To make the cut it is necessary to hold the muscle tissue with a pair of forceps. **Caution:** Avoid damaging the underlying viscera as you cut.

2. Cut upward along the midline to the rib cage and downward along the midline to the genitalia.

3. To completely expose the abdominal organs make two lateral cuts near the base of the rib cage—one to the left and the other to the right. See figure 3.6. The cuts should extend all the way to the pinned-back skin.

4. Fold out the flaps of the body wall and pin them to the wax as shown in figure 3.7. The abdominal organs are now well exposed.

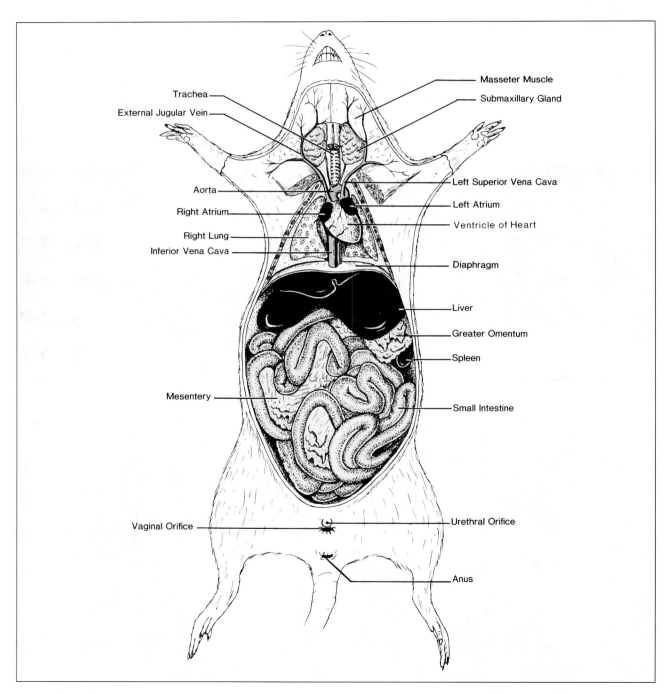

Figure 3.8 **Viscera of a female rat.**

5. Using figure 3.8 as a reference, identify all the labeled viscera without moving the organs out of place. Note in particular the position and structure of the **diaphragm.**

Examination of Thoracic Cavity

1. Using your scissors, cut along the left side of the rib cage as shown in figure 3.9. Cut through all of the ribs and connective tissue. Then, cut along the right side of the rib cage in a similar manner.
2. Grasp the xiphoid cartilage of the sternum with forceps as shown in figure 3.10 and cut the diaphragm away from the rib cage with your scissors. Now you can lift up the rib cage and look into the thoracic cavity.
3. With your scissors, complete the removal of the rib cage by cutting off any remaining attachment tissue.

4. Now examine the structures that are exposed in the thoracic cavity. Refer to figure 3.8 and identify all the structures that are labeled.
5. Note the pale-colored **thymus gland,** which is located just above the heart. Remove this gland.
6. Carefully remove the thin **pericardial membrane** that encloses the **heart.**
7. Remove the heart by cutting through the major blood vessels attached to it. Gently sponge away pools of blood with Kimwipes or other soft tissues.
8. Locate the **trachea** in the throat region. Can you see the **larynx** (voice box), which is located at the anterior end of the trachea? Trace the trachea posteriorly to where it divides into two **bronchi** that enter the **lungs.** Squeeze the lungs with your fingers, noting how elastic they are. Remove the lungs.

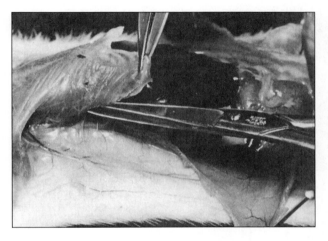

Figure 3.9 Rib cage is severed on each side with scissors.

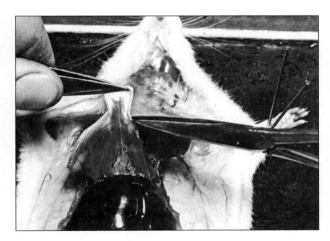

Figure 3.10 The diaphragm is cut free from the edge of the rib cage.

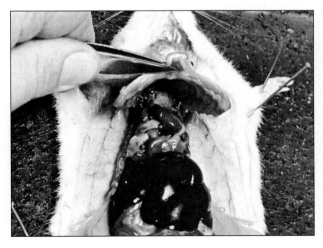

Figure 3.11 The thoracic organs are exposed as the rib cage is lifted out.

Figure 3.12 Specimen with heart, lungs, and thymus gland removed.

9. Probe under the trachea to locate the soft tubular **esophagus** that runs from the oral cavity to the stomach. Excise a section of the trachea to reveal the esophagus as illustrated in figure 3.13.

Deeper Examination of Abdominal Organs

1. Lift up the lobes of the reddish brown liver and examine them. Note that rats lack a **gallblad-** **der.** Carefully excise the liver and wash out the abdominal cavity. The stomach and intestines are now clearly visible.

2. Lift out a portion of the intestines and identify the membranous **mesentery,** which holds the intestines in place. It contains blood vessels and nerves that supply the digestive tract. If your specimen is a mature healthy animal, the mesenteries will contain considerable fat.

3. Now lift the intestines out of the abdominal cavity for a better view of the organs. Note the

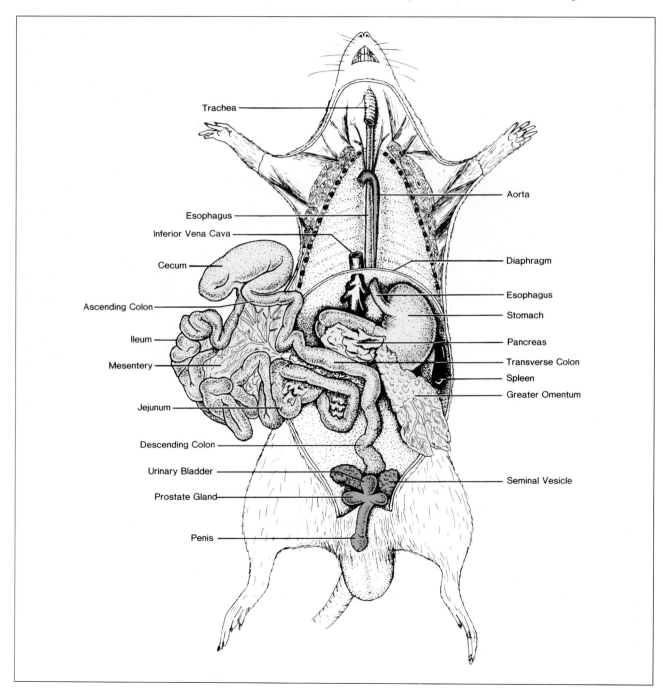

Figure 3.13   Viscera of a male rat with heart, lungs, and thymus gland removed.

great length of the **small intestine.** Its name refers to the diameter, not its length.

The first portion of the small intestine, which is connected to the stomach, is called the **duodenum.** At its distal end the small intestine is connected to a large saclike structure, the **cecum.** The *appendix* in humans is a vestigial portion of the cecum.

The cecum communicates with the **large intestine.** The latter structure consists of the **ascending, transverse, descending,** and **sigmoid** divisions. The sigmoid colon, in turn, empties into the **rectum.**

4. Try to locate the **pancreas,** which is embedded in the mesentery alongside the duodenum. It is often difficult to see. Pancreatic enzymes enter the duodenum via the **pancreatic duct.** See if you can locate this minute tube.

5. Locate the **spleen,** which is situated on the left side of the abdomen near the stomach. It is red-dish brown and held in place with mesentery. Do you recall the functions of this organ?

6. Remove the stomach, small intestine, and large intestine after severing the distal end of the esophagus and the distal end of the sigmoid colon.

Removal of these organs enables you to see the **descending aorta** and the **inferior vena cava.** The aorta carries blood posteriorly to the body tissues. The inferior vena cava is a vein that returns blood from the posterior regions to the heart.

7. Peel away the peritoneum and fat from the posterior wall of the abdominal cavity. *Removal of this fat will require special care to avoid damaging important structures.* This will make the kidneys, blood vessels, and reproductive structures more visible.

Locate the two **kidneys** and **urinary bladder.** Trace the two **ureters,** which extend from the kidneys to the bladder. Examine the anterior surfaces of the kidneys and locate the

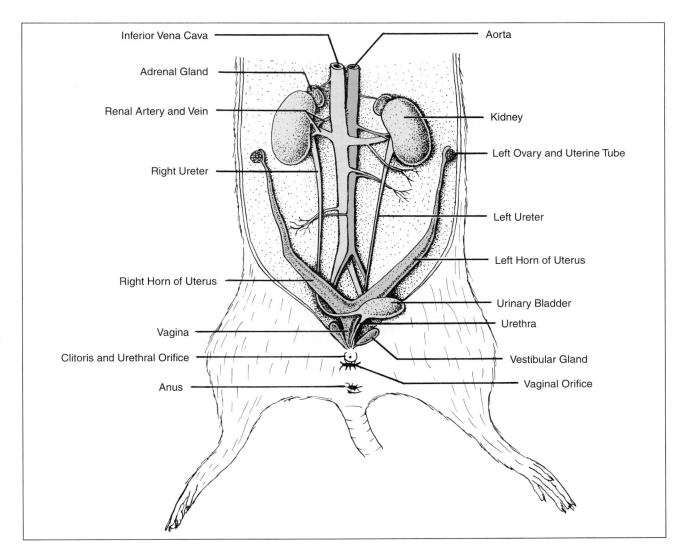

Figure 3.14 Abdominal cavity of a female rat with intestines and liver removed.

adrenal glands, which are important components of the endocrine gland system.

8. **If your specimen is a female:** Compare it with figure 3.14. Locate the two **ovaries,** which lie lateral to the kidneys. From each ovary a **uterine tube** leads posteriorly to join the **uterus.** Note that the uterus is a Y-shaped structure joined to the vagina.

   If your specimen appears to be pregnant, open up the uterus and examine the developing embryos. Note how they are attached to the uterine wall.

9. **If your specimen is a male:** Compare your specimen with figure 3.15. The **urethra** is located in the **penis.** Apply pressure to one of the testes through the wall of the **scrotum** to see if it can be forced up into the **inguinal canal.**

   Open up the scrotum on one side to expose a **testis,** the **epididymis,** and the **vas deferens.** Trace the vas deferens over the urinary bladder to where it penetrates the **prostate gland** to join the urethra.

Summarization   In this cursory dissection you have become acquainted with the respiratory, cardiovascular, digestive, urinary, and reproductive systems. Portions of the endocrine system have also been observed. Five systems (integumentary, skeletal, muscular, lymphatic, and nervous) have been omitted at this time. These will be studied later.

If you have done a careful and thoughtful rat dissection, you should have a good general understanding of the basic structural organization of the human body. Much that we have seen here has its human counterpart.

Cleanup   Dispose of the specimen as directed by your instructor. Scrub your instruments with soap and water, rinse, and dry them.

Wash your hands with soap and water, rinse, and dry thoroughly.

Laboratory Report

Answer the questions on Laboratory Report 3 to test your understanding of the organ systems.

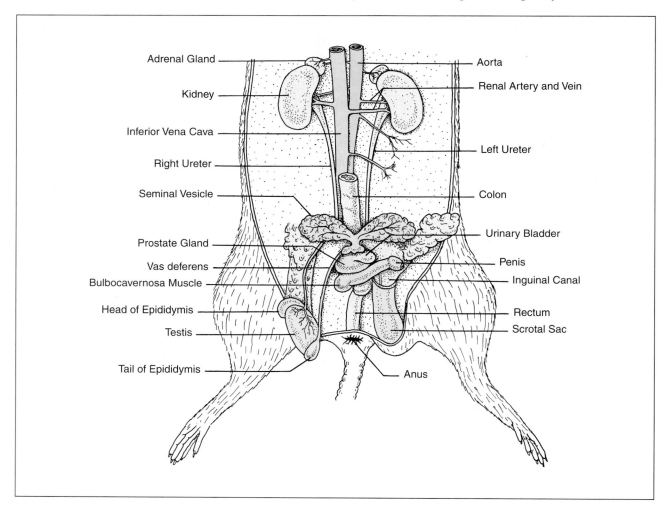

Figure 3.15   Abdominal cavity of a male rat with intestines and liver removed.

Microscopy

4

Since the next few exercises pertain to microscopic studies of various types of cells, it is essential that some orientation be provided concerning acceptable procedures in microscopy. It is very likely that you have had some previous experience using a microscope, so we will provide here only a brief summary of what you should know.

Microscopy can be fascinating or frustrating. Success in this endeavor depends to a large extent on how well you understand the mechanics and limitations of your microscope. It is the purpose of this exercise to outline procedures that should minimize difficulties and enable you to get the most out of your subsequent attempts at cytological microscopy.

Before entering the laboratory to do any of the studies on cells and tissues, it would be desirable if you answered all the questions on the Laboratory Report after reading over this entire exercise. Your instructor may require that the Laboratory Report be handed in prior to doing any laboratory work.

Care of the Instrument

Microscopes represent considerable investment and can be damaged rather easily if certain precautions are not observed. The following suggestions cover most hazards:

Transport   When carrying your microscope from one part of the room to another, use both hands when holding the instrument (see figure 4.1). If it is carried with only one hand and allowed to dangle at your side, there is always the possibility of collision with furniture or some other object. And, incidentally, under no circumstances should one attempt to carry two microscopes at one time.

Clutter   It is very important that you keep your workstation uncluttered at all times while doing microscopy. Nonessential books, lunches, and other unneeded supplies should be stashed away in a drawer or some other out-of-the-way place. A clear work area promotes efficiency and results in fewer accidents.

Electric Cord   Unbelievably, dangling microscope cords can result in catastrophic accidents. In crowded quarters a dangling cord has been known to become entangled in a student's foot. Result: a damaged microscope on a concrete floor. The best precaution is to make sure your microscope cord is kept out of harm's way.

Lens Care   At the beginning of each laboratory period check the lenses to make sure they are clean. At the end of each lab session be sure to wipe any immersion oil off the immersion lens if it has been used. More specifics about lens care are provided on pages 19 and 20.

Dust Protection   In most laboratories dustcovers are used to protect the instruments during storage. If one is available, place it over the microscope at the end of the period.

Components

Before we discuss the procedures for using a microscope, let's identify the principal parts of the instrument as illustrated in figure 4.4.

Framework   All microscopes have a basic frame structure that includes the **arm** and **base.** To this framework all other parts are attached. On many of the older microscopes the base is not rigidly attached to the arm as is the case in figure 4.4; instead, a pivot point is present that enables one to tilt the arm backward to adjust the eyepoint height.

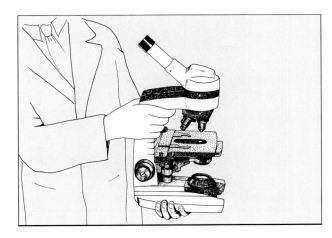

Figure 4.1   The microscope should be held firmly with both hands when it is carried.

Stage The horizontal platform that supports the microscope slide is called the *stage*. Note that it has a clamping device, the **mechanical stage,** which is used for holding and moving the slide around on the stage. Note, also, the location of the **mechanical stage control** in figure 4.4.

Light Source In the base of most microscopes is positioned some kind of light source. Ideally, the lamp should have a **voltage control** to vary the intensity of light. The microscope in figure 4.4 has a knurled wheel on the right side of its base to regulate the voltage.

Most microscopes have some provision for reducing light intensity with a **neutral density filter.** Such a filter is often needed to reduce the intensity of light below the lower limit allowed by the voltage control. On microscopes such as the Olympus CH-2, one can simply place a neutral density filter over the light source in the base. On some microscopes a filter is built into the base (see figure 4.2).

Lens Systems All microscopes have three lens systems: the oculars, the objectives, and the condenser.

The **ocular,** or eyepiece, which is at the top of the instrument, consists of two or more internal lenses and usually has a magnification of 10×. Although the microscope in figure 4.4 has two oculars, a microscope often has only one.

Three or more **objectives** are usually present. Note that they are attached to a rotatable **nosepiece,** which makes it possible to move them into position over a slide. Objectives on most laboratory microscopes have magnifications of 10×, 45×, and 100×, designated as **low power, high-dry,** and **oil immersion,** respectively.

The third lens system is the **condenser,** which is located under the stage. It collects and directs the light from the lamp to the slide being studied. The condenser can be moved up and down by a knob under the stage. A **diaphragm** within the condenser regulates the amount of light that reaches the slide. On the Olympus microscope in figure 4.4 the diaphragm is controlled by turning a knurled ring. On some microscopes a diaphragm lever is present.

Focusing Knobs The concentrically arranged **coarse adjustment** and **fine adjustment knobs** on the side of the microscope are used for bringing objects into focus when studying an object on a slide.

Ocular Adjustments On binocular microscopes one must be able to change the distance between the oculars and to make diopter changes for eye differences. On most microscopes the interocular distance is changed by simply pulling apart or pushing together the oculars.

To make diopter adjustments, one focuses first with the right eye only. Without touching the focusing knobs, diopter adjustments are then made on the left eye by turning the knurled **diopter adjustment ring** (figure 4.4) on the left ocular until a sharp image is seen. One should now be able to see sharp images with both eyes.

Resolution

The resolution limit, or **resolving power,** of a microscope lens system is a function of its numerical aperture, the wavelength of light, and the design of the condenser. The maximum resolution of the best microscopes with oil immersion lenses is around 0.2 μm. This means that two small objects that are 0.2 μm apart will be seen as separate entities; objects closer than that will be seen as a single object.

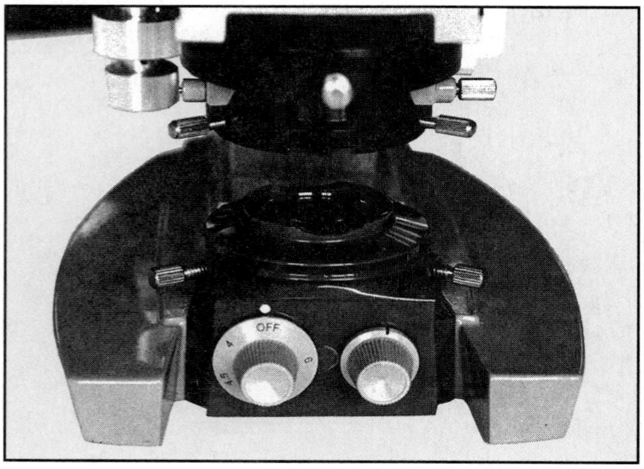

Figure 4.2 On this older type of microscope the left knob controls voltage; the other knob controls the neutral density filter.

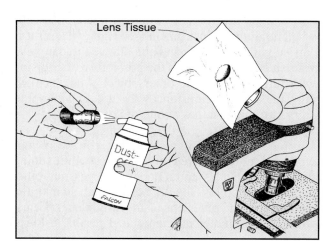

Figure 4.3 When oculars are removed for cleaning, cover the ocular opening with lens tissue. A blast from an air syringe or gas cannister removes dust and lint.

To get the maximum amount of resolution from a lens system, the following factors must be taken into consideration:

- A **blue filter** should be in place over the light source because the short wavelength of blue light provides maximum resolution.
- The **condenser** should be kept at its highest position where it allows a maximum amount of light to enter the objective.
- The **diaphragm** should not be stopped down too much. Although stopping down improves contrast, it reduces the numerical aperture.
- **Immersion oil** should be used between the slide and the 100× objective.

Lens Care

Unless all lenses are kept scrupulously clean, maximum resolution cannot occur. When cleaning lenses observe the following suggestions:

Cleaning Tissues   Only lint-free, optically safe tissues should be used to clean lenses. Tissues free of abrasive grit fall in this category. Booklets of lens tissue are most widely used for this purpose. Although several types of boxed tissues are also safe, *use only the type of tissue that is recommended by your instructor.*

Solvents   Various liquids can be used for cleaning microscopic lenses. Green soap with warm water

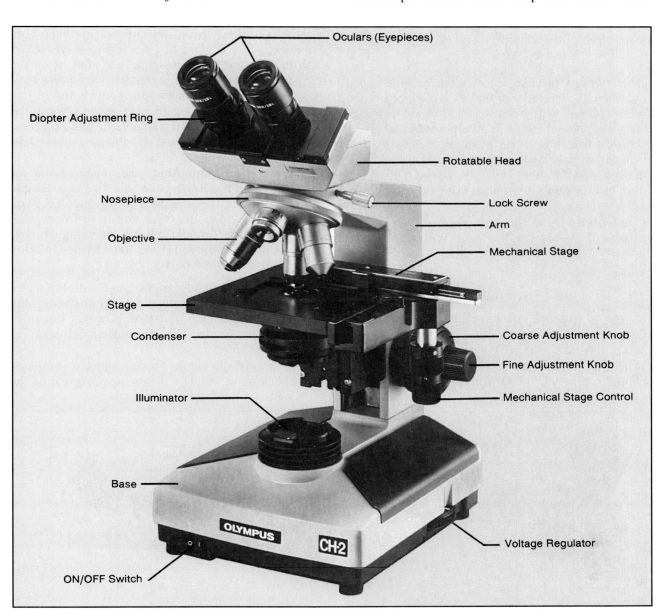

Figure 4.4   The compound microscope.
Courtesy of the Olympus Corporation, Lake Success, N.Y.

works very well. Xylene is universally acceptable. Alcohol and acetone are also recommended, but often with some reservations. Your instructor will inform you as to what solvents are best for the lenses of your microscope.

Oculars The best way to determine if your eyepiece is clean is to rotate it between the thumb and forefinger as you look through the microscope. A rotating pattern will be evidence of dirt.

If cleaning the top lens of the ocular with lens tissue fails to remove the debris, one should try cleaning the lower lens with lens tissue and blowing off any excess lint with an air syringe or gas cannister. *Whenever the ocular is removed from the microscope, it is imperative that a piece of lens tissue be placed over the open end of the microscope as illustrated in figure 4.3.*

Objectives Objective lenses often become soiled by materials from slides or fingers. A piece of lens tissue moistened with green soap and water, or one of the acceptable solvents will usually remove whatever is on the lens. Sometimes a cotton swab with a solvent will work better than lens tissue. At any time that the image on the slide is unclear or cloudy, assume at once that the objective you are using is soiled.

Condenser Dust often accumulates on the top surface of the condenser; thus, wiping it off occasionally with lens tissue is desirable.

Procedures

If your microscope has three objectives you have three magnification options: (1) low-power, or 100× magnification, (2) high-dry magnification, which is 450× with a 45× objective, and (3) 1000× magnification with a 100× oil immersion objective.

Note that the total magnification seen through an objective is calculated by simply multiplying the power of the ocular by the power of the objective.

Whether you use the low-power objective or the oil immersion objective will depend on how much magnification is necessary. Usually, it is best to start with the low-power objective and progress to the higher magnifications as your study progresses. Consider the following suggestions for setting up your microscope and making microscopic observations:

Viewing Setup If your microscope has a rotatable head, such as the ones being used by the two students in figure 4.5, there are two ways that you can use the instrument. Note that the student on the left has the arm of the microscope *near* him, and the other student has the arm *away from* her. With this type of microscope, the student on the right has the advantage in that the stage is easier to observe. Note also that when focusing the instrument she is able to rest her arm on the table. If the microscope head is not rotatable, it will be necessary to use the instrument as indicated on the left.

Low-power Examination The main reason for starting with the low-power objective is to enable you to explore the slide to look for the object you are planning to study. Once you have found what you are looking for, you can proceed to higher magnifications. Use the following steps when exploring a slide with the low-power objective:

1. Position the slide on the stage with the material to be studied on the *upper* surface of the slide. Figure 4.6 illustrates how the slide must be held in place by the mechanical stage retainer lever.
2. Turn on the light source, using a *minimum* amount of voltage. If necessary, reposition the

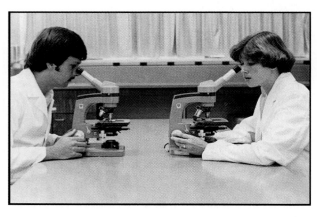

Figure 4.5 The microscope position on the right has the advantage of stage accessibility.

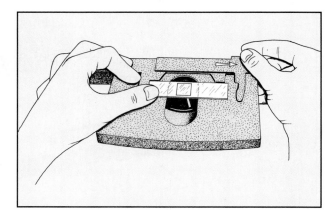

Figure 4.6 The slide must be properly positioned as the retainer lever is moved to the right.

slide so that the stained material on the slide is in the *exact center* of the light source.

3. Check the condenser to see that it has been raised to its highest point.

4. If the low-power objective is not directly over the center of the stage, rotate it into position. Be sure that as you rotate the objective into position it clicks into its locked position.

5. Turn the coarse adjustment knob to lower the objective *until it stops.* A built-in stop will prevent the objective from touching the slide.

6. While looking down through the ocular (or oculars), bring the object into focus by turning the fine adjustment focusing knob. Don't readjust the coarse adjustment knob. If you are using a binocular microscope it will also be necessary to adjust the interocular distance and diopter adjustment to match your eyes.

7. Manipulate the diaphragm to reduce or increase the light intensity to produce the clearest, sharpest image. As you close down the diaphragm the contrast improves and the depth of field increases. Stopping down the diaphragm, however, decreases resolution.

8. Once an image is visible, move the slide about to search out what you are looking for. The slide is moved by turning the knobs that move the mechanical stage.

9. Check the cleanliness of the ocular, using the procedure outlined.

10. Once you have identified the structures to be studied and wish to increase the magnification, you may proceed to either high-dry or oil immersion magnification. However, before changing objectives, *be sure to center the object you wish to observe.*

High-Dry Examination   To proceed from low-power to high-dry magnification, all that is necessary is to rotate the high-dry objective into position and open up the diaphragm somewhat. It may be necessary to make a minor adjustment with the fine adjustment knob to sharpen up the image, but *the coarse adjustment knob should not be touched.* If a microscope is of good quality, only minor focusing adjustments are needed when changing magnifications because all the objectives will be **parfocalized.** Nonparfocalized microscopes do require considerable refocusing when changing objectives.

High-dry objectives should only be used on slides that have cover glasses; without them, images are usually unclear. When increasing the lighting, be sure to open up the diaphragm first instead of increasing the voltage on your lamp; reason: *lamp life is greatly extended when used at low voltage.* If

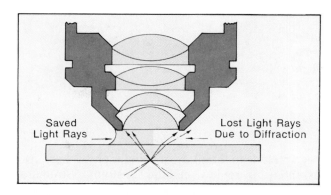

Figure 4.7   Immersion oil, having the same refractive index as glass, prevents light loss due to diffraction.

the field is not bright enough after opening the diaphragm, feel free to increase the voltage. A final point: keep the condenser at its highest point.

Oil Immersion Techniques   The oil immersion lens derives its name from the fact that a special mineral oil is interposed between the lens and the microscope slide. The oil is used because it has the same refractive index as glass, which prevents the loss of light due to the bending of light rays as they pass through the air. The use of oil in this way enhances the resolving power of the microscope. Figure 4.7 illustrates this phenomenon.

With parfocalized objectives one can go to oil immersion from either low-power or high-dry. Once the microscope has been brought into focus at one magnification, the oil immersion lens can be rotated into position without fear of striking the slide.

Before rotating the oil immersion lens into position, however, a drop of immersion oil is placed on the slide. If the oil appears cloudy it should be discarded.

When using the oil immersion lens it is best to open the diaphragm as much as possible. Stopping down the diaphragm tends to limit the resolving power of the optics. In addition, the condenser must be kept at its highest point. If different-colored filters are available for the lamp housing, it is best to use blue or greenish filters to enhance the resolving power.

Using this lens takes a little practice. The manipulation of lighting is critical. Before returning the microscope to the cabinet, all oil must be removed from the objective and stage. Slides with oil should be wiped clean with lens tissue, also.

Laboratory Report

Before entering the laboratory for your first encounter with the microscope, answer the questions on Laboratory Report 4. Preparation on your part prior to going to the laboratory will greatly facilitate your understanding.

5

Basic Cell Structure

As a prelude to the study of cellular physiology, it is essential to review cellular anatomy. Structural visualization of the plasma membrane, nucleus, and organelles such as the endoplasmic reticulum, centrosome, and lysosomes go hand in hand with understanding the physiological activities of the cell as a whole.

In the first portion of this exercise, a summary of the present knowledge of cellular structure and function is presented. This is followed by laboratory studies of epithelial cells and living microorganisms.

Since much of the discussion in this exercise pertains to cellular structure as seen with an electron microscope, and since our laboratory observations will be made with a light microscope, there will be some cellular structures that cannot be seen in great detail in the laboratory.

To get the most out of this exercise *it is recommended that the questions on the Laboratory Report be answered prior to entering the laboratory for microscopic studies.*

The Basic Design

The study of cellular structure (**cytology**) has advanced at a rapid rate over the past decades with the use of electron microscopy and special techniques in the study of cellular physiology. The old notion of the cell as a "sac of protoplasm" has been supplanted by a much more dynamic model.

A cell is now viewed as a highly complex miniature computer with an integrated, compartmentalized ultrastructure capable of communicating with its environment, altering its shape, processing information, and synthesizing a great variety of substances. It is largely through our advancing knowledge of the cellular organelles that this new viewpoint has emerged.

Figure 5.1 is a diagrammatic version of a pancreatic cell, incorporating the various organelles that can be seen with an electron microscope. A pancreatic cell has been chosen for our study here because it lacks the degree of specialization that is seen in many other cells. In addition, it contains a majority of the organelles that are present in most body cells.

Although it is not possible to see the fine structure of many organelles with an ordinary light microscope, we are including descriptions of their appearance and current theories of function. As the various cell structures are discussed in the following text, identify them in figure 5.1.

The Plasma Membrane

The outer surface of every cell consists of an extremely thin, delicate cell membrane that is often referred to as the *plasma membrane.* Chemical analyses of plasma membranes indicate that phospholipids, glycolipids, and proteins are the principal constituents. Lipids account for one-half the mass of plasma membranes; proteins make up the other half.

Examination with an electron microscope reveals that all plasma membranes have a similar basic trilaminar structure, the so-called *unit membrane.* The current interpretation of the molecular organization of this membrane proposes that **lipid molecules** form a double fluid layer in which **protein** and **glycoprotein molecules** are embedded. On the left side of figure 5.1 is an enlarged portion of this membrane illustrating the relationship of these various components to each other. Note that some of the protein molecules (color code: brown) protrude externally, others protrude internally, and some extend through both sides of the membrane. This asymmetrical architecture appears to account for the external and internal differences in receptor sites that affect the permeability characteristics of the membrane. The tubular nature of protein molecules that extend completely through the membrane provides a convenient passageway for certain types of ions and molecules.

Cell membranes are *selectively* permeable in that they allow certain molecules to pass through easily and prevent other molecules from gaining entrance to the cell. Although cell membranes play both active and passive roles in molecular movements, experimental evidence seems to indicate that the movement of all molecules, including water, is assisted by the membrane. In Exercise 7 we will study some of the forces that are involved in permeability.

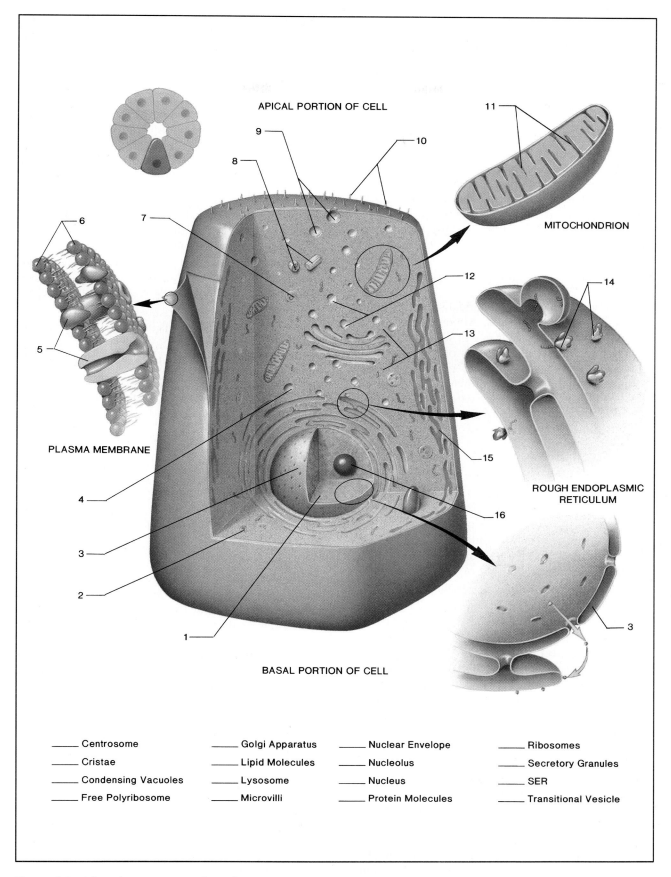

APICAL PORTION OF CELL

MITOCHONDRION

PLASMA MEMBRANE

ROUGH ENDOPLASMIC
RETICULUM

BASAL PORTION OF CELL

_____ Centrosome   _____ Golgi Apparatus   _____ Nuclear Envelope   _____ Ribosomes

_____ Cristae   _____ Lipid Molecules   _____ Nucleolus   _____ Secretory Granules

_____ Condensing Vacuoles _____ Lysosome   _____ Nucleus   _____ SER

_____ Free Polyribosome   _____ Microvilli   _____ Protein Molecules   _____ Transitional Vesicle

Figure 5.1   The microstructure of a cell.

The Nucleus

Near the basal portion of the cell is a large spherical body, the *nucleus.* It is shown in figure 5.1 with a portion cut out of it. The inner substance *(nucleoplasm)* of this body is surrounded by a double-layered **nuclear envelope** that, unlike other membranes, is perforated by pores of significant size. These openings provide a viable passageway between the nucleoplasm and cytoplasm.

A cell that has been stained with certain dyes will exhibit darkly stained regions in the nucleus called *chromatin granules* (not shown in figure 5.1). These granules are visible even with a light microscope. They represent highly condensed DNA molecules that comprise part of the chromosomes. Current theories suggest that increased condensed chromosomal material indicates a metabolically less active cell.

The most conspicuous spherical structure within the nucleus is the **nucleolus.** Although nucleoli of different cells vary considerably in size, number, and structure, they all consist primarily of RNA and function chiefly in the production of ribonucleoprotein for the ribosomes. The arrow shown in the magnified view in the lower right-hand corner of figure 5.1 illustrates the passage of RNA from the nucleoplasm through a pore of the nuclear envelope.

The Cytoplasmic Matrix

Between the plasma membrane and nucleus is an area called the *cytoplasmic matrix,* or *cytoplasm.* This region is a heterogeneous aggregation of many components involved in cellular metabolism. The cytoplasm is structurally and functionally compartmentalized by the numerous organelles, many of which are hollow and enclosed by unit membranes. A description of each of these cytoplasmic organelles follows:

Endoplasmic Reticulum The most extensive structure within the cytoplasm is a complex system of tubules, vesicles, and sacs called the *endoplasmic reticulum* (ER). It is a double-layered unit membrane that is somewhat similar to the nuclear and plasma membranes; in some instances it is continuous with these membranes (see lower right-hand enlarged view of figure 5.1). It appears to be a manufacturing organelle where many of the essential proteins of the cell are produced. It also acts as a microcirculatory system in that it transports newly formed molecules to the Golgi apparatus.

Detailed studies of cells have shown that ER may have surfaces that are smooth or rough. ER that is designated as **rough endoplasmic reticulum (RER)** is a fluid-filled canalicular system studded with ribosomes. It is the type illustrated in the magnified view on the right side of figure 5.1. RER is believed to be an important site of protein synthesis and storage. It is more developed in cells that are primarily secretory in function.

ER that lacks ribosomes is designated as **smooth endoplasmic reticulum (SER).** Some layers of SER are shown in the cytoplasm near the plasma membrane. It is thought that SER is active in various types of biosynthesis.

Ribosomes As stated above, ribosomes are small bodies attached to the surface of the RER. They are also scattered throughout the cytoplasm in rosettes and chains called **free polyribosomes** (label 2). Since they are only 170 Ångstrom units (17 nanometers) in diameter they can be resolved only by electron microscopy.

The components for each ribosome originate in the nucleolus of the nucleus. They pass from the nucleolus through the pores of the nuclear membrane into the cytoplasm, where they unite to form the completed ribosomal particle. Each ribosome consists of about 60% RNA and 40% protein. Their proposed structure is shown in the enlarged view of the RER in figure 5.1.

Ribosomes are sites of protein synthesis and are particularly numerous in cells that are actively synthesizing proteins. Free polyribosomes appear to be involved in synthesis of proteins for endogenous use; attached ribosomes on RER, on the other hand, are implicated in the synthesis of protein to be transported extracelluarly (i.e., secretion).

Mitochondria These large (1×2–3 μm) bean-shaped organelles have double-layered membranous walls with inward-protruding partial partitions called **cristae** that serve to increase the total surface area.

Mitochondria contain DNA and are able to replicate themselves. They also contain their own ribosomes. They are sometimes referred to as the "power plants" of the cell, since it is here that the energy-yielding reactions of oxidative respiration occur. Simply stated, these reactions oxidize glucose ($C_6H_{12}O_6$) to yield energy in the form of ATP:

$$C_6H_{12}O_6 + 6O_2 \rightarrow 6CO_2 + 6H_2O + ATP + Heat$$

Golgi Apparatus This unit membrane organelle is quite similar in basic structure to SER. In figure 5.1 it appears as a layered stack of vesicles between the nucleus and the apical end of the cell.

Its primary role relates to the packaging, movement, and completed synthesis of products to be released by secretory cells. The protein molecules produced by the RER are transported to the Golgi apparatus via **transitional vesicles** (color code: blue). These vesicles form on the surface of the RER, move toward the Golgi apparatus, and then coalesce with the surface membrane of the Golgi apparatus to deliver the protein molecules. Inside the Golgi apparatus additional biosynthesis occurs, often with the addition of sugars.

The completely processed materials accumulate at the apical end of the organelle and form **condensing vacuoles.** These membrane-bound vesicles move to the apex of the cell, where, as **secretory granules,** they are ultimately exocytosed. Figure 5.1 illustrates how the granules fuse with the plasma membrane. The fate of the exocytosed product depends upon the cell type.

Lysosomes Saclike structures in the cytoplasm that act like recycling centers are called *lysosomes.* These sacs (color code: turquoise) have membranes that consist of a single membrane and are believed to form by pinching off of sacs from the Golgi apparatus. The contents of lysosomes vary from cell to cell, but typically they include cellular debris and enzymes that hydrolyze proteins and nucleic acids.

Portions of mitochondria, dead microorganisms, worn-out red blood cells, and other debris have been identified in these sacs. It appears that the expired proteins of these waste products are refashioned here into proteins needed by the cell. Lysosomes that cease to be of value to a cell are exocytosed from the cell through the cell membrane.

When cells die, the membranes surrounding the lysosomes disintegrate, releasing enzymes into the cytoplasm. The hydrolytic action of the enzymes on the cell hastens the death of the cell. It is for this reason that these organelles have been called "suicide bags."

The Centrosome Near the apex of the cell is seen a pair of microtubule bundles that are arranged at right angles to each other. Each bundle, called a *centriole,* consists of nine triplets of microtubules arranged in a circle. The two centrioles are collectively referred to as the *centrosome.* Because of their small size, centrosomes appear as very tiny dots when observed with a light microscope. In most cells the centrosome is located closer to the nucleus than shown in figure 5.1.

During mitosis and meiosis the centrioles play a role in the formation of spindle fibers that aid in the

Figure 5.2 A computer-generated visualization of microtubules, vesicles, and mitochondria in the axon of a nerve cell. Vesicles and mitochondria are being propelled by the microtubules.

separation of chromatids. They also function in the formation of cilia and flagella in certain types of cells.

Cytoskeletal Elements

It now appears that a living cell has an extensive permeating network of fine fibers that contribute to cellular properties such as contractility, support, shape, and translocation of materials. Microtubules and microfilaments fall into this category.

Microtubules Molecules of protein *(tubulin),* arranged in submicroscopic cylindrical hollow bundles are called *microtubules.* They are found dispersed throughout the cytoplasm of most cells and converge especially around the centrioles to form *aster fibers* and *spindles* that are seen during cell division (Exercise 6). They also play a significant role in the transport of vesicles from the Golgi apparatus to other parts of the cell.

Nerve cells that have long processes, called axons, utilize microtubules to transport vesicles and mitochondria. Movement of these organelles is from the cell body to the axon terminal, as well as from the terminal back to the cell body.

Figure 5.2 is a computer-generated image of how these microtubules might look where they enter an axon of a nerve cell. Note that various-sized vesicles are propelled along the microtubules and that these vesicles move along the microtubules in both directions *simultaneously.* The energy for vesicle transport is supplied by mitochondria: a portion of one is seen clearly in the lower right quadrant; another is barely discernible in the upper left quadrant.

Microfilaments All cells show some degree of contractility. Cytoskeletal elements responsible for this phenomenon are the *microfilaments*. These filaments are composed of protein molecules arranged in solid parallel bundles. They are most highly developed in muscle cells that are adapted specifically for contraction. In other types of cells, they are present as interwoven networks of the cytoplasm and are relatively inconspicuous even in electron photomicrographs. Besides aiding in contractility, they also appear to be involved in cell motility and support.

Cilia and Flagella

Hairlike appendages of cells that function in providing some type of movement are either *cilia* or *flagella*. Cilia are usually less than 20 μm long, but flagella may be thousands of micrometers in length. Cells that line the respiratory tract and uterine tubes have cilia (see illustration D in figure HA-2 and illustration C in figure HA-35 of the Histology Atlas). The only human cells with flagella are the spermatozoa. Both cilia and flagella originate from centrioles, which explains why they both have microtubular internal structure similar to the centrioles.

The Microvilli

Certain cells, such as the one in figure 5.1, have tiny protuberances known as *microvilli* on their apical or free surfaces. They may appear as a fine **brush border** (illustration C, figure HA-2) or as **stereocilia** (illustration B, figure HA-38). Both of these may be mistaken for cilia with a light microscope.

Each microvillus is an extension of cytoplasm enclosed by the plasma membrane. Microvilli are very common on cells lining the intestine and kidney tubules, where they function to increase surface area of the cells for absorption of water and nutrients. Microvilli are not motile.

Laboratory Assignment

After labeling figure 5.1 and answering the questions on the Laboratory Report pertaining to cell structure and function, do the following two cellular studies in the laboratory. The epithelial cell will be removed from the inner surface of the cheek and examined with high-dry optics. The study of microorganisms (primarily protozoa) is performed here to observe the activity of living cells. Ciliary and flagellar action, amoeboid movement, secretion and excretion of cellular products, and cell division can be observed in viable cultures containing a mixture of protozoans.

Materials:
> microscope slides
> cover glasses
> toothpicks
> IKI solution
> mixed cultures of protozoa
> medicine droppers
> microscope

Epithelial Cell Prepare a stained wet mount slide of some cells from the inside surface of your cheek as follows:

1. Wash a microscope slide and cover glass with soap and water.
2. Gently scrape some cells loose from the inside surface of your cheek with a clean toothpick. It is not necessary to draw blood!
3. Mix the cells in a drop of IKI solution on the slide.
4. Cover with a cover glass.
5. After locating a cell under low power of your microscope, make a careful examination of the cell under high-dry magnification. Identify the *nucleus, cytoplasm,* and *cell membrane.*
6. Draw a few cells on the Laboratory Report, labeling the above structures.

Microorganisms Prepare a wet mount of some protozoans by placing a drop of the culture on a slide and covering it with a cover glass. Be sure to insert the medicine dropper all the way to the bottom of the jar for your study sample since most protozoans will be found there.

Examine the slide first under low power, then under high-dry. Study individual cells carefully, looking for nuclei, cilia, flagella, vacuoles of ingested food, excretory vacuoles, etc.

Laboratory Report

Answer the questions on Laboratory Report 5,6 that pertain to this exercise.

Mitosis

6

Growth of body structures and the repair of specific tissues occur by cell division. While many cells in the body are able to divide, many do not beyond a certain age. Cell division is characterized by the equal division of all cellular components to form two daughter cells that are identical in genetic composition to the original cell. Cleavage of the cytoplasm is referred to as *cytokinesis;* nuclear division is *karyokinesis,* or *mitosis.* Cells that are not undergoing mitosis are said to be in a "resting stage," or *interphase.*

During this laboratory period we will study the various phases of mitosis on a microscope slide of hematoxylin-stained embryonic cells of *Ascaris,* a roundworm. The advantage of using *Ascaris* over other organisms is that it has so few chromosomes.

To facilitate clarity in understanding this process it has been necessary to combine photomicrographs and diagrams in figures 6.1 and 6.2; however, before we get into the mitotic stages, let's see what takes place in the interphase.

The Interphase

Although a cell in the interphase doesn't appear to be active, it is carrying on the physiological activities that are characteristic of its particular specialization. The cell may be in one of two different phases: the G_1 phase or the S phase (G_1 for growth, S for synthesis).

After a previous division, cells in the *S phase* begin to synthesize DNA, with each double-stranded DNA molecule replicating itself to produce another identical molecule. Most cells in the S phase will accomplish this replication process within twenty hours and then divide.

If division does not occur within the twenty-hour time frame, the cell is said to be in the *G phase* and will not divide again. Note in illustration 1, figure 6.1, that cells in the interphase stage exhibit a more or less translucent nucleus with an intact nuclear membrane.

Stages of Mitosis

The process of mitotic cell division is a continuous one once the cell has started to divide; however, for discussion purposes, the process has been divided

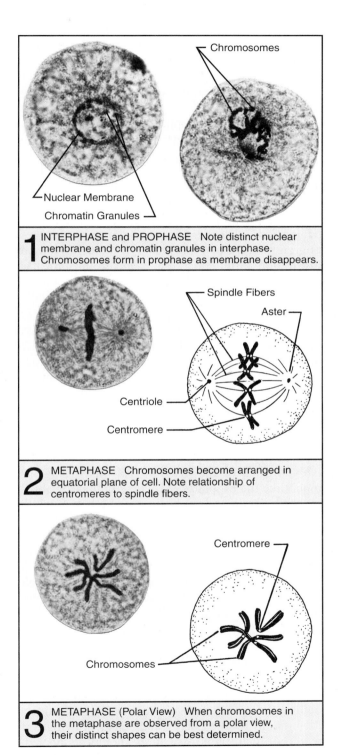

1 INTERPHASE and PROPHASE   Note distinct nuclear membrane and chromatin granules in interphase. Chromosomes form in prophase as membrane disappears.

2 METAPHASE   Chromosomes become arranged in equatorial plane of cell. Note relationship of centromeres to spindle fibers.

3 METAPHASE (Polar View)   When chromosomes in the metaphase are observed from a polar view, their distinct shapes can be best determined.

Figure 6.1   Early stages of mitosis.

27

into a series of four recognizable stages: prophase, metaphase, anaphase, and telophase. A description of each phase follows.

Prophase   In this first stage of mitosis the nuclear membrane disappears, **chromosomes** become visible, and each **centriole** of the centrosome moves to opposite poles of the cell to become an **aster.** The dark-staining chromosomes are made up of DNA and protein. Each aster consists of a centriole and astral rays (microtubules). By the end of the prophase some of the astral rays from each aster extend across the entire cell to become the **spindle fibers.**

From the early prophase to the late prophase the chromosomes undergo thickening and shortening. During this time each chromosome has a double nature, consisting of a pair of **chromatids.** This dual structure is due to the replication of the DNA molecules during the interphase.

Metaphase   During this stage the chromosomes migrate to the equatorial plane of the cell. Note in illustration 2, figure 6.1, that each chromosome has a **centromere,** which is the point of contact between each pair of chromatids and a spindle fiber.

Anaphase   As indicated in illustration 1 of figure 6.2, the individual chromatids of each chromosome are separated from each other and move apart along a spindle fiber to opposite poles of the cell.

Telophase   The telophase begins with the appearance of a **cleavage furrow** and the formation of a new **plasma membrane** between the two portions of the dividing cell. This stage is much like the prophase in reverse in that the chromosomes become less distinct and a new nuclear membrane begins to form around each set of chromosomes.

In the late telophase the spindle fibers disappear and the centrioles duplicate themselves. When the nuclear membranes and newly formed cell membrane are complete the two new structures are referred to as **daughter cells.**

Laboratory Assignment

Examine a prepared slide of *Ascaris* mitosis (Turtox slide E6.24). It will be necessary to use both the high-dry and oil immersion objectives. Identify all of the stages and make drawings, if required. Identifiable structures should be labeled. Answer the questions on Laboratory Report 5, 6.

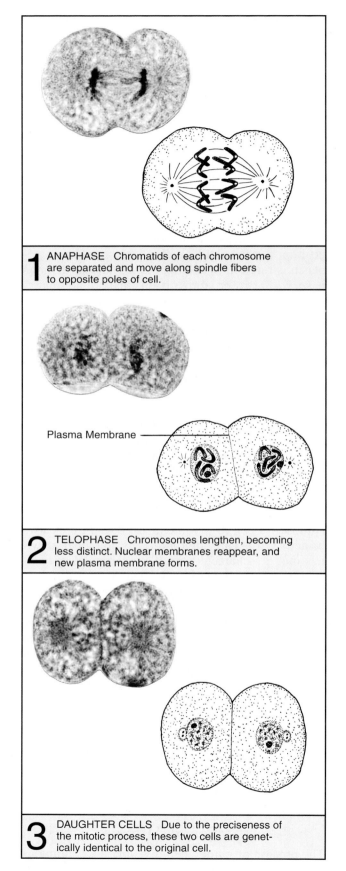

1 ANAPHASE   Chromatids of each chromosome are separated and move along spindle fibers to opposite poles of cell.

Plasma Membrane

2 TELOPHASE   Chromosomes lengthen, becoming less distinct. Nuclear membranes reappear, and new plasma membrane forms.

3 DAUGHTER CELLS   Due to the preciseness of the mitotic process, these two cells are genetically identical to the original cell.

Figure 6.2   Later stages of mitosis.

Osmosis and Cell Membrane Integrity

7

Every cell of the body is bathed in a watery fluid that contains a mixture of molecules that are essential to its survival. This fluid may be the plasma of blood or the tissue fluid in the interstitial spaces. In either case, these molecules, whether water, nutrients, gases, or ions, pass into and out of the cell through the plasma membrane.

Some molecules, usually of small size, are able to diffuse *passively* through the cell membrane from areas of high concentration to low concentration. Larger organic molecules, such as glucose and amino acids and certain ions, move through the plasma membrane either with or against a concentration gradient by *active transport,* which requires energy expenditure by the cell. Movement of all molecules occurs only if the membrane is viable and undisturbed.

The integrity of plasma membranes of cells throughout the body is due largely to the fact that body fluids are isotonic. To see what happens to cell membranes when isotonic conditions do not exist, we will study here the mechanics of **molecular diffusion** and **osmosis.**

Demonstration Setups:
Brownian movement (india ink preparation)
Osmosis (thistle tube setup)

Molecular Movement

All molecules, whether in a gas or liquid state, are in constant motion due to the release of kinetic energy. As the molecules bump into each other, directions are changed, causing random dispersal. Although direct observation of molecular movement is impossible due to their small size, one can observe this activity indirectly with both microscopic and macroscopic techniques. Two such observations are related to Brownian movement and molecular diffusion.

Brownian Movement

A Scottish botanist, Robert Brown, in 1827 observed that extremely small particles in suspension in the protoplasm of plant cells are in constant vibration. He erroneously assumed, at first, that this movement was a characteristic peculiar to living cells. Subsequent studies revealed, however, that this vibratory movement was caused by invisible water molecules bombarding the small visible particles. This type of movement became known as *Brownian movement.*

Demonstration:
The simplest way to observe this phenomenon is to examine a wet mount slide of india ink under the oil immersion objective of a microscope. India ink is a colloidal suspension of carbon particles in water, alcohol, and acetone.

Examine the demonstration setup that has been provided in the laboratory. This type of movement is of particular interest to microbiologists, since it must be differentiated from true motility in bacteria. Nonmotile bacteria are small enough to be displaced in this manner by moving water molecules.

Diffusion

If a crystal of some soluble compound is placed on the bottom of a container of water, molecules will disperse from the crystal to all parts of the container. The molecules are said to have spread throughout the water by *diffusion.* Movement of the molecules away from the original site of high concentration to low concentration is due to the fact that the reduced number of obstructing molecules in the lower concentration areas favors movement of molecules into those areas.

The rate of diffusion is variable and depends on ambient temperature and molecular size. Given sufficient time, diffusion eventually achieves even dispersal of all molecules throughout the container. This slow dispersal of molecules can be considerably augmented by convection currents caused by temperature differences.

To observe the phenomenon of diffusion, one can use crystals of colored compounds such as potassium permanganate (purple) and methylene blue. Figure 7.1 illustrates how to set up such a demonstration.

Methylene blue has a molecular weight of 320. Potassium permanganate has a molecular weight of 158. A Petri plate of 1.5% agar will be used. This agar medium of 98.5% water allows freedom of movement of molecules. Prepare such a plate as follows and record your observations on the Laboratory Report.

Materials:

> crystals of potassium permanganate and
>   methylene blue
> Petri plate with about 12 ml of 1.5% agar
> plastic ruler with metric scale

1. Select one crystal of each of the two different chemicals and place them on the surface of the agar medium about 5 cm apart. Try to select crystals of similar size.
2. Every 15 minutes, for 1 hour, examine the plate and measure the distance of diffusion (in millimeters) from the center of each crystal. Use a small plastic metric ruler.
3. Record the measurements on Laboratory Report 7.

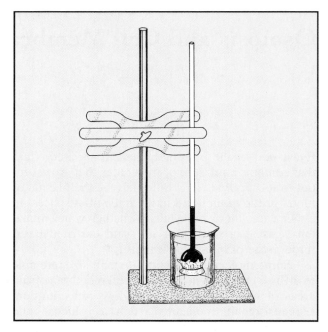

Figure 7.2   Osmosis setup.

Osmotic Effects

Water molecules, because of their small size, move freely through cell membranes into and out of the cell. Water acts as a vehicle that enables ions and other molecules to move through the membrane. This movement of water molecules through a semipermeable membrane is called *osmosis.*

Although water molecules are always moving both ways through the membrane, the predominant direction of flow is determined by the concentrations of solutes on each side of the membrane. If a funnel containing a sugar solution is immersed in a beaker of water and separated from the water with a semipermeable membrane, as in figure 7.2, more water molecules will enter the funnel through the membrane than will leave the sugar solution to enter the beaker of water. If a demonstration of this phenomenon is set up by your instructor in the laboratory, observe the upward movement that occurs during the period.

As was observed in the diffusion experiment, molecules tend to move from an area of high concentration to areas of low concentration. In this case, the water molecules are obviously of greater concentration in the beaker than in the sugar solution. The inward flow of water molecules produces an upward movement of the sugar solution in the tube. The force that would be required to restrain the upward flow is called the *effective osmotic pressure.*

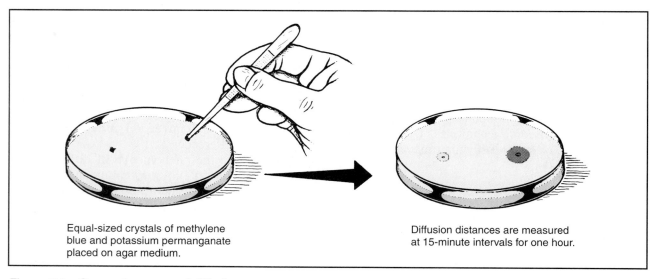

Equal-sized crystals of methylene blue and potassium permanganate placed on agar medium.

Diffusion distances are measured at 15-minute intervals for one hour.

Figure 7.1   Comparing areas of diffusion.

Solutions that contain the same concentration of solutes as cells are said to be **isotonic solutions.** Blood cells immersed in such solutions gain and lose water molecules at the same rate, establishing an *osmotic equilibrium.* See figure 7.3.

If a solution contains a higher concentration of solutes than is present in cells, water leaves the cells faster than it enters, causing the cells to shrink and develop cupped (crenated) edges. This shrinkage is called *plasmolysis,* or *crenation.* Such solutions of high solute concentration are called **hypertonic solutions** and are said to have a higher osmotic potential than isotonic solutions.

A solution that has a lower solute concentration than is present in cells is said to be a **hypotonic solution.** In such solutions water flows rapidly into the cells, causing them to swell *(plasmoptysis)* and burst *(lyse)* as the plasma membrane disintegrates. Lysis of red blood cells is called *hemolysis.*

To observe the effects of the various types of solutions on red blood cells, we will follow the procedures outlined in figure 7.4. Blood cells will be added to various concentrations of solutions. The effects of the solutions on the cells will be determined macroscopically and microscopically. Proceed as follows:

Materials:

  5 serological test tubes and tube rack
  small beaker of distilled water (50 ml size)
  2 depression microscope slides, cover glasses
  wax pencil, Vaseline, and toothpicks
  1 serological pipette (5 ml size)
  cannister for used pipettes

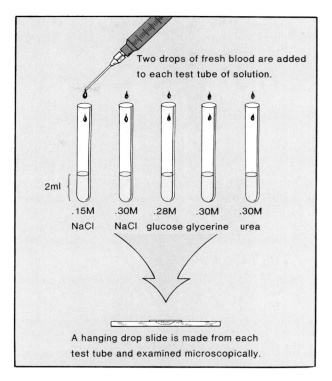

Figure 7.4   Routine for making cell suspensions.

  syringes, needles, Pasteur pipettes
  mechanical pipetting device
  fresh blood
  solutions of: 0.15M NaCl, 0.30M NaCl, 0.28M glucose, 0.30M glycerine, 0.30M urea

1. Label five clean serological tubes **1** to **5** and arrange them sequentially in a test-tube rack.
2. With a 5 ml pipette, deliver 2 ml of each solution to the appropriate tube. Refer to figure 7.4

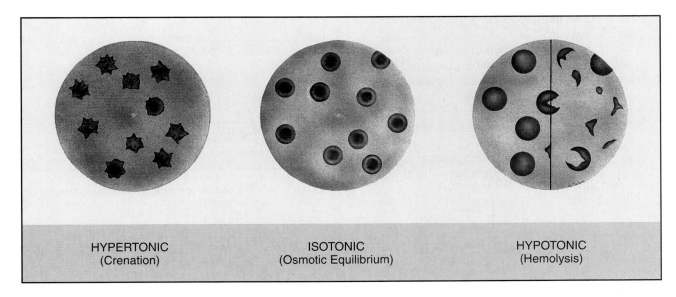

Figure 7.3   Effects of solutions on red blood cells.

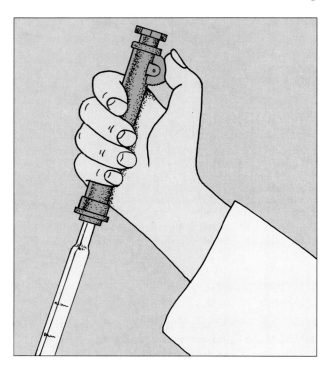

Figure 7.5 Use a mechanical pipetting device for all pipetting.

to determine which solution goes into which tube and refer to figure 7.5 for delivery method. Be sure to rinse out the pipette with distilled water between each delivery.

3. Dispense 2 drops of blood to each tube, using a syringe. Shake each tube from side to side to mix, then let stand for **5 minutes.**

4. Hold the rack of tubes up to the light and compare them. If the solution is transparent, **hemolysis** has occurred. If you are unable to see through the tube, **no hemolysis** has occurred and the cells should be intact.

   If crenation has occurred, the appearance will be somewhat between the clarity of hemolysis and the opacity of osmotic equilibrium.

   Record your results on the Laboratory Report.

5. Make a hanging drop slide from each tube as follows:

   a. Place a tiny speck of Vaseline near each corner of a cover glass. See illustration 1, figure 7.6.

   b. With a Pasteur pipette transfer a drop of the cell suspension to the center of the cover glass.

   c. Place a depression slide on the cover glass with the depression facing downward.

   d. With the cover glass held to the slide by the Vaseline, quickly invert.

6. Examine each slide under high-dry and record your results on the Laboratory Report.

Laboratory Report

After recording your results on the Laboratory Report, complete the report by answering all the questions.

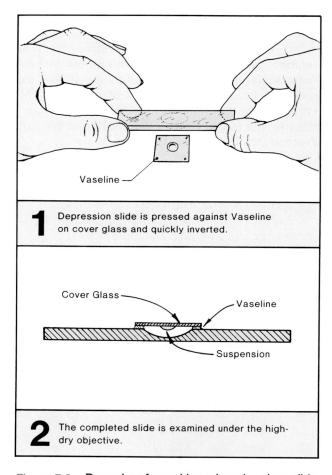

Vaseline

1 Depression slide is pressed against Vaseline on cover glass and quickly inverted.

Cover Glass — Vaseline

Suspension

2 The completed slide is examined under the high-dry objective.

Figure 7.6 Procedure for making a hanging drop slide.

Epithelial Tissues

8

Although all cells of the body share common structures such as nuclei, centrosomes, Golgi, etc., they differ considerably in size, shape, and structure according to their specialized functions. An aggregate of cells that are similar in structure and function is called a **tissue.** The science that relates to the study of tissues is called **histology.**

In this exercise we will study the different types of epithelial tissues. Prepared microscope slides from portions of different organs will be available for study. Learning how to identify specific epithelial tissues on slides that have several types of tissue will be part of the challenge of this exercise.

Epithelial tissues are aggregations of cells that perform specific protective, absorptive, secretory, transport, and excretory functions. They often serve as coverings for internal and external surfaces and they rest upon a bed of connective tissue. Characteristics common to all epithelial tissues are as follows:

- The individual cells are closely attached to each other at their margins to form tight sheets of cells lacking in extracellular matrix and vascularization.
- The cell groupings are oriented in such a way that they have an apical (free) surface and a

basal (bound) region. The basal portion is closely anchored to underlying connective tissue. This thin adhesive margin between the epithelial cells and connective tissue is called the **basal lamina.** Although this structure is also referred to as the "basement membrane," it is not a true membrane. Unlike true membranes, the basal lamina is acellular. In reality, it is a colloidal complex of protein, polysaccharide, and reticular fibers.

Differentiation of the various epithelia is illustrated in the separation outline, figure 8.1. The basic criteria for assigning categories are cell shape, surface specializations, and layer complexity. The three divisions *(simple, pseudostratified,* and *stratified)* are based on layer complexity. A discussion of each follows:

Simple Epithelia

Epithelial tissues that fall in this category are composed of a single cell layer that extends from the basement lamina to the free surface. Figure 8.2 illustrates three basic kinds of simple epithelia.

Simple Squamous Epithelia   These cells are very thin, flat, and irregular in outline. An example of

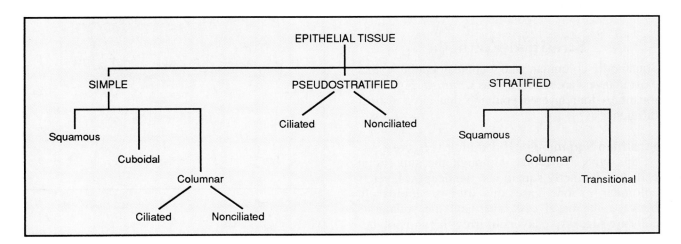

Figure 8.1   A morphological classification of epithelial types.

this type is seen in illustration 1, figure 8.2. They form a pavementlike sheet in various organs that perform filtering or exchange functions. Capillary walls, alveolar walls in the lungs, the peritoneum, the pleurae, and blood vessel linings consist of simple squamous epithelia.

Cuboidal Epithelia   Cells of this type (illustration 2, figure 8.2) are stout and blocklike in cross section and hexagonal from a surface view.

This type of tissue is seen in many glands, such as the thyroid, salivary, and pancreas. It is also seen in the ovaries and the capsule surrounding the lens of the eye.

Simple Columnar Epithelia   Illustration 3, figure 8.2, reveals that cells of this type are elongated between their apical and basal surfaces. When the free surfaces are observed, the cells have a polygonal configuration.

Functionally, this type of tissue may be specialized for protection, secretion, or absorption. Note in illustration 3 that certain secretory cells, called **goblet cells,** may be present that produce **mucus,** a protective glycoprotein. Also, note that the cells may be *ciliated* or not ciliated *(plain).*

Plain columnar epithelial tissue lines the stomach, intestines, and kidney collecting tubules. Ciliated columnar epithelia can be found in the lining of the respiratory tract, uterine tubes, and portions of the uterus. Goblet cells can be present in both the plain and ciliated types.

Another modification of the free surfaces of columnar cells is the presence of **microvilli.** These structures are seen as a **brush border** when observed with a light microscope under oil immersion. Columnar cells that line the small intestine and make up the walls of the kidney collecting tubules exhibit distinct brush borders.

Stratified Epithelia

Squamous, columnar, and cuboidal epithelia that exist in layers are referred to as *stratified.* Figure 8.3 illustrates four layered epithelia that have distinct differences.

Stratified Squamous   Illustration 1, figure 8.3 is of this type. Note that, although the superficial cells are distinctly squamous, the deepest layer is columnar; in some cases this layer is cuboidal. In between the basal cell layer and the squamous cells are successive layers of irregular and polyhedral cells.

Protection is the chief function of this type of tissue. Exposed inner and outer surfaces of the

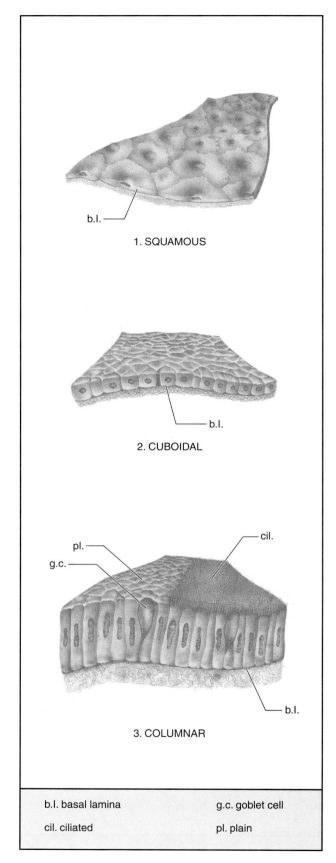

| b.l. basal lamina | g.c. goblet cell |
| cil. ciliated | pl. plain |

Figure 8.2   Simple epithelia.

body, such as the skin, oral cavity, esophagus, vagina, and cornea consist of stratified squamous epithelia.

Stratified Columnar This type of epithelium is shown in illustration 2, figure 8.3. Observe that, although the superficial cells are distinctly columnar, the deeper cells are irregular or polyhedral. Note also that the superficial columnar cells are variable in height.

Protection and secretion are the chief functions of this type of tissue. Distribution of stratified columnar is limited to some glands, the conjunctiva, the pharynx, a portion of the urethra, and lining of the anus.

Stratified Cuboidal Although no stratified cuboidal tissue is listed in figure 8.1, tissue of this configuration is seen in the ducts of sweat glands of the skin. The ducts in illustration C, figure HA-9, show what stratified cuboidal epithelium looks like.

Transitional A unique characteristic of this type of stratified epithelium is that the surface layer consists of large, round, dome-shaped cells that may be binucleate. The deeper strata are cuboidal, columnar, and polyhedral.

Note in illustration 3, figure 8.3, that the deeper cells are not as closely packed as in other stratified epithelia. Distinct spaces can be seen between the cells. This looseness of cells imparts a certain

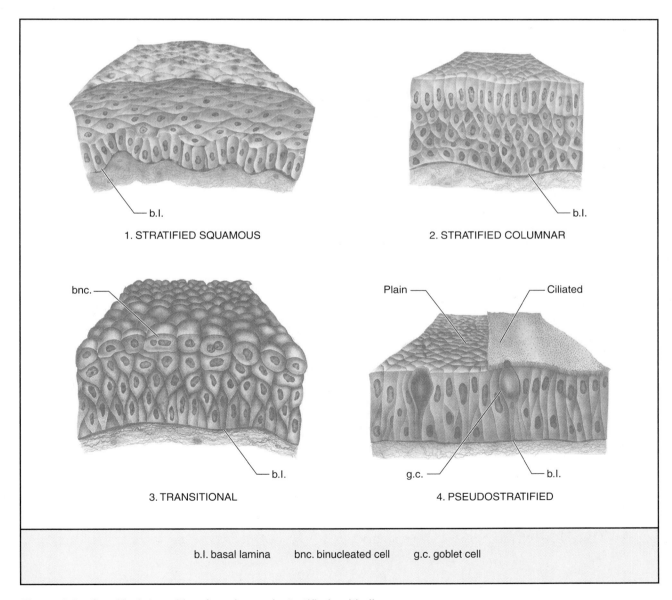

b.l. basal lamina   bnc. binucleated cell   g.c. goblet cell

Figure 8.3 Stratified, transitional, and pseudostratified epithelia.

degree of elasticity to the tissue. Organs such as the urinary bladder, ureters, and kidneys (the calyces), contain transitional epithelium that enables distension due to urine accumulation.

Pseudostratified Epithelia

This epithelium presents a superficial stratified appearance because of the staggered nuclei as well as two different cell orientations. It is only one layer thick, however. Close examination will reveal that every cell is in contact with the basal lamina, but only the columnar types extend to the free surface. The smaller cells wedged between them have no free surface.

Two types are shown in illustration 4 of figure 8.3: plain and ciliated. The *plain,* or nonciliated, is found in the male urethra and parotid gland. The *ciliated* type lines the trachea, bronchi, auditory tube, and part of the middle ear. Since both types frequently produce mucus, they possess goblet cells.

Laboratory Assignment

Do a systematic study of each kind of epithelial tissue by examining prepared slides that should show the kinds of tissue you are looking for. Remembering that epithelial cells have a free surface and a basement lamina, always look for the tissue near the edge of a structure.

Note in the materials list below that preferred and optional lists of slides are given. The *preferred* list is of slides that are limited to the type of tissue being studied. If these are available, use them. If they are not available, use the *optional* slides, which are very good but more difficult to use because they usually have several kinds of tissue on each slide.

The numbers listed in parentheses after each tissue are Turtox code numbers. Since most of the photomicrographs in the Histology Atlas were made of Turtox slides, the code numbers are provided so that instructors can provide slides that most closely match the illustrations.

Materials:
preferred slides:
    squamous (H1.1)
    stratified squamous (H1.14)
    cuboidal (H1.21)
    pseudostratified ciliated (H1.32)
    transitional (H1.41)

optional slides:
    skin (H11.12 or H11.14)
    trachea (H6.41)
    stomach (H5.415 or H5.425)
    ileum (H5.531)
    kidney (H9.11 or H9.15)
    thyroid gland (H14.11, H14.12, or H14.13)

Squamous Epithelium Look for the flattened cells that are representative of this kind of tissue. Use figure HA-1 in the Histology Atlas for reference. The exfoliated cells are the same ones you studied in Exercise 5.

Explore each slide first with the low-power objective before using the high-dry or oil immersion objectives. Make drawings if required.

Columnar Epithelium Consult figure HA-2 in the Histology Atlas for representatives of this group of epithelial cells. To see the **brush border** or **cilia** on these cells it will be necessary to study the cells with the high-dry or oil immersion objectives. Can you identify the **basal lamina** on each tissue?

The **lamina propria** consists of connective tissue, vascular and lymphatic channels, lymphocytes, plasma cells, eosinophils, and mast cells. All these structures will be studied later.

Cuboidal Epithelium Illustrations A and B, figure HA-3, reveal the appearance of cuboidal tissue. The preferred slide (H1.21) is usually made from the uterine lining of a pregnant guinea pig. If the preferred slide is unavailable, use a slide of the thyroid gland for this type of tissue.

Transitional Tissue Illustrations C and D of figure HA-3 and illustration C, figure HA-33, provide good examples of this tissue. Note that the cells seem to be loosely arranged. Look for **binucleate cells.**

Ciliated Pseudostratified Columnar Epithelium The best place to look for this type of tissue is in a cross section of the trachea. Illustration D, figure HA-2, is the epithelium of the trachea.

Laboratory Report

Answer the questions on Laboratory Report 8,9 that pertain to the epithelial tissues.

9

Connective Tissues

Connective tissues include those tissues that perform binding, support, transport, and nutritive functions for organs and organ systems. Characteristically, all connective tissues have considerable amounts of nonliving extracellular substance that holds and surrounds various specialized cells.

The extracellular material, or **matrix,** is composed of fibers, fluid, organic ground substance, and/or inorganic components intimately associated with the cells. This matrix is a product of the cells in the tissue. The relative proportion of cells to matrix will vary from one tissue to another.

Several systems of classification of connective tissues have been proposed. Figure 9.1 reveals the system that we will use here. It is based on the nature of the extracellular material; in some cases there is a certain amount of overlapping of categories.

Connective Tissue Proper

Histologically, *connective tissue proper* is composed, primarily, of protein fibers, special cells, and a ground substance that varies among the several types. The fibers may differ in protein composition and density. From the standpoint of composition, fibers consist of either *collagenous* or *elastin* protein.

There are three basic types of fibers: collagenous, reticular, and elastic. **Collagenous** (white) **fibers,** are relatively long, thick bundles seen in most ordinary connective tissues in varying amounts. **Reticular fibers** constitute minute networks of very fine threads. Although both collagenous and reticular fibers are composed of collagen, the reticular fibers stain more readily with silver dyes (i.e., they are *argyrophilic*).

Elastic (yellow) **fibers** are often found in connective tissue stroma of organs that must yield to changes in shape. These are the only fibers to contain elastin protein. All three types of fibers are seen in loose (areolar) tissue (figure 9.2).

The ground substance usually consists of complex peptidoglycans that form an amorphous solution or gel around the cells and fibers. Chondroitin sulfate and hyaluronic acid are frequent components.

Various types of cells are seen in the matrix of connective tissue proper. Examples include fibroblasts, adipose cells, mast cells, plasma cells, macrophages, and other types of blood cells.

Loose Fibrous (Areolar) Connective Tissue

This tissue is found throughout the body, interwoven into the stroma of many organs. The cells

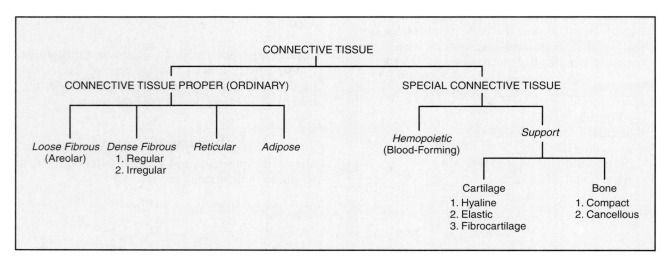

Figure 9.1   Types of connective tissue.

and extracellular substances of this tissue are very loosely organized. Because of the preponderance of spaces in this tissue, it is often designated as *areolar* (a small space). It is highly flexible and capable of distension when excess extracellular fluid is present. The left-hand illustration in figure 9.2 depicts its generalized structure. Note the presence of **mast cells, macrophages,** and **fibroblasts.**

Three important functions are served by loose connective tissue: (1) it provides flexible support and a continuous network within organs, (2) it furnishes nutrition to cells in adjacent areas due to its capillary network, and (3) it provides an arena for activities of the immune system.

This type of tissue is found beneath epithelia, around and within muscles and nerves, and as part of the serous membranes. The deep and superficial *fasciae* encountered in the rat dissection are of this type.

Adipose Tissue

Fat or adipose cells can be found in small groupings throughout the body. However, as they constitute a storage depot for fat, they frequently accumulate in large areas to make up the bulk of body fat. These latter areas are made up, essentially, of *adipose tissue*. There are various types of adipose tissue, but *ordinary adipose* is the most common type found in humans. Refer to figure 9.2.

Fat cells are very large and are characterized by a spherical or polygonal shape. As much as 95% of their mass may be stored fat. As the process of fat deposition ensues, the cell cytoplasm becomes reduced and thin, and the vacuole, filled with lipids, appears as a large open space with a thin periphery of cytoplasm. The nucleus is displaced to one side, producing a signet ring appearance. A fine network of reticular fibers exists between the cells.

Fat tissue serves as a protective cushion and insulation for the body, in addition to being a potential source of energy and heat generator. Its rich vascular supply points to its relatively high rate of metabolism and turnover.

Reticular Tissue

This tissue (figure 9.3) is generally regarded as a network of reticular fiber elements within certain organs. Since the fibers show a unique pattern and staining reaction, they can be identified as the supporting framework of many vascular organs, such as the liver, lymphatic structures, hemopoietic tissue, and basal lamina. The left-hand illustration in figure 9.3 reveals the appearance of the reticulum of a lymph node.

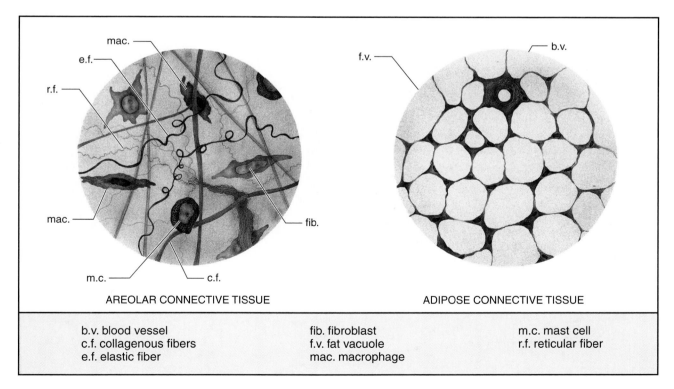

AREOLAR CONNECTIVE TISSUE

ADIPOSE CONNECTIVE TISSUE

b.v. blood vessel	fib. fibroblast	m.c. mast cell
c.f. collagenous fibers	f.v. fat vacuole	r.f. reticular fiber
e.f. elastic fiber	mac. macrophage	

Figure 9.2 Loose fibrous and adipose tissues.

Dense Fibrous (White) Connective Tissue

Tissue in this category differs from the loose variety in having a predominance of fibrous elements and a sparseness of cells, ground substance, and capillaries. According to the arrangement of fibers this tissue can be separated into two groups: *regular dense* and *irregular dense* tissues. The right-hand illustration in figure 9.3 reveals both types.

Dense regular connective tissue is the type found in tendons, ligaments, fascia, and aponeuroses. It consists of thick groupings of longitudinally organized collagenous fibers and some elastic fibers. These tough strands have enormous tensile strength and are capable of withstanding strong pulling forces without stretching. Fibroblasts are the principal cells, although they are quite scarce.

Dense irregular connective tissue has elastic fibers that are woven into flat sheets to form capsules around certain organs. The sheaths that encircle nerves and tendons are of this type. A large portion of the dermis also consists of the dense irregular type.

Special Connective Tissue

As indicated in the categorization chart in figure 9.1, this type of tissue includes two divisions:

hemopoietic and support tissues. *Hemopoietic tissue* consists of red bone marrow connective tissue that produces the formed elements of blood (red blood cells, white blood cells, and platelets). Although this tissue will not be studied here, we will study the products of this tissue in Exercises 40 and 41.

Our principal concern here will pertain to the *support tissues*, which include cartilage and bone. These two tissues are adapted for the bearing of weight. Structural characteristics shared by these tissues include: (1) a solid, flexible, yet *strong extracellular matrix*, (2) cells contained in matrix cavities called *lacunae*, and (3) an external covering *(periosteum* or *perichondrium)* that is capable of generating new tissue.

Cartilage

In general, cartilage consists of a stiff, plastic matrix that has lubricating as well as weight-bearing capability. As a result, it is found in areas that require support and movement (skeleton and joints).

Basically, all three types of cartilage in humans consist of cartilage cells (**chondrocytes**), embedded within a matrix that contains ground substance and fibers. The proportion devoted to matrix is much greater than that for chondrocytes. Unlike other connective tissues, cartilage is devoid of a

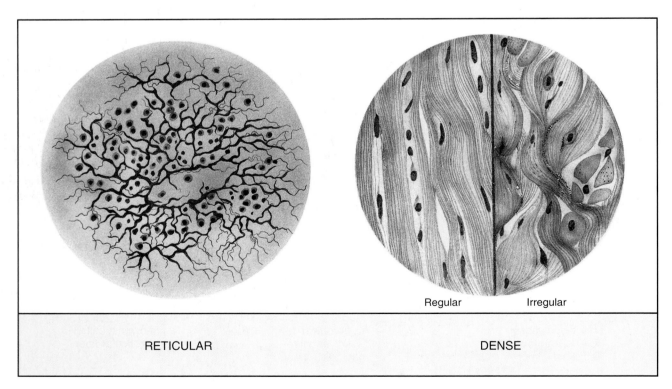

Regular   Irregular

RETICULAR   DENSE

Figure 9.3   Reticular and dense connective tissue.

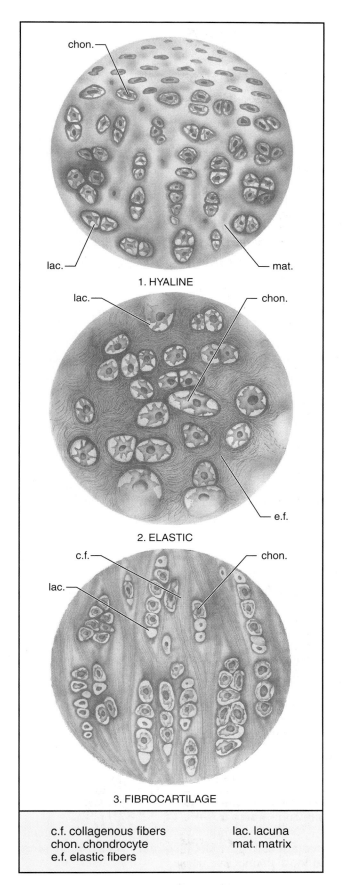

chon.

lac. ——— mat.

1. HYALINE

lac. ——— chon.

e.f.

2. ELASTIC

c.f. ——— chon.

lac.

3. FIBROCARTILAGE

c.f. collagenous fibers	lac. lacuna
chon. chondrocyte	mat. matrix
e.f. elastic fibers	

Figure 9.4   Types of cartilage.

vascular supply and receives all nutrients through diffusion. The three classes of cartilage are shown in figure 9.4.

Hyaline cartilage (illustration 1) is a pearly glasslike tissue that makes up a very large part of the fetal skeleton. During fetal development it is gradually replaced by bone, except for certain areas in the joints, ear, larynx, trachea, and ribs.

The matrix is a firm homogenous gel made up of chondroitin sulfate and collagen. Cells occur singly or in "nests" of several cells, the sides of which may be flattened. Often the areas directly around lacunae appear denser in sections.

Elastic cartilage (illustration 2) is similar in basic structure to hyaline cartilage, but differs in that it contains significant amounts of elastic fibers in the matrix. The result is a tissue that has highly developed flexibility and elasticity. The pinna of the external ear, the epiglottis, and the auditory tube are reinforced with this type of cartilage.

Fibrocartilage (illustration 3) differs from the other two types in that its chondrocytes are arranged in groupings between bundles of collagenous fibers. It serves a useful cushioning function in strategic joint ligaments and tendons. It is the major component, for instance, of the intervertebral disks of the vertebral column and the pubic symphysis.

Bone Tissue

The tissue that makes up the bones of the skeleton meets all the criteria of connective tissue, yet it has many striking and unique features. It is hard, unyielding, very strong, and relatively lightweight. It is found in those parts of the anatomy that require maximum weight-bearing capacity, protection, and storage of certain materials.

In a sense, bone can be visualized as a living organic cement. It consists of specialized cells, blood vessels, and nerves that are reinforced within a hard ground substance made up of organic secretions impregnated with mineral salts. It arises during embryonic and fetal development from cartilage and/or fibrous connective tissue precursors. The initial organic matrix consists of collagenous fibers that serve as a framework for the gradual deposition of calcium and phosphate salts by special bone cells, the *osteoblasts*. Far from being an inactive tissue, it is continuously being modified and reconstructed by both metabolic and external influences.

In macroscopic sections of bones, two frameworks are apparent: the solid, dense **compact bone,** which makes up the outermost layer; and the more porous **spongy** *(cancellous)* **bone** that is located internally. Each of these variations has a distinctive histological character.

Compact Bone   Upon close inspection of a cut section of the sternum viewed from a three-dimensional perspective (figure 9.5, in illustration A), it is apparent that compact bone (label 2) is permeated by a microscopic framework of tunnels, channels, and interconnecting networks that are surrounded by a hard matrix. Within this hollow network exist the living substances of bone that facilitate its nourishment and maintenance.

The functional and structural unit of compact bone is a cylindrical component called the **osteon,** or Haversian system (label 9, illustration B). In the center of each osteon is a hollow space, the **central** (Haversian) **canal,** which contains one or two blood capillaries. Surrounding each central canal are several concentric rings of matrix called the **lamellae. Osteocytes,** which are responsible for secreting the lamellae, can be seen within small hollow cavities, the **lacunae.** Note that these cavities, which are oriented between the lamellae, have many minute hollow tunnels, called **canaliculi,** that radiate outward, imparting a spiderlike appearance. The canaliculi contain protoplasmic processes of the osteocytes.

If one follows the various levels of structure shown in figure 9.5, it should become evident that the network comprises a continuous communication system from the central canal to the lacunae and between adjacent lacunae via the canaliculi. The entrapped osteocytes, thus, can receive nourishment and exchange materials within the hard space of the matrix. Intimate contact between the protoplasmic processes of adjacent osteocytes through the canaliculi makes all of this possible.

Groups of osteons lie in vertical array with adjacent lamellae separated at lines of demarcation called **cement lines.** Continuity between the central canals of adjacent osteons is achieved by **perforating** (Volkmann's) **canals** that penetrate the bone obliquely or at right angles.

Spongy Bone   This type of bone lies adjacent to compact bone and is continuous with it; there is no distinct line of demarcation between the two regions. Histologically, it presents a lesser degree of organization than compact bone. Illustration A reveals that its outstanding feature is a series of branching, overlapping plates of matrix called **trabeculae.** These plates are oriented so as to produce large, interconnecting cavelike spaces. These spaces function well in storage and as pockets to hold the blood-forming cells (hemopoietic tissue) of the bone marrow. They also function in weight reduction.

Note that the trabeculae are also randomly punctuated by the osteocyte-holding spaces, or lacunae, and that blood vessels meander through the large spaces between the trabeculae, bringing nourishment to nearby osteocytes.

The Periosteum   Contiguous with the outer layer of compact bone, and tightly adherent to it, is a thick, tough membrane called the *periosteum.* It is composed of an outer layer of fibrous connective tissue and an inner *osteogenic layer,* which serves as a source of new bone-forming cells and provides an access for blood vessels. The periosteum is anchored tightly to compact bone by bundles of collagen fibers that perforate and become firmly embedded within the outer lamellae. These minute attachments are called **Sharpey's fibers.**

Assignment:
Label figure 9.5.

Histological Study

Do a systematic study of each type of connective tissue by examining prepared slides that are available. Note that Histology Atlas references are indicated for each type of tissue.

Materials:
prepared slides of:
    areolar (H2.13)
    adipose (H2.51)
    white fibrous (H2.115)
    yellow (elastic) fibrous (H2.125)
    reticular (H2.31)
    hyaline cartilage (H2.61)
    fibrocartilage (H2.63)
    elastic cartilage (H2.62)
    bone, x.s. (H2.735)
    developing bone (H2.79)

Connective Tissue Proper   (figure HA-4) Examine slides of areolar, adipose, white fibrous, and reticular connective tissues, identifying all the structures shown in figure HA-4.

Note the comments that are made in the legend pertaining to the function of mast cells that are seen in loose connective tissue.

Cartilage (figure HA-5) Study the three different types of cartilage, noting their distinct differentiating characteristics.

Bone (figure HA-6) When studying a slide of developing membrane bone try to differentiate the osteoblasts from the osteoclasts. Note that an osteo-

clast is a large multinucleate cell with a clear area between it and the bony matrix.

Laboratory Report

Answer all the questions on combined Laboratory Report 8,9.

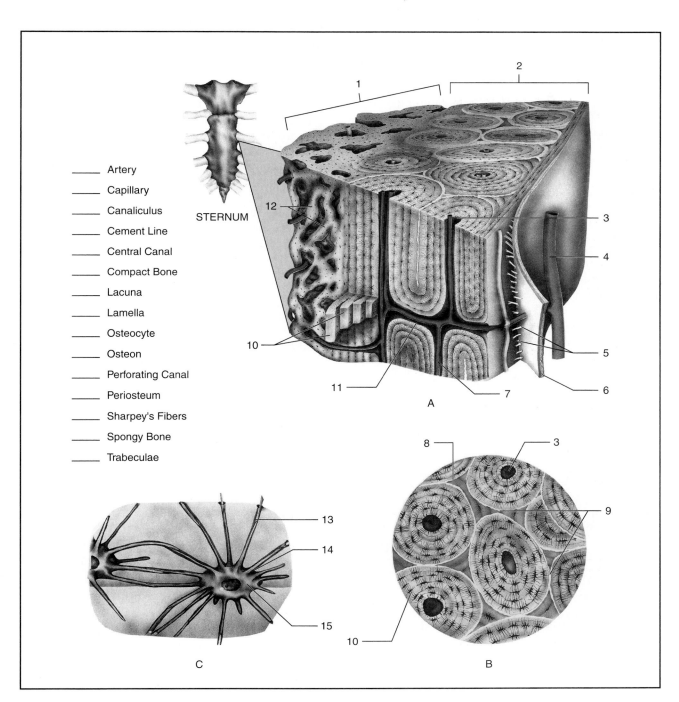

_____ Artery
_____ Capillary
_____ Canaliculus
_____ Cement Line
_____ Central Canal
_____ Compact Bone
_____ Lacuna
_____ Lamella
_____ Osteocyte
_____ Osteon
_____ Perforating Canal
_____ Periosteum
_____ Sharpey's Fibers
_____ Spongy Bone
_____ Trabeculae

Figure 9.5   Bone tissue.

10

The Integument

Since the skin is constructed of epithelial and connective tissues, a study of it at this time provides one with an opportunity to review experiences of the last two laboratory periods. During this laboratory period prepared slides of the skin will be available for study. Prior to examining the slides, however, label figures 10.1 and 10.2. Note that the skin consists of two layers: an outer multilayered **epidermis** and a deeper **dermis.**

The Epidermis

The enlarged section of skin on the right side of figure 10.1 is the *epidermis*. Note that it consists of four distinct layers: an outer **stratum corneum,** a thin translucent **stratum lucidum,** a darkly stained **stratum granulosum,** and a multilayered **stratum spinosum** (stratum mucosum).

All four layers of the epidermis originate from the deepest layer of cells of the stratum spinosum. It is called the **stratum basale** (stratum germinativum). The columnar cells of this deep layer are constantly dividing to produce new cells that move outward to undergo metamorphosis at different levels. The brown skin pigment *melanin,* which is produced by stellate *melanocytes* of the stratum basale, is responsible for skin color. Skin color differences are due to the amount of melanin present.

The stratum corneum of the epidermis consists of many layers of the scaly remains of dead

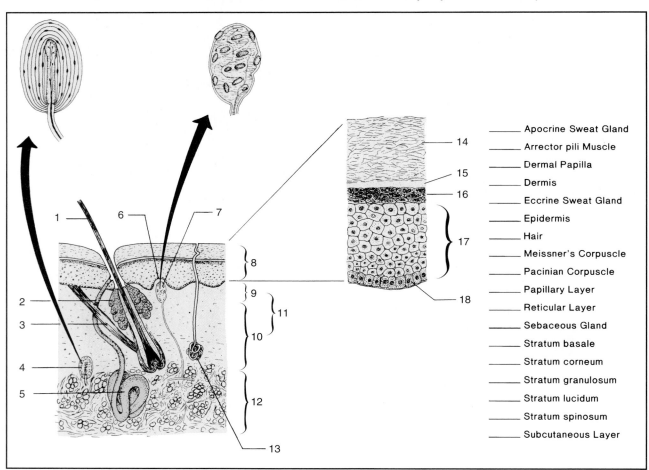

_____ Apocrine Sweat Gland

_____ Arrector pili Muscle

_____ Dermal Papilla

_____ Dermis

_____ Eccrine Sweat Gland

_____ Epidermis

_____ Hair

_____ Meissner's Corpuscle

_____ Pacinian Corpuscle

_____ Papillary Layer

_____ Reticular Layer

_____ Sebaceous Gland

_____ Stratum basale

_____ Stratum corneum

_____ Stratum granulosum

_____ Stratum lucidum

_____ Stratum spinosum

_____ Subcutaneous Layer

Figure 10.1   Skin structure.

epithelial cells. This protein residue of dead cells is primarily *keratin,* a water-repellent material. As the cells of the stratum spinosum are pushed outward, they move away from the nourishment of the capillaries, die, and undergo *keratinization.* The *eleidin* granules of the stratum basale are believed to be an intermediate product of keratinization. The translucent stratum lucidum consists of closely packed cells with traces of flattened nuclei.

The Dermis

This layer is often referred to as the "true skin." It varies in thickness of less than a millimeter to over 6 millimeters. It is highly vascular and provides most of the nourishment for the epidermis. It consists of two strata, the papillary and reticular layers.

The outer portion of the dermis, which lies next to the epidermis, is the **papillary layer.** It derives its name from numerous projections, or **dermal papillae,** which extend into the upper layers of the epidermis. In most regions of the body these papillae form no pattern; however, on the fingertips, palms, and soles of the feet they form regularly arranged patterns of parallel ridges that improve frictional characteristics in these areas.

The deeper portion, or **reticular layer,** contains more collagenous fibers than the papillary layer. These fibers greatly enhance the strength of the skin. The surface texture of suede leather is, essentially, the reticular layer of animal hides.

Subcutaneous Tissue

Beneath the dermis lies the **subcutaneous layer,** or *hypodermis;* it is also referred to as the *superficial fascia.* It consists of loose connective tissue, nerves, and blood vessels. One of its prime functions is to provide attachment for the skin to underlying structures.

Hair Structure

Hair *(pili)* consists of keratinized cells that are compactly cemented together. Each shaft of hair (label 1, figures 10.1 and 10.2) is surrounded by a tube of epithelial cells, the **hair follicle.** The terminal end of the hair shaft, or *root,* is enlarged to form an onion-shaped region called the **bulb** (see figure 10.2). Within the bulb is an involution of loose connective tissue called the **follicular (hair) papilla.** It is through the latter structure that nourishment enters the shaft. The root of the hair is encased in an **internal root sheath** and an **external root sheath.**

Extending diagonally from the wall of the hair follicle to the epidermis is a band of smooth muscle fibers, the **arrector pili muscle.** Contraction of these muscle fibers causes the hair to move to a more perpendicular position, causing elevations on the skin surface, commonly referred to as "goose pimples."

Glands

Two kinds of glands are present in the skin: sebaceous and sweat.

Sebaceous Glands These glands are located within the epithelial tissues that surround each hair follicle. An oily secretion, called *sebum,* is secreted by these glands into the hair follicles and out onto the skin surface. Secretion is facilitated to some extent by the force of the arrector pili muscles during contraction. Sebum keeps hair pliable and helps to waterproof the skin.

Sweat Glands Sweat glands are of two types: eccrine and apocrine. The small sweat glands that empty directly out through the surface of the skin are **eccrine sweat glands.** These glands are simple tubular structures that have their coiled basal

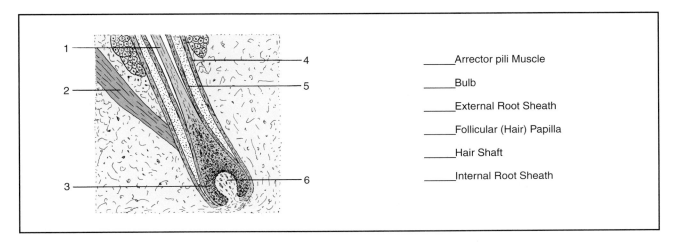

_____ Arrector pili Muscle

_____ Bulb

_____ External Root Sheath

_____ Follicular (Hair) Papilla

_____ Hair Shaft

_____ Internal Root Sheath

Figure 10.2 Structure of the hair root.

portions located deep in the dermis. Except for the lips, glans penis, and clitoris, they are widely distributed throughout the body.

Although all eccrine gland secretions are similar in composition, there are two different controlling stimuli. Almost everyone is aware that the sweat glands of some parts of the body, such as the palms and axillae, are affected by emotional factors. Glands in other regions, however, such as the forehead, neck, and back are regulated, primarily, by thermal stimuli.

Apocrine sweat glands are much larger than the eccrine type and have their secretory coiled portions located in the subcutaneous layer. Instead of emptying out onto the surface of the epidermis, all apocrine glands empty directly into a hair follicle canal.

These glands are found in the axillae, scrotum of the male, female perigenital region, external ear canal, and nasal passages. While eccrine sweat is watery, the secretion of apocrine glands is thick white, gray, or yellowish. Malodorous substances in apocrine sweat are the principal contributors to body odors. Psychic factors, rather than temperature changes, contribute mostly to apocrine secretions.

Receptors

The receptors shown in figure 10.1 are Meissner's and Pacinian corpuscles. **Meissner's corpuscles** are located in the papillary layer of the dermis, projecting up into papillae of the epidermis. They function as receptors of touch.

Pacinian corpuscles are spherical receptors with onionlike laminations. They lie deep in the reticular layer of the dermis. Pacinian corpuscles are sensitive to variations in sustained pressure.

Laboratory Assignment

After labeling figures 10.1 and 10.2 and answering the questions on Laboratory Report 10, proceed as follows:

Materials:
  prepared slide of the skin (H11.11, H11.12, or H11.14)

Examine prepared slides of sections through the skin and identify the structures seen in figures 10.1, HA-9, HA-10, and HA-11. Make drawings if required.

The Skeletal Plan

In this exercise we will study the structure of the skeleton as a whole and the anatomy of a typical long bone. Detailed examination of individual parts will follow in subsequent exercises.

Materials:
> fresh beef bones, sawed longitudinally
> articulated human skeleton

Long Bone Structure

We see in figure 11.1 a diagram of the femur cut open, longitudinally, to reveal its internal structure. Note that it has a long shaft called the **diaphysis** and two enlarged ends, the **epiphyses.** Where the epiphyses meet the diaphysis are growth zones called **metaphyses.**

During the growing years a plate of hyaline cartilage, the *epiphyseal disk*, exists in each of these growth areas. As new cartilage forms on the epiphyseal side, it is destroyed and replaced by bone on the diaphyseal side. While the metaphysis during the growing years consists of the epiphyseal disk and calcified cartilage, at maturity the area becomes completely ossified, and linear growth ceases.

Note that the central portion of the diaphysis is a hollow chamber, the **medullary cavity.** Lining this cavity is a thin membrane called the **endosteum.** This membrane is continuous with the linings of the central canals of the osteons.

The entire medullary cavity and much of the cancellous tissue of the bone extremities contain **yellow marrow,** a fatlike substance. The cancellous bone of the epiphyses of the femur (and humerus) contains **red marrow** in the adult. Other long bones of the skeleton contain only yellow marrow. Most of the red marrow in adults is contained in the ribs, sternum, and vertebrae.

A tough covering, the **periosteum,** envelops the surfaces of the entire bone except for the areas of articulation. This covering consists of fibrous connective tissue that is quite vascular. The surfaces of each epiphysis that contact adjacent bones are covered with smooth **articular cartilage** that is of the hyaline type.

Assignment:
Identify the labels in figure 11.1.

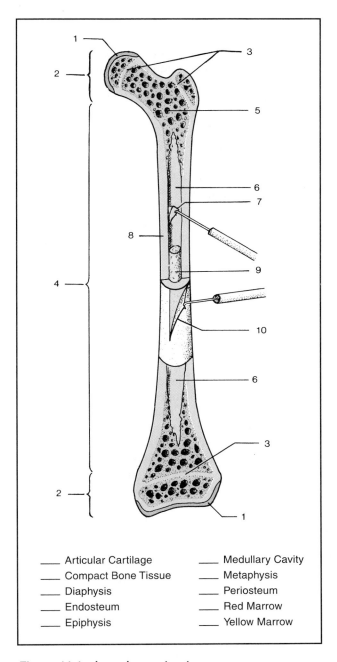

_____ Articular Cartilage   _____ Medullary Cavity
_____ Compact Bone Tissue   _____ Metaphysis
_____ Diaphysis   _____ Periosteum
_____ Endosteum   _____ Red Marrow
_____ Epiphysis   _____ Yellow Marrow

Figure 11.1   Long bone structure.

Beef Bone Study   Examine a freshly cut section of bone. Identify all structures shown in figure 11.1. Probe into the periosteum near a torn ligament or tendon; note the continuity of fibers between the periosteum and these structures. Probe into the marrow and note its texture.

Bone Processes, Depressions, and Openings

In addition to the structures seen in figure 11.1, the following terms pertaining to processes, depressions, openings, and canals will be encountered as this skeletal study progresses:

Foramen: An opening in a bone that provides a passageway for nerves and blood vessels.

Fossa: A shallow depression in a bone. In some instances the fossa is a socket into which another bone fits.

Sulcus: A groove or furrow.

Meatus: A canal or long tubelike passageway.

Fissure: A narrow slit.

Sinus (antrum): A cavity in a bone.

Condyle: A rounded knucklelike eminence on a bone that articulates with another bone.

Tuberosity: A large roughened process on a bone that serves as a point of anchorage for a muscle.

Tubercle: A small rounded process.

Trochanter: A very large process on a bone.

Head: A portion of a bone supported by a constricted part, or *neck.*

Crest: A narrow ridge of bone.

Spine: A sharp slender process.

Parts of the Skeleton

The adult skeleton is made up of 206 named bones and many smaller unnamed ones. They are classified as being long, short, flat, irregular, or sesamoid.

The bones of the skeleton fall into two main groups: those that make up the axial skeleton and those forming the appendicular skeleton. Identify the following in figure 11.2.

The Axial Skeleton The parts of the axial skeleton are the **skull, hyoid bone, vertebral column** (spine), and **rib cage.** The hyoid bone is a horseshoe-shaped bone that is situated in the neck under the lower jaw. The rib cage consists of twelve pairs of **ribs** and a **sternum** (breastbone).

The Appendicular Skeleton This portion of the skeleton includes the upper and lower extremities.

Each upper extremity consists of a pectoral girdle, arm, and hand. The **pectoral girdle** consists of a **scapula** (shoulder blade) and **clavicle** (collar bone). Each arm consists of an upper portion, the **humerus,** and two forearm bones, the **radius** and **ulna.** The radius is lateral to the ulna. The **hand** includes the bones of the fingers and wrist.

The lower extremities consist of the pelvic girdle and legs. The **pelvic girdle** is formed by two bones, the **os coxae,** which are attached posteriorly, to the sacrum of the vertebral column and anteriorly, to each other. The anterior joint where the os coxae are united on the median line is the **symphysis pubis.**

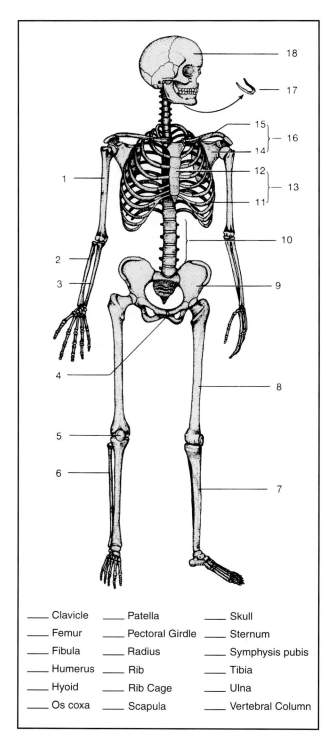

___ Clavicle	___ Patella	___ Skull
___ Femur	___ Pectoral Girdle	___ Sternum
___ Fibula	___ Radius	___ Symphysis pubis
___ Humerus	___ Rib	___ Tibia
___ Hyoid	___ Rib Cage	___ Ulna
___ Os coxa	___ Scapula	___ Vertebral Column

Figure 11.2   **The human skeleton.**

Each leg consists of four bones: the **femur** in the upper leg, a **tibia** (shinbone), a thin, long **fibula** parallel to the tibia, and the **patella,** or kneecap.

Assignment:
Label the parts of the skeleton in figure 11.2.

Bone Fractures

Various terms are used to describe different kinds of bone fractures. Fractures that do not penetrate the skin or mucous membranes are said to be **closed,** or **simple,** fractures. On the other hand, those that do break through are said to be **open,** or **compound** fractures.

Fractures may also be complete or incomplete. **Incomplete** fractures are the type in which the bone is split, splintered, or only partially broken. Illustrations A, B, and C in figure 11.3 are of this type. When a bone breaks through on only one side as a result of bending, it is often referred to as a **greenstick** fracture. Linear splitting of a long bone may be referred to as a **fissured** fracture.

Complete fractures are those in which the bone is broken clear through. If the break is at right angles to the long axis, it is considered to be a **transverse** fracture. Breaks that are at an angle to the long axis are termed **oblique** fractures. If a fracture results from torsional forces, it may be referred to as a **spiral** fracture.

If a piece of bone is broken out of the shaft it is a **segmental** fracture. More extensive fractures, in which two or more fragments are seen, are designated as **comminuted** fractures. When bone fragments have been moved out of alignment, as in illustration G, the fracture may also be referred to as being **displaced.**

Severe vertical forces can result in compacted or compression bone fractures. If a broken portion of bone is driven into another portion of the same bone, it is referred to as a **compacted** fracture. This is often seen in femur fractures, such as in illustration J. **Compression** fractures (not shown) often occur in the vertebral column when vertebrae are crushed due to falls from excessive heights.

Assignment:
Identify the types of fractures shown in figure 11.3.
Complete Laboratory Report 11 for this exercise.

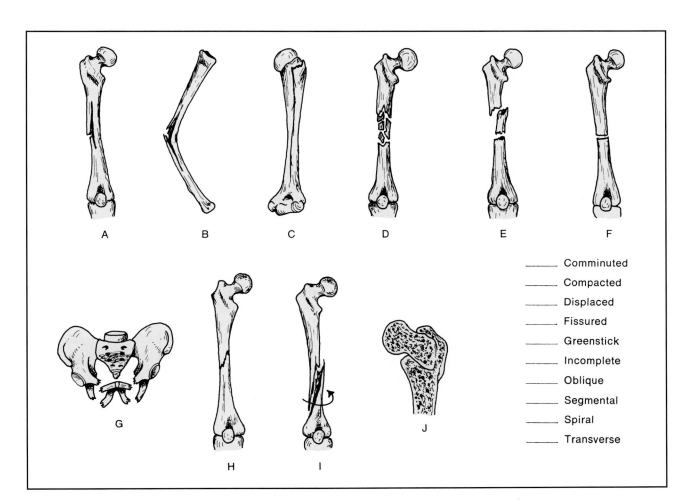

_____ Comminuted
_____ Compacted
_____ Displaced
_____ Fissured
_____ Greenstick
_____ Incomplete
_____ Oblique
_____ Segmental
_____ Spiral
_____ Transverse

Figure 11.3 **Types of bone fractures.**

12

The Skull

For this study of the skull, specimens will be available in the laboratory. As you read through the discussion of the various bones, identify them first in the illustrations and then on the specimens. Compare the specimens with the illustrations to note the degree of variance.

Care of Skulls

When handling laboratory skulls be very careful to avoid damaging them. **Never use a pencil as a pointer.** Pencil marks must not be made on the bones. A metal probe or a pipe cleaner should be used instead. If a metal probe is used, **touch the bones very gently to avoid bone perforation** where bone is thin.

Materials:
  whole and disarticulated skulls
  fetal skulls
  metal probe or pipe cleaner

The Cranium
(Front and Lateral Views)

That portion of the skull that encases the brain is called the *cranium*. It consists of the following bones: a single frontal, two parietals, one occipital, two temporals, one sphenoid, and an ethmoid. All these bones are joined together at their margins by irregular interlocking joints called *sutures*. The lateral and anterior aspects of the cranium are illustrated in figures 12.1 and 12.2. A sagittal section is seen in figure 12.8.

Frontal   The anterior superior portion of the skull consists of the *frontal* bone (color code: yellow). It forms the eyebrow ridges and the ridge above the nose. The most inferior edge of this bone extends well into the orbit of the eye to form the **orbital plates** of the frontal bone. On the superior ridges of the eye orbits are a pair of foramina, the **supraorbital foramina.** See figure 12.2.

Parietals   Directly posterior to the frontal bone on the sides of the skull are the *parietal* bones (c.c.: white). The lateral view of the skull actually shows only the left parietal bone. The right parietal is on the other side of the skull.

The right and left parietals meet on the midline of the skull to form the **sagittal suture.** Between the

frontal and each parietal bone is another suture, the **coronal suture.** Two semicircular bony ridges that extend from the forehead (frontal bone) and over the parietal bone are the **superior temporal line** and **inferior temporal line.** These ridges form the points of attachment for the longest muscle fibers of the *temporalis* muscle. Figure 26.1, page 127, shows the position of the muscle (label 2). It is the upper extremity of the muscle that falls on the superior temporal line.

Temporals   On each side of the skull, inferior to the parietal bones, are the *temporals* (c.c.: yellow). Each temporal is joined to its adjacent parietal by the **squamosal suture.** A depression, the **mandibular** *(glenoid)* **fossa,** on this bone provides a recess into which the lower jaw articulates. Pull the jaw away from the skull to note the shape of this fossa. The rounded eminence of the mandible that fits into this depression is the **mandibular condyle.** Just posterior to the mandibular fossa is the ear canal, or **external acoustic meatus** *(acoustic:* hearing; *meatus:* canal or passage).

The temporal bone has three significant processes: the zygomatic, styloid, and mastoid. The **zygomatic process** is a long slender process that extends forward to articulate with the zygomatic bone. The zygomatic process and a portion of the zygomatic bone constitute the *zygomatic arch.* The **styloid process** is a slender spinelike process that extends downward from the bottom of the temporal bone to form a point of attachment for some muscles of the tongue and pharynx. This process is often broken off on laboratory specimens. The **mastoid process** is a rounded eminence on the inferior surface of the temporal just posterior to the styloid process. It provides anchorage for the *sternocleidomastoideus* muscle of the neck. Middle ear infections that spread into the cancellous bone of this process are referred to as *mastoiditis.*

Sphenoid   The pink-colored bone seen in the lateral view of the skull, figure 12.1, is the *sphenoid* bone. Note in the inferior view, figure 12.5, that this bone extends from one side of the skull to the other.

Identify the two greater wings, the orbital surfaces, and the pterygoid processes of the sphenoid.

The portions that are on the sides of the skull in the "temple" region are the **greater wings of the sphenoid.** On the ventral surface (figure 12.5) are the **pterygoid processes of the sphenoid** to which the pterygoid muscles of mastication are attached. The **orbital surfaces of the sphenoid** (figure 12.2) make up the posterior walls of each eye orbit.

Ethmoid On the medial surface of each orbit of the eye is seen the *ethmoid* bone (label 6, figure 12.2, c.c.: pale yellow). The bone forms a part of the roof of the nasal cavity and closes the anterior portion of the cranium.

Examine the upper portion of the nasal cavity of your skull. Note that the inferior portion of the ethmoid has a downward extending **perpendicular plate** on the median line. Refer to figure 12.8 (label 3). This portion articulates anteriorly with the nasal and frontal bones. Posteriorly, it articulates with the sphenoid and vomer. On each side of the perpendicular plate are irregular curved plates, the **superior** and **middle nasal conchae.** They provide bony reinforcement for the fleshy **upper nasal conchae** of the nasal cavity.

Occipital The posterior inferior portion of the skull consists, primarily, of the *occipital* bone (c.c.: magenta). It is joined to the parietal bones by the **lambdoidal**

suture. Further details of this bone of the cranium will be described after the facial bones have been studied.

The Face

The face of the skull consists of thirteen bones fused together, plus a movable mandible. Of the thirteen fused bones, only one bone, the vomer, is not paired. Figure 12.2 reveals a majority of the facial bones.

Maxillae The upper jaw consists of two maxillary bones (*maxillae*) that are joined by a suture on the median line. Its color code is magenta. Remove the mandible from your laboratory skull and examine the hard palate. Compare it with figure 12.5. Note that the anterior portion of the hard palate consists of two **palatine processes of the maxillae.** A **median palatine suture** joins the two bones on the median line.

The maxillae of an adult support sixteen permanent teeth. Each tooth is contained in a socket, or *alveolus.* That portion of the maxillae that contains the teeth is called the *alveolar process.* The alveolar process consists of two compact tissue bony plates, the *external* and *internal alveolar plates.* These two plates of bone are joined by partitions or *septae,* that lie between the teeth and make up the transverse walls of the alveoli.

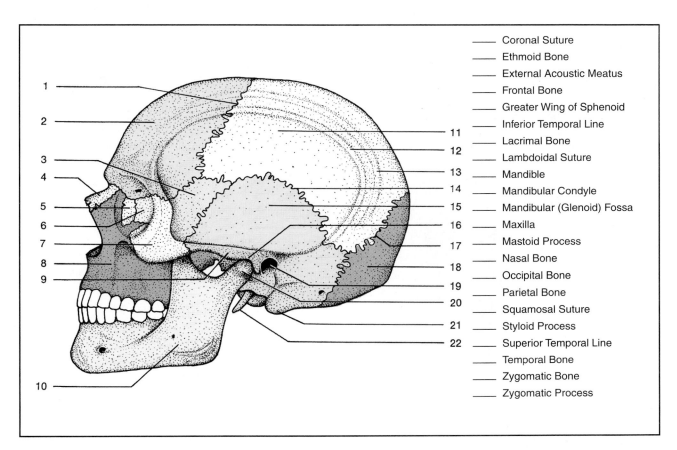

_____ Coronal Suture
_____ Ethmoid Bone
_____ External Acoustic Meatus
_____ Frontal Bone
_____ Greater Wing of Sphenoid
_____ Inferior Temporal Line
_____ Lacrimal Bone
_____ Lambdoidal Suture
_____ Mandible
_____ Mandibular Condyle
_____ Mandibular (Glenoid) Fossa
_____ Maxilla
_____ Mastoid Process
_____ Nasal Bone
_____ Occipital Bone
_____ Parietal Bone
_____ Squamosal Suture
_____ Styloid Process
_____ Superior Temporal Line
_____ Temporal Bone
_____ Zygomatic Bone
_____ Zygomatic Process

Figure 12.1 Lateral view of skull.

Three significant foramina are seen on the maxillae: two infraorbital and one anterior palatine. The **infraorbital foramina** are situated on the front of the face under each eye orbit. Nerves and blood vessels emerge from each of these foramina to supply the nose. The **anterior palatine** (incisive) **foramen** is seen in the anterior region of the hard palate just posterior to the central incisors. Refer to figure 12.5.

Palatines   In addition to the horizontal palatine processes of the maxillae, the hard palate also consists of two *palatine* bones (c.c.: orange). See figures 12.5 and 12.8. The two palatines form the posterior third of the palate and are joined to the maxillae by the **transverse palatine suture.**

Note that each palatine bone has a large **greater palatine foramen** and two smaller **lesser palatine foramina** (not labeled in figure 12.5).

Zygomatics   On each side of the face are two *zygomatic* (malar) bones (uncolored). They form the prominence of each cheek and the inferior, lateral surface of each eye orbit. Each zygomatic has a small foramen, the **zygomaticofacial foramen.**

Lacrimals   Between the ethmoid and upper portion of the maxillary bones is a pair of *lacrimal* bones (Latin: *lacrima:* tear)—one in each orbit. These bones are uncolored in figures 12.1 and 12.2.

Locate a groove on the surface of each lacrimal that is continuous with a groove on the maxilla. This groove provides a recess for the lacrimal duct through which tears flow from the eye into the nasal cavity.

Nasals   The bridge of the nose is formed by a pair of thin, rectangular *nasal* bones (uncolored).

Vomer   This thin bone is located in the nasal cavity on the median line. Its posterior upper edge articulates with the back portion of the perpendicular plate of the ethmoid and the rostrum of the sphenoid. This bone is uncolored.

The lower border of the vomer is joined to the maxillae and palatines. The *septal cartilage* of the nose extends between the anterior margin of the vomer and the perpendicular plate of the ethmoid. Locate the vomer on figures 12.2, 12.5, and 12.8 as well as on your laboratory skull.

Inferior Nasal Conchae   The *inferior nasal conchae* are curved bones attached to the walls of the nasal fossa. They are situated beneath the superior and middle nasal conchae, which are part of the ethmoid bone.

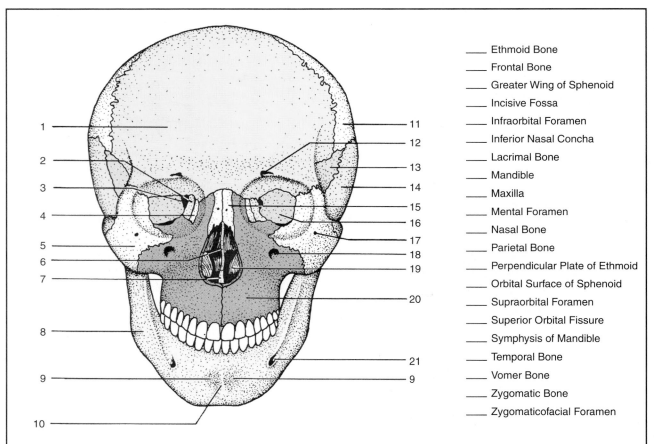

_____ Ethmoid Bone
_____ Frontal Bone
_____ Greater Wing of Sphenoid
_____ Incisive Fossa
_____ Infraorbital Foramen
_____ Inferior Nasal Concha
_____ Lacrimal Bone
_____ Mandible
_____ Maxilla
_____ Mental Foramen
_____ Nasal Bone
_____ Parietal Bone
_____ Perpendicular Plate of Ethmoid
_____ Orbital Surface of Sphenoid
_____ Supraorbital Foramen
_____ Superior Orbital Fissure
_____ Symphysis of Mandible
_____ Temporal Bone
_____ Vomer Bone
_____ Zygomatic Bone
_____ Zygomaticofacial Foramen

Figure 12.2   Anterior aspect of the skull.

The Paranasal Sinuses

Some of the bones of the skull contain cavities, the *paranasal sinuses,* which reduce the weight of the skull without appreciably weakening it. All of the sinuses have passageways leading into the nasal cavity and are lined with a mucous membrane similar to the type that lines the nasal cavities.

The paranasal sinuses are named after the bones in which they are situated. Figure 12.4 shows the location of these cavities. Above the eyes in the forehead are the **frontal sinuses.** The largest sinuses are the **maxillary sinuses,** which are in the maxillary bones. These sinuses are also referred to as the *antrums of Highmore.* The **sphenoidal sinus** is the most posterior sinus shown in figure 12.4. Between the frontal and sphenoidal sinuses are a group of small spaces called the **ethmoid air cells.**

Assignment:
Label figures 12.1 and 12.4.

The Mandible

The only bone of the skull that is not fused as an integral part of the skull is the lower jaw, or mandible. Figures 12.2 and 12.3 reveal the anatomical details of this bone.

The mandible consists of a horizontal portion, the **body,** and two vertical portions, the **rami.** Embryologically, the mandible forms from two centers of ossification, one on each side of the face. As the bone develops toward the median line, the two

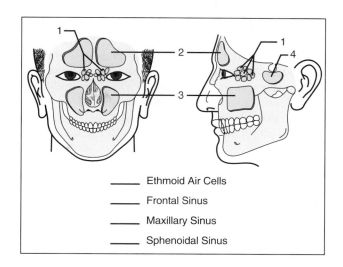

_____ Ethmoid Air Cells
_____ Frontal Sinus
_____ Maxillary Sinus
_____ Sphenoidal Sinus

Figure 12.4 **The paranasal sinuses.**

halves finally meet and fuse to form a solid ridge. This point of fusion on the midline is called the **symphysis** (label 10, figure 12.2). On each side of the symphysis are two depressions, the **incisive fossae.**

Like the maxillae, that portion of the mandible that supports sixteen teeth is an **alveolar process** (label 4, figure 12.3). Each alveolus in the mandible is constructed like the alveoli in the maxillae.

The superior portion of each ramus has a condyle, a coronoid process, and a notch. The **mandibular condyle** occupies the posterior superior terminus of the ramus. The process on the superior anterior portion of the ramus is the **coronoid**

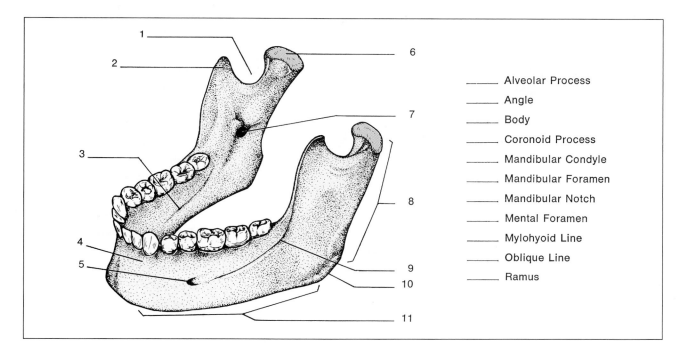

_____ Alveolar Process
_____ Angle
_____ Body
_____ Coronoid Process
_____ Mandibular Condyle
_____ Mandibular Foramen
_____ Mandibular Notch
_____ Mental Foramen
_____ Mylohyoid Line
_____ Oblique Line
_____ Ramus

Figure 12.3 **The mandible.**

process. This tuberosity provides attachment for the *temporalis* muscle (see illustration A, figure 26.1). Between the mandibular condyle and the coronoid process is the **mandibular notch.** At the posterior inferior corners of the mandible, where the body and rami meet, are two protuberances, the **angles.** The angles provide attachment for the *masseter* and *internal pterygoid* muscles (figure 26.1, labels 1 and 15).

A ridge of bone, the **oblique line,** extends at an angle from the ramus down the lateral surface of the body to a point near the mental foramen. This bony elevation is strong and prominent in its upper part, but gradually flattens out and disappears, as a rule, just below the first molar.

On the internal (medial) surface of the mandible is another diagonal line, the **mylohyoid line.** It extends from the ramus down to the body. To this crest is attached a muscle, the *mylohyoid,* which forms the floor of the oral cavity. The bony portion of the body that exists above this line makes up a portion of the sides of the oral cavity proper.

Note also that on the medial surfaces of the rami are two openings, the **mandibular foramina.** On the external surface of the body are two prominent **mental foramina** (*mental:* chin).

Assignment:
Label figures 12.2 and 12.3.

Bottom of the Cranium

Now that you are familiar with the front and lateral aspects of the skull, let's study the bottom of the cranium to identify the principal processes, foramina, and sutures in this region. Use figures 12.5 and 12.6 for reference.

External Features

Note that the occipital bone on the bottom of the skull has a large central opening, the **foramen magnum.** It is through this opening that the brain stem protrudes downward. On each side of this opening is seen a pair of **occipital condyles** (label 18, figure 12.5). These two condyles rest on fossae of the *atlas,* the first cervical vertebra of the spinal column. Note that a **hypoglossal canal** passes through each condyle. The right occipital condyle is shown with a piece of wire passing through it. This canal provides a passageway for the hypoglossal (12th cranial) nerve to pass through it. On the median line posterior to the foramen magnum is a prominent process called the **external occipital protuberance.**

Identify the foramen lacerum and carotid canal on the bottom of your skull. The **foramen lacerum**

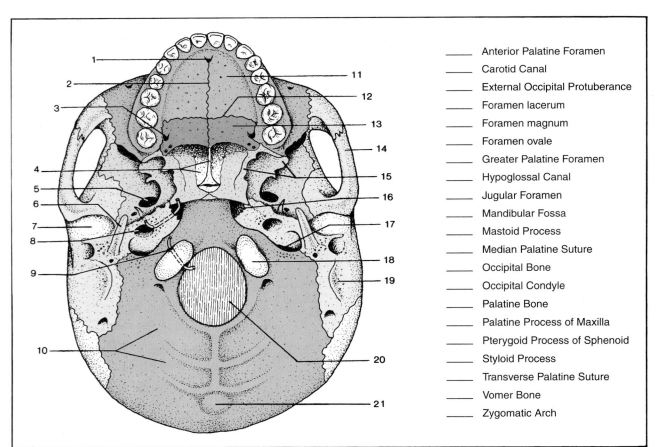

_____ Anterior Palatine Foramen
_____ Carotid Canal
_____ External Occipital Protuberance
_____ Foramen lacerum
_____ Foramen magnum
_____ Foramen ovale
_____ Greater Palatine Foramen
_____ Hypoglossal Canal
_____ Jugular Foramen
_____ Mandibular Fossa
_____ Mastoid Process
_____ Median Palatine Suture
_____ Occipital Bone
_____ Occipital Condyle
_____ Palatine Bone
_____ Palatine Process of Maxilla
_____ Pterygoid Process of Sphenoid
_____ Styloid Process
_____ Transverse Palatine Suture
_____ Vomer Bone
_____ Zygomatic Arch

Figure 12.5 Inferior surface of the skull.

is a large jagged-edged opening between the borders of the temporal and occipital bones. Note that if a piece of wire (bent paper clip) is carefully inserted into this foramen, as shown in figure 12.5, a passageway, the **carotid canal,** can be demonstrated. It is through this canal that the internal carotid artery carries blood from the neck into the brain.

Identify the foramen ovale on the sphenoid bone. The **foramen ovale** (label 5, figure 12.5) is a large elliptical foramen, which provides a passageway for the mandibular branch of the trigeminal nerve.

Just posterior to the carotid canal opening is an irregular slitlike opening, the **jugular foramen,** which allows drainage of blood from the cranial cavity via the inferior petrosal sinus.

Internal Structure

Examine the inner bottom of a laboratory skull by removing the upper half of a sectioned skull and compare it to figure 12.6. Identify the following structures:

Cranial Fossae As you look down on the entire floor of the cranium note that it is divided into three large depressions called *cranial fossae.* The one formed by the orbital plates of the frontal is called the **anterior cranial fossa.** The large depression formed within the occipital bone is the **posterior cranial fossa.** It is the deepest fossa. In between these two fossae is the **middle cranial fossa,** which is not labeled in figure 12.6.

Ethmoid Identify the *ethmoid* bone (c.c.: pale yellow) in the anterior cranial fossa. It lies between the orbital plates of the frontal bone. Note that it consists of a perforated horizontal portion, the **cribriform plate,** and an upward projecting process, the **crista galli** (cock's comb). The holes in the cribriform plate allow branches of the olfactory nerve to pass from the brain into the nasal cavity. The crista galli serves as an attachment for the *falx cerebri* (label 1, figure 34.1).

Sphenoid Note the batlike configuration of this bone, with the greater wings extending out on each side. On the median line of the sphenoid there is a

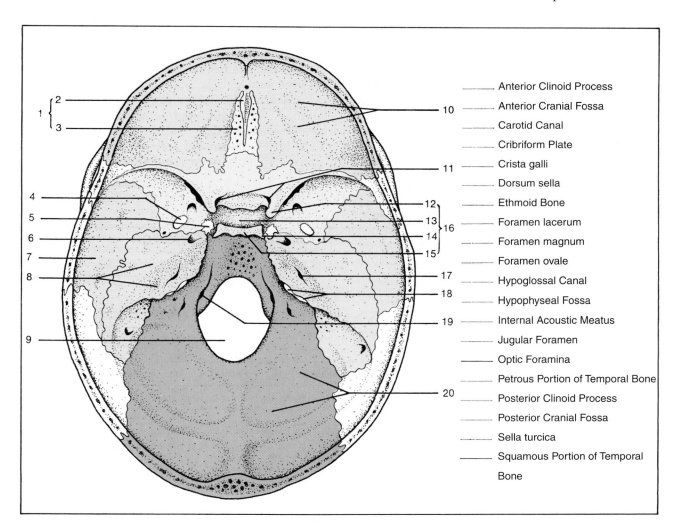

Anterior Clinoid Process
Anterior Cranial Fossa
Carotid Canal
Cribriform Plate
Crista galli
Dorsum sella
Ethmoid Bone
Foramen lacerum
Foramen magnum
Foramen ovale
Hypoglossal Canal
Hypophyseal Fossa
Internal Acoustic Meatus
Jugular Foramen
Optic Foramina
Petrous Portion of Temporal Bone
Posterior Clinoid Process
Posterior Cranial Fossa
Sella turcica
Squamous Portion of Temporal Bone

Figure 12.6 **Floor of the cranium.**

deep depression called the **hypophyseal fossa.** This depression contains the pituitary gland *(hypophysis)* in real life. Posterior to this fossa is an elevated ridge called the **dorsum sella.** The two spine-like processes anterior and lateral to the hypophyseal fossa that project backward are the **anterior clinoid processes.** The outer spiny processes of the dorsum sella are the **posterior clinoid processes.** The hypophyseal fossa, dorsum sella, and clinoid processes, collectively, make up the **sella turcica,** or *Turkish saddle.*

Temporals The significant parts of the temporal bone to identify in figure 12.6 are the petrous, squamous, and mastoid portions. The **squamous portion** of the temporal is that thin portion that forms a part of the side of the skull. The **petrous portion** (label 8) is probably the hardest portion of the skull. The medial sloping surface of the petrous portion has an opening to the **internal acoustic meatus.** This canal contains the facial and vestibulocochlear cranial nerves. The latter nerve passes from the inner ear to the brain. The most posterior part of the temporal bone is the **mastoid portion.** Although it is not labeled in figure 12.6, it is shown in figure 12.5.

Occipital Identify the large **foramen magnum** in the center of the bone and the **jugular foramen,** where the internal jugular vein takes its origin. The jugular foramen appears as an irregular slit between the anterolateral margin of the occipital bone and the petrous portion of the temporal bone. Cranial nerves IX, X, and XI also pass through this foramen.

Identify the openings to the **hypoglossal canals.** Note that in figure 12.5 this is the foramen that passes through the occipital condyles. Explore these foramina with a slender probe or pipe cleaner.

Assignment:
Label figures 12.5 and 12.6.

Foramina Summary

To review the principal foramina of the skull, consult table 12.1. Use this table for quick reference.

The Fetal Skull

The human skull at birth is incompletely ossified. Figure 12.7 reveals its structure. These unossified membranous areas, called *fontanels,* facilitate compression of the skull at childbirth.

There are six fontanels joined by five areas where future sutures of the skull will form. The largest fontanel is the **anterior fontanel,** a somewhat diamond-shaped membrane that lies on the median line at the juncture of the frontal and parietal bones. The **posterior fontanel** is somewhat smaller and lies on the median line at the juncture of the parietal and occipital bones. Between these two fontanels on the median line is a membranous area where the **future saggital suture** of the skull will form.

On each side of the skull, where the frontal, parietal, sphenoid, and temporal bones come together behind the eye orbit, is an **anterolateral fontanel.** Between the anterior and anterolateral fontanels can be seen a membranous line that is

Table 12.1 Location and function of skull foramina.

FORAMEN	LOCATION	PASSAGEWAY FOR
Anterior palatine foramen	maxillae	nasopalatine nerves and descending palatine vessels
Carotid canal	temporal bone	internal carotid artery
Foramen lacerum	between sphenoid and temporal bone	internal carotid artery and plexus of sympathetic nerves
Foramen magnum	occipital bone	brain stem
Foramen ovale	sphenoid bone	mandibular nerve and accessory meningeal artery
Greater palatine foramen	palatine bone	anterior palatine nerve and descending palatine vessels
Hypoglossal canal	occipital bone	12th cranial nerve (hypoglossal)
Jugular foramen	between temporal and occipital bones	inferior petrosal sinus, vagus n., glossopharyngeal n., accessory n., etc.
Mandibular foramen	mandible	inferior alveolar nerve (branch of 5th cranial n.) and blood vessels
Mental foramen	mandible	mental nerve and blood vessels

the area where the **future coronal suture** will develop.

The most posterior fontanel on the side of the skull is the **posterolateral fontanel,** which lies at the juncture of the parietal, temporal, and occipital bones. Between the anterolateral and posterolateral fontanels is a membranous line that will develop into the **future squamosal suture.** Ossification of these fontanels and membranous future sutures is usually completed in the two-year-old child.

Assignment:
Examine a fetal skull to identify all of the structures shown in figure 12.7.

Label figure 12.8 and complete Laboratory Report 12 for this exercise.

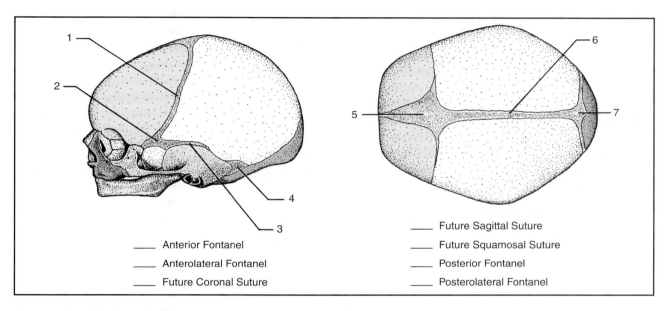

_____ Anterior Fontanel

_____ Anterolateral Fontanel

_____ Future Coronal Suture

_____ Future Sagittal Suture

_____ Future Squamosal Suture

_____ Posterior Fontanel

_____ Posterolateral Fontanel

Figure 12.7   The fetal skull.

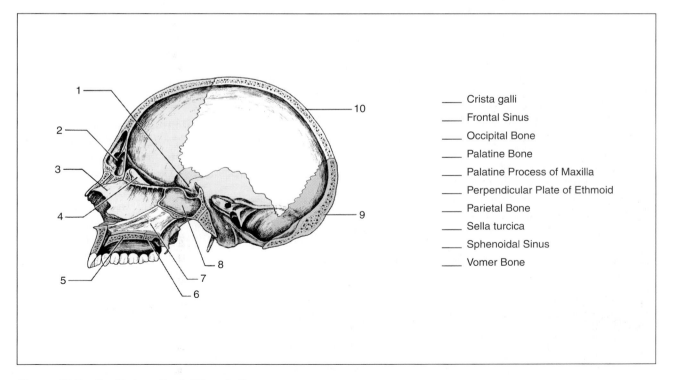

_____ Crista galli

_____ Frontal Sinus

_____ Occipital Bone

_____ Palatine Bone

_____ Palatine Process of Maxilla

_____ Perpendicular Plate of Ethmoid

_____ Parietal Bone

_____ Sella turcica

_____ Sphenoidal Sinus

_____ Vomer Bone

Figure 12.8   Sagittal section of the skull.

13

The Vertebral Column and Thorax

The skeletal structure of the trunk of the body will be studied in this exercise. The vertebral column, ribs, sternum, and hyoid bone make up this part of the skeleton.

Materials:
> skeleton, articulated
> skeleton, disarticulated
> vertebral column, mounted

The Vertebral Column

The vertebral column consists of thirty-three bones, twenty-four of which are individual movable vertebrae. Figure 13.1 illustrates its structure. Note that the individual vertebrae are numbered from the top.

The Vertebrae

Although the vertebrae in different regions of the vertebral column vary considerably in size and configuration, they do have certain features in common. Each one has a structural mass, the **body,** which is the principal load-bearing contact area between adjacent vertebrae. The space between the surfaces of adjacent vertebral bodies is occupied by a fibrocartilaginous **intervertebral disk.** The collective action of these twenty-four disks imparts a vital cushion effect to the spinal column.

Note that in addition to the bodies, adjacent vertebrae contact each other on articular surfaces, which are located on the **transverse processes** (label 20). The **superior articular surfaces** (label 21) of each vertebra contact the **inferior articular surfaces** of the vertebra above it. At these three points, adjacent vertebrae are secured together by ligaments, uniting the entire column into a single functioning unit.

In the center of each vertebra is an opening, the **spinal** (vertebral) **foramen,** which contains the spinal cord. Note that this foramen gets smaller as one progresses down the vertebral column.

In addition to the two transverse processes, each vertebra has a **spinous process** that projects out on its posterior surface. These spinous processes provide points of attachment for various muscles of the neck and back.

Extending backward from the body of each vertebra are two processes, the **pedicles,** which form a portion of the bony arch around the spinal foramen (see illustrations C and D). The openings between the pedicles of adjacent vertebrae allow spinal nerves to exit from the spinal cord. These openings are the **intervertebral foramina** (label 29) of the vertebral column. Between the transverse processes and the spinous process are two broad plates, called **laminae** (label 24). The two pedicles and two laminae constitute the *neural arch* of each vertebra.

Cervical Vertebrae The upper seven bones are the *cervical vertebrae* of the neck. The first of these seven is the **atlas** (illustration A). Note that the spinal foramen is much larger here to accommodate a short portion of the brain stem, which extends down into the vertebral column.

Note in illustrations A, B, and C that the cervical vertebrae have a small **transverse foramen** in each transverse process. Collectively, these foramina form a passageway on each side of the spinal column for the vertebral artery and vein.

Observe, also, that the second cervical vertebra, or **axis** (illustration B), is unique in that it has a vertical protrusion, the **odontoid process** (dens), which provides a pivot for the rotation of the atlas. When the head is turned from side to side, movement occurs between the axis and atlas around this process.

Thoracic Vertebrae Below the seven cervical vertebrae are twelve *thoracic vertebrae.* Illustration D reveals the structure of a typical thoracic vertebra. Note that these bones are larger and thicker than ones in the neck. A distinguishing feature of these vertebrae is that all twelve of them have **articular facets for ribs** on their transverse processes.

Lumbar Vertebrae Inferior to the thoracic vertebrae lie five *lumbar vertebrae.* The bodies of these bones are much thicker than those of the other

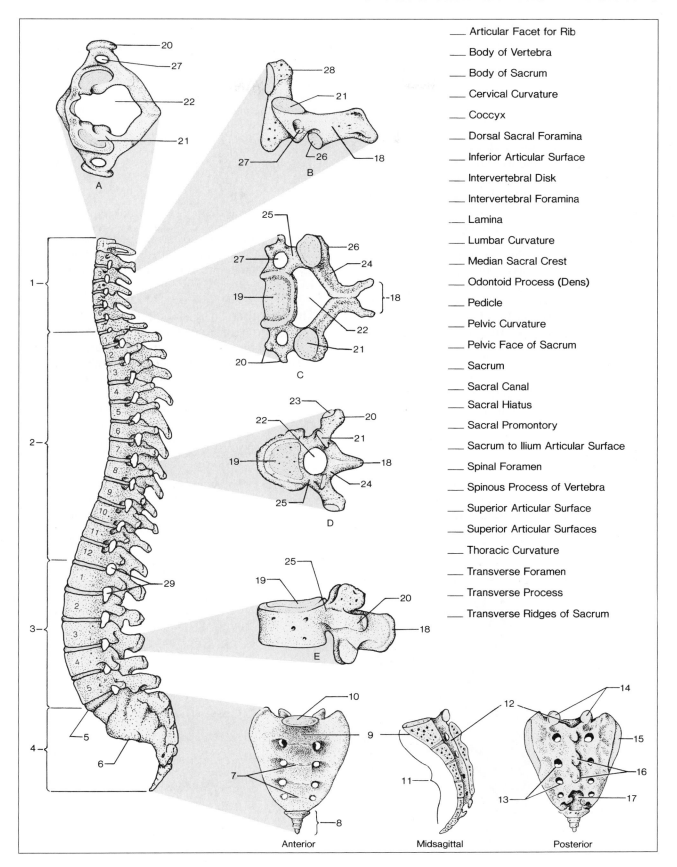

___ Articular Facet for Rib
___ Body of Vertebra
___ Body of Sacrum
___ Cervical Curvature
___ Coccyx
___ Dorsal Sacral Foramina
___ Inferior Articular Surface
___ Intervertebral Disk
___ Intervertebral Foramina
___ Lamina
___ Lumbar Curvature
___ Median Sacral Crest
___ Odontoid Process (Dens)
___ Pedicle
___ Pelvic Curvature
___ Pelvic Face of Sacrum
___ Sacrum
___ Sacral Canal
___ Sacral Hiatus
___ Sacral Promontory
___ Sacrum to Ilium Articular Surface
___ Spinal Foramen
___ Spinous Process of Vertebra
___ Superior Articular Surface
___ Superior Articular Surfaces
___ Thoracic Curvature
___ Transverse Foramen
___ Transverse Process
___ Transverse Ridges of Sacrum

Figure 13.1   The vertebral column.

vertebrae due to the greater stress that occurs in this region of the vertebral column. Illustration E is of a typical lumbar vertebra.

The Sacrum

Inferior to the fifth lumbar vertebra lies the *sacrum.* It consists of five fused vertebrae. Note that there are two oval **superior articular surfaces** (facets) that articulate with the inferior articular surfaces of the fifth lumbar vertebra. On each lateral surface of the sacrum is a **sacrum to ilium articular surface.** Note that the part of the sacrum that articulates with the intervertebral disk superior to it is called the **body of the sacrum.**

Observe that the anterior surface of the sacrum, or **pelvic face,** is curved backward and that the body of the first sacral vertebra forms a protrusion called the **sacral promontory.** Note also that four **transverse ridges** can be seen on the pelvic face that reveal where the five vertebrae are fused together.

Identify the **median sacral crest** (label 16) and the **dorsal sacral foramina** on the posterior surface. The neural arches of the fused sacral vertebrae form the **sacral canal,** which exits at the lower end as the **sacral hiatus.**

The Coccyx

The "tailbone" of the vertebral column is the **coccyx.** It consists of four or five rudimentary vertebrae. It is triangular in shape and is attached to the sacrum by ligaments.

Spinal Curvatures

Four curvatures of the vertebral column, together with the intervertebral disks, impart considerable springiness along its vertical axis. Three of them are identified by the type of vertebrae in each region: the **cervical, thoracic,** and **lumbar curves.** The fourth curvature, which is formed by the sacrum and coccyx, is the **pelvic curve.**

Assignment:
Label figure 13.1

The Thorax

The sternum, ribs, costal cartilages, and thoracic vertebrae form a cone-shaped enclosure, the *thorax.* Its components are illustrated in figures 13.2 and 13.3.

The Sternum   The sternum, or breastbone, consists of three separate bones, an upper **manubrium,** a middle **body** *(gladiolus),* and a lower **xiphoid** *(ensiform)* **process.** A **sternal angle** is formed where the inferior border of the manubrium articulates with the body.

On both sides of the sternum are notches (facets) where the sternal ends of the costal cartilages are attached. Note that the second rib fits into a pair of *demifacets (demi,* half) at the sternal angle.

The Ribs   There are twelve pairs of ribs. The first seven pairs attach directly to the sternum by costal cartilages and are called **vertebrosternal,** or **true, ribs.** The remaining pairs are called **false ribs.** The

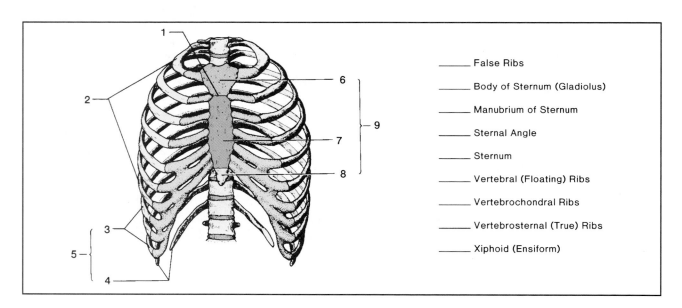

_____ False Ribs

_____ Body of Sternum (Gladiolus)

_____ Manubrium of Sternum

_____ Sternal Angle

_____ Sternum

_____ Vertebral (Floating) Ribs

_____ Vertebrochondral Ribs

_____ Vertebrosternal (True) Ribs

_____ Xiphoid (Ensiform)

Figure 13.2   **The thorax.**

upper three pairs of false ribs, the **vertebrochon-dral ribs,** have cartilaginous attachments on their anterior ends but do not attach directly to the sternum. The lowest false ribs, the **vertebral,** or **floating ribs,** are unattached anteriorly.

Figure 13.3 illustrates the structure of a central rib, a lateral view of a thoracic vertebra, and articulation details. Although considerable vari-

ability exists in size and configuration of the various ribs, the central rib reveals structures common to most ribs.

The principal parts of each rib are a head, neck, tubercle, and body. The **head** (label 15) is the enlarged end of the rib that articulates with the vertebral column. The **tubercle** (label 17) consists of two portions: an **articular portion** and a **nonarticular**

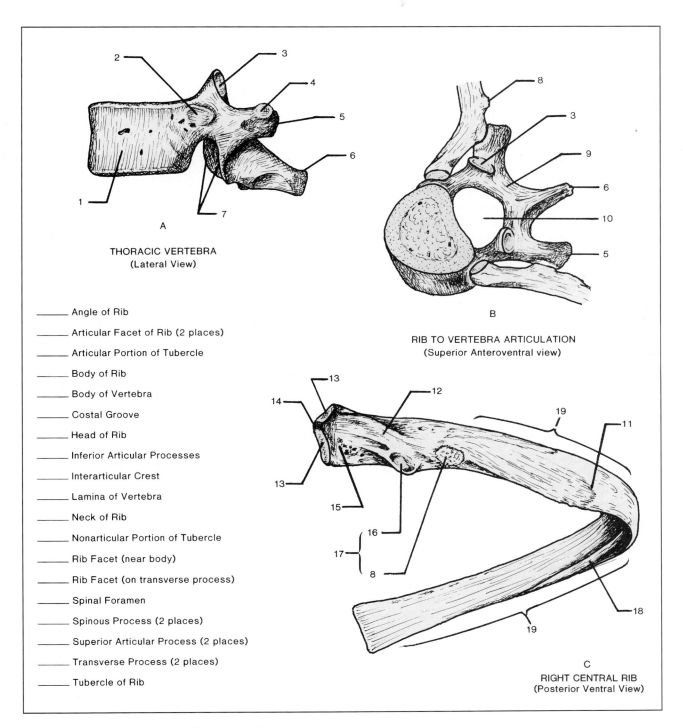

THORACIC VERTEBRA
(Lateral View)

RIB TO VERTEBRA ARTICULATION
(Superior Anteroventral view)

RIGHT CENTRAL RIB
(Posterior Ventral View)

_____ Angle of Rib

_____ Articular Facet of Rib (2 places)

_____ Articular Portion of Tubercle

_____ Body of Rib

_____ Body of Vertebra

_____ Costal Groove

_____ Head of Rib

_____ Inferior Articular Processes

_____ Interarticular Crest

_____ Lamina of Vertebra

_____ Neck of Rib

_____ Nonarticular Portion of Tubercle

_____ Rib Facet (near body)

_____ Rib Facet (on transverse process)

_____ Spinal Foramen

_____ Spinous Process (2 places)

_____ Superior Articular Process (2 places)

_____ Transverse Process (2 places)

_____ Tubercle of Rib

Figure 13.3   Rib anatomy and its articulation.

portion. The **neck** is a flattened portion, about 2.5 cm long, between the head and tubercle. The **body** is the flattened curved remainder of the rib.

Note that the head of this rib has two **articular facets** on its medial surface that make contact with two adjacent vertebrae. Between these two facets is a roughened **interarticular crest** to which ligamentous tissue is anchored. Illustration B shows how the rib articulates with two points on a thoracic vertebra. Although the rib shown in illustration C has two articular facets, the 1st, 2nd, 10th, 11th, and 12th ribs exhibit only a single articular facet on their heads.

The body of the rib in illustration C has two landmarks: an angle and a costal groove. The **angle** is a ridge on the external surface that provides anchorage for the *iliocostalis* muscle of the back. Note that the portion between the angle and tubercle is rough and irregular; it is on this surface that another back muscle, the *longissimus dorsi,* is attached. Both of these muscles are shown in figure 27.2 (labels 8 and 9). The **costal groove** is a depression on the ventral side of the rib, which provides a recess for the intercostal nerve and blood vessels.

In addition to revealing the location of the two rib facets, the lateral view of the thoracic vertebra (illustration A) reveals the **superior articular process,** which contacts the inferior articular process of the vertebra above it. On its underside are seen the two **inferior articular processes.**

Assignment:
Label figures 13.2 and 13.3.

The Hyoid Bone

The *hyoid* bone is a horseshoe-shaped bone located in the neck region between the mandible and larynx. Although it does not articulate directly with any other bone, it is held in place by various ligaments and muscles. Figure 13.4 illustrates its structure.

The bone consists of five segments: a body, two greater cornua, and two lesser cornua. The massive central portion of the bone is the **body.** The two long arms that extend out from each side of the body are the **greater cornua** (*cornu,* singular). Note that the distal end of each greater cornu terminates in a tubercle.

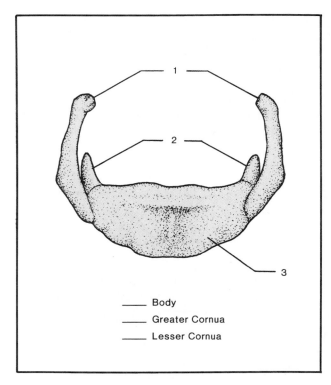

_____ Body

_____ Greater Cornua

_____ Lesser Cornua

Figure 13.4   The hyoid bone.

The **lesser cornua** are the conical eminences that are located on the sides of the body superior to the bases of the greater cornua. These structures are separate bone entities that are connected to the body, and occasionally to the greater cornua, by fibrous connective tissue. These joints are usually diarthrotic (freely movable), but occasionally become fused (ankylosed) in later life.

The following muscles have points of attachment on the hyoid bone: *mylohyoideus, sternohyoideus, omohyoideus, thyrohyoideus, digastricus, hyoglossus, genioglossus,* and *constrictor pharyngis medius.*

Assignment:
Label figure 13.4.

Laboratory Report

Complete Laboratory Report 13 for this exercise.

The Appendicular Skeleton

14

In this exercise a study will be made of the individual bones of the upper and lower extremities.

Materials:
> skeleton, articulated
> skeleton, disarticulated
> male pelvis and female pelvis

The Upper Extremities

The upper extremities are held in place by a pair of *pectoral girdles,* each of which includes a scapula and clavicle. Figure 14.2 illustrates the upper arm and its shoulder girdle.

The Clavicle   Note in figure 14.2 that the *clavicle* is a slender S-shaped bone that articulates with the manubrium on its medial end. Its lateral end articulates with the scapula. Important muscles of the shoulder attach to this bone.

The Scapula   As shown in figure 14.1, the *scapula* is a flattened triangular bone with a cartilage-lined socket, the **glenoid cavity,** on its upper lateral edge. Instead of being firmly attached to the axial skeleton, this bone is loosely held in place by muscles that are attached to the upper arm, back, and chest.

Observe in figure 14.1 that its posterior surface has a pronounced elongated diagonal ridge, the **scapular spine.** This ridge divides the back surface of the scapula into two depressions: an upper **supraspinous fossa** and a larger **infraspinous fossa** below it. The anterior surface of the scapula has a single large concavity, the **subscapular fossa** (label 13). These three fossae provide anchorage for various shoulder muscles.

Two prominent processes, the acromion and coracoid, distinguish themselves on the lateral aspect of figure 14.1. The **acromion process** is the large one that is formed at the lateral terminus of the scapular spine. It is to this process that the lateral end of the clavicle is attached. The **coracoid process** is anterior to the acromion and glenoid cavity.

Note that the perimeter of the scapula consists of three margins. The convex edge on the left side of the posterior aspect (figure 14.1) is the **vertebral (medial) margin.** This border extends from the **superior angle** (label 2) down to the **inferior angle** at the bottom. The **axillary (lateral) margin** makes up the border opposite to the vertebral border, extending from the inferior angle to the glenoid cavity. The third border is the **superior margin,** which extends from the superior angle to a deep depression called the **scapular notch.**

Assignment:
Label figure 14.1.

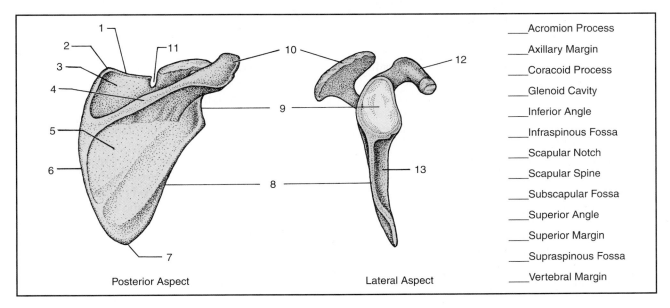

____Acromion Process	
____Axillary Margin	
____Coracoid Process	
____Glenoid Cavity	
____Inferior Angle	
____Infraspinous Fossa	
____Scapular Notch	
____Scapular Spine	
____Subscapular Fossa	
____Superior Angle	
____Superior Margin	
____Supraspinous Fossa	
____Vertebral Margin	

Posterior Aspect   Lateral Aspect

Figure 14.1   The scapula.

The Upper Arm The skeletal structure of the upper arm consists of a single bone, the **humerus.** Its smooth rounded proximal end, the **head,** fits into the glenoid cavity of the scapula. Just below the head is a narrowing section called the **anatomic neck.**

Inferior to the head and anatomic neck are two prominent eminences, the greater and lesser tubercles. The **greater tubercle** is the larger process that is lateral to the **lesser tubercle.** The narrowing section below the two tubercles is the **surgical neck,** so named because of the frequency of bone fractures in this area.

The surface of the diaphysis of the humerus has a roughened raised area near its midregion that is called the **deltoid tuberosity.** In this same general area there is also a small opening, the **nutrient foramen.**

The distal terminus of the humerus has two condyles, the capitulum and trochlea, which contact the bones of the forearm. The **capitulum** is the lateral condyle that articulates with the radius. The **trochlea** is the medial condyle that articulates with the ulna. Superior and lateral to the capitulum is an eminence, the **lateral epicondyle.** On the opposite side is a larger tuberosity, the **medial epicondyle.** Above the trochlea on the anterior surface is a depression, the **coronoid fossa.** Note that the posterior surface of this end of the humerus has a depression also, the **olecranon fossa.**

The Forearm The radius and ulna constitute the skeletal components of the forearm. Figure 14.2 shows the relationship of these two bones to each other and the hand.

The **radius** is the lateral bone of the forearm. The proximal end of this bone has a disk-shaped **head,** which articulates with the capitulum of the humerus. The disklike nature of the head makes it possible for the radius to rotate at the upper end when the palm of the hand is changed from palm down to palm up (pronation to supination).

A few centimeters below the head of the radius on its medial surface is an eminence, the **radial tuberosity.** This process is the point of attachment for the *biceps brachii,* a flexor muscle of the arm. The region between the head and the radial tuberosity is the **neck** of the radius. The distal lateral prominence of the radius that articulates with the wrist is the **styloid process.** Locate this process on your own wrist.

The **ulna,** or elbow bone, is the largest bone in the forearm. Note in the cutaway section of the elbow that the proximal end of the ulna has a large prominence, the **olecranon process.** On the anterior surface of this process is a large depression, the **semilunar notch,** which articulates with the trochlea of the humerus. The eminence just below

the semilunar notch on its anterior surface is the **coronoid process** (label 24). Where the head of the radius contacts the ulna is a depression, the **radial notch.**

The lower end of the ulna is small and terminates in two eminences: a large portion, the **head,** and a small **styloid process.** The head articulates with a fibrocartilaginous disk that separates it from the wrist. The styloid process is a point of attachment for a ligament of the wrist joint.

The Hand Each hand consists of a **carpus** (wrist), a **metacarpus** (palm), and **phalanges** (fingers). The **carpus** consists of eight small bones arranged in two rows of four bones each. The **metacarpus** consists of five metacarpal bones that are numbered one to five, the thumb-side being one.

The phalanges are the skeletal elements of the fingers distal to the metacarpal bones. There are three phalanges in each finger and two in the thumb.

Assignment:
Label figure 14.2.

The Lower Extremities

The Pelvic Girdle The two hipbones *(ossa coxae)* articulate in front to form a bony arch called the *pelvic girdle.* The back of this arch is formed by the union of the ossa coxae with the sacrum and coccyx. The enclosure formed within this bony ridge is the **pelvis** (Latin: *pelvis,* basin). Figure 14.4 illustrates one half of the pelvis and the right leg.

Figure 14.3 illustrates the lateral and medial views of the right os coxa. The large circular depression into which the head of the femur fits is the **acetabulum.** Note that each os coxa consists of three fused bones (ilium, ischium, and pubis) that are joined together in the center of the acetabulum. Although these ossification lines are easy to discern in young children, they are generally obliterated in the adult.

Two articulating surfaces (labels 15 and 16) are visible on the medial aspect of the os coxa in figure 14.3. The upper one is the **sacrum articulating surface,** which interfaces with the sacrum. The lower one is the **symphysis pubis articulating surface,** which joins with the other os coxa to form the symphysis pubis. The line of juncture between the ilium and sacrum is the **sacroiliac joint.**

The **ilium** is the yellow portion of the bone. On the greater portion of its medial surface is a large concavity called the **iliac fossa.** The **arcuate line** (label 11) is a ridge of demarcation below the iliac fossa.

The upper margin of the ilium, which is called the **iliac crest,** extends from the **anterior superior spine** (label 9) to the **posterior superior spine** (label 2). Just below the latter process is the **posterior inferior spine.** An **anterior inferior spine** is seen on the opposite margin below the anterior superior spine.

The reddish colored portion of the os coxa is the **ischium.** Note that it has two distinct processes: a small eminence, the **ischial spine,** and a larger one, the **tuberosity of the ischium,** which makes up the lower bulk of the bone. It is on this tuberosity that one sits. Just below the ischial spine is a depression, the **lesser sciatic notch.** The larger depression superior to it is the **greater sciatic notch.**

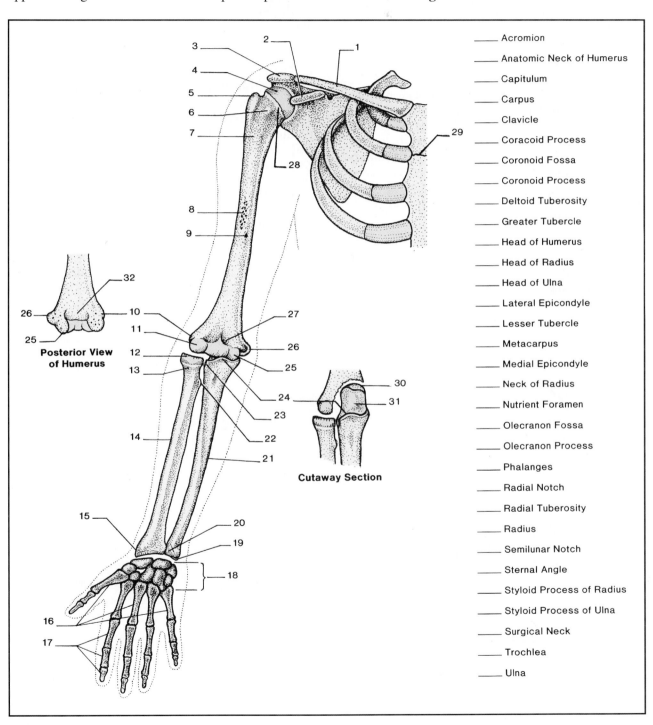

_____ Acromion
_____ Anatomic Neck of Humerus
_____ Capitulum
_____ Carpus
_____ Clavicle
_____ Coracoid Process
_____ Coronoid Fossa
_____ Coronoid Process
_____ Deltoid Tuberosity
_____ Greater Tubercle
_____ Head of Humerus
_____ Head of Radius
_____ Head of Ulna
_____ Lateral Epicondyle
_____ Lesser Tubercle
_____ Metacarpus
_____ Medial Epicondyle
_____ Neck of Radius
_____ Nutrient Foramen
_____ Olecranon Fossa
_____ Olecranon Process
_____ Phalanges
_____ Radial Notch
_____ Radial Tuberosity
_____ Radius
_____ Semilunar Notch
_____ Sternal Angle
_____ Styloid Process of Radius
_____ Styloid Process of Ulna
_____ Surgical Neck
_____ Trochlea
_____ Ulna

Posterior View of Humerus

Cutaway Section

Figure 14.2   **The arm and shoulder girdle.**

The blue-colored bone is the **pubis.** The opening surrounded by the pubis and ischium is the **obturator foramen.** Note that the **inferior ramus of the ischium** and **inferior ramus of the pubis** form the lower bony arch around this foramen.

Assignment:

Label figure 14.3.

Compare a male pelvis with a female pelvis and answer questions on the Laboratory Report that pertain to their anatomical differences.

The Upper Leg   Skeletal support of the thigh is achieved with one bone, the **femur.** Its upper end consists of a hemispherical **head,** a **neck,** and two processes, the greater and lesser trochanters.

The **greater trochanter** is the large process on its lateral surface. The **lesser trochanter** is located farther down on its medial surface. A ridge, the **intertrochanteric line,** extends obliquely between these two trochanters on the anterior surface. On the posterior surface a ridge between these two trochanters is called the **intertrochanteric crest.**

The lower extremity of the femur is larger than the upper end and is divided into two condyles: the **lateral** and **medial condyles.**

The Lower Leg   The tibia and fibula constitute the skeletal structures of the lower leg. The **tibia** is the stronger bone of the two. Its upper portion is expanded to form two condyles and one tuberosity. The condyles (labels 8 and 20) are named according to their location: **medial** and **lateral condyles.** The **tibial tuberosity** is located just below these condyles on the anterior surface of the tibia.

The distal extremity of the tibia is smaller than the upper portion. A strong process, the **medial malleolus,** forms the inner prominence of the ankle. Along its anterior surface the tibia has a ridge, the **anterior crest.**

The **fibula** is lateral to the tibia and parallel to it. The upper extremity, or **head,** articulates with the tibia, but it does not form a part of the knee joint. Below the head is the **neck** of the fibula. The lower extremity of the fibula terminates in a pointed process, the **lateral malleolus,** which lies just under the skin forming the outer anklebone. Locate this process on your own leg. Like the tibia, the fibula has an **anterior crest** extending down its anterior surface.

The Foot   Each foot consists of a tarsus, metatarsus, and phalanges. The **tarsus,** or ankle, consists of seven tarsal bones. The *calcaneus,* or heel bone, is the largest tarsal bone. The tibia of the leg articulates with the *talus,* a tarsal bone on top of the foot.

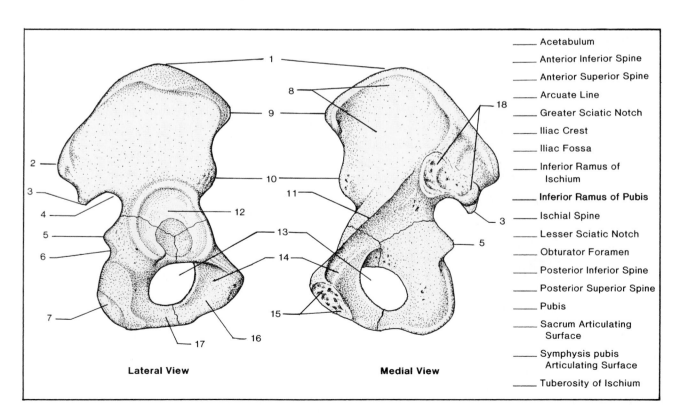

_____ Acetabulum
_____ Anterior Inferior Spine
_____ Anterior Superior Spine
_____ Arcuate Line
_____ Greater Sciatic Notch
_____ Iliac Crest
_____ Iliac Fossa
_____ Inferior Ramus of Ischium
_____ **Inferior Ramus of Pubis**
_____ Ischial Spine
_____ Lesser Sciatic Notch
_____ Obturator Foramen
_____ Posterior Inferior Spine
_____ Posterior Superior Spine
_____ Pubis
_____ Sacrum Articulating Surface
_____ Symphysis pubis Articulating Surface
_____ Tuberosity of Ischium

Lateral View   **Medial View**

Figure 14.3   The os coxa.

The **metatarsus,** or instep, consists of five elongated metatarsal bones. They are numbered one through five, number one being on the medial side of the foot. The **phalanges** are the bones of the toes. There are two phalanges in the great toe and three in each of the other toes.

Assignment:
Label figure 14.4.

Answer the remaining questions on the first part of the combined Laboratory Report 14, 15 that pertain to this exercise.

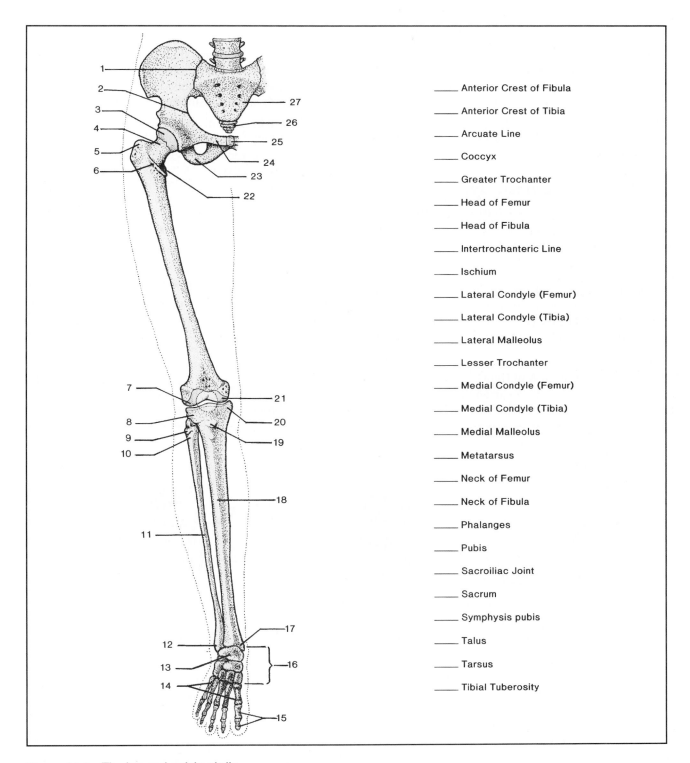

_____ Anterior Crest of Fibula

_____ Anterior Crest of Tibia

_____ Arcuate Line

_____ Coccyx

_____ Greater Trochanter

_____ Head of Femur

_____ Head of Fibula

_____ Intertrochanteric Line

_____ Ischium

_____ Lateral Condyle (Femur)

_____ Lateral Condyle (Tibia)

_____ Lateral Malleolus

_____ Lesser Trochanter

_____ Medial Condyle (Femur)

_____ Medial Condyle (Tibia)

_____ Medial Malleolus

_____ Metatarsus

_____ Neck of Femur

_____ Neck of Fibula

_____ Phalanges

_____ Pubis

_____ Sacroiliac Joint

_____ Sacrum

_____ Symphysis pubis

_____ Talus

_____ Tarsus

_____ Tibial Tuberosity

Figure 14.4   The leg and pelvic girdle.

15

Articulations

During this laboratory period we will study the basic characteristics of three types of joints and the specific anatomical details of the shoulder, hip, and knee joints.

Materials:
   fresh knee joint of cow or lamb, sawed through longitudinally

Three Basic Types

All joints in the body fall into one of three categories illustrated in figure 15.1. Note that classification is based on the degree of movement.

Immovable Joints

Immovable joints lack mobility because the adjacent bones in the joint are bonded to each other by fibrous connective tissue or cartilage. There are two kinds of immovable joints: sutures and synchondroses.

Sutures   Illustration A in figure 15.1 illustrates the structure of a *suture*. These irregular joints are seen between the flat bones of the skull.

The "glue" between these bones is **fibrous connective tissue,** which is continuous with the **periosteum** (label 2) on the outside of the skull and the **dura mater** on the inside of the skull.

Synchondroses   This type of immovable joint utilizes cartilage instead of fibrous tissue as the bonding agent. The metaphyses of long bones in children are joints of this type. As was noted in Exercise 11, these growth areas in children consist of hyaline cartilage during the growing years.

Slightly Movable Joints

Joints that exhibit slight movement are either symphyses or syndesmoses.

Symphyses   Intervertebral joints of the spine and the symphysis pubis are representative joints of this type. The most characteristic feature of this type of joint is the presence of a pad of **fibrocartilage** (label 6, figure 15.1), which provides a cushion in the joint. **Articular cartilage** (hyaline type) on the bone surfaces provide a frictionless surface adjacent to the fibrocartilage. The joints are held together with a fibroelastic **capsule.**

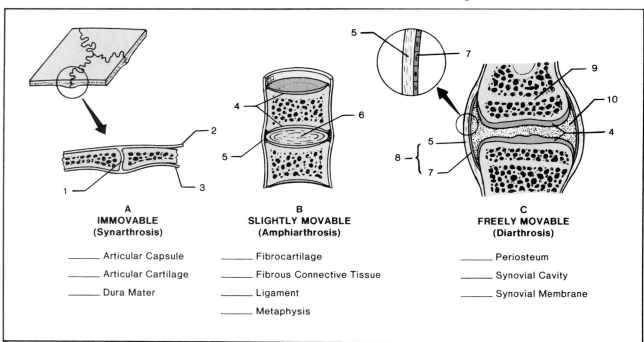

A IMMOVABLE (Synarthrosis)	B SLIGHTLY MOVABLE (Amphiarthrosis)	C FREELY MOVABLE (Diarthrosis)
_____ Articular Capsule	_____ Fibrocartilage	_____ Periosteum
_____ Articular Cartilage	_____ Fibrous Connective Tissue	_____ Synovial Cavity
_____ Dura Mater	_____ Ligament	_____ Synovial Membrane
	_____ Metaphysis	

Figure 15.1   Types of articulations.

Syndesmoses Slightly movable joints that lack fibrocartilage and are held together by an inter-osseous ligament are of this type. A good example is the attachment of the fibula to the tibia, as seen in figure 15.4.

Freely Movable Joints

Articulations that move easily are designated as *diarthrotic,* or *synovial,* joints. Note in illustration C, figure 15.1, that the bone ends are covered with smooth **articular cartilage.** Holding the joint together is a fibrous **articular capsule,** which consists of an outer layer of **ligaments** and an inner lining of **synovial membrane.** The latter membrane produces a viscous fluid called *synovium,* which lubricates the joint. These joints also contain sacs, or **bursae,** of synovial tissue. Fibrocartilaginous pads may also be present.

Types of Freely Movable Joints

Diarthrotic joints of the body are classified into six different categories.

Gliding Joints The articular surfaces in these joints are nearly flat or slightly convex. Bones that exhibit gliding action are seen in between carpal bones of the wrist and tarsal bones of the ankle.

Hinge Joints Movement through the ankle, elbow, and knee is hingelike, moving primarily in one plane. Movement between the occipital condyles of the skull and the atlas is also hingelike.

Condyloid Joints Joints that have bones with oval-shaped heads, or condyles, that move in elliptical cavities can move in two directions. The wrist is a typical condyloid joint.

Saddle Joints Like the condyloid joints, these joints allow movement in two directions. The ends of the bones differ, however, in that each bone end is convex in one direction and concave in the other direction. The joint between the thumb metacarpal and trapezium is of this type.

Pivot Joints A joint in which movement is rotational around an axis is a pivot type. Rotation of the atlas around the odontoid process of the axis is a good example.

Ball and Socket Joints These joints have angular movement in all directions. The shoulder and hip joints are representative.

Assignment:
Label figure 15.1.

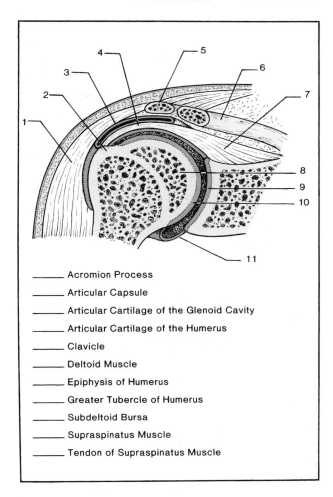

_____ Acromion Process

_____ Articular Capsule

_____ Articular Cartilage of the Glenoid Cavity

_____ Articular Cartilage of the Humerus

_____ Clavicle

_____ Deltoid Muscle

_____ Epiphysis of Humerus

_____ Greater Tubercle of Humerus

_____ Subdeltoid Bursa

_____ Supraspinatus Muscle

_____ Tendon of Supraspinatus Muscle

Figure 15.2   The shoulder joint.

The Shoulder Joint

The shoulder joint is the most freely movable articulation of the body. Its structure is revealed in figure 15.2. Observe that the articulating surfaces of the humerus and glenoid cavity are covered with **articular cartilage.**

Although there are usually six or seven bursae in the neighborhood of this joint, only one, the **subdeltoid bursa,** is shown in figure 15.2. Note that the **tendon of the supraspinatus muscle** inserts on the **greater tubercle of the humerus.** Bony sections seen in the upper portion of the joint are the **acromion process** (label 5) and a portion of the **clavicle.** The membranous fold beneath the head of the humerus is a portion of the **articular capsule.**

Assignment:
Label figure 15.2.

The Hip Joint

The hip joint, another ball and socket joint, is shown in figure 15.3. The rounded head of the femur is confined in the acetabulum of the os coxa by the acetabular labrum and the transverse acetabular ligament. The **acetabular labrum** (label 8) is a fibrocartilaginous rim attached to the margin of the acetabulum. The **transverse acetabular ligament** (label 11) is an extension of the acetabular labrum.

Note in the sectional view that a structure, the **ligamentum teres femoris,** is attached to the middle of the curved condylar surface of the femur. The other end of this ligament is attached to the surface of the acetabulum. This ligament adds nothing to the strength of the joint; instead, it contributes to nourishment of the head of the femur and supplies synovial fluid to the joint.

The entire joint, including the above three structures, is enclosed in an articular capsule. This capsule consists of longitudinal and circular fibers surrounded by three external accessory ligaments. The three accessory ligaments are the iliofemoral, pubocapsular, and ischiocapsular ligaments.

The **iliofemoral ligament** (label 7) is a broad band on the anterior surface of the joint. This ligament is attached to the **anterior inferior iliac spine** at its upper margin and the **intertrochanteric line** on its lower margin. Adjacent and medial to the iliofemoral ligament lies the **pubocapsular ligament.** The posterior surface of the capsule is reinforced by the **ischiocapsular ligament.** Like the knee joint, the capsule is lined with a **synovial membrane,** which provides lubrication for the joint.

Movements of flexion, extension, abduction, adduction, rotation, and circumduction of the thigh are readily achieved through this joint. This is made possible by the unique angle of the neck of the femur and the relationship of the condyle to the acetabulum. Excessive backward movement of the body at the joint is limited to some extent by the iliofemoral ligament, which takes some of the strain off certain muscles.

Assignment:
Label figure 15.3.

The Knee Joint

Two views and a sagittal section of the knee are shown in figure 15.4. Although the action of this joint has been described earlier as being essentially hingelike, it is by no means a simple hinge. The curved surfaces of the condyles of the femur allow rolling and gliding movements within the joint. There is also some rotary movement due to the nature of the hip and foot alignment.

The knee joint is probably the most highly stressed joint in the body. To absorb some of this stress are two *semilunar cartilages,* or *menisci,* in each joint. As revealed in the sagittal section, these fibrocartilaginous pads are thick at the periphery and thin in the center of the joint, providing a deep recess for the condyles.

The **lateral meniscus** lies between the lateral condyle of the femur and the tibia; the **medial meniscus** is between the medial condyle and the tibia. These menisci are best seen in the anterior and posterior views of figure 15.4. Anteriorly and peripherally, they are connected by a **transverse ligament.**

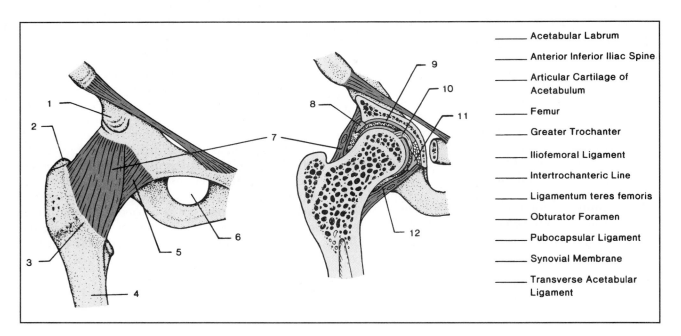

_____ Acetabular Labrum

_____ Anterior Inferior Iliac Spine

_____ Articular Cartilage of Acetabulum

_____ Femur

_____ Greater Trochanter

_____ Iliofemoral Ligament

_____ Intertrochanteric Line

_____ Ligamentum teres femoris

_____ Obturator Foramen

_____ Pubocapsular Ligament

_____ Synovial Membrane

_____ Transverse Acetabular Ligament

Figure 15.3   The hip joint.

The entire joint is held together by several layers of ligaments. Innermost are two cruciate and two collateral ligaments. The **posterior** and **anterior cruciate ligaments** form an × on the median line of the posterior surface, with the posterior cruciate ligament being outermost. The anterior cruciate ligament is the outermost one seen on the anterior surface when the leg is flexed. The **fibular collateral ligament** is on the lateral surface, extending from the lateral epicondyle of the femur to the head of the fibula. The **tibial collateral ligament** extends from the medial epicondyle of the femur to the upper medial surface of the tibia. In addition to these four ligaments are the *oblique* and *arcuate popliteal ligaments* on the posterior surface. These are not shown in figure 15.4.

Encompassing the entire joint is the *fibrous capsule,* a complicated structure of special ligaments, united with muscle tendons that pass over the whole joint.

Observe in the sagittal section that the kneecap is held in place by an upper **quadriceps tendon** and a lower **patellar ligament.** The innermost surfaces of these latter structures are lined with **synovial membrane.** The space between the synovial membrane and the femur is the **suprapatellar bursa.**

It is significant that although this joint is capable of sustaining considerable stress, it lacks bony reinforcement to prevent dislocations in almost any direction. It relies almost entirely on soft tissues to hold the bones in place. It is because of this fact that knee injuries are so commonplace in vigorous sport activities. It is a vulnerable joint particularly to lateral and rotational forces. An understanding of its anatomy should alert one to its limitations.

Assignment:
Label figure 15.4.

Laboratory Assignment

Animal Joint Study Examine the knee joint of a cow or lamb that has been sawed through longitudinally. Identify as many of the structures as possible that are shown in figure 15.4.

Laboratory Report Complete the Laboratory Report for this exercise.

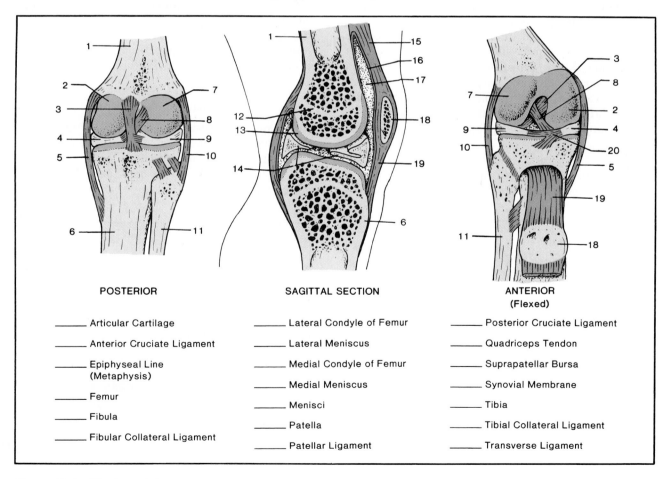

Figure 15.4 The knee joint.

POSTERIOR — SAGITTAL SECTION — ANTERIOR (Flexed)

_____ Articular Cartilage
_____ Anterior Cruciate Ligament
_____ Epiphyseal Line (Metaphysis)
_____ Femur
_____ Fibula
_____ Fibular Collateral Ligament

_____ Lateral Condyle of Femur
_____ Lateral Meniscus
_____ Medial Condyle of Femur
_____ Medial Meniscus
_____ Menisci
_____ Patella
_____ Patellar Ligament

_____ Posterior Cruciate Ligament
_____ Quadriceps Tendon
_____ Suprapatellar Bursa
_____ Synovial Membrane
_____ Tibia
_____ Tibial Collateral Ligament
_____ Transverse Ligament

16

Electronic Instrumentation

Various types of electronic equipment will be used in subsequent experiments that pertain to muscle contraction, heart action, respiration, etc. It is the purpose of this exercise to present a basic understanding of how this equipment is set up and how it works.

Figure 16.1 illustrates the arrangement of equipment that might be used in a typical instrumentation setup. Note that the biological phenomenon being studied, whether it produces an electrical or mechanical signal, must be picked up by either an electrode or an input transducer. If the signal is electrical, an **electrode** will be used. Nonelectrical signals, however, must be converted to electrical signals before they can be measured by the setup. For conversion of nonelectrical signals, such as temperature or pressure, an **input transducer** must be used.

Since biological phenomena often produce weak signals, the next unit in the system to receive the signal is the **amplifier,** where amplification takes place. Finally, the amplified signals are converted to some form of visual or audio display with an **output transducer,** that may be a chart recorder, oscilloscope, computer monitor, or some other such type of instrument.

The considerable variety of components that one might encounter in a college laboratory precludes a complete description here of all types of equipment.

Our main concern in this exercise will be to make a statement concerning the principal types that you are likely to encounter in this laboratory manual.

Electrodes

Two basic types of electrodes will be used in our experiments: pickup and stimulating electrodes. **Pickup electrodes** are the type discussed above that are used to *receive* a signal. EEG (electroencephalograph) and GSR (galvanic skin response) electrodes are of this type (see figure 16.2). Both of these electrodes work best when an electrode jelly, such as Biogel, is used on the skin to improve conductivity.

Electrodes that *induce* electrical stimulation into tissues to elicit a response are referred to as **stimulating electrodes.** This type will be used in some of our muscle experiments. An electronic stimulator must be used with this type of electrode.

Input Transducers

An *input transducer,* by definition, is any device that converts a nonelectrical form of energy to an electrical signal. The principal input transducers that we will use will be for measurements of force, pulse, and sound.

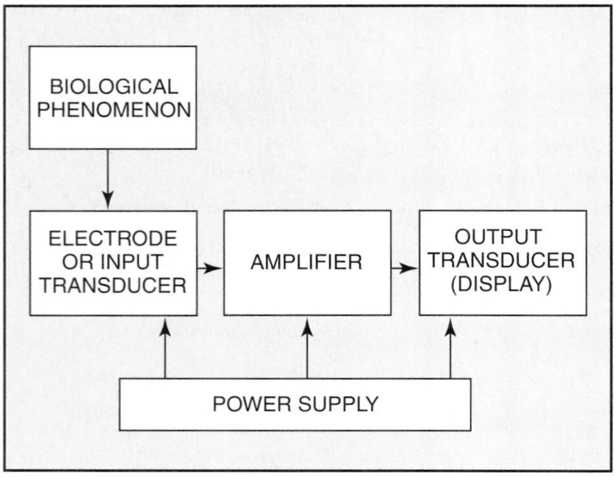

Figure 16.1 Equipment setup for monitoring various biological phenomena.

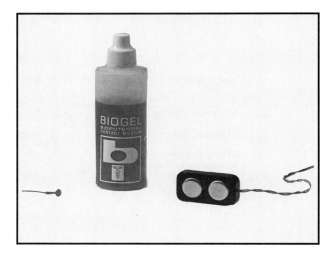

Figure 16.2 EEG and GSR electrodes with a dispenser of electrode jelly.

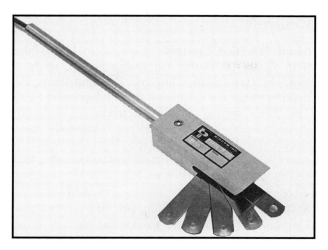

Figure 16.3 Bending of the steel leaves on this type of force transducer changes the sensitivity in a stress-sensitive resistor in the transducer.

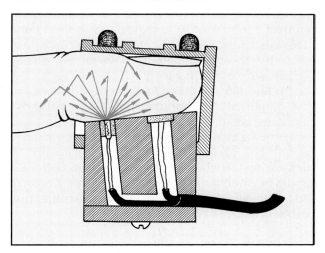

Figure 16.4 This pulse transducer utilizes a photo sensitive resistor to detect differences in light intensity as blood surges through the soft tissues.

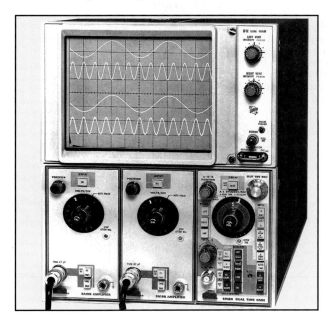

Figure 16.5 A dual-beam oscilloscope, manufactured by Tektronix, works well for two-channel experiments.

Force Transducers

Transducers of this type are able to convert the mechanical force exerted by a muscle into an electrical signal. Figure 16.3 illustrates one of the types that can be used in our muscle experiments. Since the force exerted by a muscle may vary from a very weak pull to a very strong one, the transducer has a number of flexible leaves that can be used individually or together. Weak forces may require only a single leaf; strong forces can be matched with two or more leaves. Figure 22.1 illustrates a setup in which this type of transducer is used.

The element in transducers of this type that makes this possible is a resistor that is affected by the stress applied to the leaves. As deformation occurs in the spring leaves, due to bending, proportional bending also occurs in the resistor. The induced stretching, or compression of the resistor changes its resistivity. Resistivity changes, in turn, result in changing electrical signals that pass to the amplifier, accomplishing the conversion of mechanical activity to electrical activity.

One complication with this type of transducer is that the resistor is part of a Wheatstone bridge that must be electronically balanced before each experiment is performed. The procedure for balancing the transducer resistor is provided in appendix B and will be utilized in those experiments where it is required.

Pulse Transducers

The pulse transducer illustrated in figure 16.4 utilizes a photoresistor and a small light source to detect the presence or absence of the pulse in the finger. As blood surges through the finger, the amplitude of light is altered, resulting in changes of reflected light that strike the photoresistor. The variations in light intensity striking the photoresistor generate a variable signal that is fed to the amplifier and output transducer for interpretation.

Microphones

Microphones that pick up sound waves and convert them to electronic signals are used for studying heart and breathing sounds. An audio monitor (loud speaker) is often used in such setups.

Output Transducers

Chart recorders, oscilloscopes, audio monitors, and computer monitors are the principal kinds of output transducers used in the physiology laboratory today. In the past, chart recorders have been the instrument of choice; however, the computer with

73

its monitor and a good printer is rapidly replacing many of the chart recorder applications. Three popular types of recorders will be discussed here as well as the cathode ray tube, which is the basis for the oscilloscope and computer monitor.

The Cathode Ray Tube

The cathode ray tube (CRT) is an output transducer that visually displays input voltage signals on a fluorescent screen. Television screens, oscilloscopes (figure 16.5), and computer monitors are, basically, cathode ray tubes. Figure 16.6 illustrates the mechanics of a CRT. Note that the electrical potential difference, or **voltage,** of an input signal is expressed vertically, and the time base, or **sweep,** is shown horizontally. A stream of electrons, generated by an electron gun (cathode) at the back of the tube, is fed first between two horizontal (green) deflection plates and then between two vertical (blue) plates. The deflection of the electron beam by the Y and X plates determines the trace that is shown on the screen by the stream of electrons that causes fluorescence on the screen of the tube.

The Unigraph

The Gilson Unigraph, figure 16.7, is a compact chart recorder that can be used for monitoring blood pressure, pulse rate, heart action, and many other physiological activities. Because of its small size (portability) it has been used in many of the experimental setups in this manual.

Controls

Mode Selection Control   This control knob (label 6 in figure 16.7) has the following six settings: EEG, ECG, CC-Cal, CD-CAL, DC, and Trans. For making electrocardiograms, the knob is set at ECG. The EEG setting is for electroencephalography (brain waves). The Trans setting is used for transducers. The Cal setting is used when one wishes to calibrate the instrument for certain recordings, such as electrocardiograms.

Styluses   Instead of using ink, the stylus of a Unigraph is heated and marks the moving paper with a blue line as the heat acts on a heat sensitive chemical in the paper.

Two styluses are present: a **recording stylus** (label 4) and an **event marker** (label 3). Since stylus temperature is critical, there is a **stylus heat control** knob (label 8) to regulate the temperature. This control is usually set at the *two o'clock position.* If the styluses are too warm the tracing becomes excessively wide.

While the recording stylus is used to record the amplified input signal, the event marker is used to record when an event begins, or when some new pertinent influence affects the recorded event. The event may be recorded manually by depressing the white **event push button** (label 12) at the beginning of the event, or it may be recorded automatically if an **event synchronization cable** is used.

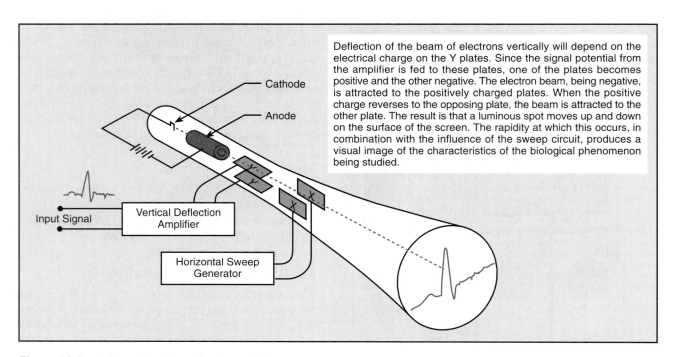

Deflection of the beam of electrons vertically will depend on the electrical charge on the Y plates. Since the signal potential from the amplifier is fed to these plates, one of the plates becomes positive and the other negative. The electron beam, being negative, is attracted to the positively charged plates. When the positive charge reverses to the opposing plate, the beam is attracted to the other plate. The result is that a luminous spot moves up and down on the surface of the screen. The rapidity at which this occurs, in combination with the influence of the sweep circuit, produces a visual image of the characteristics of the biological phenomenon being studied.

Figure 16.6   Schematic of a cathode ray tube.

Centering Control (Zero Offset Control) Note on the left side of the control panel in figure 16.7 that there is a knob labeled CENTERING. This control is used for moving the baseline of the recording stylus to the most desirable position. It should be kept in mind that the most desirable location of the baseline is not always the center of the paper; it may be near one of the edges.

Chart Speed There are two speeds at which the paper can move on the Unigraph: 2.5 mm and 25 mm per second. The speed is controlled by the **chart speed control lever** (label 1). When pointed toward the end of the instrument, it is set at the faster speed. To change the speed to slow, the lever is rotated 180° in the opposite direction. For the chart to move at all, however, the chart control switch has to be in the Chart On position.

Toggle Switches The Unigraph has four toggle switches, two of which are extensively used in its operation. The **main power switch** (label 7) is a small one with ON and OFF labels near it. It is located near the stylus heat control.

The large gray plastic switch (label 9) is the **chart control switch.** It has three settings: STBY, Chart On, and Stylus On. In the STBY (standby) position, the chart does not move and the stylus is deactivated. In the Stylus On position, the stylus heats up, but the chart does not move. At the Chart On position the chart moves and the stylus is activated.

The other two small toggle switches at the other end of the unit come into play only occasionally. They should be set at the NORM position for most experiments.

Sensitivity Controls Sensitivity of the Unigraph is controlled by the **gain control** (label 5) and the **sensitivity knob,** which has the label SENS near it. The gain control is calibrated in MV/CM (millivolts per centimeter) and it has a range of 2, 1, .5, .2, and .1 millivolts. The least sensitive position is 2; the greatest sensitivity is .1. When the sensitivity knob is turned clockwise the gain at any setting is increased toward the next level.

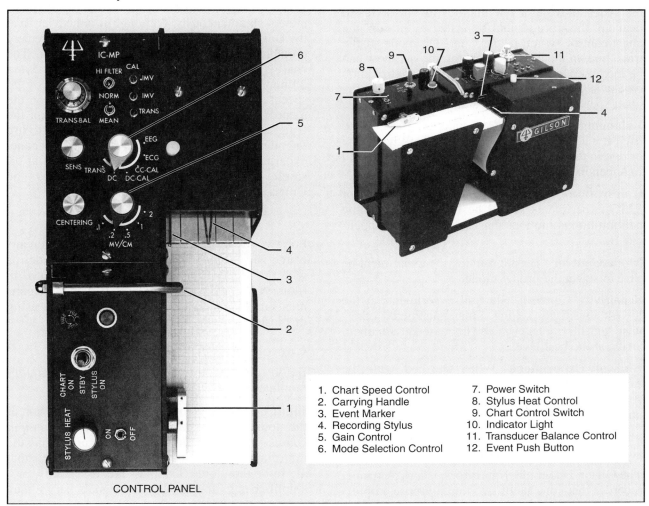

1. Chart Speed Control	7. Power Switch
2. Carrying Handle	8. Stylus Heat Control
3. Event Marker	9. Chart Control Switch
4. Recording Stylus	10. Indicator Light
5. Gain Control	11. Transducer Balance Control
6. Mode Selection Control	12. Event Push Button

CONTROL PANEL

Figure 16.7 The Gilson Unigraph.

Operational Procedures

When setting up the Unigraph in an experiment the following points should be kept in mind.

Cable Hookup The cable end of electrodes or the transducer must be inserted into an input receptacle in the end of the Unigraph. The cable end must be compatible with the Unigraph. For some experiments it will be necessary to use an adapter unit between the cable end and the Unigraph.

Synchronization Cable If an electronic stimulator is to be used, it is desirable, but not mandatory, to install an event synchronization cable between the stimulator and the Unigraph.

Power Source The end of the power cord on this instrument has three prongs. This indicates that it must be used only in a 110V *grounded* outlet. Before plugging it in, however, the power switch should be OFF and the chart control switch at STBY.

Stylus Heat Once the unit is plugged in, the power should be turned ON and the stylus heat control set at the two o'clock position.

After allowing a few minutes for the stylus to heat up, the trace should be tested by putting the chart control switch in the Chart On mode. If the trace line on the chart is too light, the darkness can be increased by turning the stylus heat control knob clockwise. This testing should be done with the chart speed set at the slow rate.

Balancing If a force transducer is going to be used it will be necessary to balance the Wheatstone bridge in the circuitry, using the procedure outlined in appendix B.

Calibration When monitoring EEG, ECG, or EMG experiments it will be necessary to follow calibration instructions that are provided in appendix B.

Mode Control Before running the experiment be sure to select the proper mode.

Sensitivity Control Start all experiments at the lowest sensitivity setting, unless instructed otherwise. Sensitivity level can be increased gradually to arrive at the best level.

End of Experiment When finished with the experiment always return all controls to their least sensitive settings, or to the OFF position. The chart control switch should be left at STBY.

The Duograph

The Gilson Duograph, figure 16.8, is a two-channel recorder that consists of two multipurpose amplifier modules. The modules (IC-MPs) are identical to

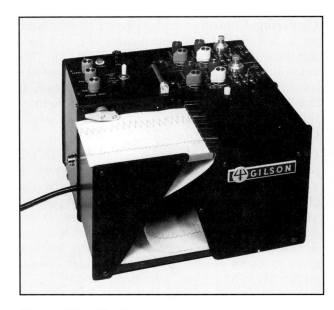

Figure 16.8   The Duograph.

the unit used in the Unigraph. This chart recorder enables one to monitor the interaction of two parameters simultaneously. Although it is heavier than the Unigraph (25 vs. 18 pounds), it is still extremely portable and can be carried with one hand. The chart is 127 mm wide and has two speeds (2.5 and 25 mm/sec) on the standard unit. An option of ten speeds is also available. Since the controls on the recorder are essentially the same as those on the Unigraph, learning to use one makes it easy to use the other. This unit can be used for the two channel experiments in Exercises 57 and 58 as well as all single channel experiments that utilize the Unigraph.

The Narco Physiograph®

For two or more channel recording experiments Narco Bio-Systems of Houston, Texas, provides us with the Mark IIIS Physiograph®, as illustrated in figure 16.9. All single and multichannel experiments in this manual provide instructions for using this instrument.

Components

Before utilizing this instrument in the laboratory familiarize yourself with its components.

Input Couplers Note that on the upper portion of the sloping control panel are four units called *input couplers* (label 3). It is into these units that the transducers or electrodes feed the input signals that are to be monitored. Couplers are, essentially, preamplifier units that modify the signals before they enter the amplifier units (label 2).

Couplers of various types are available from the manufacturer. At the present time Narco Biosystems produces thirteen different kinds of couplers. The Universal Coupler is probably the most widely used type. It is equivalent to the Gilson Unigraph in that it can be used for monitoring ECG, EEG, EMG, and other DC and AC potentials. Other couplers that are available include the Strain Gage Coupler, GSR Coupler, Transducer Coupler, and Temperature Coupler.

Amplifier Controls Note in figure 16.9 that each input coupler has its individual cluster of amplifier controls located below it. There are two knobs and one push button on each of these clusters.

The control on the left, with nine settings, adjusts the sensitivity of the channel. It has an outer portion that enables one to select the desired millivolt setting: 1, 2, 5, 10, 20, 50, 100, 200, 500, or 1000. At setting 1 it takes only 1 millivolt to produce 1 centimeter of pen movement; when set at 1000, it takes 1000 millivolts to produce the same amount of pen travel. Thus, we see that the 1 millivolt setting is the most sensitive setting available here. To adjust the sensitivity between any of these settings, one rotates the inner portion of this control. When the inner knob is completely clockwise, the sensitivity is identical to the reading on the dial.

Counterclockwise rotation of the inner knob decreases the sensitivity.

The knob on the right with POSITION under it is used for repositioning the pen to locate the baseline of your recording to the most desirable spot. The location of the baseline will be determined by the degree of pen travel. For example, if the pen deflection is to be as much as 6 centimeters, it would be desirable to set the baseline substantially below the center of the arc of the pen. The total range of deflection of the pen is approximately 8 centimeters.

The push button below the positioning knob is used for starting the recording process. When depressed, this button is in the ON mode, which activates the pen. To stop the recording process one merely presses the button again, causing it to come up to a higher level, which is the OFF position.

Ink Pens Note that the Mark IIIS has five ink pens: one for each of the four couplers and one to record one-second time intervals at the bottom of the chart. Each pen is supplied by ink from five separate ink **reservoirs** (label 5). These pens can be raised from the paper with the **pen lifter lever** (label 7). This lever actuates a metal bar that lifts all the pens simultaneously off the paper. *The pens should always be kept in the raised position when recording is not taking place.*

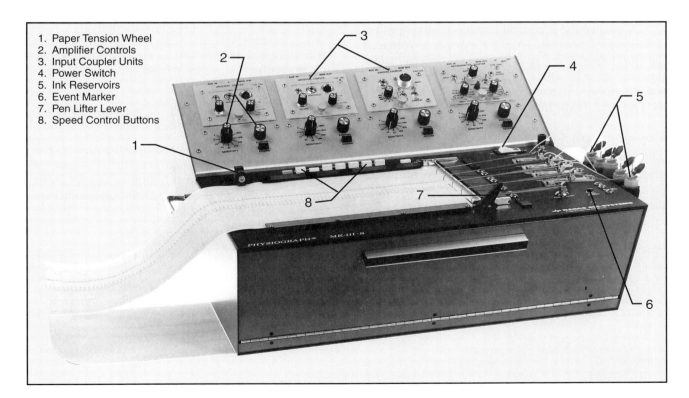

1. Paper Tension Wheel
2. Amplifier Controls
3. Input Coupler Units
4. Power Switch
5. Ink Reservoirs
6. Event Marker
7. Pen Lifter Lever
8. Speed Control Buttons

Figure 16.9 The Physiograph® Mark IIIS.

Ink reaches each pen through a small tube that leads from its ink reservoir. To get the ink to flow to the pen it is necessary to first raise the reservoir 2–3 centimeters, and then squeeze the rubber bulb at the top of the reservoir as the index finger is held over the small hole at the tip of the rubber bulb. Once the tube is full of ink the fingertip is released before the bulb is released. When it becomes necessary to increase or decrease the flow of ink during recording, one simply raises or lowers the reservoir slightly.

To prevent clogging of pens it is necessary to remove all ink from the pens at the end of each laboratory period. With the pens in the raised position, the ink is drawn back into the reservoir by squeezing the reservoir bulb, placing the index finger over the hole of the bulb, and then releasing pressure on the bulb. The vacuum thus created will draw all the ink back into the reservoir.

Chart Speed The Mark IIIS Physiograph® has seven push buttons (figure 16.10) that allow the following chart speeds: 0.05, 0.1, 0.25, 0.50, 1.0, 2.5, and 5.0 cm/sec. It also has an OFF push button on the left and a TIMER push button on the right end of the series.

To set the desired speed, one depresses the TIME button first (the up position is OFF), then the speed selection button is depressed, and finally, the OFF button is depressed again, putting the button in the ON position.

To change the speed, all one has to do is push a different speed button. To stop the paper movement, the OFF button is pressed. As soon as the paper stops moving, the pens should be raised with the pen lifter lever. At the end of the period it is important that both the OFF and TIME buttons be in the OFF position.

Physiograph® Operational Procedures

Although each experiment performed on the Physiograph® will differ in certain respects, there are some procedures that should always be followed. It is these general procedures that are outlined here.

Dustcover The dustcover should be removed from the unit, neatly folded, and placed somewhere where it will not be in the way or damaged. Because laboratory dust can affect some electronic components over a period of time, it is important that the unit be covered during storage.

Recording Paper If there is no recording paper in the paper compartment, place a stack of paper in the compartment that is accessed through the hinged front panel of the instrument. Feed the paper up through the opening provided in the top right margin of the paper compartment and then slide the paper under the paper guides on each side.

Draw the paper across the top of the recording surface and slide it under the paper tension wheel (label 1, figure 16.9) after lifting the wheel slightly by pulling the black lever straight up. After the paper is in place, release the lever. The paper should now be in firm contact with the paper drive mechanism and will move when the proper controls are activated.

Each small block on the paper measures 0.5 cm by 0.5 cm. Knowing this, one can determine the exact speed of the paper by the time marks.

Power Cord The power cord is a loose component that should be inserted first into the right side of the Physiograph®, and then into a grounded electric outlet.

Power Switch Turn on the white power switch (label 4). Note that a small lighted indicator in the switch reveals that the unit has been turned on. This switch should be left in the ON position during the entire laboratory period and turned OFF at the end of the period.

Ink Pens Fill the ink pens, observing the procedure outlined on the preceding page.

Paper Speed Set the speed of the paper by first setting the TIMER push button in the ON mode and depressing the speed push button called for in the experiment. Do not press the OFF button until you are ready to start recording.

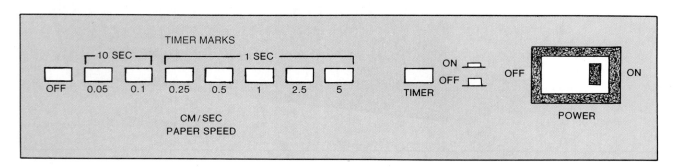

Figure 16.10   Push-button control panel of the Mark IIIS.

Coupler Attachment Insert the cable from the experimental setup into the receptacle of the proper input coupler.

Amplifier Setting Set the Sensitivity control at the sensitivity level recommended in the experiment. In most experiments a low sensitivity will be used first.

Transducer Balancing If a transducer is to be used it will have to be balanced as per appendix B.

Calibration In certain experiments, such as Exercise 47 (Electrocardiogram monitoring), it will have to be balanced as per appendix B.

Monitoring Once all the hookups have been completed, lower the pens to the paper and press the OFF button, which will start the paper moving, and then press the RECORD buttons on each amplifier cluster that is involved in the experiment. The events occurring in the experiment will now record.

The Electronic Stimulator

Stimulation of nerve and muscle cells in the laboratory is most readily accomplished with an electronic stimulator. Although we will focus our attention on the Narco and Grass stimulators, it is the general principles of stimulator operation that we will be primarily concerned with here.

The electronic stimulator accomplishes several things that are essential to controlled experiments. First of all, it converts alternating current (AC) to direct current (DC). As stated previously, DC moves continuously in one direction only. AC, on the other hand, reverses its direction repeatedly and continuously.

The advantage of DC stimulation is that there is an abrupt rise of voltage from zero to maximum and then a sudden return to zero when a pulse of elec-

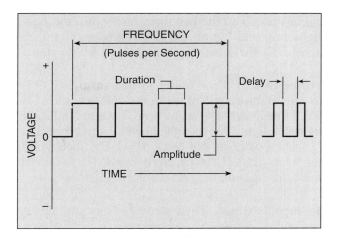

Figure 16.12 Square wave characteristics.

tricity is applied to a nerve or muscle. This type of voltage control produces a **square wave** instead of a sine wave. The differences are seen in figure 16.11.

In addition to converting AC to DC, the stimulator, through a variety of controls and switches, can vary the voltage, duration, and frequency of pulses that are fed to a specimen. With the increase of **voltage,** the **amplitude** or height, of the wave form is increased. The length of time from start to end of a single pulse of electrical stimulus is the **duration,** and the number of pulses delivered per second is called the **frequency.** Some stimulators, such as the Grass SD9, can also vary the time between twin pulses; this time interval is designated as the **delay.** Figure 16.12 illustrates these various characteristics.

Stimulator Control Functions

Figures 16.13 and 16.14 are of the Grass SD9 and Narco SM-1 electronic stimulators. Although the

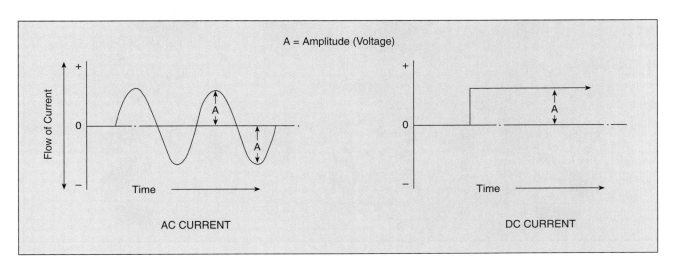

Figure 16.11 Comparison of AC and DC current.

control panels of the two instruments differ in configuration and terminology, they both accomplish essentially the same functions.

Power Switch   The power switch for the SD9 is located at the bottom of the front panel near the middle. Immediately above the switch is a red indicator light that lights up when the power is turned on.

The power switch for the SM-1 is located on the back panel. Like the SD9, the SM-1 has a red indicator light on the horizontal front panel.

Voltage Settings   The intensity of any stimulus is varied by increasing or decreasing the voltage. Note that voltage on the SD9 is determined by two settings: a round knob in the upper right-hand corner and a decade (multiplier) switch below it. The voltage range here is from 0.1 to 100 volts.

Voltage adjustments on the SM-1 are made with the VOLTAGE RANGE switch and the VARIABLE control knob. The voltage range switch has three settings: 0–10, 0–100, and OFF. When the voltage range switch is set at the 0–100 setting, each digit on the variable dial is multiplied by 10; thus, the range on this scale is from 0 to 100 volts. When the voltage range switch is set on 0–10, the voltage output is between 0 and 10 volts. In the OFF mode no electrical stimulus can occur. The lowest voltage output for this stimulator is 50 millivolts.

Duration Settings   The duration of each electrical pulse is measured in milliseconds (msec). On the SD9 this stimulus characteristic is controlled by the DURATION knob, which has a four-position decade switch below it. The combination of these two controls yields a range of 0.02 to 200 milliseconds (msec). The duration setting should not exceed 50% of the interval between pulses.

On the SM-1 duration is regulated by the WIDTH control, which is the left-hand knob on the horizontal panel. Its range is from 0.1 to 2.0 msec.

Mode Settings   Since stimulators can produce single stimuli, continuous pulses, and even twin pulses, there has to be a way of selecting the mode preferred for a particular experiment.

The two mode selection switches on the SD9 are located under the STIMULUS label. The right-hand MODE switch has three positions: repeat, single, and off. In the REPEAT position repetitive impulses are produced according to the setting on the Frequency control. To produce single pulses, the switch lever is depressed manually to the SINGLE position and released. One pulse is administered for each downward press of the lever; a spring returns the switch lever to the OFF position.

The other three-position switch under the stimulus label on the SD9 is used for twin pulses and hooking up with another stimulator. The MOD position is used when another stimulator is used in

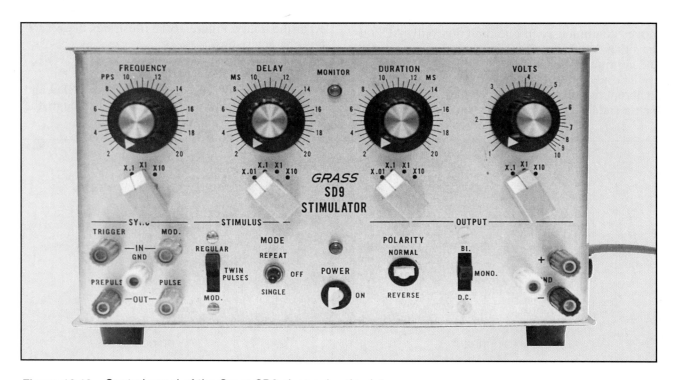

Figure 16.13   Control panel of the Grass SD9 electronic stimulator.

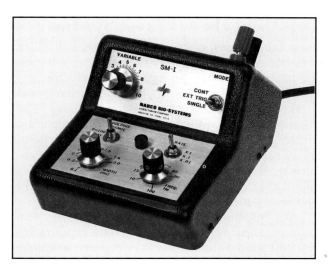

Figure 16.14 The Narco Bio-Systems SM-1 stimulator.

Frequency Settings The frequency control on a stimulator is set only when repetitive pulses are administered. In other words, the mode must be set at REPEAT on the SD9, or CONT on the SM-1.

On the SD9 the frequency control with its decade switch has a range of 0.20 to 200 pulses per second (PPS).

On the SM-1 the frequency range is from 0.1 Hz to 100 Hz. A RATE switch, which is calibrated ×1, ×.1, and ×.01, must be used in conjunction with the round knob to get the proper setting. Incidentally, the Hz (Hertz) unit means essentially the same thing as PPS. One Hz represents one cycle per second.

Delay Settings Stimulators that can produce twin pulses have a DELAY control to regulate the time interval between the pairs of pulses. The delay on the SD9 can be varied from 0.02 to 200 milliseconds, utilizing the control knob and four-place decade switch. The delay control can be used in either a single or repetitive mode. When the REPEAT mode is used, the delay time should not exceed 50% of the period between each set of pulses to avoid overlap.

Since the Narco SM-1 stimulator does not produce twin pulses, there is no need for a delay control.

Output Switches on SD9 In addition to the aforementioned controls, the Grass SD9 has two switches positioned under the OUTPUT label that can modify the wave form. Note that the three-position slide switch on the right is for selecting monophasic (MONO), biphasic (BI), or DC. Figure 16.15, middle

tandem with this one to modulate its output. When it is desirable to have the stimulator produce two pulses close together, the TWIN PULSES setting is used. When twin pulses or a second stimulator are not used, the setting should be on REGULAR. For most of our experiments this switch should be set at the regular setting.

The MODE switch on the SM-1 has three positions: continuous, single, and external triggering. When in the CONT position the stimulator will produce repetitive pulses according to the setting on the frequency control. When the switch lever is pressed to SINGLE, a single pulse will be delivered. The EXT TRIG position is used when an external triggering device is used. Such a device is plugged into a receptacle in the back of the stimulator.

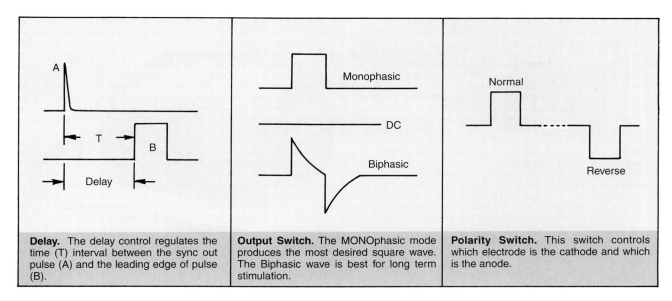

Delay. The delay control regulates the time (T) interval between the sync out pulse (A) and the leading edge of pulse (B).

Output Switch. The MONOphasic mode produces the most desired square wave. The Biphasic wave is best for long term stimulation.

Polarity Switch. This switch controls which electrode is the cathode and which is the anode.

Figure 16.15 Wave form modifications (Grass SD9 stimulator).

illustration, shows the differences between these three types of current. The proper setting for most experiments will be in the monophasic position. However, when tissues are stimulated for a long period of time with the monophasic waveform, hydrolysis occurs, causing gas bubbles to form around the electrodes. For the longer duration experiments, the biphasic wave form will be preferable.

The DC position is seldom used due to the formation of electrolytes and gas bubbles in the preparation. Activation of the DC position overrides all other controls except voltage and mode.

The POLARITY switch to the left of the other switch is used for reversing direction of electron flow. In the NORMAL position the red output binding post is positive (+) with respect to the black binding post. In the REVERSE position the red binding post becomes negative (−) with respect to the black binding post. See the right-hand illustration in figure 16.15.

Operational Procedures

Due to the fact that each experiment will require different settings of controls, no operational procedure will be provided here at this time. In any experiment where a stimulator is used, the control settings will be given at the beginning of the experiment.

Using the Intelitool System

Four experiments in this laboratory manual utilize transducers with a computer to monitor physiological activities. The hardware and software for performing these experiments are manufactured and marketed by Intelitool, Inc., of Batavia, Illinois. The four exercises that utilize Intelitool equipment are Exercise 23 (the *Physiogrip*™ for muscle physiology), Exercise 33 (the *Flexicomp*™ for reflex monitoring), Exercise 49, (the *Cardiocomp*™ for ECG monitoring), and Exercise 62 (the *Spirocomp*™ for spirometry measurements).

An Intelitool data acquisition/analysis system is an excellent way to perform certain experiments if a compatible computer and printer are available. Since the equipment is designed to be used only on the human body (with safety), no animals are needed or sacrificed.

Although student computer knowledge is helpful in performing these experiments, it is not absolutely required. The reason is that the software instructions carefully guide the experimenter along, one step at a time, to the completion of the experiment. A great advantage of these systems over the use of chart recorders is that as the experiment is performed, data are accumulated, automatically, and calculations are instantly made by the software. Another great advantage is that with a printer that has graphic capability, the results of

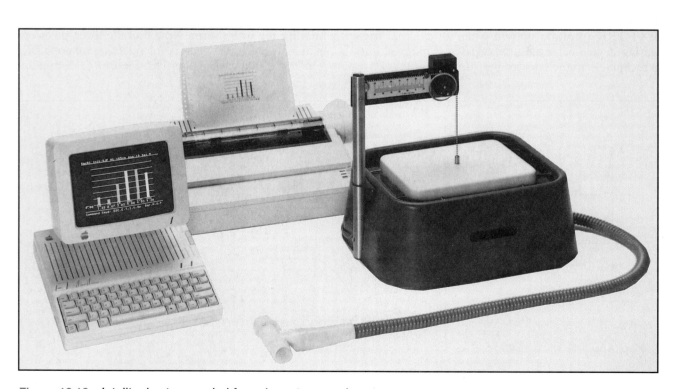

Figure 16.16 Intelitool setup needed for spirometry experiments.

the experiment can be printed out to produce a permanent record.

There is one limitation to these systems, however: they are all single channel systems. If one wishes to monitor two pysiological activities, simultaneously, such as ECG and respiration, one must use a Duograph or polygraph.

All of the four systems can be operated on the following computer platforms*:

- IBM or compatible PC (MS-DOS 3.3 or higher)
- Macintosh (System 6.0.8 or higher)
- Apple II (ProDOS)

The instructions for each experiment are written with the assumption that you have a basic knowledge of the computer platform that you will be using. To accommodate the different platforms, separate instructions are provided at each step when the computer is used.

To see how an Intelitool experiment is set up, refer to figure 16.16. It illustrates the equipment that is used in Exercise 62 to perform breathing experiments (spirometry). A wet spirometer with hose attached, mouthpieces, computer, printer, and software are all the equipment that is needed. The Spirocomp transducer is an interface box that is bolted to the scale arm that has a pulley on it. As the subject breathes into the spirometry hose, respiratory volumes are seen on the monitor of the computer, as the air chambered bell moves. Mathematical computations are automatically graphed by the software to determine pertinent information such as

---

*Platform refers to a specific combination of computer hardware and operating system software.

tidal volume, expiratory reserve volume, and vital capacity.

Although setup instructions are included in each experiment, the actual setup process will probably be performed for you prior to your entering the laboratory. Even the software will be loaded and ready to use. Assembling the various components for each experiment consists primarily of connecting the Intelitool transducer unit to the computer with the proper cable. Only one system, the Physiogrip, uses an electronic stimulator that is coupled to the subject's arm via a fitting on the front of the Physiogrip (see figure 23.1). All other systems have a single cable that leads to the computer.

No attempt will be made here to identify the manner in which one accesses the procedures for doing an Intelitool experiment. Once you get into the program, prompts on the monitor screen and instructions in this manual will guide you through the experiment.

In addition to the instructions in this manual there are two manuals produced by Intelitool that will be available to you. For each experiment there are the Intelitool *Lab Manual* and *User Manual*. Each manual is limited to the specific experiment being performed. There is no single manual that includes instructions for all four of the Intelitool systems. If some of the instructions in this manual seem confusing, you are encouraged to refer to the two Intelitool manuals for clarification.

Laboratory Report

Answer all the questions on Laboratory Report 16 for this exercise.

17

Muscle Structure

The movement of limbs and other structures by skeletal muscles is a function of the contractility of muscle fibers and the manner of attachment of the muscles to the skeleton. In this exercise our concern will be with the detailed structure of whole muscles and their mode of attachment to the skeleton.

Figure 17.1 reveals a portion of the arm and shoulder girdle with only two muscles shown. Several other muscles of the upper arm have been omitted that would obscure the points of attachment of these muscles. The longer muscle that is attached to the scapula and humerus at its upper end is the **triceps brachii.** The shorter muscle on the anterior aspect of the humerus is the **brachialis.**

Muscle Attachment

Each muscle of the body is said to have an origin and insertion. The **origin** of the muscle is the immovable end. The **insertion** is the other end, which moves during contraction. Contraction of the muscle results in shortening of the distance between the origin and the insertion, causing movement of the insertion end.

Muscles may be attached to bone in three different ways: (1) directly to the periosteum, (2) by means of a tendon, or (3) with an aponeurosis. The upper end of the brachialis (the origin) is attached by the first method, i.e., directly to the periosteum. The insertion end of this muscle, however, does have a short tendon for attachment.

A **tendon** is a band or cord of white fibrous tissue that provides a durable connection to the skeleton. The tendon of the triceps insertion is much longer and larger due to the fact that the muscle exerts a greater force in moving the forearm.

An example of the aponeurosis type of attachment is seen in figure 29.1, page 148. Label 5 in this figure is the aponeurosis of the external oblique muscle. An **aponeurosis** is a broad flat sheet of glistening pearly-white fibrous connective tissue that attaches a muscle to the skeleton or another muscle.

Although most muscles attach directly to the skeleton, there are many that attach to other muscles or soft structures (skin) such as the lips and eyelids. In Exercises 26 through 30 the precise origins and insertions of various muscles will be studied.

Microscopic Structure

Illustrations B and C in figure 17.1 reveal the microscopic structure of a portion of the brachialis muscle. Note that the muscle is made up of bundles, or **fasciculi,** of skeletal muscle cells. A single fasciculus is shown, enlarged, in illustration C.

Protruding upward from the upper cut surface of the fasciculus is a portion of an individual muscle cell, or **muscle fiber.** Note that between the individual muscle fibers of each fascicle is a thin layer of connective tissue, the **endomysium.** This connective tissue enables each muscle fiber to react independently of the others when stimulated by nerve impulses. Surrounding each fascicle is a tough sheath of fibrous connective tissue, the **perimysium.**

Surrounding the fasciculi of the muscle is another layer of coarser connective tissue that is called the **epimysium.** Exterior to the epimysium is the **deep fascia,** which covers the entire muscle. Since this muscle lies close to the skin, the deep fascia of this muscle would have fibers that intermesh with the **superficial fascia** (not shown) that lies between the muscle and the skin. The connective tissue fibers of the deep fascia are also continuous with the tissue in the tendons, ligaments, and periosteum.

Illustration A reveals the continuity of the connective tissue of the endomysium, perimysium, and epimysium. A fasciculus of muscle fibers is shown bracketed in this illustration.

Assignment:
Label figure 17.1.

Examine some muscles on a freshly killed or embalmed animal, identifying as many of the aforementioned structures as possible.

Laboratory Report

Complete the first portion of Laboratory Report 17,18.

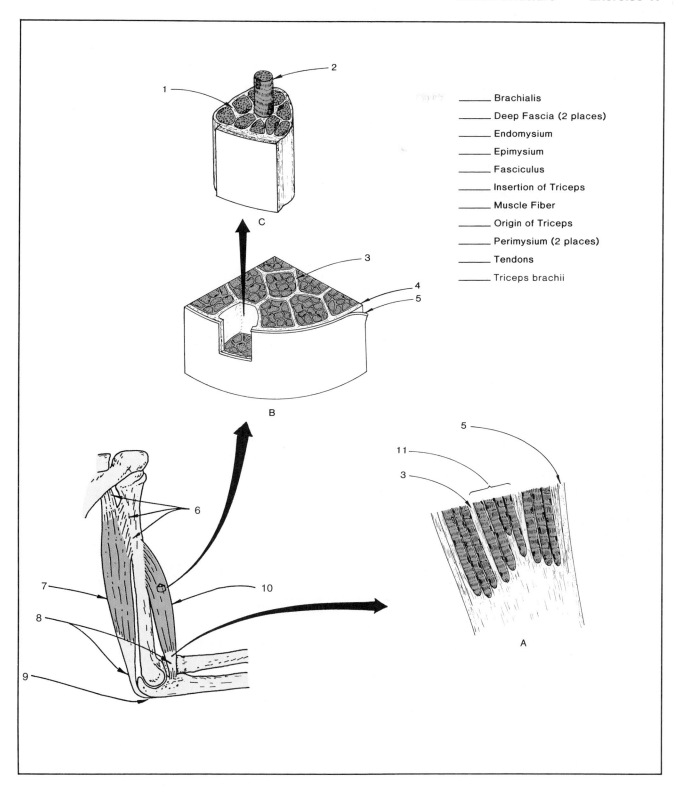

_____ Brachialis

_____ Deep Fascia (2 places)

_____ Endomysium

_____ Epimysium

_____ Fasciculus

_____ Insertion of Triceps

_____ Muscle Fiber

_____ Origin of Triceps

_____ Perimysium (2 places)

_____ Tendons

_____ Triceps brachii

Figure 17.1   Muscle anatomy and attachments.

18

Body Movements

The action of muscles through diarthrotic joints results in a variety of types of movement. The nature of movement is dependent on the construction of the individual joint and the position of the muscle. Muscles working through hinge joints result in movements that are primarily in one plane. Ball and socket joints, on the other hand, will have many axes through which movement can occur; thus, movement through these joints is in many planes. The different types of movement follow.

Flexion   When the angle between two parts of a limb is decreased the limb is said to be "flexed." Flexion of the arm takes place at the elbow when the antebrachium is moved toward the brachium. The term *flexion* may also be applied to movement of the head against the chest and of the thigh against the abdomen.

When the foot is flexed upward, the flexion is designated as being **dorsiflexion.** Movement of the sole of the foot downward, or flexion of the toes, is called **plantar flexion.**

Extension   Increasing the angle between two portions of a limb or two parts of the body is *extension.* Straightening the arm from a flexed position is an example of extension. Extension of the foot at the ankle is essentially the same as plantar flexion. When extension goes beyond the normal posture, as in leaning backward, the term **hyperextension** is used.

Abduction and Adduction   The movement of a limb away from the median line of the body is called *abduction.* When the limb moves toward the median line, *adduction* occurs. These terms may also be applied to parts of a limb, such as the fingers and toes, by using the longitudinal axes of the limbs as points of reference.

Rotation and Circumduction   The movement of a bone or limb around its longitudinal axis without lateral displacement is *rotation.* Rotation of the arm occurs through the shoulder joint. The head can be rotated through movement in the cervical vertebrae. If the rotational movement of a limb through a freely movable joint describes a circle at its termi-

nus, the movement is called *circumduction.* The arm, leg, and fingers can be circumducted.

Supination and Pronation   Rotation of the antebrachium at the elbow affects the position of the palm. When the palm is raised upward from a downward-facing position, *supination* occurs. When the rotation is reversed so that the palm is returned to a downward position, *pronation* takes place.

Inversion and Eversion   These two terms apply to foot movements. When the sole of the foot is turned inward or toward the median line, *inversion* occurs. When the sole is turned outward, *eversion* takes place.

Sphincter Action   Circular muscles such as those around the lips *(orbicularis oris)* and the eye *(orbicularis oculi)* are called *sphincter muscles.* When the fibers of these muscles contract, the openings they encircle are closed.

Muscle Grouping

For reasons of simplicity, muscles are often studied individually, as in figure 18.1. It should be kept in mind, however, that seldom, if ever, do they act alone. Rather, muscles are arranged in groups with specific functions to perform: e.g., flexion and extension, abduction and adduction, supination and pronation.

The flexors are the prime movers, or **agonists.** The opposing muscles, or **antagonists,** contribute to smooth movements by their power to maintain tone and give way to movement by the flexor group. Variance in the tension of flexor muscles results in a reverse reaction in the extensor muscles.

Muscles that assist the agonists to reduce undesired action or unnecessary movement are called **synergists.** Other groups of muscles that hold structures in position for action are called **fixation muscles.**

Laboratory Report

Identify the types of movement in figure 18.1 and complete the last portion of Laboratory Report 17,18.

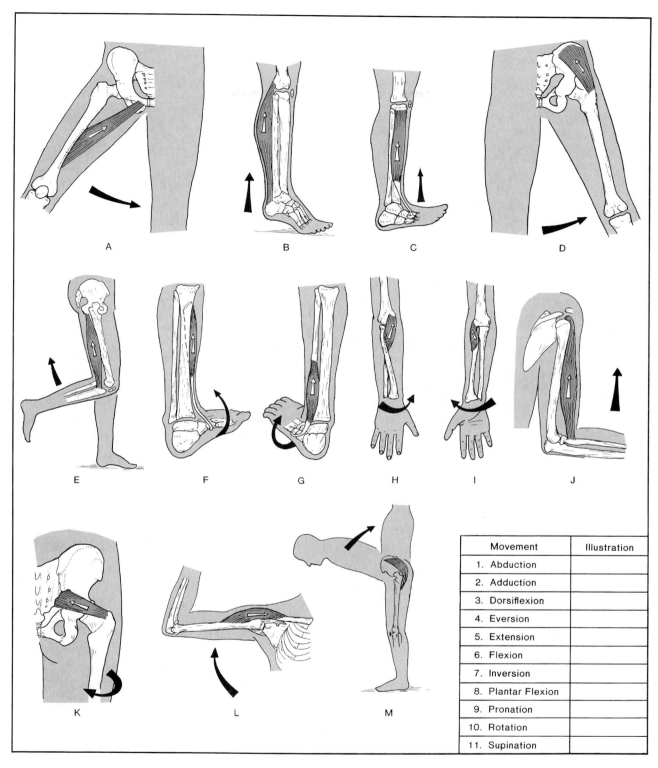

Movement	Illustration
1. Abduction	
2. Adduction	
3. Dorsiflexion	
4. Eversion	
5. Extension	
6. Flexion	
7. Inversion	
8. Plantar Flexion	
9. Pronation	
10. Rotation	
11. Supination	

Figure 18.1 Body movements.

19

Nerve and Muscle Tissues

The characteristic of nerve and muscle cells that sets them apart from other cells is that they are capable of transmitting electrochemical impulses along their membranes. This property is called **excitability.** Excitability is a function of the cell membrane's electrical potential. Although all cells in the body possess membrane potentials, only nerve and muscle cells utilize this characteristic to generate action potentials (impulses) along their membranes. The end result is that nerve cells provide for neural coordination and muscle cells shorten to provide for movement.

During this laboratory period you will have an opportunity to study different kinds of neurons and three kinds of muscle tissue. Prior to examining the tissues under the microscope, label figure 19.1 and answer the questions on the Laboratory Report.

Neurons

Since neurons perform different roles in different parts of the nervous system, they exist in a variety of sizes and configurations. In spite of this diversity, all neurons have much in common. Figures 19.1 and 19.2 illustrate the principal types.

The Cell Body and Its Processes

The large peripheral neurons illustrated in figure 19.1 are the principal neurons associated with skeletal muscle physiology. The cross section through the spinal cord illustrates the relationship of these neurons to the spinal cord and to each other. Three kinds of neurons are shown: sensory, motor, and interneuron. These three neurons, together with the receptor and effector, compose a **reflex arc,** which is described more in detail in Exercise 31.

The cell body of a neuron is called a **perikaryon.** It usually contains a rather large nucleus surrounded by Nissl bodies and neurofibrils in the cytoplasm. **Nissl bodies** are dense aggregations of rough endoplasmic reticulum. They stain readily with basic aniline dyes such as toluidine blue or cresyl violet. They represent sites of protein synthesis. Some Nissl bodies are shown in the perikaryon of the motor neuron in figure 19.1.

Neurofibrils are slender protein filaments that extend throughout the cytoplasm parallel to the long axis. They resemble roadways that circumvent the Nissl bodies. It is significant that these minute fibers converge into the axon to form the core of the nerve fiber.

The most remarkable features of neurons are their cytoplasmic processes: the axons and dendrites. Traditional terminology once categorized these processes purely from a morphological sense; however, numerous exceptions require more accurate and universal definitions that emphasize their *functional* attributes.

A **dendrite,** or **dendritic zone,** functions in receiving signals from receptors or other neurons and plays an important part in integration of information.

An **axon,** on the other hand, is a single elongated extension of the cytoplasm that has the specialized function of transmitting impulses away from the dendritic zone.

The perikaryon can be located at any position along the conduction pathway. In the motor neuron of figure 19.1, the dendrites are the short branching processes of the perikaryon; the axon is the long single process. Although a neuron will usually have several dendrites, there will be only one axon. Some neurons, however, such as the *amacrine cells* of the retina, have no axon at all.

Fiber Construction

The neural fibers of all peripheral neurons are enclosed by a covering of **neurolemmocytes** (also called *Schwann cells*). This covering extends from the perikaryon to the peripheral termination of the fiber. In the larger peripheral neurons, such as those seen in figure 19.1, the neurolemmocytes are wrapped around the axon in a unique manner to form a myelin sheath of lipoprotein. The outer portion of the neurolemmocytes that contain nuclei and cytoplasm make up the **neurolemma** (*neurolemmal sheath*).

Those axons that have a thick myelin sheath with a neurolemma are designated as being *myelinated*. The smaller peripheral neurons that have fibers enclosed in neurolemmocytes, but lack the myelin

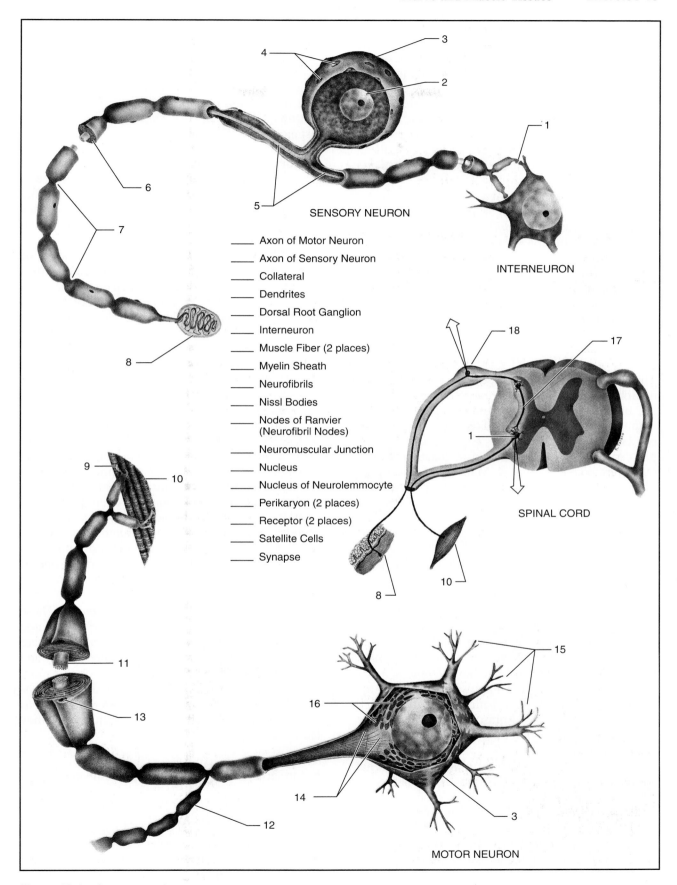

_____ Axon of Motor Neuron
_____ Axon of Sensory Neuron
_____ Collateral
_____ Dendrites
_____ Dorsal Root Ganglion
_____ Interneuron
_____ Muscle Fiber (2 places)
_____ Myelin Sheath
_____ Neurofibrils
_____ Nissl Bodies
_____ Nodes of Ranvier
(Neurofibril Nodes)
_____ Neuromuscular Junction
_____ Nucleus
_____ Nucleus of Neurolemmocyte
_____ Perikaryon (2 places)
_____ Receptor (2 places)
_____ Satellite Cells
_____ Synapse

SENSORY NEURON

INTERNEURON

SPINAL CORD

MOTOR NEURON

Figure 19.1   Sensory and motor neurons.

sheath, are said to have *unmyelinated* fibers. One of the most important functions of neurolemmocytes is to facilitate the regeneration of damaged fibers. When a fiber is damaged, regeneration occurs along the pathway formed by the neurolemmocytes.

The myelin sheath and neurolemma are formed during embryological development by the folding of the neurolemmocytes around the axons. The axon of the motor neuron in figure 19.1 reveals an enlarged cross section of the lipoprotein sheath and neurolemma.

Note in figure 19.1 that the myelin sheath and neurolemma are indented at regular intervals by **nodes of Ranvier** *(neurofibril nodes),* which are points of discontinuity between successive neurolemmocytes. Each internode consists of a single neurolemmocyte, and each node of Ranvier represents a place where the exposed axon core lacks the myelin and neurolemma. The nodes play a significant role in facilitating the speed of nerve impulse transmission.

Types of Neurons

The variability in size, shape, and arborization of neurons is considerable. Neurons may be multipolar, bipolar, unipolar, or pseudounipolar. Some of them have perikarya as small as 4 μm, others may be as large as 150 μm. Some neurons have perikarya that are encapsulated with satellite cells; others are not encapsulated. Many have myelinated fibers; others do not. More detailed structural features of spinal cord and brain neurons follows.

Neurons of the Spinal Cord

As stated above, the principal neurons of the spinal cord are motor, sensory, and interneuron. Each type has unique characteristics.

Motor Neurons As indicated in figure 19.1, these multipolar neurons have their perikarya located in the gray matter of the central nervous system (CNS). They are referred to as *efferent neurons* because they carry nerve impulses away from the CNS. The axons of these neurons are myelinated and terminate in skeletal muscle fibers. They function in the maintenance of voluntary control over skeletal muscles.

The perikarya of these neurons are somewhat angular or star-shaped rather than oval. The cytoplasm in a properly stained perikaryon reveals a nucleus, many Nissl bodies, and neurofibrils. Many short branching dendrites are present that make synaptic connections with axons of interneurons or sensory neurons. Illustrations C and D of figure HA-13 in the Histology Atlas reveal the

appearance of these cells when observed under oil immersion optics.

The myelinated axons of these neurons often have **collaterals** emanating at right angles from a node of Ranvier. Collaterals are branches that provide innervation to additional muscle fibers, other neurons, and even the same neuron in some instances.

Sensory Neurons These neurons are also called *ganglion cells* because their perikarya are located in dorsal root ganglia outside of the CNS (see label 18, figure 19.1). Since they carry nerve impulses from a receptor toward the CNS, they are also often referred to as *afferent neurons.*

As illustrated in figure 19.1, the perikaryon of a sensory neuron is ovoid and is covered with a capsule of **satellite cells.** These cells have the same embryological origin and function as neurolemmocytes. Structurally, satellite cells resemble certain neuroglial cells (oligodendrocytes); however, the two types of cells do not have the same embryological origin. Illustration E, figure HA-13 of the Histology Atlas reveals what satellite cells look like in a sectioned sensory ganglion.

Although sensory neurons appear to be unipolar, histologists classify them as *pseudounipolar* on the basis of their embryological development. Note that the entire fiber is myelinated from the receptor in the skin to the synaptic connection (synapse) in the spinal cord. The entire myelinated fiber is designated as an **axon** even though part of it functions like a dendrite in carrying nerve impulses toward the cell body. A **synapse** (label 1, figure 19.1) is the point where the terminus of an axon of one neuron adjoins the dendritic zone of another neuron.

Interneurons *(Associated* or *Internuncial)* These multipolar unmyelinated neurons exist in large numbers in the gray matter of the CNS. They are essentially the integrator cells between sensory and motor neurons. Most reflexes involve one or more interneurons between the sensory and motor neurons; however, some automatic responses, such as the knee-jerk reflex, are monosynaptic in that they lack interneurons.

Assignment:
Before going on to the study of the brain cells, label figure 19.1.

Neurons of the Brain

Four different types of neurons that are found in various parts of the brain are shown in figure 19.2. They are pyramidal, Purkinje, stellate, and basket cells.

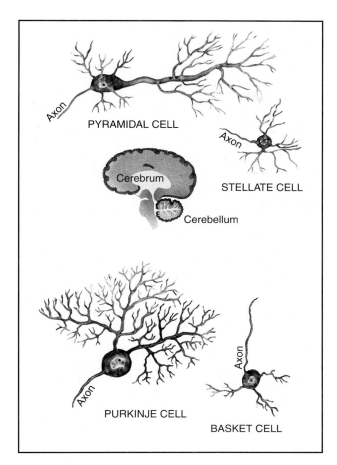

Figure 19.2 **Neurons of the brain.**

Pyramidal Cells These multipolar neurons are the principal cells of the cortex of the cerebrum. The upper neuron in figure 19.2 is of this type. Figure HA-14 of the Histology Atlas has three photomicrographs of these cells that were made from three different slide preparations. The perikarya of these neurons are somewhat triangular to impart a pyramidlike shape.

A unique characteristic of these cells is that they have two separate sets of dendrites: (1) one long trunklike structure with many branches that ascends vertically in the cortex and (2) several basal branching dendrites that extend outward from the basal portion of the perikaryon. The single axon is distinguishable from the basal dendrites by its smoother surface.

Purkinje Cells The large multipolar neuron in the lower left-hand corner of figure 19.2 is a typical Purkinje cell. The perikarya of these neurons are located deep within the molecular layer of the cerebellum (see figure HA-15, Histology Atlas). The large branching dendritic processes of these cells fill up most of the space in the molecular layer of the cerebellum. The axons of these neurons extend inward into the granule cell layer of the cerebellum.

Stellate Cells *(Golgi Type II)* These small neurons are found in both the cerebrum and cerebellum, where they act as linkage (association) cells for the pyramidal cells of the cerebrum and the Purkinje cells of the cerebellum. Both multipolar and bipolar types exist. See figures 19.2, HA-14, and HA-15.

Basket Cells These multipolar neurons are a variant of large stellate cells that are located deep in the molecular layer of the cerebellum in close association with the Purkinje cells (figures 19.2 and HA-15). They have short, thick, branching dendrites and a long axon. The axons of these neurons usually have five or six collaterals that make synaptic connections with dendrites of the Purkinje cells. Like the stellate cells, these cells function as association neurons.

Neuronal Histological Study

After answering the questions on the Laboratory Report that pertain to cells of the nervous system examine prepared slides of the nervous system using oil immersion optics whenever necessary. The following procedure will involve most of the cells discussed in this exercise.

Materials:
prepared slides of:
    x.s. spinal cord (H10.341 or H10.342)
    spinal cord smear (H4.11)
    spinal ganglion (H4.25)
    cerebral cortex (H10.11 or H10.12)
    cerebellum (H10.23)

Spinal Cord Examine the cross section of a rat spinal cord under low power and identify the structures labeled in figure HA-13. Study several of the large motor neurons in the gray matter with a high-dry objective.

Also, study a smear of ox spinal cord tissue and compare the cells with the photomicrograph in illustration D, figure HA-13.

Ganglion Cells Study a slide of a section through a spinal ganglion and look for *satellite cells* that surround the sensory neurons. Refer to illustration E, figure HA-13.

Cerebral Cortex Examine a slide of the cerebral cortex. Look for *pyramidal* and *stellate cells.* Refer to figure HA-14. Try to differentiate dendrites from axons on the pyramidal cells.

Note the presence of spiny processes called "gemmules" that are seen on the branches of the dendrites. These processes greatly increase the effective surface area of the dendrites, allowing large neurons to receive as many as 100,000 separate axon terminals, or synapses.

Cerebellum Examine a slide of the cerebellum and look for *Purkinje cells, stellate cells,* and *basket cells.* Refer to figure HA-15.

Muscle Fibers

As stated above, excitability in muscle cells manifests itself in contractility. This property is made possible by the presence of contractile protein fibers within the cells. The shortening of muscle cells due to this property is responsible for the various movements of appendages and organs of the body. Because of their elongated structure, muscle cells are often referred to as **muscle fibers.**

On the basis of cytoplasmic differences, there are two categories of muscle fibers: striated and smooth. **Striated muscle** fibers are characterized by regularly spaced transverse bands called *striae.* **Smooth muscle** fibers lack these transverse bands and have other distinctions to set them apart.

Striated muscle cells are further subdivided into skeletal and cardiac muscle. **Skeletal muscle** fibers are multinucleated, or *syncytial.* Attached primarily to the skeleton, they are responsible for voluntary body movements. **Cardiac muscle** cells, which make up the muscle of the heart wall, are not syncytial and have less prominent striations than skeletal muscle.

Smooth Muscle Tissue

Smooth muscle fibers are long, spindle-shaped cells as illustrated in figure 19.3. Each cell has a single elongated nucleus that usually has two or more nucleoli. When studied with electron microscopy, the cytoplasm reveals the presence of fine linear filaments of the same protein composition that exists in striated muscle. The fact that both smooth and striated fibers contain actin, myosin, troponin, and tropomyosin indicates that the chemistry of contraction is probably similar or identical in all types of muscle fibers.

Smooth muscle cells are found in the walls of blood vessels, walls of organs of the digestive tract, the urinary bladder, and other internal organs. They are innervated by the autonomic nervous system and, thus, are *under involuntary nervous control.*

Skeletal Muscle Tissue

Muscle fibers of this type are long, cylindrical, and multinucleated. Large numbers of these fibers are grouped together into bundles called **fasciculi.** The individual cells within each fascicle are separated from each other by a thin layer of connective tissue called **endomysium.** This separateness of muscle fibers enables each cell to respond independently to nerve stimuli.

Examination of the cytoplasm, or **sarcoplasm,** with a light microscope reveals a series of distinct dark and light **striae** (cross stripes). Surrounding each cell is a cell membrane called the **sarcolemma.**

Figure 19.3   Smooth muscle fibers, teased.

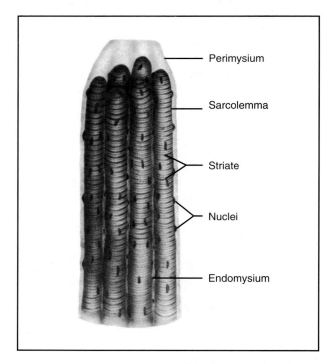

Figure 19.4   A fascicle of skeletal muscle fibers.

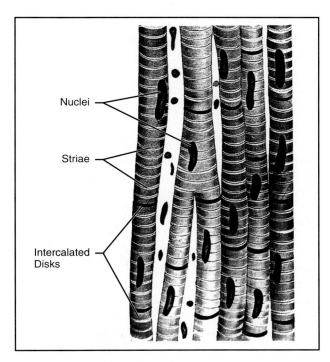

Figure 19.5 **Cardiac muscle fibers.**

Note in figure 19.4 that the nuclei of each cell lie in close association with the sarcolemma.

Although striated muscle is referred to as being "skeletal" and "voluntary," there are certain striated muscles that are not attached to bone or subject to voluntary control. Two examples: (1) striated muscles of the tongue and external anal sphincter have no skeletal attachment, and (2) striated muscles of the pharynx and upper esophagus have no skeletal attachment and are not voluntarily controlled. In spite of these differences, these aberrational examples are morphologically indistinguishable from other skeletal muscle.

Cardiac Muscle Tissue

Muscle cells of this type are found in the wall of the heart and in the walls of the venae cavae, where they enter the right atrium of the heart. A unique characteristic of cardiac muscle is its ability to contract rhythmically and continuously as a result of intrinsic cellular activity.

Morphologically, cardiac muscle is readily distinguished from skeletal and smooth fibers, yet it shares some characteristics of each type. The principal distinguishing characteristics of cardiac muscle fibers are as follows:

- Instead of being anatomically syncytial, cardiac muscle is made up of separate cellular units that are separated from each other by **intercalated disks** (cell membranes). These disks stain somewhat darker than transverse striae.

- Some fibers of cardiac cells are bifurcated to form a branched three-dimensional network instead of forming only long straight cylinders.
- The elongated nuclei of each cellular unit lie deep within the cells instead of near the surface as in skeletal muscle.
- Cardiac tissue is not normally subject to voluntary control.

Histological Study

After answering the questions on the Laboratory Report that pertain to the characteristics of the various types of nerve and muscle tissue, examine prepared slides to identify the various kinds of cells.

Materials:
prepared slides of:
striated muscle, teased (H3.11)
striated muscle, x.s. and l.s. (H3.13)
smooth muscle, teased (H3.21)
muscle to tendon (H7.13)
smooth muscle, sectioned (H3.22)
cardiac muscle, teased (H3.31)
cardiac muscle, l.s. (H3.32 or H3.33)

Skeletal Muscle   Scan a slide of striated muscle tissue with low power to find a muscle fiber with the most pronounced striae. Increase the magnification to 1000× (oil immersion) and see if you can differentiate the A and I bands.

Refer to illustration A, figure HA-7. Note the position of the nuclei in these fibers. Also examine a cross section of this tissue to identify the endomysium.

If a slide of muscle to tendon (H7.13) is available examine it under high-dry magnification and refer to illustration C, figure HA-7.

Cardiac Muscle   Scan a slide of this tissue with low power, looking for pronounced striae and intercalated disks. Increase the magnification to 1000× to observe the nature of the striae and disks. Can you find a place on the slide where bifurcation of the fibers is seen? Consult illustration A, figure HA-8, of the Histology Atlas.

Smooth Muscle   Study slides of this type of muscle tissue in the same manner as above. Finding individual fibers, separated from others, is not always as easy as it might seem. Compare your observations with illustrations B and C in figure HA-8.

Laboratory Report

Complete Laboratory Report 19 for this exercise.

20

The Neuromuscular Junction

It was observed in Exercise 19 that motor neurons emerge from the spinal cord through the anterior roots of spinal nerves to innervate the skeletal muscle fibers throughout the body. Each large myelinated nerve fiber branches many times through collaterals to stimulate from three to two thousand skeletal muscle fibers. At the end of each branch is a **neuromuscular junction,** where the nerve fiber contacts the muscle fiber. This juncture is usually made at approximately the midpoint of the fiber so that the action potential generated in the fiber will reach the opposite ends of the fiber at about the same time. Normally there is only one junction per muscle fiber.

Figure 20.1 reveals the structure of a neuromuscular junction as determined by electron microscopy. Chemical studies that have taken place over the past several decades have revealed a series of events that occur here. Our present understanding of these events is somewhat different from what it was only a few years ago, and it is certain that future research will reveal a slightly different picture as present questions are answered.

After you have labeled the structures on this illustration and answered the questions on the Laboratory Report, you will have an opportunity to study microscope slides of the neuromuscular junction. The configuration revealed in figure 20.1 is the result of electron microscopy, don't expect to see this degree of detail on your slides.

Note in figure 20.1 that the end of the neuron consists of a cluster of knoblike structures that lie within invaginations of the **sarcolemma** (label 3) of the muscle fiber. These invaginations are called *synaptic gutters;* the knoblike branching terminals of the neuron are called *axon terminals.* Each knob of an axon terminal is called an **axon terminal branch.** The lower enlarged illustration in figure 20.1 reveals the structure of a single axon terminal branch.

Between the axon terminal and lining of the synaptic gutter is a gap, the **synaptic cleft** (label 11), which is approximately 20 μm wide and is filled with a gelatinous ground substance. The surface of the axon terminal branch within this cleft is called the **presynaptic membrane** (label 13). That

portion of the infolded sarcolemma that is on the other side of the synaptic cleft is the **postsynaptic membrane.** Note that the postsynaptic membrane has many infoldings that increase its surface area. These infoldings are called **junctional folds,** or *subneural clefts.*

The transmission of the nerve impulse (action potential) from the axon terminal to the muscle fiber is accomplished by chemical rather than electrical means. This is accomplished by the production of acetylcholine, a neurotransmitter, in the axon terminal that passes across the synaptic cleft. The sequence of events involved in action potential activation to bring about muscle fiber contraction is as follows:

- Depolarization of the neurolemma occurs with sodium ions flowing into the neuron and potassium ions flowing outward. This ionic interchange produces an action potential in the nerve fiber.
- Calcium ions flow into the neuron.
- Acetylcholine (ACh), the neurotransmitter, is secreted into the synaptic cleft by the presynaptic membrane.
- ACh molecules hook up with receptors on the postsynaptic membrane.
- An action potential develops in the sarcolemma of the muscle fiber, causing muscle fiber shortening.

Since the early 1950s it was believed by most neuroscientists that all the ACh that bridges the synaptic cleft was *exocytosed* by the **synaptic vesicles** (color code: yellow). This conclusion was based on two facts: (1) chemical analysis revealed that the vesicles do contain ACh, and (2) electron microscopy showed that the vesicles do fuse with the presynaptic membrane. Thus, the belief prevailed for many years that the action potential in the neuron caused synaptic vesicle membranes to fuse with the presynaptic membrane to release the acetylcholine.

Subsequent studies by Yves Dunant, Maurice Israel, and others, however, contradict this simple picture. The present concept is that ACh passes directly from the cytoplasm of the axon terminal through the presynaptic membrane. The releasing

94

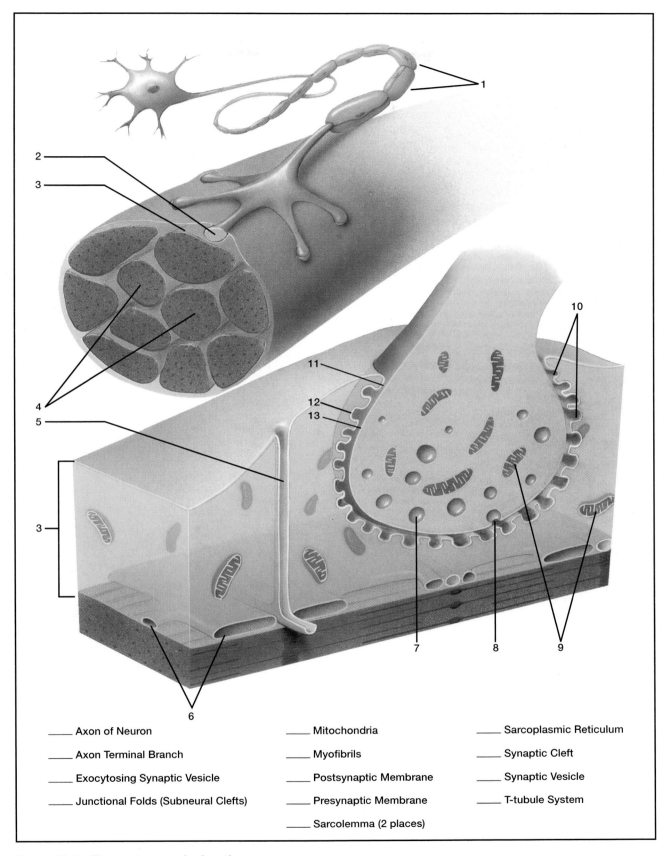

__ Axon of Neuron

__ Axon Terminal Branch

__ Exocytosing Synaptic Vesicle

__ Junctional Folds (Subneural Clefts)

__ Mitochondria

__ Myofibrils

__ Postsynaptic Membrane

__ Presynaptic Membrane

__ Sarcolemma (2 places)

__ Sarcoplasmic Reticulum

__ Synaptic Cleft

__ Synaptic Vesicle

__ T-tubule System

Figure 20.1   The neuromuscular junction.

mechanism appears to be a compound, most likely a protein, that is embedded in the presynaptic membrane. The protein may act as a valve, enabling ACh to pass through when the action potential reaches the area.

This raises the question of the role of the synaptic vesicles. According to Dunant and Israel, it appears that these vesicles do two things: (1) they release stored ACh only when the cytoplasmic pool of the neurotransmitter has been exhausted by sustained nervous activity, and (2) they accumulate calcium ions that assist in the termination of the release of ACh.

Once the action potential is initiated in the sarcolemma of the muscle fiber, the following events occur within the muscle fiber:

- The action potential moves inward to the **T-tubule system** (label 5).
- Calcium ions, which permit contraction of the myofibrils, are released by the **sarcoplasmic reticulum** (color code: blue outside, green inside).
- ACh is inactivated by the enzyme *acetylcholinesterase* within 1/500 of a second.
- All energy for muscle fiber contraction is supplied by ATP from mitochondria in the area.

The action of acetylcholinesterase on ACh is to break it down into acetate and choline. This enzyme is found on the surfaces of the presynaptic and postsynaptic membranes as well as in the ground substance of the synaptic cleft.

Once the choline and acetate diffuse back into the axon terminal they are resynthesized back into ACh. The enzyme that catalyzes the synthesis of ACh is *choline acetyltransferase*. This enzyme is produced in the perikaryon of the neuron and travels down microtubules of the axon to the axon terminal, where it is stored in the cytoplasm.

These events, which depict how the nerve impulse bridges the gap from the axon terminal to a skeletal muscle fiber, resemble nerve to nerve transmission, and nerve to secretory cell, as well. One thing to keep in mind about neuron to neuron transmission in the brain, however, is that there are many neurotransmitters other than acetylcholine that bridge the synaptic clefts between different kinds of neurons.

Reference:
Dunant, Y., and M. Israel. The release of acetylcholine. *Scientific American* 252(4) (April 1985).

Laboratory Assignment

After labeling figure 20.1 examine a microscope slide of motor nerve endings (H4.31) under low-power and high-dry magnification. Consult illustration D, figure HA-7, for reference. Complete the part of Laboratory Report 20, 21 for this exercise.

Histology Self-Quiz No. 1

If at this point you have completed all the histology assignments in the first twenty exercises in this manual, you should be ready to test your knowledge of the histology of these areas by taking the self-quiz on page 407. After recording your answers on a sheet of paper, check your answers against the key that you will find on page 196.

21

The Physicochemical Nature of Muscle Contraction

The microscopic changes that occur during skeletal muscle contraction will be observed here in a controlled experimental setup. The physicochemical nature of muscle contraction will be reviewed first for clarification of this complex phenomenon.

Skeletal muscle contraction involves many components within the muscle cell. The entire process is initiated by an action potential that reaches the neuromuscular junction. As we learned in Exercise 20, acetylcholine initiates depolarization of the muscle fiber. The depolarization proceeds along the surface (sarcolemma) of the muscle fiber and deep into the sarcoplasm by way of the T-tubule system of the sarcoplasmic reticulum.

Figure 21.2 illustrates the ultrastructure of a single muscle fiber. Note that the sarcoplasm of the fiber consists of many long tubular **myofibrils,** the **sarcoplasmic reticulum,** and **mitochondria. Nuclei** can also be seen beneath the **sarcolemma.** The abundance of mitochondria and their close association to the myofibrils indicate an availability of ATP for contraction.

Illustrations C and D in figure 21.2 reveal the molecular structure of the contractile **myofilaments** of the myofibrils. While myofibrils can be seen with a good light microscope, myofilaments are visible only with electron microscopy.

Note that there are two types of myofilaments: actin and myosin. **Myosin** filaments are the thickest filaments and are the principal constituents of the dark **A band.** They are slightly thicker in the middle and taper toward both ends. The thinner **actin** filaments extend about 1 μm in either direction from the **Z line** and make up the lighter **I band.**

The shortening of myofibrils to achieve muscle fiber contraction is due to the interaction of actin and myosin in the presence of calcium, magnesium, and ATP. Two other protein molecules plus troponin and tropomyosin are also involved.

As the depolarization wave (action potential) moves along the sarcolemma and T-tubule system, large quantities of calcium ions are released by the sarcoplasmic reticulum. As soon as these ions are

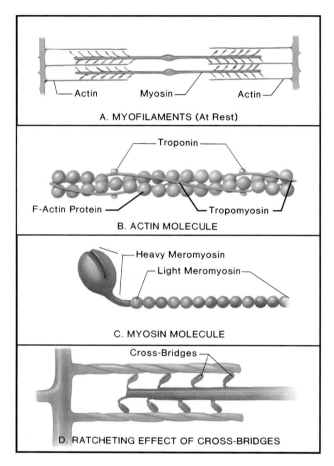

Figure 21.1   Myofilament structure.

released, attractive forces develop between the actin and myosin filaments that cause the ends of the actin fibers to be pulled toward each other, as shown in illustration D, figure 21.2. The exact mechanism that causes this shortening of the myofibrils is unknown, but the following explanation is the most acceptable theory we have at the present time.

It is in the nature of the actin and myosin molecules that we seek an answer to the attractive forces. The myosin molecule is made up of **heavy meromyosin** and **light meromyosin** (see figure 21.1). The heavy meromyosin forms a head that is flexible and able to pivot. The actin filament consists of three components: a double-stranded helix of **F-actin protein, tropomyosin,** and **troponin.** The tropomyosin is a long protein strand that lies within the groove of the F-actin protein. The third protein, troponin, occurs at various active sites along the helix. The cross-bridges of heavy meromyosin are able to bridge the gap between the myosin and actin molecules.

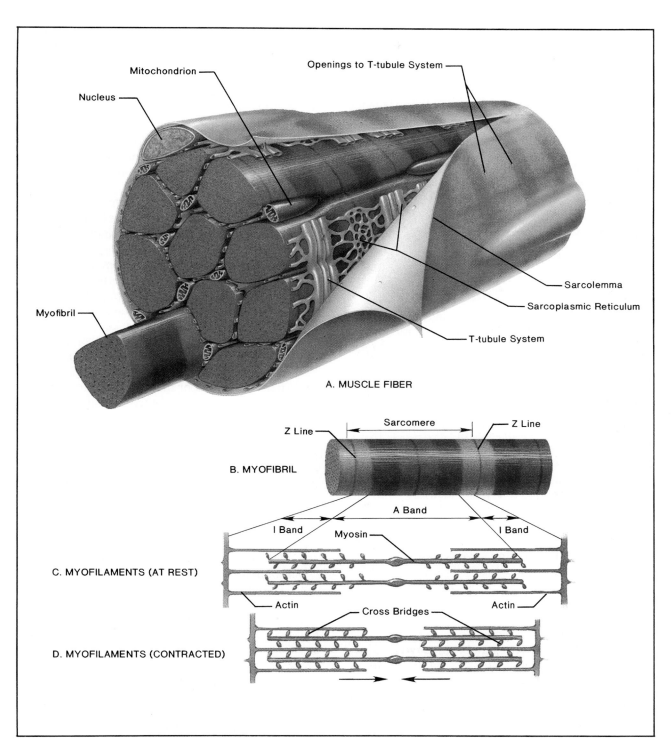

Figure 21.2   The ultrastructure of a muscle fiber.

The *ratchet theory of contraction* proposes that the following series of events takes place:

1. Calcium ions combine with troponin.
2. Calcium-bound troponin causes the tropomyosin to be pulled deeper into the groove of the two actin strands, exposing the active sites on the actin filament.
3. The heads of the cross-bridges form a bond with the active sites and pull on the actin filament with a ratchetlike power stroke. The energy for the first power stroke is derived from the meromyosin molecule that was previously activated by ATP.
4. After the first power stroke, ATP combines with the head of the cross-bridge, causing the head to separate from the active site to which it has been attached.
5. The ATP on the head is split by ATPase, which is contained in the heavy meromyosin. The energy released by this cleavage is used for cocking the head again, attaching the head to another active site, and performing another power stroke.

Microscopic Examination

To observe the nature of muscle contraction at the microscopic level, we will induce contraction in glycerinated rabbit muscle fibers with ATP and ions of magnesium and potassium. Measurements of the muscle fibers will be made before and after contraction to determine the percentage of contraction that occurs. Microscopic comparisons will be made of fibers in both the relaxed and contracted state to determine the changes that occur.

Materials:
test tube of glycerinated rabbit psoas muscle (muscle is tied to a small stick)
dropping bottle of ATP in triple distilled water
dropping bottle of 0.25%ATP, 0.05M KCl, and 0.001M MgCl$_2$
3 microscope slides and cover glasses
sharp scissors
forceps
dissecting needle
Petri dish
plastic ruler, metric scale
microscopes, dissecting and compound

1. Remove the stick of glycerinated muscle tissue from the test tube and pour some of the glycerol from the tube into a Petri dish.
2. With scissors, cut the muscle bundle into segments about 2 cm long, allowing the pieces to fall into the Petri dish of glycerol.
3. With forceps and dissecting needle, tease one of the segments into very thin groups of muscle fibers; single fibers, if obtained, will demonstrate the greatest amount of contraction. Strands of muscle fibers exceeding 0.2 mm in cross-sectional diameter are too thick and should not be used.
4. Place one of the strands on a clean microscope slide and cover with a cover glass. Examine under a high-dry or oil immersion objective. Sketch appearance of striae on the Laboratory Report.
5. Transfer three or more of the thinnest strands to a minimal amount of glycerol on a second microscope slide. Orient the strands straight and parallel to each other.
6. Measure the lengths of the fibers in millimeters by placing a plastic ruler underneath the slide. Use a dissecting microscope, if available. Record lengths on the Laboratory Report.
7. With the pipette from the dropping bottle, cover the fibers with a solution of ATP and ions of potassium and magnesium. Observe the reaction.
8. After 30 seconds or more, remeasure the fibers and calculate the percentage of contraction. Has width of the fibers changed? Record the results on the Laboratory Report.
9. Remove one of the contracted strands to another slide. Examine it under a compound microscope and compare the fibers with those seen in step 4. Record the differences on the Laboratory Report.
10. Place some fresh fibers on another clean microscope slide and cover them with a solution of ATP and distilled water (no potassium or magnesium ions). Does contraction occur?

Laboratory Report

Record all data on Laboratory Report 20, 21 and answer all the questions.

Muscle Contraction Experiments:
Using Chart Recorders

22

The following phenomena of skeletal muscle contraction will be studied here: the **simple twitch, recruitment,** and **wave summation.** All of these physiological activities will be observed on frog muscle tissue of a freshly killed frog. Although the frog will be dead, the tissues will still be viable and respond to electrical stimulation. Frog tissue is used here in place of mammalian tissue because of its tolerance to temperature change and adverse handling. The results are similar to what would be seen in a more carefully controlled mammalian experiment.

The study of skeletal muscle physiology requires the use of considerable equipment and careful handling of tissue. First of all, a frog must be carefully prepared to keep the tissues viable. Second, several pieces of equipment must be properly arranged and hooked up. Finally, stimuli must be administered to the skeletal muscle in a manner that will elicit the desired responses.

Because of the complexity of the overall procedure, and the need to complete all tests within a limited-time laboratory period, it will be necessary to perform the experiment as a team effort. The class will be divided into four or five teams of five or six students. When the teams are determined, the members of each group can decide among themselves which roles they prefer to perform on the team. The responsibilities of the various members on each team are outlined in table 22.1 on page 104.

Note that figures 22.1 and 22.2 illustrate two different setups for this experiment. Although one setup utilizes Gilson and Grass equipment, the other is essentially a Narco Bio-Systems setup. The type of equipment that is available in your laboratory will determine which setup is to be used. It is also possible to interchange some of the compatible components so that your particular setup is not exactly like either illustration. Your instructor will indicate the actual arrangement that is to be used.

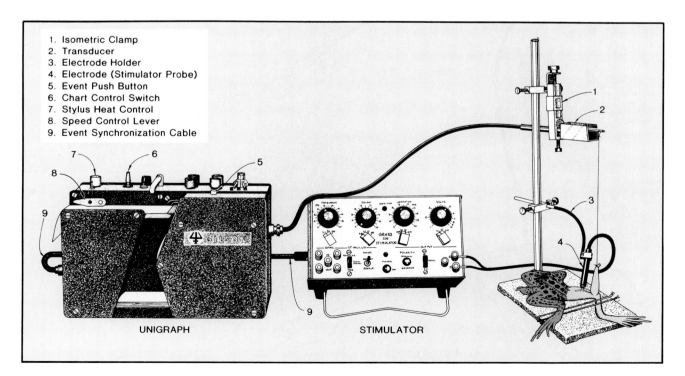

1. Isometric Clamp
2. Transducer
3. Electrode Holder
4. Electrode (Stimulator Probe)
5. Event Push Button
6. Chart Control Switch
7. Stylus Heat Control
8. Speed Control Lever
9. Event Synchronization Cable

UNIGRAPH

STIMULATOR

Figure 22.1   Equipment setup with Gilson recorder.

Materials:

for dissection:
small frog (one per team)
decapitation scissors
scalpel, small scissors, dissecting needle, and
    forceps
squeeze bottle of frog Ringer's solution
spool of cotton thread

for Gilson setup:
Unigraph, Duograph, or polygraph
electronic stimulator
event synchronization cable
electrode and electrode holder
T/M Biocom #1030 force transducer
ring stand
2 double clamps
isometric clamp (Harvard Apparatus, Inc.)
galvanized iron wire and pliers

for Narco setup:
Physiograph® IIIS with transducer coupler
SM-1 electronic stimulator
stimulus switch and myograph
transducer stand and frog board
myograph tension adjuster
event marker cable

sleeve electrode and pin electrodes
dissecting pins

Preparations

While one team is setting up the equipment, another team will decapitate the frog and perform the dissection.

Frog Dissection

To hold the frog for decapitation, grip it with the left hand so that the forefinger is under the neck and the thumb is over the neck. Hold the frog under a water tap, allowing cool water to flow onto its head and body. This tends to calm the animal and helps to wash away the blood.

Insert one blade of a sharp heavy-duty scissors into the mouth, well into the corner; the other scissors blade should be poised over a line joining the tympanic membranes (eardrums). With a swift clean action, snip off the head. Immediately after this, force a probe down into the spinal canal to destroy the spinal cord. The frog should become limp and show no signs of reflexes.

Strip the skin off the right or left hind leg following the instructions in figures 22.3 through 22.5.

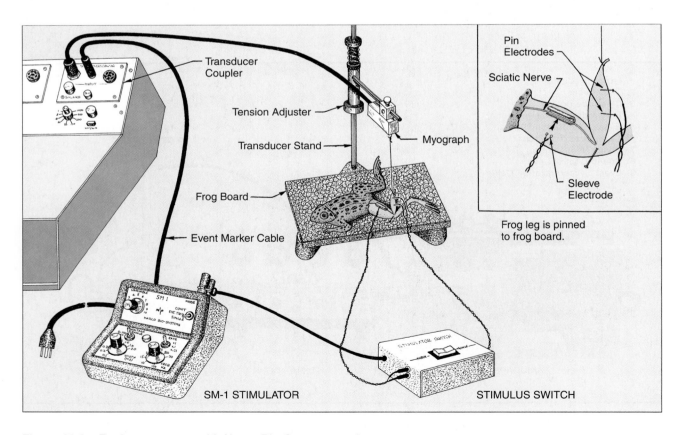

Figure 22.2 Equipment setup with Narco Bio-Systems equipment.

Figure 22.3   The first incision through the skin is made at the base of the thigh.

Figure 22.4   After the skin is cut around the leg, it is pulled away from the muscles.

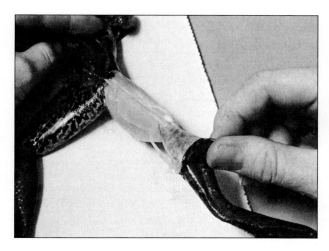

Figure 22.5   While the body is held firmly with the left hand, the skin is stripped off the entire leg.

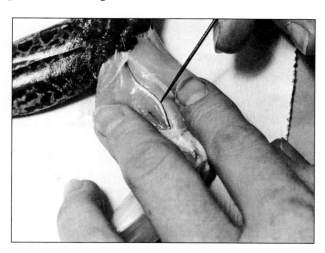

Figure 22.6   The muscles of the thigh are separated with a dissecting needle to expose the sciatic nerve.

Figure 22.7   Forceps are inserted under the nerve to grasp the end of the string on the other side.

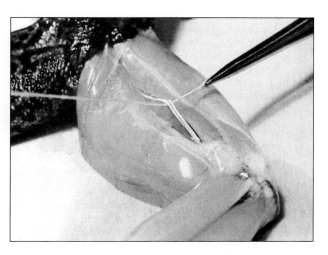

Figure 22.8   The thread used for manipulating the nerve is pulled through.

If the Gilson setup is used, use the right hind leg. If the Narco setup is used, use the left leg.

With a sharp dissecting needle, tear through the fascia and muscle tissue of the thigh to expose the sciatic nerve and insert a short length of cotton thread (10 inches long) under it. This thread will be used for manipulation of the nerve. Figures 22.6 through 22.8 outline the desired procedure. **Be careful in the way you handle the nerve.** The less it is traumatized or comes in contact with metal, the better. Keep the exposed tissues moist with frog Ringer's solution.

Equipment Hookup *(Gilson Arrangement)*

Set up the various pieces of equipment as shown in figure 22.1. The power cords of both the Unigraph and stimulator must be plugged into a grounded 110-volt outlet. Note that an event synchronization cable connects the stimulator to the Unigraph.

Attach the Harvard isometric clamp to the ring stand with a double clamp, keeping the adjustment screw of the isometric clamp upward. Attach the shaft of the transducer to the lower part of the isometric clamp and plug the transducer cord into the Unigraph.

Put another double clamp on the ring stand shaft at the lower level and clamp the shaft of the electrode holder (label 3, figure 22.1) in place.

Plug the jacks of the electrode into the stimulator at the position shown in the illustration, but do not fix the electrode to the clamp at this time since the electrode will have to be maneuvered about in the first tests.

With the completion of all the necessary connections, balance the transducer and calibrate the Unigraph according to the instructions in appendix B.

Equipment Hookup *(Narco Arrangement)*

Set up the various pieces of equipment as shown in figure 22.2. In this setup a stimulus switch is used between the stimulator and the frog. This switch makes it possible to switch from direct muscle stimulation to nerve stimulation without disturbing the setup.

Note that an event marker cable connects the stimulator with the Physiograph®. The receptacle for this cable is on the stimulator back; it is labeled MARKER OUTPUT. The other end of the cable is inserted into the middle receptacle of the transducer coupler.

After affixing the myograph tension adjuster to the transducer stand, attach the myograph to the support rod, and tighten the thumbscrew to hold it in place. Insert the free end of the myograph cable into the transducer coupler.

Plug the power cords for the Physiograph® and SM-1 stimulator into grounded 110-volt outlets. Insert the jacks for the pin electrodes into the right side of the stimulus switch and the jacks for the sleeve electrode into the left side of the stimulus switch. Also, attach the cable from the stimulus switch to the output posts of the SM-1 stimulator.

With the completion of all the hookups, balance the channel and calibrate the myograph according to instructions in appendix B. You are now ready to begin muscle and nerve stimulations.

Table 22.1   Team member responsibilities.

Equipment Engineer
Hooks up the various components (recorder, stimulator, transducer, etc.). Balances the transducer and calibrates recorder. Operates the stimulator. Responsible for correct use of equipment to prevent damage. Dismantles equipment and returns all components to proper places of storage at end of period.

Engineer's Assistant
Works with engineer in setting up equipment to ensure procedures are correct. Makes necessary tension adjustments before and during experiment. Applies electrode to nerve and muscle for stimulation.

Recorder
Labels events produced on tracings. Keeps a log of the sequence of events as experiment proceeds. Sees that each member of team is provided with chart records for attachment to Laboratory Report.

Surgeon
Decapitates and piths frog. Removes skin on leg and exposes sciatic nerve. Responsible for maintaining viability of tissue. Attaches string to nerve and muscle end.

Head Nurse
Assists surgeon as needed during dissection. Keeps exposed tissues moist with Ringer's solution. Helps to keep work area clean.

Coordinator
Oversees entire operation. Communicates among team members. Reports to instructor any problems that seem to be developing that seem insurmountable. Double-checks the setup and cleanup procedures. Keeps things moving.

The Threshold Stimulus

Before severing the Achilles tendon on the muscle and attaching the muscle with string to the transducer, we are going to stimulate the intact muscle and nerve to determine the minimum voltage that will elicit a simple twitch. A stimulus that barely induces a muscle twitch is called the **threshold stimulus.** A stimulus of less than threshold strength is designated as being **subliminal.** Proceed as follows to compare the threshold differences between muscle and nerve stimulation. *Be sure to keep the muscle and nerve moist with frog Ringer's solution.*

Gilson Procedure

1. Set the Duration control at 15 msec and the Voltage control at 0.1 volt. The stimulus switches should be set on REGULAR and OFF. The Polarity switch should be on NORMAL and the Output switch on MONO.
2. Turn on the Power switch and depress the Mode switch to SINGLE, while holding the electrode against the belly of the gastrocnemius muscle as illustrated in figure 22.9. You are now administering a single pulse of 0.1 volt of 15 msec duration to the muscle. This is the minimum voltage possible with the Grass stimulator.

   If the muscle twitches at this minimum voltage, reduce the Duration to the lowest value that will produce a twitch. In this manner we can record 0.1 volt as the threshold stimulus on the Laboratory Report.
3. If the muscle does not twitch at 0.1 volt, increase the voltage at increments of 0.1 volt (Duration at 15 msec) until a twitch is seen.

Record this voltage as the threshold stimulus on the Laboratory Report.

4. Cut the nerve at the point indicated by the white arrow in figure 22.10, and position the stimulator probe under the nerve as shown in the illustration. The nerve is cut to prevent extraneous reflexes.
5. Place the foot in a flexed position.
6. Return the Voltage control to 0.1 volt and stimulate by pressing the Mode lever to SINGLE. If no twitch is seen, proceed as previously to stimulate the nerve at increasing increments of 0.1 volt until a twitch is visible.

   Record this voltage as the threshold stimulus by nerve stimulation.
7. Now, determine the minimum voltage that causes extension of the foot by stimulating the nerve. Record this voltage on the Laboratory Report. If extension occurs with a stimulus of 0.1 volt, try reducing the Duration in the same manner that you used for direct muscle stimulation.

Narco Procedure

1. Set the Width control at 2 msec, and the Voltage at 0.1 volt (Range switch at 0–10, Variable control at its lowest value).
2. Insert the pin electrodes into the muscle about 1 centimeter apart.
3. Press the MUSCLE side of the button on the Stimulus switch to ensure that the muscle will get stimuli from the stimulator.
4. Now, deliver a single stimulus to the muscle by depressing the Mode switch on the stimulator to SINGLE. If the muscle twitches, reduce the

Figure 22.9 To determine direct muscle response to stimulation, the belly of the gastrocnemius muscle is stimulated with hand-held electrode (Gilson setup).

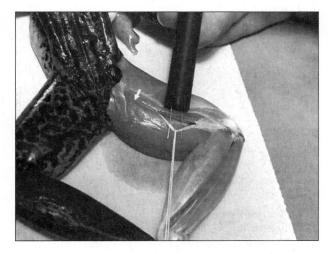

Figure 22.10 Nerve is cut at point indicated by white arrow in both setups. Note that the probe is under the nerve for Gilson setup.

width to the lowest value that will produce a twitch. In this manner we can record 0.1 volt as the threshold stimulus on the Laboratory Report.

If the muscle does not twitch at 0.1 volt, 2 msec, increase the voltage at increments of 0.1 volt until a twitch is seen. Record this voltage as the threshold stimulus on the Laboratory Report.

5. Now, place the foot in a flexed position and proceed to stimulate the muscle by gradually increasing the voltage until the foot extends completely due to muscle contraction. Record the voltage on the Laboratory Report.
6. Carefully position the rubber tubing sleeve around the nerve and insert the sleeve electrode within the tubing next to the nerve. Use the string to manipulate the nerve, and *be sure to bathe the nerve in frog Ringer's solution.*
7. Stimulate the nerve by pressing the NERVE side of the button on the Stimulus switch.
8. Cut the nerve at the point indicated by the white arrow in figure 22.10, to prevent extraneous reflexes.
9. Place the foot in a flexed position.
10. Return the voltage to 0.1 volt and stimulate by pressing the Mode lever to SINGLE. If no twitch is seen, proceed as previously to stimulate the nerve by increasing increments of 0.1 volt until a twitch is visible.

Record this voltage as the threshold stimulus by nerve stimulation.
11. Now, determine the minimum voltage that causes extension of the foot by stimulating the nerve. Refer to illustration B, figure SR-1 in appendix C for Physiograph® sample record.

Record this voltage on the Laboratory Report.

Myogram of Muscle Twitch

Now that we have determined the threshold stimulus that will induce a visible twitch on the muscle, we are ready to produce a *myogram,* or tracing, of the twitch using a transducer and recorder. Figure 22.11 illustrates what a myogram of a human muscle, the soleus, looks like. Although the frog muscle twitch will have the same general configuration, it will have a different amplitude and spread (msec).

Note in figure 22.11 that the myogram of a muscle twitch has three distinct phases: a latent period, the contraction phase, and the relaxation phase.

The **latent period** is a very short time lapse between the time of stimulation and the start of contraction. In most muscles of the body, it lasts about 10 milliseconds.

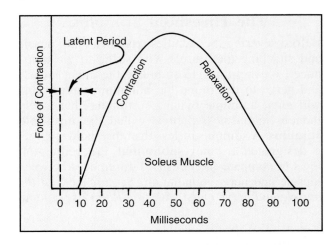

Figure 22.11 **A simple twitch of the human soleus muscle.**

In the **contraction phase** the muscle shortens due to the chemical changes that occur within the muscle fibers. Different muscles have different durations in this phase.

The **relaxation phase** follows the contraction phase. During this phase the muscle returns to its former length.

Proceed as follows to attach the frog's leg to the transducer and to record a myogram.

Thread Hookup

Free the connective tissue that holds the gastrocnemius muscle to adjacent muscles with a dissecting needle. Insert a pair of forceps under the muscle near the Achilles tendon (figure 22.12) to grasp one end of an 18-inch length of cotton thread. Pull the thread through with forceps and tie it to the tendon a short distance away from where the tendon will be severed. Now, with a pair of scissors, cut off the tendon distal to where it is attached to the string (figure 22.13).

If the Gilson setup is being used, tie the free end of the thread to two leaves (the stationary one and the one next to it) of the transducer. If the Narco myograph is being used, tie a loop in the thread and slip it onto the little hook.

To prevent the leg from moving during contraction, pin the leg down to the cork board with a pin (Narco setup, figure 22.2). For the Gilson setup, it will be necessary to secure the leg to the base with wire. *When handling these transducers, do so gently. Undue stress to the leaves or hook can cause internal damage.*

Adjust the tension on the thread with the fine adjustment knob on the tension adjuster just enough to take the slack off the thread.

Electrode Positioning

For the Narco setup, check the sleeve electrode to make sure the electrode is in the proper position.

For the Gilson setup, position the electrode prongs so that the nerve rests on the prongs. Maneuver the electrode holder in such a way that the electrode is held firmly in place. Note that the stem of the holder consists of a soft metal that can be bent into any position.

Although the pin electrodes are shown in figure 22.2, they will not be used in the remainder of the experiment.

Recorder Settings

Unigraph Turn on the power switch. Place the chart control (c.c.) lever at STBY, the stylus heat control knob at the two o'clock position, the speed selector level at the slow position (opposite of position shown in figure 22.1), the gain control on 2 MV/CM, and the mode control on TRANS (transducer).

Now, place the c.c. lever at Chart On and note the position of the line produced on the chart. The line should be about 1 centimeter from the nearest margin of the paper. If it is not at this position, relocate the stylus with the centering knob. Now, stop the chart by placing the c.c. lever at STBY. Proceed to Stimulator Settings below.

Physiograph® Turn on the power switch. Make sure that the myograph is calibrated so that 100 grams of tension is equivalent to 4 centimeters of pen deflection (appendix B).

After the initial attachment of the muscle to the myograph and with minimal tension on the myo-graph, set the baseline to the centerline with the position control knob.

Place the record button in the ON position and gradually increase the tension by moving the myo-graph upward with the myograph tension adjuster. Tension should be increased until the pen has moved upward 0.5 centimeters (you have applied 12.5 grams of tension).

By using the Balance control, return the pen to centerline.

Now, by using the position control, move the pen to any desired baseline; normally, this would be 2 centimeters below the centerline.

Be sure to periodically bathe the muscle and nerve preparation with frog Ringer's solution.

Stimulator Settings

Grass Stimulator Set the Duration control at 15 msec and the Voltage at the level that produced flexion of the leg by nerve stimulation in the previous experiment. Check the switches to make sure that the Mode switch is set at OFF, pulses at REGULAR, Polarity at NORMAL, and the right-hand slide switch at MONO.

SM-1 Stimulator Set the Width control at 2 msec and the Voltage at the level that produced flexion of the leg by nerve stimulation in the previous experiment.

Recording

You are now ready to produce a simple muscle twitch myogram on the chart of your recorder and determine the duration of each of the three phases in the contraction cycle. Proceed as follows:

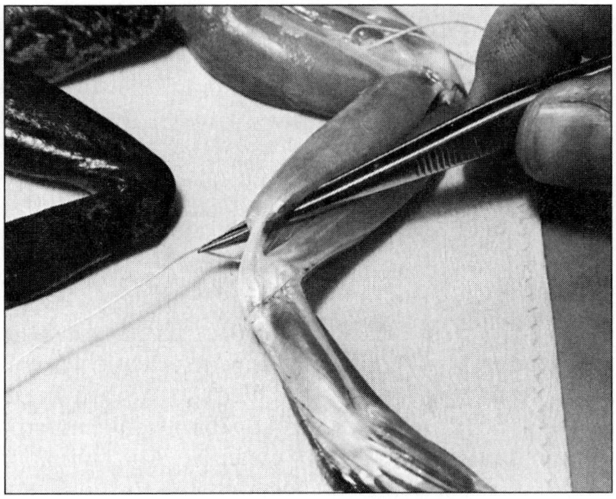

Figure 22.12 Thread is drawn through the space behind the Achilles tendon by inserting forceps through to grasp the thread.

Figure 22.13 After the thread is securely tied to the Achilles tendon, the tendon is severed between the thread and joint.

Unigraph

1. Place the c.c. lever at Chart On. The paper should be moving at the slow rate (2.5 mm/sec).
2. Depress the Mode switch to SINGLE and observe the tracing on the chart. The stylus travel should be approximately 1.5 centimeters. Adjust the sensitivity control to produce the desired stylus displacement.
3. If the stylus travel remains insufficient by adjusting the sensitivity knob, increase the sensitivity by changing the Gain control to 1 MV/CM and readjusting the sensitivity knob again.
4. Change the speed of the paper to 25 mm per second by repositioning the speed control lever 180° to the left.
5. Administer 25 to 30 stimuli to provide enough chart material so that each member of your team will have at least 5 inches of chart for attachment to the Laboratory Report sheet.
   Note: If your setup does not have an event sync cable it will be necessary to depress the event marker button on the Unigraph *simultaneously* with depression of the Mode lever on stimulation.
6. Calculate the duration of each of the periods, knowing that 1 millimeter on the chart is equivalent to 0.04 seconds (40 milliseconds).

Physiograph®

1. Start the paper moving at 0.1 cm/sec, and lower the pens onto the recording paper. It is not necessary that the timer be turned on.
2. Press the Mode switch to SINGLE and observe the tracing on the chart. The pen travel should be approximately 1.5 centimeters. Adjust the sensitivity control to produce the desired pen travel.
3. Change the chart speed to 2.5 cm/sec.
4. Administer 25 to 30 stimuli to provide enough chart material so that each member of your team will have enough chart to attach to the Laboratory Report sheet. Refer to illustration A, figure SR-1, appendix C, for Physiograph® sample recording.
5. Calculate the duration of each of the periods on the contraction cycle and report your results on the Laboratory Report.

Summation

The ability of whole muscles to exert different degrees of pull is achieved primarily by summation. Two types of summation are known to occur: multiple motor unit summation and wave summation.

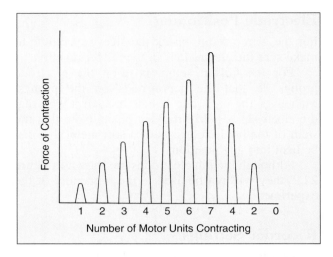

Figure 22.14 **Multiple motor unit summation.**

Multiple motor unit summation is an increased force caused by the recruitment of additional motor units. It is also known as *recruitment* or *spatial summation.*

Wave summation is due to an increase in the frequency of nerve impulses and is also referred to as *temporal summation.*

An attempt will be made in this portion of the experiment to demonstrate both types of summation.

Multiple Motor Unit Summation

Once the threshold stimulus has been determined for producing a twitch, it will be observed that increasing the strength of the stimulus (by voltage increase) will cause increasingly greater degrees of contraction. This phenomenon is schematically illustrated in figure 22.14. As the diagram reveals, the force of contraction is a function of the number of motor units that are stimulated.

A **motor unit** consists of a group of muscle fibers that is innervated by a single motor neuron (motoneuron). Small muscles that react quickly and precisely may have as few as two or three muscle fibers per motor unit. Large muscles that lack a fine degree of control, such as the gastrocnemius, may have as many as 1000 muscle fibers per motor unit. These large muscles are well adapted to a posture-sustaining function.

The muscle fibers of adjacent motor units are arranged in an overlapping fashion so that each unit lends support to its neighbors. When some nerve fibers to a muscle are destroyed, as in poliomyelitis, **macromotor units** are formed by extensive arborization of the ends of surviving motoneurons to provide neural connections to all muscle fibers. These large motor units may be four or five times as large as normal units.

As voltage increases are administered, a degree of contraction is reached that cannot be exceeded by further voltage increase. This maximum contraction, called the **maximal response,** is due to the fact that all motor units are activated and no further contraction by individual stimuli can be produced. The lowest voltage that produces a maximal response is called the **maximal stimulus.**

Proceed as follows to demonstrate recruitment, maximal stimulus, and maximal response.

1. **Unigraph:** For this unit set the speed control at the slow speed (2.5 mm/sec) and the c.c. switch at STYLUS ON. All other settings should be left as they were for the previous experiment.
   Physiograph®: Set the paper speed at 0.25 cm/sec, and lower the pens onto the paper, but do not start the paper at this time.
2. **Stimulator:** Except for the voltage, leave all other settings as they were in the previous experiment. Return the voltage controls to 0.1 volt.
3. **Electrodes:** Check the electrodes on the frog's nerve to see that they are in good contact with the nerve. *Remember to keep the tissues moist with frog Ringer's solution.*
4. With the chart still, administer a stimulus to the nerve by depressing the Mode lever to SINGLE. Look for movement of the stylus on the chart.
5. Advance the chart approximately 5 mm, and stop it again.
6. Raise the voltage by 0.2 volt and produce another single stimulus.
7. Continue this process of advancing the chart 5 mm and increasing the voltage by 0.2 volt until a maximal response has been achieved. *Be sure to record all voltages on the chart.*
8. Repeat this overall process five or more times to provide enough chart material for all members of your team.

Wave Summation

If motor unit summation were the only way in which maximum contraction could be achieved by muscles they would have to be much larger than they are to accomplish the work that they are able to do. Fortunately, the phenomenon of wave summation augments multiple motor unit summation.

Wave summation in skeletal muscle occurs when the muscle receives a series of stimuli in very rapid succession, as illustrated in figure 22.15. If the rate of stimulation is kept very slow only single twitches will occur; however, as seen in figure

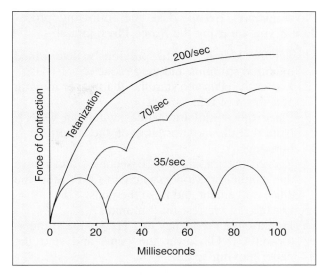

Figure 22.15 **Wave summation.**

22.15, the single twitches produced at 35 pps will produce some summation. Note that as the frequency is stepped up to 70 pulses per second, considerable summation occurs, and at 200 pps no relaxation of the muscle takes place; instead, *complete tetanization* occurs.

In this portion of the experiment, we will use the Frequency control to regulate the pps or Hz.

Unigraph Setup Use the following procedure if you are using the Unigraph-Grass setup.

1. Set the Voltage at the previously determined maximal stimulus, the Frequency at 1 pps and the stimulator Mode switch at OFF.
2. Put the c.c. switch at Chart On with chart speed at slow setting (2.5 mm/sec).
3. Move the Mode switch on the stimulator to REPEAT. Note that the monitor lamp on the stimulator is blinking.
4. Record the tracing for 10 seconds, return the stimulator Mode switch to OFF, and the c.c. switch to STBY. Let the muscle rest for 2 minutes.
5. Increase the Frequency to 2 pps, start the paper moving, and record for another 10 seconds. Stop the chart (STBY), and rest the muscle for another 2 minutes.
6. Now, double the Frequency to 4 pps for another 10 seconds. Rest again for 2 minutes.
7. Continue to double the frequency, recording for 10 seconds, and resting for 2 minutes, until the muscle goes into complete tetany.
8. After resting the muscle for a few minutes, repeat the experiment enough times to provide a record for each member of your team.

Physiograph® Setup Use the following procedure if you are using the Narco Physiograph®.

1. Set the Voltage at the previously determined maximal stimulus and the Frequency at 1 Hz.
2. Lower the pens and start the chart paper moving at a speed of 0.25 cm/sec.
3. Place the stimulator Mode switch on CONT. Note that the monitor lamp on the stimulator is blinking.
4. Record the tracing for 10 seconds, and return the stimulator Mode switch to OFF. Stop the chart movement and raise the pens.
5. Let the muscle rest for 2 minutes.
6. Increase the Frequency to 2 Hz, set the stimulator on CONT, lower the pens, and start the paper moving again.
     Record for another 10 seconds, return the stimulator Mode switch to OFF, stop the paper and raise the pens.
7. Allow the muscle to rest another 2 minutes.

8. Now, double the frequency to 4 Hz, turn on the stimulator to CONT again, lower the pens and start the paper moving. After recording for another 10 seconds, shut down the system again and rest the muscle for another 2 minutes.
9. Continue to double the frequency, recording for 10 seconds, and resting the muscle each time for 2 minutes until the muscle goes into complete tetany.
10. Refer to illustration C, figure SR-1, appendix C, for sample Physiograph® records.
11. After resting the muscle for a few minutes, repeat the experiment enough times to provide a record for each member of your team.

Laboratory Report

Complete the first portion of Laboratory Report 22, 24 by attaching required charts and answering all questions.

Muscle Contraction Experiments:
Using The Intelitool Physiogrip™

23

The same phenomena of muscular contraction studied in Exercise 22 will be studied here utilizing an Intelitool Physiogrip™ that is controlled by computer software. The setup is illustrated in figure 23.1. Note that electrical stimuli will be applied to nerves in the right arm that control flexion of the third and fourth fingers. With either the third or fourth finger on the trigger of the Physiogrip, flexor response is fed into the computer for data accumulation and evaluation. The following phenomena of muscle contraction will be studied: (1) threshold stimulus, (2) motor unit (spatial) summation, (3) wave (temporal) summation, (4) tetany, (5) single muscle twitch, and (6) fatigue.

This exercise is actually divided into four experiments. This is accomplished by combining two or more of the six experiments into single experiments.

You will be working with a laboratory partner: one of you will be the subject, and the other individual will read the instructions and operate the stimulator and computer.

The instructions provided here assume that you have a basic knowledge of the computer platform you are using. Instructions will be provided for three platforms: (1) IBM or compatible PC (MS-DOS 3.3 or higher), (2) Macintosh (System 6.0.8 or higher), and (3) Apple II (ProDOS).

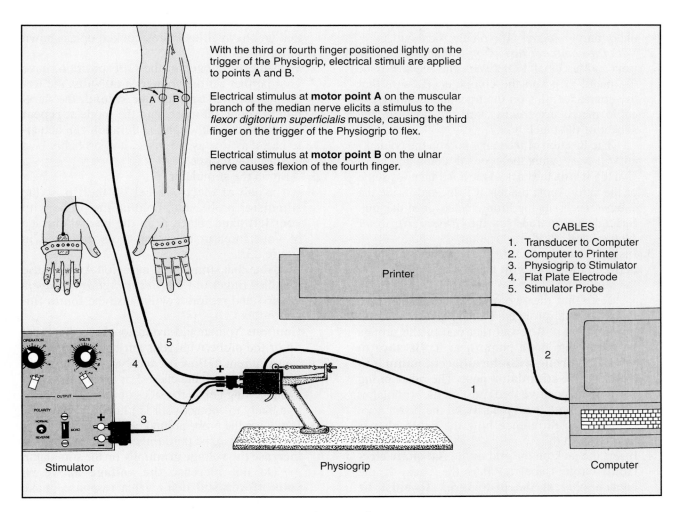

With the third or fourth finger positioned lightly on the trigger of the Physiogrip, electrical stimuli are applied to points A and B.

Electrical stimulus at **motor point A** on the muscular branch of the median nerve elicits a stimulus to the *flexor digitorium superficialis* muscle, causing the third finger on the trigger of the Physiogrip to flex.

Electrical stimulus at **motor point B** on the ulnar nerve causes flexion of the fourth finger.

CABLES
1. Transducer to Computer
2. Computer to Printer
3. Physiogrip to Stimulator
4. Flat Plate Electrode
5. Stimulator Probe

Printer

Stimulator

Physiogrip

Computer

Figure 23.1 Equipment hookup for muscle contraction experiments.

Note in the materials list below that among the items used in this experiment are copies of two Physiogrip manuals that are supplied with the equipment. Although you may not need them, the manuals will be available for reference.

Materials:
Physiogrip transducer on wooden base
Physiogrip program disk
Physiogrip Lab Manual
Physiogrip User Manual
blank disk for saving data
electronic stimulator, computer, and printer
flat-plate electrode cable and rubber strap
stimulator probe cable
dual banana cable, gameport cable
electrode gel

Equipment Hookup

Arrange the various components illustrated in figure 23.1 in a manner that will make it convenient to use all pieces of equipment. Hook up the electronic components with the appropriate cables according to the following sequence:

1. Plug the cables for the computer, printer, and stimulator into *grounded* 110 volt electrical outlets.
2. *With the computer turned off,* connect the **game port cable** (label 1) between the black box of the pistol grip and the computer. Be sure that the connector pins on the transducer end of the cable match up *exactly* with the holes in the socket of the black box.

   The location of the game port on the computer will vary with the type of computer used. Usually it is on the back of the computer. On some of the early Apple models it is located inside the cabinet, necessitating lifting the cover of the unit. Special instructions in the *Physiogrip User Manual* will clarify its location.
3. Install the **dual banana cable** (label 3) between the pistol grip and the output posts on the stimulator.

   Note that the two ends of this cable have identical dual-pronged banana plugs. Careful examination will reveal that one of the prongs on each plug has a bump near it. **Be sure to insert the prong with the adjacent bump into the negative stimulator post.** The other prong fits into the positive (red) post on the stimulator.

   Since there is no polarity on the pistol grip, it makes no difference how the cable plug is inserted into it.
4. Insert the jack on the end of the **flat-plate electrode cable** (label 4) into the back of the banana plug at the pistol grip. **It must be inserted into the negative side of the plug,** **which has the bump near it.** Note that this plate electrode, which is the ground, is strapped to the back surface of the right hand. Before strapping it to the hand, place an ample amount of electrode gel to the plate so that good electrical contact occurs.
5. Insert the jack of the **stimulator probe cable** (label 5) into the other hole in the back of the banana plug at the pistol grip.

Locating the Motor Point

Before getting into the actual experiments it is necessary that you locate a precise motor point and become comfortable while stimulating it. To find the motor point it will be necessary to probe around on the arm surface with the stimulator set at a certain voltage, frequency, and duration.

Once you locate the motor point, almost all discomfort from the stimulator will disappear and a good strong flexion of the middle finger will occur. Proceed as follows to locate a sensitive motor point on the median nerve (point A, figure 23.1):

1. Double-check your connections to make sure that you have all the wires hooked up as shown in figure 23.1.
2. With the stimulator in the OFF position, have your partner set the voltage at **60 volts,** the frequency at **one stimulus per second,** the duration to about **2.0 msec,** and the mode at **repeat** (continuous). Although the duration can actually be set as low as 0.5 msec, it is probably best to start at 2.0 msec.
3. Turn on the stimulator.
4. Put a dab of electrode gel on the tip of the stimulator probe, and, holding the probe with your left hand, place it on the medial surface of your forearm in region A as indicated in figure 23.1.

   Note that stimulation in region A will cause the third finger to flex (*flexor digitorum superficialis*) and region B will cause the fourth finger to flex.
   Caution: You should avoid touching the metal tip of the probe with your left hand as this will cause current to flow between your left and right arm: i.e., through the chest. For the same reason, avoid getting gel on the sides of the probe, where it is likely to contact your left hand.
5. If moving the probe around point A does not elicit a response in the third finger, have your partner **increase the voltage gradually** on the stimulator.
   Do not increase the voltage above 90 volts. If you still don't get a response, make sure that the stimulator is set up correctly. In

particular, pay attention to the settings of any multiplier switches.

6. Once the motor point is located, experiment with it; for instance, you may have to press rather hard with the probe to get maximum muscle flexion.

Try to relax! Get used to the sensation so that when you are actually doing an experiment you are relaxed with it. When you are doing an experiment you want to avoid distorting the data due to body jerking, voluntary finger flexion, or extension.

7. For additional information on locating the motor points consult the *Physiogrip Lab Manual.*

Threshold Stimulus, Motor Unit Summation, and Maximal Stimulus

Our first experiment will be to determine the minimum electrical stimulus that will elicit a contraction. This amount of stimulus is called the *threshold* or *liminal stimulus.*

Next we will demonstrate *motor unit summation,* also known as *spacial summation* or *recruitment.* Motor unit summation is increased contraction due to an increase in the number of muscle fibers that are stimulated to contract. Figures 23.2 and 23.3 illustrate what this phenomenon looks like.

Finally, the third objective of this experiment will be to determine the magnitude of stimulus above which there is no further increase in the strength of contraction. This is known as the *maximal stimulus.*

Proceed as follows to demonstrate these three phenomena using the instructions that apply to your computer platform:

1. Begin Data Collection.

MS-DOS: Select "Experiment Menu" from the PHYSIOGRIP MENU. Then select "Alter Sweep Speed" and choose "15 Seconds per Frame."

Next, select "Continuous Experiment Mode" from the "Experiment Menu." Make sure that Autostop is turned ON. Press the A key to toggle its status.

Macintosh: Start a new data file by selecting **New** from the **File** menu. Disable stimulus marks by clicking the OFF button in the upper right portion of the window. Set the controls on the left of the window so that **AutoStop** is on (checked). **Samples/Sec** is 50, **Sec/Data Set** is 100, and **Metronome** is off (click the Metronome icon to toggle status). Click the **Start** button.

Apple II: Select "Change Sweep" from the MAIN MENU and choose "15 Seconds per Sweep." Next, select "Experiment Mode Menu" from the MAIN MENU and then choose "Run Experiment Mode."

2. Position Your Hand on Transducer.

Without applying a stimulus, grasp the pistol grip so that when you are relaxed (no stimulus) the natural tonus in your finger is putting a very slight pressure on the trigger. The trigger should be displaced 3–5 mm (3–5 increments on the screen). In other words, before a stimulus is applied, you should have enough pressure on the trigger to lift the plot up off the bottom of the screen. If you are using your third finger, rest your second finger on the shelf on the right side of the pistol grip. **Try to relax the muscles in your arm** so that you don't dampen the response.

3. Determine the Threshold Stimulus.

With the frequency set to 1 stimulus per second and the duration at 2.0 msec, have your lab partner turn the

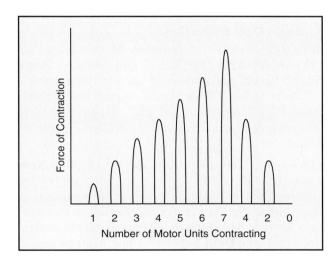

Figure 23.2 Diagrammatic representation of multiple motor unit summation (spatial summation) and recruitment of motor units.

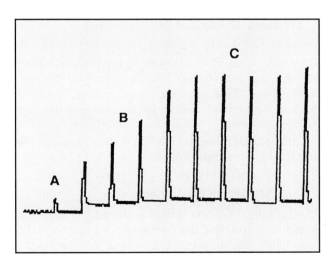

Figure 23.3 The threshold (liminal) stimulus (A), motor unit (spatial)) summation (B), and maximal stimulation (C) using Intelitool software.

stimulator voltage down to well below the point where you get a response. This will be about 30 volts.

While stimulating the motor point location on your forearm, have your lab partner gradually increase the voltage until you get a slight response. This is the *threshold* (liminal) *stimulus*.

Record the stimulator voltage setting here:
_____ Volts

4. Gradually Ramp Up the Voltage.

Ask your partner to increase the stimulus voltage gradually. On the screen you should see a stair stepping in the strength of responses corresponding to the increase in stimulus strength.

When the point where all motor units are responding is reached, turning up the voltage further will only increase your discomfort but will not increase the strength of response.

Do not continue to turn the voltage up to the point of real pain. If the responses are so strong that the trigger is reaching the end of its travel, increase the spring tension or use a stronger spring and start over.

Record the maximal stimulus here:
_____ Volts

5. Try Again, If Necessary.

If data collection stops (memory is full) before you are satisfied with your results, just begin data collection again as described below. If you are satisfied with results, proceed to step 6.

MS-DOS: Press "C" to clear memory and restart data acquisition. All data currently in memory will be lost.

Macintosh: Click the **Start** button and then click **Erase** to confirm that all data currently in memory will be lost.

Apple II: Press "1" to clear memory and restart data acquisition. All data currently in memory will be lost.

6. Measure Amount of Displacement.

To measure the displacement, follow these steps:

MS-DOS: Press ESC to return to the PHYSIOGRIP MENU, and then select "Review/Analyze Menu." Next choose "Displacement/Time Analysis," and use the S, Z, and arrow (cursor) keys to position the data markers.

Measure the height of both the threshold and the maximal contraction. Results:

Threshold height:_____mm

Maximal contraction height:_____mm

Macintosh: Select **Displacement** from the **Analysis** menu. Use the mouse to scroll through the plot. Click and drag in the plot area to position the data markers. Measure the height of both threshold and maximal contractions.

Threshold height:_____mm

Maximal contraction height:_____mm

Apple II: Press ESC to return to the MAIN MENU, and then select "Review/Analyze." From the REVIEW/ANALYZE MENU, select "Review/Analyze" again, and use the cursor keys to position the data marker. Measure the height of both the threshold and maximal contractions:

Threshold height:_____mm

Maximal contraction height:_____mm

7. Printout.

Upon completion of this portion of the exercise print enough copies for you and your partner to hand in.

8. Save Data to Disk.

To incorporate this data into memory do as follows:

MS-DOS: Press ESC to return to the PHYSIOGRIP MENU and select "Data File/Disk Commands." Then choose SAVE File and give the file an appropriate name and press the Return (Enter) key.

Macintosh: Select **Save** from the **File** menu and name the file appropriately.

Apple II: From the MAIN MENU select "Save Current Data" and then select "Catalog/Save Data File." On single floppy drive systems, you will be prompted to insert a data disk. Give the file an appropriate name and press the Return key.

Wave Summation and Tetanizing Contractions

Wave (temporal) *summation* in skeletal muscle occurs when the muscle receives a series of stimuli in very rapid succession. If the frequency of stimulation is increased enough, *tetanization* will occur. See figures 23.4 and 23.5. Proceed as follows to demonstrate these two phenomena:

1. Begin Data Collection.

MS-DOS: Select "Experiment Menu" from the PHYSIOGRAPH MENU. Then select "Alter Sweep Speed" and choose "5 Seconds per Frame."

Next select "Continuous Experiment Mode" from the "Experiment Menu." Make sure that Autostop is turned ON. Press the A key to toggle its status.

Macintosh: Start a new data file by selecting **New** from the **File** menu. Enable stimulus marks by clicking the ON button in the upper right portion of the window. Set the controls on the left of the window so that **AutoStop** is on (checked). **Samples/Sec** is 50, **Sec/Data Set** is 100, and **Metronome** is off (click the metronome icon to toggle status). Click the **Start** button.

Apple II: Select "Change Sweep Speed" from the MAIN MENU and choose "3 Seconds per Sweep."

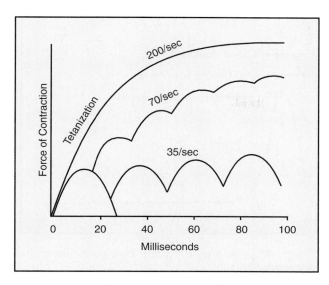

Figure 23.4 Diagrammatic representation of wave (temporal) summation and tetanization caused by increasing the frequency of stimulation.

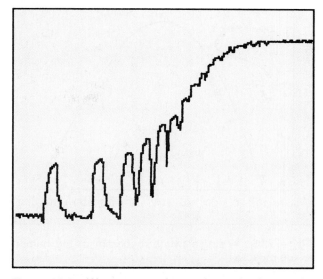

Figure 23.5 Wave summation and tetanization on the monitor screen. The first two waves before summation should occupy about 25% of vertical screen height.

Next, select "Experiment Mode Menu" from the MAIN MENU and then choose "Run Experiment Mode."

2. Position Your Hand on Transducer.

Without applying a stimulus, grasp the pistol grip so that when you are relaxed (no stimulus) the natural tonus in your finger is putting a slight pressure on the trigger. The trigger should be displaced 3–5 mm (3–5 increments on the screen). In other words, before a stimulus is applied, you should have enough pressure on the trigger to lift the plot up off the bottom of the screen. If you are using your third finger, rest your second finger on the shelf on the right side of the pistol grip.

3. Set the Initial Voltage.

With the stimulator in repeat (continuous) mode, the duration between 0.5 and 2 msec (previously determined), and the frequency at 1 pps, set the voltage to a level where the response from each stimulus takes up about 25% of the plotting screen vertically.

4. Gradually Ramp Up the Frequency.

Have your partner gradually increase the frequency of stimulation up to a tetanic contraction. This should be accomplished within 5 seconds or so. Do this several times, sustaining tetanic contraction for 1–2 seconds.

If the tetanic contraction causes a contraction that is so strong that the pistol grip is reaching the end of its travel, increase the spring tension or use a stronger spring.

5. Try Again If Necessary.

If data collection stops (memory is full) before you are satisfied with your results, just begin data collection again as follows. If you are satisfied with results, proceed to step 6.

MS-DOS: Press "C" to clear memory and restart data acquisition. All data currently in memory will be lost.

Macintosh: Click the **Start** button and then click **Erase** to confirm that all data currently in memory will be lost.

Apple II: Press "1" to clear memory and restart data acquisition. All data currently in memory will be lost.

6. Measure the Stimulus Frequency.

As described below, determine and record both the stimulus frequency and the magnitude of the response at the following two points on the plot: (1) just before summation occurred and (2) at the onset of tetany.

Also, determine the percent increase in the displacement caused by the tetanic contraction as compared with a complete single twitch.

MS-DOS: Press the ESC key to return to the PHYSIOGRIP MENU, and then select "Review/Analyze Menu." Next choose "Displacement/Time Analysis," and measure the interval between stimuli (i.e., the stimulus frequency). Do so by lining up the data markers with the stimulus marks under the "Stim on" indicator. Use the S, Z, and arrow keys to position the markers.

Macintosh: Select **Displacement** from the **Analysis** menu. Use the mouse to scroll through the plot. Click and drag in the plot area to position the data markers. Measure the interval between stimuli (i.e., the stimulus frequency) by lining up the data markers with the stimulus marks on the stimulus indicator.

Apple II: Press ESC to return to the MAIN MENU, and then select "Review/Analyze." From the REVIEW/ANALYZE MENU, select "Review/

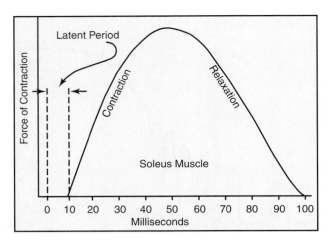

Figure 23.6 A single twitch myogram of the human soleus muscle.

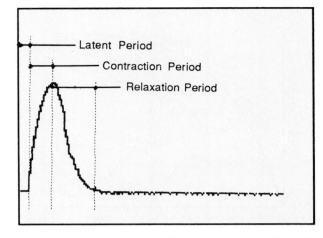

Figure 23.7 A single twitch myogram using the Physiogrip™.

Analyze" again, and use the D key and cursor keys to position the data markers.

The best approximation for determining stimulus frequency is to measure the interval between the very beginning of each response. As the frequency of contractions occur closer together, it may be easier to measure peak to peak.

7. Printout.

Upon completion of this portion of the exercise print enough copies for you and your partner to hand in.

8. Save Data to Disk.

To incorporate these data into memory do as follows:

MS-DOS: Press ESC to return to the PHYSIOGRIP MENU and select "Data File/Disk Commands." Then choose SAVE File and give the file an appropriate name and press the Return (Enter) key.

Macintosh: Select **Save** from the **File** menu and name the file appropriately.

Apple II: From the MAIN MENU select "Save Current Data" and then select "Catalog/Save Data File." On single floppy drive systems, you will be prompted to insert a data disk. Give the file an appropriate name and press the Return key.

Single Muscle Twitch

We are now ready to study the myogram (tracing) of a single muscle twitch and determine its characteristics. All muscle twitches display three phases: a latent period, a contraction phase, and a relaxation phase. Each of these phases has a characteristic time duration, which may vary from muscle to muscle in the body.

The **latent period** is a very short time lapse between the time of stimulation and the start of con-

traction. In most muscles of the body it lasts about 10 milliseconds.

In the **contraction phase** the muscle shortens due to chemical changes that occur within the muscle fibers. The time duration here varies from muscle to muscle.

After contraction the muscle goes into a **relaxation phase** during which the muscle returns to its former length. The duration of this phase will also vary to some extent from muscle to muscle.

Figure 23.6 illustrates a myogram of the soleus muscle. Figure 23.7 illustrates what the trace will look like with a Physiogrip. Proceed as follows to demonstrate a single muscle twitch with the Physiogrip:

1. Adjust the Stimulator Settings.

Without applying a stimulus, set the stimulator on SINGLE (momentary) mode. The voltage and duration should be set to the values that were determined for the maximal stimulus on page 114, typically, about 2.0 msec and 70 to 90 volts.

You may find it interesting to repeat the maximal stimulus experiment, using a longer stimulus duration. Increasing the duration will lower the voltage setting needed to get a true maximal stimulus. To find the maximal stimulus voltage for a different stimulus duration, do that portion over again.

2. Position Your Hand on Transducer.

Grasp the pistol grip so that when you are relaxed (no stimulus) the natural tonus in your finger is putting a very slight pressure on the trigger. The trigger should be displaced 3–5 mm (3–5 increments on the screen). In other words, before a stimulus is applied, you should have enough pressure on the trigger to lift the plot up off the bottom of the screen. If you are using your third finger, rest your second finger on the shelf on the right side of the pistol grip.

3. Begin Sampling Trial Data.

MS-DOS: Select "Experiment Mode" from the PHYSIOGRIP MENU, and then select "Single Twitch Mode." This screen is simply for setting up. None of what you see at this time can be saved and/or reviewed.

Macintosh: Start a new data file by selecting **New** from the **File** menu. Disable stimulus marks by clicking the OFF button in the upper right portion of the window.

Set the controls on the left of the window so that **AutoStop** is off (unchecked), **Samples/Sec** is 50, **Sec/Data Set** is 100, and **Metronome** is off (click the metronome icon to toggle status). Click the **Start** button.

Apple II: Select "Single Twitch Mode" from the program's MAIN MENU, and then select "Run Single Twitch Mode." This screen is simply for setting up. None of what you see at this time can be saved and/or reviewed.

4. Verify a Good Twitch Response.

While stimulating your motor point, have your partner repeatedly depress the stimulus switch, and, if necessary, gradually turn up the stimulator voltage until you are getting a good response that fills about 75% of the screen vertically.

If you have increased the duration, you will notice that you can now get a response at a lower voltage, and even the threshold response was at a lower voltage. This demonstrates that even a subliminal stimulus has an effect on an excitable membrane, so that stimulus voltage that is subliminal, if only applied for a short period of time, may become significant if applied for a longer period of time.

5. Collect Single Twitch Data.

MS-DOS: Have your lab partner press the B key on the computer. The computer is now waiting for stimulus application in order to acquire data.

Have your partner press the stimulus switch to evoke a twitch response. The "Waiting for Stimulus" message disappears as the data are acquired and drawn on the screen. Repeat this process, acquiring as many frames of data as you like.

Macintosh: Click the mouse to end the trial data sampling. Enable stimulus marks by clicking the ON button in the upper right portion of the window.

Set the controls on the left of the window so that **AutoStop** is on (unchecked). **Samples/Sec** is 500, **Sec/Data Set** is 20, and **Metronome** is off (click the metronome icon to toggle status). Click the **Start** button and discard the data currently in memory.

Have your partner press the stimulus switch every one to two seconds to evoke a twitch response. Repeat this process until you are satisfied with the results.

Apple II: Have your lab partner press the T (Twitch) key on the computer. The computer is now waiting for stimulus application in order to acquire data.

Next, have your partner press the stimulus switch every one to two seconds to evoke a twitch response. Since the data are being acquired faster than they can be plotted, you will not see a trace on the screen. However, when you look at the data in one of the review modes, the resolution of the data will be about as high as is possible using the computer's built-in input capabilities.

6. Measure Twitch Velocity.

When viewing the velocity plot, it should be very apparent that during the contraction period of the muscle twitch the muscle was moving (contracting) faster than it was moving (relaxing) during the relaxation period. This can be observed by noting that, corresponding to each muscle twitch, the peak of the positive velocity (contraction) is greater (farther from the zero line) than is the peak of the negative velocity (relaxation). View the velocity plot as follows:

MS-DOS: Press ESC to return to the PHYSIOGRIP MENU, and then select "Review/Analyze Menu." Next, choose "Velocity Plot."

Study the double plot. The left edge of the screen represents the exact moment when the stimulator fired. Use the S, Z, and arrow keys to position the data markers and take readings.

Macintosh: Click the mouse to stop data acquisition if it is not already stopped. Select **Disp. & Vel.** from the **Analysis** menu.

Study the double plot. The stimulator bar will indicate the exact moment the stimulator fired. Use the mouse to scroll through the plot. Click and drag in the plot area to position the data markers and take reading.

Apple II: When you have finished collecting single twitch data, the computer will process the data so that they are in a form that can be plotted when you analyze them.

Press ESC to return to the MAIN MENU, select "Velocity Plot," and then choose "Velocity Plot" mode.

Study the double plot. The left-hand edge of the screen represents the exact moment when the stimulator fired. Use the D and cursor keys to position the data markers and read the values.

7. Measure Twitch Time Intervals.

Several characteristic time intervals are readily identified on the time-displacement plot of a single

117

muscle twitch. These are the periods of *latency, contraction, relaxation,* and the *entire twitch.*

Follow the procedures below to view the time-displacement plot, and average the periods from several different twitches.

MS-DOS: Press ESC to return to the "Review/ Analyze Menu," and then select "Displacement/ Time Analysis." Again, the left edge of the screen represents the exact moment when the stimulator fired. Use the S, Z, and arrow keys to position the data markers and measure the various twitch intervals.

Macintosh: Select **Displacement** from the **Analyze** menu. Again, the stimulator bar will indicate the exact moment the stimulator fired. Use the mouse to scroll through the plot. Click and drag in the plot area to position the data markers and measure the various twitch intervals.

Apple II: Press ESC to return to the MAIN MENU. Next, select REVIEW/ANALYZE MENU and then choose "Review/Analyze." Again, the left-hand edge of the screen is the exact moment when the stimulator fired. If a long duration was used, a bar will also appear at the bottom of the screen showing the stimulus duration. Use the D key and cursor keys to position the data markers and measure the various twitch intervals.

8. Printout.
Upon completion of this portion of the exercise print enough copies for you and your partner to hand in.

9. Save Data to Disk.
To incorporate these data into memory do as follows:

MS-DOS: Press ESC to return to the PHYSIOGRIP MENU and select "Data File/Disk Commands." Then choose SAVE File and give the file an appropriate name and press the Return (Enter) key.

Macintosh: Select **Save** from the **File** menu and name the file appropriately.

Apple II: From the MAIN MENU select "Save Current Data" and then select "Catalog/Save Data File." On single floppy drive systems, you will be prompted to insert a data disk. Give the file an appropriate name and press the Return key.

Fatigue and Muscle Contraction

In this experiment we will study the effect of blood supply on fatigue in a muscle. Muscle fatigue is caused by the depletion of oxygen, glucose, and other raw materials, and/or the accumulation of

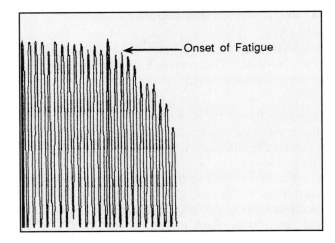

Figure 23.8   Onset of muscle fatigue.

waste products such as lactic acid. All of these factors have a depressing effect on the neuromuscular junction.

Since all of these materials are brought to or removed from the muscle by blood, the rate of fatigue for a given rate and force of contraction will vary with the quantity and quality of the blood supply to the muscle. To regulate the blood flow to the hand, we will apply a sphygmomanometer (blood pressure cuff) to the upper arm. Figure 23.8 illustrates the appearance of the onset of fatigue.

Materials:
Physiogrip setup from previous experiment
sphygmomanometer
metronome (for Apple II version only)

1. Change Trigger Tension.
Install the strongest spring that is available for the Physiogrip.

2. Position Hand on the Transducer.
Grasp the pistol grip so that your *index finger* is positioned on the trigger.

3. Set Acquisition Parameters.
MS-DOS: Select "Experiment Mode" from PHYSIOGRIP MENU. Then select "Alter Sweep Speed" and choose "25 Seconds per Frame."

Macintosh: Start a new data file by selecting **New** from the **File** menu. Disable stimulus marks by clicking the OFF button in the upper right portion of the window.

Set the controls on the left of the window so that **AutoStop** is on (checked). **Samples/Sec** is 10, **Sec/Data Set** is 500, **Metronome** is on (click the metronome icon); metronome rate is **2 Hz,** and **Sound** (if available) is on (click the speaker icon).

Apple II: Select "Change Sweep Speed" from the MAIN MENU and choose "20 Seconds per Sweep."

4. Begin Data Collection.

MS-DOS: Select "Continuous Experiment Mode" from the EXPERIMENT MENU. Immediately press the F key to turn on the flashing asterisk in the upper right portion of the screen and *at the same time* start rhythmically squeezing the trigger *all the way in and out* at a rate of about two times per second, or at least twice as fast as the asterisk flashes.

When you can no longer squeeze the trigger through its full range of travel, proceed to step 5.

Macintosh: Click the **Start** button and immediately begin squeezing the trigger *all the way in and out* in cadence with the metronome (2 times per second).

When you can no longer squeeze the trigger through its full range of travel, click the mouse to end acquisition and proceed to step 5.

Apple II: Select "Experiment Mode Menu" from the MAIN MENU and then choose "Run Experiment Mode." Immediately begin rhythmically squeezing the trigger *all the way in and out* at a rate of about two times per second. If a metronome is available, use it for cadence.

When you can no longer squeeze the trigger through its full range of travel, press ESC to stop acquisition and proceed to step 5.

5. Save Data to Disk.

To incorporate these data into memory do as follows:

MS-DOS: Press ESC to return to the PHYSIOGRIP MENU and select "Data File/Disk Commands." Then choose SAVE File and give the file an appropriate name and press the Return (Enter) key.

Macintosh: Select **Save** from the **File** menu and name the file appropriately.

Apple II: From the MAIN MENU select "Save Current Data" and then select "Catalog/Save Data File." On single floppy drive systems, you will be prompted to insert a data disk. Give the file an appropriate name and press the Return key.

6. Repeat with Reduced Blood Flow.

Rest your arm for about 5 minutes, and then place a sphygmomanometer cuff on your arm just above the elbow. Inflate the cuff to 140 mm Hg and repeat steps 2 through 5.

7. Measure Fatigue Rate.

MS-DOS: Press ESC to return to the PHYSIOGRIP MENU and then select "Review/Analyze Menu." Choose "Displacement/Time Analysis" and use the S, Z, and arrow keys to determine the elapsed time before fatigue sets in. Open the data file from the first trial (i.e., with normal blood flow) and repeat the measurement.

To open the first data file, select "Data File/Disk Commands" from the PHYSIOGRIP MENU and then choose OPEN File.

Macintosh: Using **Displacement** mode under the **Analyze** menu, measure the time it took for fatigue to set in for each data file. Use the mouse to scroll through the plot. Click and drag in the plot area to position the data marker and determine fatigue time.

Apple II: Press ESC to return to the MAIN MENU. Next select REVIEW/ANALYZE MENU, and then choose "Review/Analyze." Use the arrow keys to determine the elapsed time before fatigue set in. Open the data file from the first trial (i.e., with normal blood flow) and repeat the measurement. To open the first data file, select "Load Old Data File" from the MAIN MENU and then enter the name of the file you wish to load. On single floppy drive systems you will be prompted to insert the data disk.

Your partner may wish to reverse the procedure: that is, perform the reduced blood flow experiment first. Would you expect the magnitude of difference in fatigue rate to be as great if the procedure were reversed for the same individual?

8. Printout.

Upon completion of this portion of the exercise print enough copies for you and your partner to hand in.

Laboratory Report

Since there is no Laboratory Report for this exercise in the back of the manual, your instructor will indicate how these various muscle experiments should be written up.

24

Electromyography

Since the early experiments of the 1920s, when muscle contraction proved to be accompanied by bioelectricity, electromyography (hereafter EMG) has become firmly established in physiology. Electrical activity associated with muscle contraction arises from nerve and muscle action potentials of the motor units. Depolarizations initiate contraction while repolarizations accompany relaxation. Together, this activity produces voltage changes that are detectable at the body surface with skin electrodes or intramuscularly with needle electrodes.

The type of recordings obtained in EMG depends upon the selection of electrodes. Recordings with needle electrodes inserted into muscle tissue yield precise information concerning individual motor unit action potentials. Such a procedure permits the diagnostician to distinguish between myopathic and neurogenic disorders. Because of the skill required, not to mention the hazard of infection and legal implications, student applications must be limited to surface electrodes. Although only a general impression of whole muscle groups is feasible with surface electrodes, single muscles may sometimes be detected.

Individual action potentials cannot be readily distinguished with skin electrodes since the region beneath the electrode often consists of several motor units. Figure 24.1 illustrates the nature of a typical EMG. The high-frequency, irregular, overlapping spikes of an EMG of this type are called an **interference pattern.** The voltage spread of these potentials may be as great as 50 millivolts.

With sufficient amplification, an interference pattern may be detectable even in muscles that are completely relaxed. This spontaneous activity is associated with the maintenance of **normal muscle tone.** As such a muscle is activated, however, the interference pattern becomes exaggerated.

To display an EMG, one might use an oscilloscope, audio monitor, or a chart recorder. Although the Unigraph is the recorder of choice in this experiment because of its portability, a polygraph with an EMG module is excellent. A Unigraph in which the IC-MP module has been replaced by an IC-EMG module can also be used.

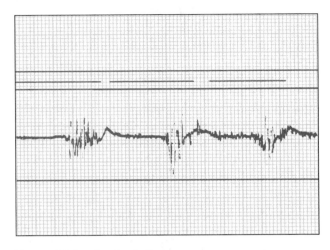

Figure 24.1   An electromyogram.

In fact, the latter module is better adapted to this type of experiment.

The class will be divided up into a number of teams of three or four students each. While one individual prepares the Unigraph, another will attach the electrodes to the subject and assist in recording information on the chart.

Materials:

Unigraph, Duograph, or polygraph
handgrip dynamometer (Stoelting #19117)
3-lead patient cable for Unigraph
3 skin electrodes (Gilson self-adhering #E1081K)
adhesive pads for electrodes
electrode gel (ECG or other)
Scotchbrite pads (grade #7447)
alcohol or alcohol swabs

Preparations

Equipment Hookup

While the subject is being prepped for EMG recording, connect the 3-lead cable to the Unigraph. Plug the power cord into an outlet. Turn on the power switch and the stylus heat control, and check out the intensity of the stylus line on the paper at slow speed.

If the Unigraph is not provided with a special EMG module, set the Mode control on the EEG setting. To be able to quantify the EMG in terms of millivolts, it will be necessary to calibrate the instrument according to instructions in appendix B.

If a Gilson polygraph is used, the following modules can be used for EMG monitoring: IC-MP, IC-UM, and IC-EMG. If all modules are available, use the IC-EMG module.

Subject Preparation

Figure 24.2 illustrates the placement of three electrodes on the arm. The two recording red-wired electrodes are placed within an inch of each other on the belly of the major flexor muscle of the forearm (*flexor digitorum superficialis*). The third electrode is a ground and should be placed some distance away over a bony area.

Prior to placing the electrodes on the arm, scrub the skin areas gently with a Scotchbrite pad and disinfect the skin with 50%–70% alcohol. Removing some of the dead cells from the skin surface will greatly facilitate conductivity.

Consult figures 24.3 through 24.6 to see how the self-adhesive pads and electrode gel are placed on the electrodes prior to attaching them to the skin. There are several other kinds of electrodes that can also be used. Figure 24.7 reveals how one might substitute three ECG plate electrodes on the arm. As in the case of the other electrodes, it is necessary to scrub first with Scotchbrite and apply electrode gel before strapping on the electrodes.

When the electrodes are firmly attached to the skin, connect the wires to the appropriate wire of the 3-lead cable. These cables usually have alligator clips and are color coded to accept the correct electrodes. Be sure that the ground electrode is attached to the black ground clip of the 3-lead cable.

Monitoring

Three phenomena of muscle activity will be studied with this setup: spontaneous activity, recruitment, and fatigue. If a Duograph, dual-beam oscilloscope, or polygraph is available, it will also be possible to monitor EMGs of agonist and antagonist muscles, simultaneously.

Spontaneous Activity

With the arm completely relaxed and fully supported on the tabletop, turn on the Unigraph (high speed) and observe if any evidence of motor unit activity is discernible at the lowest gain. Increase the sensitivity by changing the gain settings and turning the sensitivity control. Determine the magnitude and frequency of peaks.

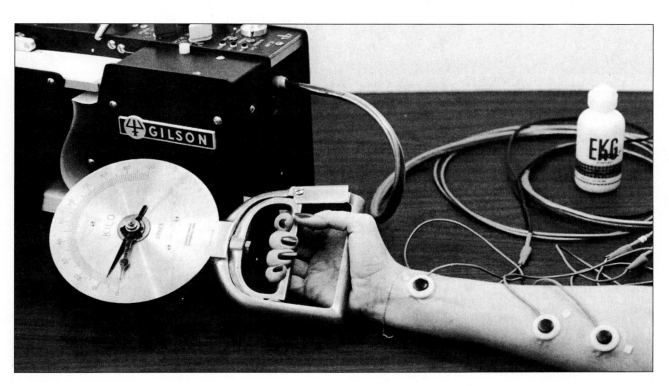

Figure 24.2   EMG monitoring setup.

Recruitment

Instruct the subject to clench his or her fingers to form a fist and note the burst of activity to form a typical EMG interference pattern on the chart.

To demonstrate recruitment, have the subject grip a hand dynamometer, as in figure 24.2. Record, first, a few bursts at 5 kilograms. Then double the force to 10 kilograms. Continue to increase the force by 5- or 10-kilogram increments to maximum, recording each force magnitude on the chart with a pen or pencil.

Do you see an increase in amplitude on the chart as evidence of recruitment of motor units? Produce enough tracings for all members of your team.

Fatigue

With the Unigraph turned off, direct the subject to perform some handgripping work on the dynamometer until the hand muscles are thoroughly fatigued.

When the subject cannot continue, take the dynamometer away and allow the subject's arm to recover on the tabletop. Start the recorder and monitor any electrical activity. How does the EMG of fatigue compare with that of spontaneous activity?

Agonist vs. Antagonist Muscles

Opposing movements of muscle pairs are caused by *agonist* or *antagonist* muscles. For example, flexion

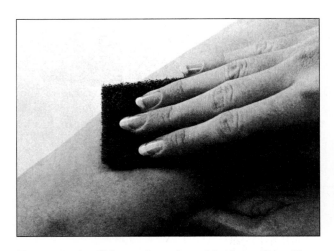

Figure 24.3 Skin surface is rubbed gently with a Scotchbrite pad to improve skin conductivity.

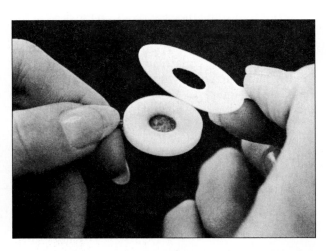

Figure 24.4 An adhesive disk is applied to the clean, dry undersurface of the skin electrode.

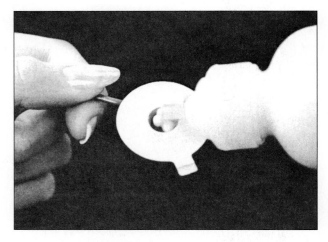

Figure 24.5 Electrode gel is applied to the depression of the skin electrode.

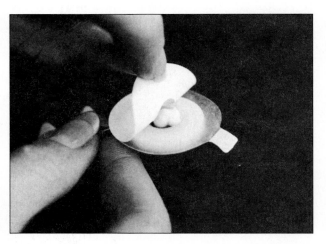

Figure 24.6 The top covering of the adhesive disk is removed, and the electrode is placed on the skin.

122

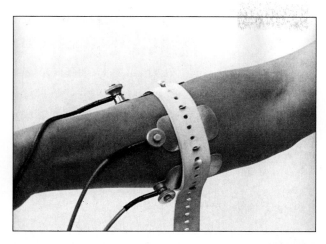

Figure 24.7 An alternative method of electrode application for an EMG demonstration.

of the forearm against the upper arm is achieved by contraction of the *biceps brachii,* the agonist. Extension occurs when the *triceps brachii,* its antagonist, contracts. The action of these two muscles must be coordinated so that when one contracts the other relaxes.

Activity in flexors cannot be accompanied by simultaneous activity in extensors, or rigidity will occur instead of flexion. Coordination may be demonstrated by recording the activity of both muscles simultaneously with a Duograph, polygraph, or dual-beam oscilloscope.

To demonstrate coordination, hook up three electrodes to each muscle of one arm through separate channels of whatever type of equipment that is available. Have the subject execute the following maneuvers:

1. Flexion of arm against resistance; assistant restrains subject's fist while flexion is attempted
2. Extension against resistance
3. Isometric tension (contraction) of both agonist and antagonist, simultaneously

Laboratory Report

Complete the last portion of combined Laboratory Report 22, 24.

123

25

Cat Dissection:
Skin Removal

In preparation for the muscle studies in the next few exercises it will be necessary to remove the skin from your cat. Once removed, the skin will be used as a wrap for the carcass to prevent dehydration.

Laboratory cats are preserved with a mixture of alcohol and formaldehyde that is unpleasant to the nose, eyes, and skin; yet, this preservative is essential in preventing bacterial decomposition of the specimen. Protection of the hands is best achieved with rubber gloves. If gloves are not used, a protective cream, such as *Protek,* affords some protection. The latter is rubbed into the pores of the skin before dissection begins.

Students will work in pairs to dissect their cat. At the end of each laboratory period the specimen will be packed away in a plastic bag and sealed tightly with a name tag tied to one end. Under no circumstances should a student use any other student's cat for dissection.

When the dissection is completed, your specimen should look like the one illustrated in figure 25.1. Proceed as follows:

Materials:
    cat (preserved and latex injected)
    dissecting board or tray
    dissecting instruments
    plastic bag for cat
    rubber surgical gloves or Protek
    hand lotion

1. Place the cat ventral side down on a dissecting board. Make a short incision on the midline of the neck region with a sharp scalpel, as illustrated in figure 25.2. Cut only through the skin, which is about 1/4 inch thick here. Avoid cutting into the muscles beneath the skin.
2. With a pair of scissors or scalpel continue the incision along the back midline all the way to the tail. See figure 25.3.
3. As illustrated in figures 25.4 and 25.5, separate the skin from the body muscles by pulling on it and severing the superficial fascia with a scalpel.

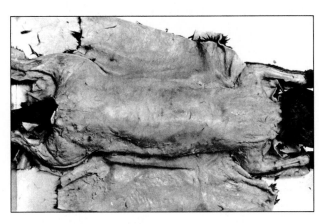

Figure 25.1   **Cat with skin removed.**

4. From the tail, make incisions down each hind leg to the ankles and cut all around each ankle. See figure 25.6.
5. As shown in figure 25.7, cut the skin around the neck and down each foreleg to the wrists. Cut the skin around each wrist.
6. Now, gripping the skin with your fingers, gently pull it away from the body. If your specimen is a female, look for the **mammary glands** on the underside of the abdomen and thorax. Remove the glands and discard them.
7. If laboratory time is still available, proceed to Exercise 26 to study the neck muscles of your cat.
8. If no more time is available, wash your instruments and tray with soap and water and scrub down your desktop to get rid of any debris. Your table should be left as clean and dry as it was when you entered the room.
9. Wrap the specimen in the skin and seal it in the plastic bag. Attach a label to the bag with your names on it. Place the cat in the assigned storage bin.
10. After washing your hands with soap and water, apply hand lotion to them.

Figure 25.2 Make the first cut in neck region with a sharp scalpel. Be careful to cut only through the skin, not into the muscles.

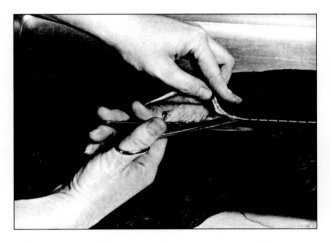

Figure 25.3 With a pair of sharp scissors (or a very sharp scalpel) cut the skin along the back from the neck to the tail region.

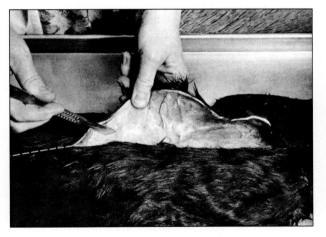

Figure 25.4 Pull the skin away from the muscles, using the scalpel to separate the superficial fascia from the muscles.

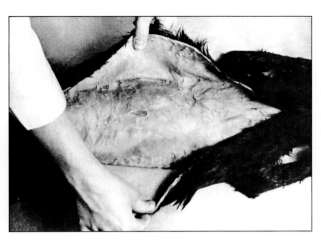

Figure 25.5 Separate as much of the skin as possible from the muscles of the body before proceeding to the next step.

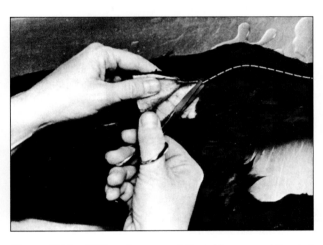

Figure 25.6 With scissors cut the skin around the tail and down each of the hind legs to the ankles. Cut around each ankle.

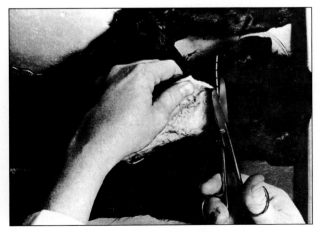

Figure 25.7 With scissors cut the skin around the neck and down each side of the front legs to the wrists. Cut around wrists.

26

Head and Neck Muscles

The principal muscles of the head and neck will be studied in this exercise. Depending on the availability of materials, cadavers, muscle manikins, or cats will be used for laboratory studies. The head muscles are grouped here according to their functions.

Scalp Movements

Removal of the scalp reveals an underlying muscle known as the **epicranial.** It consists of the **frontalis** in the forehead region, an **occipitalis** in the occiput region, and an aponeurosis extending between these two over the top of the skull.

The occipitalis has its *origin* on the mastoid process and occipital bone; its *insertion* is on the aponeurosis. The frontalis takes its *origin* on the aponeurosis and its *insertion* on the soft tissue of the eyebrows.

Action: Contraction of the frontalis causes horizontal wrinkling of the forehead and elevation of the eyebrows. Action of the occipitalis causes the scalp to be pulled backward.

Sphincter Action

Circular muscles that close openings are called *sphincters.* The eyes and mouth are surrounded by muscles of this type.

Orbicularis Oris This circular muscle lies just under the skin of the lips. It takes its *origin* in various facial muscles, the maxilla, mandible, and septum of the nose. Its *insertion* is on the lips.

Action: It closes the lips in various ways: by compression over the teeth, or by pouting and pursing them; utilized in kissing.

Orbicularis Oculi Each eye is surrounded by one of these sphincters. Each muscle *arises* from the nasal portion of the frontal bone, the frontal process of the maxilla, and the medial palpebral ligament. It *inserts* within the tissues of the eyelids (palpebrae).

Action: Blinking and squinting.

Facial Expressions

Emotions such as pleasure, sadness, fear, and anger are expressed by certain muscles of the face. The following five muscles play different roles in facial expressions.

Zygomaticus This muscle extends diagonally from the zygomatic bone to the corner of the mouth. Its *origin* is on the zygomatic; its *insertion* is on the edge of the orbicularis oris.

Action: Contraction of these muscles draws the angles of the mouth upward and backward as in smiling and laughing.

Quadratus Labii Superioris This thin muscle, consisting of three heads, lies between the eye and upper lip. Its origin is on the upper part of the maxilla and part of the zygomatic bone. Its insertion is mainly on the superior margin of the orbicularis oris, and partially on the alar (Latin: *alaris,* wing) region of the nose.

Action: Furrowing of the upper lip occurs when the entire muscle contracts; result: expression of disdain or contempt. Expression of sadness occurs if only the infraorbital head contracts.

Quadratus Labii Inferioris This small muscle extends from the lower lip to the mandible. Its *origin* is on the mandible, and its *insertion* is on the lower margin of the orbicularis oris.

Action: It pulls the lower lip down in irony.

Triangularis This muscle is lateral to the quadratus labii inferioris. Its *origin* is on the mandible. Its *insertion* is on the orbicularis oris.

Action: Since it is an antagonist of the zygomaticus, it depresses the corner of the mouth.

Platysma This broad sheetlike muscle extends from the mandible over the side of the neck. Only a portion of it is seen in figure 26.1. Figure 26.2 (label 3) reveals its entirety. Since its *insertion* is on the mandible and muscles around the mouth, it profoundly affects facial expressions. Contraction causes the lower lip to move backward and downward, expressing horror.

Masticatory Movements

The chewing of food (mastication) involves five pairs of muscles. Four pairs move the mandible and one pair helps to hold the food in place. All of these muscles are revealed in illustrations A and B, figure 26.1.

Masseter Of the four muscles that move the mandible, the masseter is the most powerful one. It extends from the angle of the mandible to the zygomatic bone. Only the lower portion of this muscle is seen in illustration A; its entirety is seen on the large illustration. Its *origin* is on the zygomatic arch, and it *inserts* on the mandible.

Action: It raises the mandible.

Temporalis The large fan-shaped muscle on the side of the skull is the temporalis. It *arises* on portions of the frontal, parietal, and temporal bones and is *inserted* on the coronoid process of the mandible.

Action: It acts synergistically with the masseter to raise the mandible. It can also cause retraction of the mandible when only the posterior fibers of the muscle are activated.

Pterygoideus Internus *(Internal Pterygoid)* The position of this muscle has led some to refer to it as the "internal masseter." It lies on the medial surface of the ramus, taking its *origin* on the maxilla, sphenoid, and palatine bones. Its *insertion* is on the medial surface of the mandible between the mylo-

hyoid line and the angle. Refer to illustration B, figure 26.1.

Action: It works synergistically with the masseter and temporalis to raise the mandible.

Pterygoideus Externus *(External Pterygoid)* This muscle is the uppermost one in illustration B. It *arises* on the sphenoid bone and *inserts* on the neck of the mandibular condyle and the articular disk of the joint.

Action: Acting together, the two external pterygoids can cause the mandible to move forward (protrusion) and downward. Individually, they can move the mandible, laterally, to the right and left.

Buccinator The horizontal muscle that partially obscures the teeth in illustration B is the buccinator. It is situated in the cheeks *(buccae)* on each side of the oral cavity. It *arises* on the maxilla, mandible, and a ligament, the **pterygomandibular raphé** (figure 26.1, label 17). This ligament joins the buccinator to the **constrictor pharyngis superioris.** The buccinator *inserts* on the orbicularis oris.

Action: It compresses the cheek to hold food between the teeth during chewing.

Assignment:
Locate all these muscles on a muscle manikin or cadaver. The cat is not a suitable specimen for most of these muscles.

Label figure 26.1.

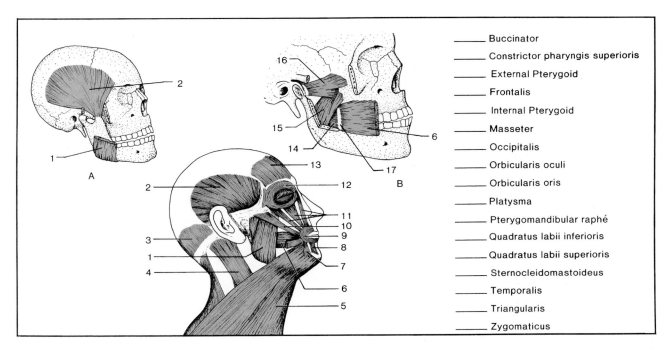

Figure 26.1 Head muscles.

Neck Muscles

Figures 26.2 and 26.3 illustrate a majority of the major muscles of the neck region. Figure 26.2 is primarily of surface and posterior deep muscles, and figure 26.3 reveals the deeper anterior muscles.

Surface Muscles

The principal surface muscles of the neck are the platysma on the anterolateral surfaces and the trapezius on the back of the neck.

Platysma   It was noted previously that this muscle takes its *insertion* on the mandible and the muscles around the mouth. Its *origin* is primarily the fascia that covers the pectoralis and deltoideus muscles of the shoulder region.

> *Action:* Primarily to draw the lower lip backward and downward; assists to some extent in opening the jaws.

Trapezius   This large triangular muscle occupies the upper shoulder region of the back. Only its upper portion functions as a neck muscle.

It *arises* on the occipital bone, the ligamentum nuchae, and the spinous processes of the seventh cervical and all the thoracic vertebrae. The **ligamentum nuchae** is a ligament that extends from the occipital bone to the seventh cervical vertebra, uniting the spinous processes of all the cervical vertebrae. The *insertion* of the trapezius is on the spine and acromion of the scapula and the outer third of the clavicle.

> *Action:* Pulls the scapula toward the median line (adduction); raises the scapula, as in shrugging the shoulder; and draws the head backward (hyperextension), if the shoulders are fixed.

Deeper Neck Muscles, Posterior

Illustration C, figure 26.2, reveals three posterior muscles of the neck as seen when the trapezius is removed. Another prominent muscle, the levator scapulae, has been omitted here for clarity. It will be studied in the next exercise.

Splenius Capitis   This muscle lies just beneath the trapezius and medial to the levator scapulae. Only the left splenius is shown in illustration C.

Its *origin* is on the ligamentum nuchae and the spinous process of the seventh cervical and upper three thoracic vertebrae. Its *insertion* is on the mastoid process.

> *Action:* Hyperextension of the head if both muscles act together. If only one muscle contracts, the head is inclined and rotated toward the contracting muscle.

Semispinalis Capitis   This muscle is the larger of two muscles shown on the right side of the neck. Its *origin* consists of the transverse processes of the upper six thoracic vertebrae and the articular processes of the lower four cervical vertebrae. Its *insertion* is on the occipital bone.

> *Action:* It works synergistically with the splenius capitis.

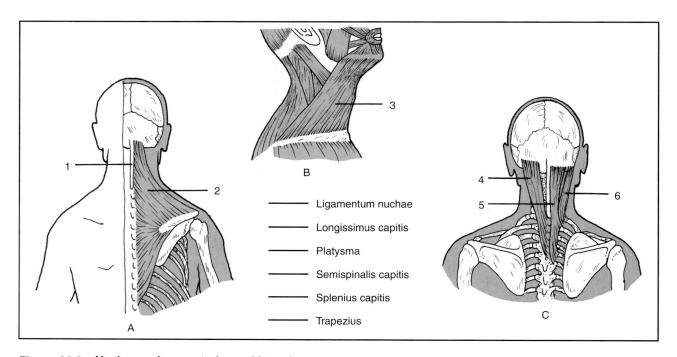

Figure 26.2   Neck muscles, posterior and lateral.

Ligamentum nuchae
Longissimus capitis
Platysma
Semispinalis capitis
Splenius capitis
Trapezius

Longissimus Capitis This muscle is lateral and slightly anterior to the semispinalis capitis. The *origin* of this muscle is on the transverse processes of the first three thoracic vertebrae and the articular processes of the lower four cervical vertebrae. The *insertion* is on the mastoid process.

Action: Same as the two muscles on page 128.

Deeper Neck Muscles, Anterior

Refer to figure 26.3 to identify the anterior muscles that are seen when the two platysma muscles are removed from the neck.

Sternocleidomastoideus This large neck muscle derives its name from the skeletal components that provide its anchorage. Its *origin* is located on the manubrium of the sternum and the sternal (medial) end of the clavicle. The *insertion* is on the mastoid process.

Action: Simultaneous contraction of both sternocleidomastoids causes the head to be flexed forward and downward on the chest. Independently, each muscle draws the head down toward the shoulder on the same side as the muscle.

Sternohyoideus This long muscle extends from the hyoid bone to the sternum and clavicle. The left sternohyoid is exposed by the removal of the lower portion of the left sternocleidomastoid. The sternohyoid *arises* on a portion of the manubrium and clavicle. It *inserts* on the lower border of the hyoid bone.

Action: Draws the hyoid bone downward.

Sternothyroideus This muscle is somewhat shorter than the sternohyoideus. It *originates* on the posterior surface of the manubrium and *inserts* on the inferior edge of the thyroid cartilage of the larynx.

Action: Draws the thyroid cartilage of the larynx downward.

Omohyoideus The long narrow curving muscle that is seen on the left side of the neck and that extends from the hyoid bone into the shoulder region is the omohyoid. It arises from the upper surface of the scapula and inserts on the hyoid bone, lateral to the sternohyoid.

Action: Draws the hyoid bone downward.

Digastricus Two of these narrow V-shaped muscles are visible under the mandible in figure 26.3. Each muscle consists of anterior and posterior bellies with the central portion attached to the hyoid bone by a fibrous loop.

The posterior belly *arises* from the medial side of the mastoid process of the temporal bone. The anterior belly *arises* on the inner surface of the body of the mandible near the symphysis. The point of *insertion* is the fibrous loop on the hyoid bone.

Action: Raises the hyoid bone and assists in lowering the mandible. Acting independently, the anterior belly can pull the hyoid bone forward; the posterior belly can pull it backward.

Mylohyoideus The two mylohyoids extend from the medial surfaces of the mandible to the median line of the head to form the floor of the mouth. They

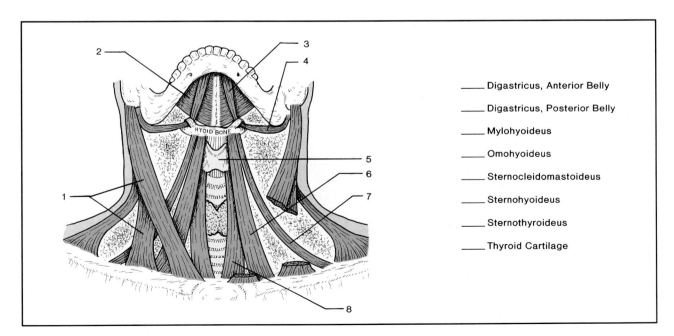

_____ Digastricus, Anterior Belly

_____ Digastricus, Posterior Belly

_____ Mylohyoideus

_____ Omohyoideus

_____ Sternocleidomastoideus

_____ Sternohyoideus

_____ Sternothyroideus

_____ Thyroid Cartilage

Figure 26.3 Anterior view of neck muscles with platysmas removed.

lie just superior to the anterior bellies of the digastric muscles. Each mylohyoid *arises* along the mylohyoid line of the mandible and *inserts* on a median fibrous raphé that extends from the symphysis menti of the mandible to the hyoid bone.

Action: Raises the hyoid bone and tongue.

Assignment:
Label figures 26.2 and 26.3 and proceed to cat dissection of the neck muscles.

Cat Dissection

Once the skin has been removed from the cat, place the animal ventral side up on a dissecting tray and identify the neck muscles by referring to figures 26.4 and 26.5. Since an embalming incision, which destroys several muscles, has been made along one side of the neck, perform your studies on the side of the neck opposite to the incision. As you read through these instructions, you will encounter three key words, which are defined as follows:

Dissect: To dissect a muscle in this laboratory means to use a probe to free the muscle from adjacent muscles, exposing its full length. *The muscle is not to be cut up.*

Transect: Transection involves cutting the muscle transversally with a scalpel or scissors.

Reflect: Once a muscle has been transected its cut ends can be lifted out of the way, or reflected, to expose deeper muscles.

It is important that the muscles on your specimen not be mutilated or accidentally removed, unless directed to do so. Muscle dissection should be methodical and careful. A good way to keep track

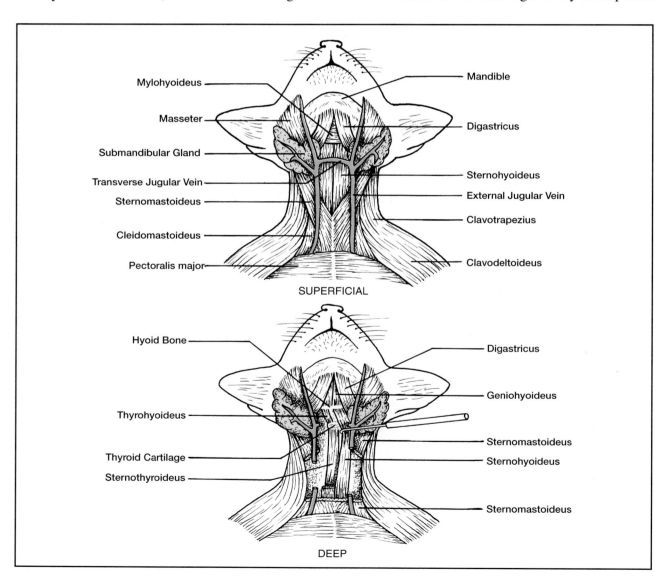

SUPERFICIAL

Mylohyoideus — Mandible
Masseter — Digastricus
Submandibular Gland
Transverse Jugular Vein — Sternohyoideus
Sternomastoideus — External Jugular Vein
Cleidomastoideus — Clavotrapezius
Pectoralis major — Clavodeltoideus

DEEP

Hyoid Bone — Digastricus
— Geniohyoideus
Thyrohyoideus
— Sternomastoideus
Thyroid Cartilage — Sternohyoideus
Sternothyroideus
— Sternomastoideus

Figure 26.4 Cat neck muscles, superficial and deep.

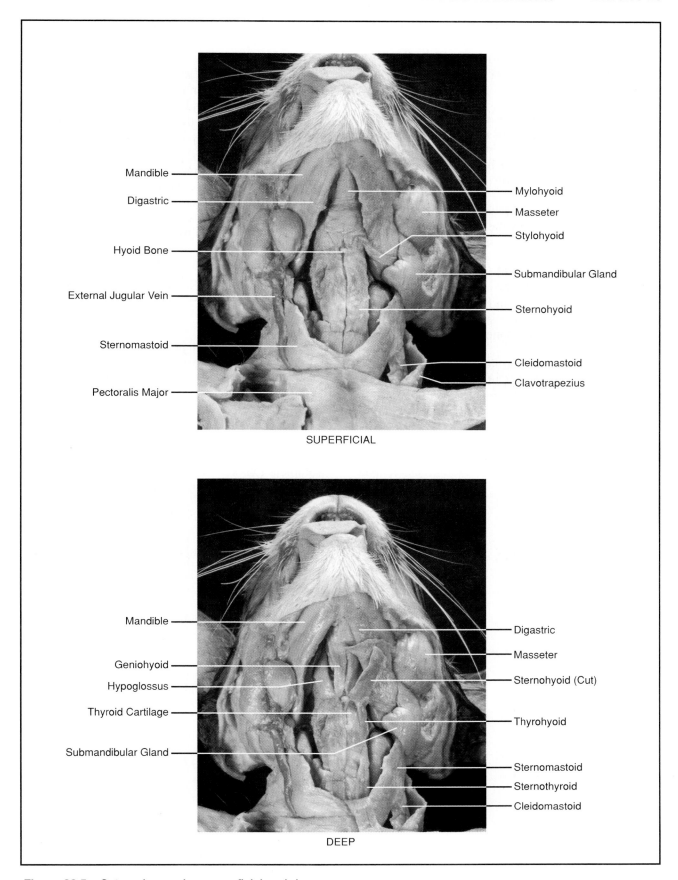

Mandible

Digastric

Hyoid Bone

External Jugular Vein

Sternomastoid

Pectoralis Major

Mylohyoid

Masseter

Stylohyoid

Submandibular Gland

Sternohyoid

Cleidomastoid

Clavotrapezius

SUPERFICIAL

Mandible

Geniohyoid

Hypoglossus

Thyroid Cartilage

Submandibular Gland

Digastric

Masseter

Sternohyoid (Cut)

Thyrohyoid

Sternomastoid

Sternothyroid

Cleidomastoid

DEEP

Figure 26.5 **Cat neck muscles, superficial and deep.**

of muscles as they are identified is to color each of the muscles on the diagrams with different-colored pencils. This procedure may also be helpful for review purposes. Proceed to identify the cat muscles in the following order:

Digastric Like the human digastric, this superficial muscle has two portions: an anterior belly and a posterior belly. The anterior belly is attached to the mandible, and the posterior belly is attached to the mastoid process of the skull.

Mylohyoid This thin, flat muscle with fibers that course horizontally below the mandible lies deep to the anterior belly of the digastric. Pull the anterior bellies of the digastric muscles laterally to see the full extent of the mylohyoid.

Sternomastoid This bandlike muscle extends from the manubrium of the sternum to the mastoid portion of the temporal bone and lies deep to the external jugular vein.

Note that the right and left sternomastoid muscles fuse on the midline. Cut through this point of fusion in the caudal direction to the manubrium. *Dissect* the sternomastoid that is on the side opposite to the embalming incision.

Cleidomastoid Note that this narrow band of muscle is located between the sternomastoid and clavotrapezius. *Dissect.* What muscle in the human takes the place of these last two muscles?

Sternohyoid This straight, narrow muscle extends from the manubrium to the hyoid bone, covering the larynx and trachea. *Dissect.* It will be necessary to cut away the transverse jugular vein at this time.

Sternothyroid This narrow band of muscle is lateral and dorsal to the sternohyoid. It extends from the manubrium to the thyroid cartilage of the larynx. To expose the sternothyroid, it will be necessary to lift up the sternohyoid and push it to one side. *Dissect.*

Thyrohyoid Cranial to the sternothyroid, and extending between the hyoid bone and the lateral portion of the thyroid cartilage, is a very small muscle, the thyroid.

Masseter This most powerful muscle of mastication is seen as a large, rounded muscle in the cheek region. Its origin is on the zygomatic arch and its insertion is on the lateral surface of the mandible.

Laboratory Report

Complete Laboratory Report 26 for this exercise.

Trunk and Shoulder Muscles

27

In this study of the trunk and shoulder muscles, all of the surface and a majority of the deeper muscles of the trunk and shoulder will be studied. These muscles function primarily to move the shoulders, upper arms, spine, and head. Some of them assist in respiration.

Anterior Muscles

Removal of the skin and subcutaneous fat from the body will reveal muscles as shown in figure 27.1.

Surface Muscles

Pectoralis Major   This muscle is the thick fan-shaped one that occupies the upper quadrant of the chest. Its *origin* is on the clavicle, sternum, costal cartilages, and aponeurosis of the external oblique. It *inserts* into the crest of the greater tubercle of the humerus.

> *Action:* Adducts, flexes, and rotates the humerus medially.

Deltoideus   This muscle is the principal muscle of the shoulder. It *originates* on the lateral third of the clavicle, the acromion, and the spine of the scapula. It *inserts* on the deltoid tuberosity of the humerus.

> *Action:* Activation of the middle portion of the deltoid causes abduction of the humerus. The anterior portion flexes the humerus and medially rotates it. Its posterior portion extends the humerus and rotates it laterally.

Serratus Anterior   The upper and lateral surfaces of the rib cage are covered by this muscle. It takes its *origin* on the upper eight or nine ribs and *inserts* on the anterior surface of the scapula near the vertebral border.

> *Action:* Rotates the scapula, raising the shoulder as in full flexion and abduction. Draws the scapula forward as in the act of pushing. Upper digitation may draw scapula downward and forward. Lower digitations draw the scapula downward.

Deeper Muscles

Pectoralis Minor   This muscle lies beneath the pectoralis major and is completely obscured by the latter. It *arises* on the third, fourth, and fifth

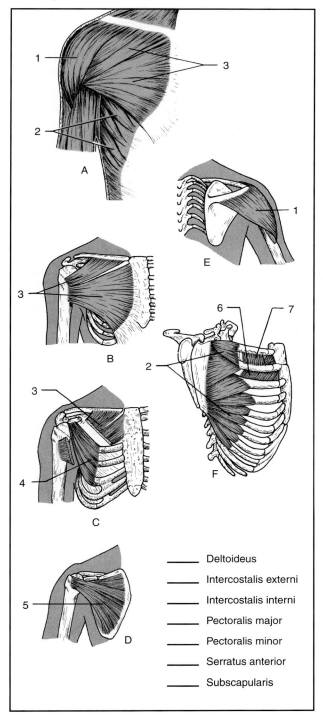

Figure 27.1   Shoulder and trunk muscles, anterior.

_____ Deltoideus

_____ Intercostalis externi

_____ Intercostalis interni

_____ Pectoralis major

_____ Pectoralis minor

_____ Serratus anterior

_____ Subscapularis

ribs, and *inserts* on the coracoid process of the scapula.

> *Action:* Draws the scapula forward and downward with some rotation.

Intercostalis Externi *(External Intercostals)* Between the ribs on both sides are eleven pairs of short muscles, the external intercostals. Each of these short muscles *arises* from the lower border of a rib and *inserts* on the upper border of the rib below it.

Note that their fibers are directed obliquely forward on the front of the ribs. Label 7 in figure 27.1 illustrates one of these muscles.

> *Action:* They pull the ribs closer to each other, causing them to be raised. Raising the ribs increases the volume of the thorax to cause inspiration of air in breathing.

Intercostalis Interni *(Internal Intercostals)* These antagonists of the external intercostals lie on the internal surface of the rib cage. The complete rib cage is lined with these muscles.

Each muscle *arises* from a ridge on the inner surface of a rib, as well as its corresponding costal cartilage, and *inserts* on the upper border of the rib below. The fibers of these muscles are oriented in a direction that is opposite to the fibers of the external intercostals.

> *Action:* They draw adjacent ribs closer together, which has the effect of lowering the ribs and decreasing the volume of the thoracic cavity. Result: expiration of air from the lungs.

Subscapularis This large triangular muscle fills the subscapular fossa of the scapula. Some fibers *originate* on the vertebral (medial) margin; others are anchored to the lower two-thirds of the axillary margin of the scapula. All fibers pass laterally to *insert* on the lesser tubercle of the humerus and the anterior portion of the joint capsule.

> *Action:* Rotates the arm medially; assists in other directional movements of the arm, depending on position of the arm.

Assignment:
Label figure 27.1.

Posterior Muscles

The muscles of the upper back region that are revealed when the skin is removed are shown in illustration A, figure 27.2. The deeper muscles of the back and shoulder are shown in illustrations B, C, and D of figure 27.2.

Latissimus Dorsi This large muscle of the back covers the lumbar area. It takes its *origin* in a broad aponeurosis that is attached to thoracic and lumbar vertebrae, the spine of the sacrum, iliac crest, and the lower ribs. Its *insertion* is on the intertubercular groove of the humerus.

> *Action:* Extends, adducts, and rotates the arm medially; draws the shoulder downward and backward.

Infraspinatus This muscle derives its name from its position. It is attached to the inferior margin of the scapular spine. Although a portion of it can be seen in illustration A, its complete structure cannot be seen unless the deltoideus and trapezius are removed as in illustration C. Its *origin* occupies the infraspinous fossa. Its *insertion* is the middle facet of the greater tubercle of the humerus.

> *Action:* Rotates the humerus laterally.

Teres Major Of the three muscles that cover most of the scapula in illustration C, the teres major is the most inferior one. Note that, although the latissimus dorsi obscures part of it, a portion of it is visible in illustration A. It *originates* near the inferior angle of the scapula and *inserts* into the crest of the lesser tubercle of the humerus.

> *Action:* Rotates the humerus medially and weakly adducts it.

Teres Minor This small muscle lies between the infraspinatus and teres major. Only a small portion of it is visible in illustration A. It takes its *origin* from the lateral margin of the scapula, and its *insertion* is on the lowest facet of the greater tubercle of the humerus.

> *Action:* Rotates the arm laterally and weakly adducts it.

Levator Scapulae This muscle lies under the trapezius in the neck region. It takes its *origin* on the transverse processes of the first four cervical vertebrae. Its *insertion* is on the upper portion of the vertebral border of the scapula.

> *Action:* Raises the scapula and draws it medially. With the scapula in a fixed position, it can bend the neck laterally.

Supraspinatus This muscle is completely covered by the trapezius and deltoideus; thus, it cannot be seen in illustration A. It *arises* from the supraspinous fossa of the scapula and *inserts* on the greater tubercle of the humerus.

> *Action:* This muscle assists the middle portion of the deltoid to abduct the humerus. It initiates the first 5-10 degrees of movement. Without this assist the deltoid has difficulty moving the humerus.

Rhomboideus Major and Minor These two muscles lie beneath the trapezius. They are flat muscles that extend from the vertebral border of the scapula to the spine (see illustration C).

The rhomboideus minor is the smaller one. It occupies a position between the levator scapulae and the rhomboideus major. It *arises* from the lower part of the ligamentum nuchae and the first thoracic vertebra. It *inserts* on that part of the scapular vertebral margin where the scapular spine originates.

The rhomboideus major *arises* from the spinous processes of the second, third, fourth, and fifth thoracic vertebrae. It *inserts* below the rhomboideus minor on the vertebral border of the scapula.

Action: These two muscles act together to pull the scapula medially and slightly upward. The lower part of the major rotates the scapula to depress the lateral angle, assisting in adduction of the arm.

Sacrospinalis *(Erector Spinae)* This long muscle extends over the back from the sacral region to the midshoulder region (see illustration B, figure 27.2). It consists of three portions: a lateral **iliocostalis,** an intermediate **longissimus,** and a medial **spinalis.** It *arises* from the lower and posterior portion of the sacrum, the iliac crest, and the lower two thoracic vertebrae. *Insertion* of the muscle is on the ribs and transverse processes of the vertebrae.

Action: This muscle is an extensor, pulling backward on the ribs and vertebrae to maintain erectness.

Assignment:
Label figure 27.2.

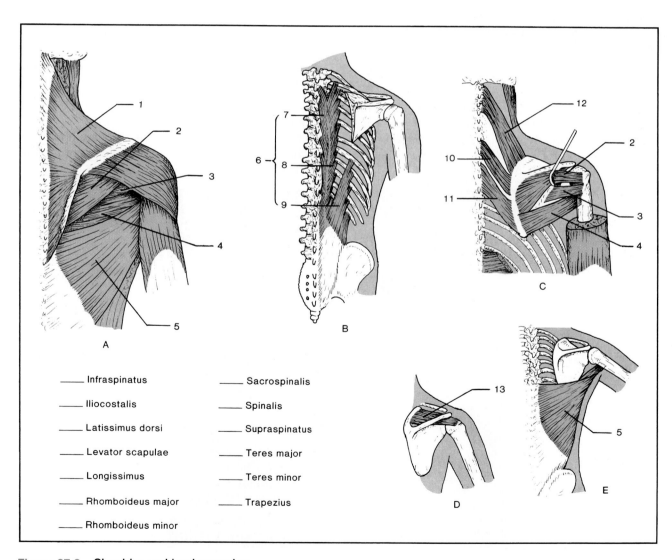

_____ Infraspinatus

_____ Iliocostalis

_____ Latissimus dorsi

_____ Levator scapulae

_____ Longissimus

_____ Rhomboideus major

_____ Rhomboideus minor

_____ Sacrospinalis

_____ Spinalis

_____ Supraspinatus

_____ Teres major

_____ Teres minor

_____ Trapezius

Figure 27.2   Shoulder and back muscles.

Cat Dissection
Thorax and Shoulder Muscles (Superficial)

Cat muscles of the thorax fall into three categories: the pectoralis, trapezius, and deltoideus groups. Use figures 27.3, 27.4, and 27.5 for reference.

The Pectoralis Group

Pectoantebrachialis This narrow ribbonlike muscle is about 1 cm wide and extends from the upper sternum to the forearm fascia. Since there is no homolog in humans, *transect* and remove this muscle.

Pectoralis Major This muscle is about 5 cm wide and lies deep to the pectoantebrachialis. With a blunt probe separate it from the clavodeltoid and pectoralis minor. *Dissect.*

Pectoralis Minor Note that this muscle is larger than the pectoralis major. It arises from the sternum, passes diagonally craniad and laterad beneath the pectoralis major to insert on both the scapula and humerus. Note that in humans, this muscle inserts only on the scapula. *Transect* and *reflect.*

Xiphihumeralis The most caudal muscle of the pectoral group, its fibers run parallel to the pectoralis minor and eventually pass deep to the pectoralis minor to insert on the humerus. There is no human homolog. Remove this muscle.

The Trapezius Group

The trapezius muscle in humans is represented by three separate muscles in the cat. Use figures 27.4 and 27.5 to identify them on your specimen.

Clavotrapezius The most cranial trapezius muscle on the cat; covers the dorsal surface of the neck. It inserts on the clavicle and unites with the clavodeltoid to form a single muscle, the **brachiocephalicus.** This muscle is homologous to the portion of the human trapezius that inserts on the clavicle.

Acromiotrapezius This muscle lies caudad to the clavotrapezius. It is comparable to the part of the human trapezius that inserts on the acromion. *Dissect, transect* by cutting through the tendon of origin on the mid-dorsal line, and *reflect.*

Spinotrapezius This muscle is caudad to the acromiotrapezius and superficial to the cranial border of the latissimus dorsi. That part of the human trapezius that inserts on the spine of the scapula is comparable to this muscle. *Dissect, transect,* and *reflect.*

The Deltoid Group

The cat has three deltoid muscles that are homologs of the single human deltoid. Locate each of the following muscles.

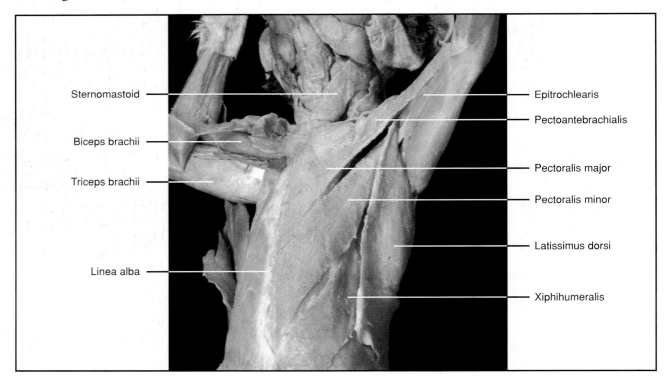

Figure 27.3 Ventrolateral aspect of superficial thoracic muscles of the cat.

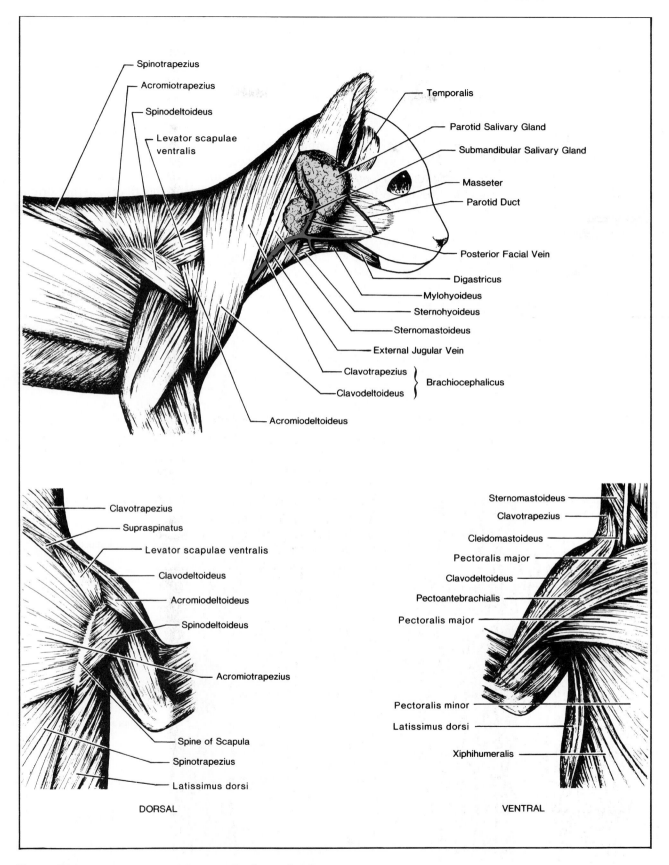

Figure 27.4 Trunk and shoulder muscles (superficial).

Clavodeltoideus *(Clavobrachialis)* The position of this muscle is best seen on the ventral view, figure 27.4, and on figure 27.5. This muscle is an extension of the clavotrapezius. It originates on the clavicle and inserts, primarily, on the proximal end of the ulna; some of its fibers, however, merge with the pectoantebrachialis to insert on the fascia of the forearm. *Dissect.*

Acromiodeltoideus This muscle is comparable to the acromial portion of the human deltoid. *Dissect.* Although freeing up this muscle is difficult, at least the cranial and caudal borders need to be clearly separated from adjacent muscles.

Spinodeltoideus This thick muscle lies caudad to the acromiodeltoid and arises from the spine of the scapula. Use the dorsal view, figure 27.4, and figure 27.5 for reference. *Dissect, transect,* and *reflect.*

Other Superficial Thoracic Muscles
In addition to the above three groups there are two more superficial muscles of importance:

Levator Scapulae Ventralis The location of this muscle is shown in the lateral and dorsal views, figure 27.4. Note that this muscle is caudad to the clavotrapezius and passes beneath the clavotrape-zius. It joins the craniolateral edge of the acromio-trapezius and inserts on the acromial part of the scapular spine. It arises from the transverse process of the atlas and the occipital bone. Since it has no human homolog, identify it and move on to the next muscle.

Latissimus Dorsi This broad, flat, triangular muscle has an extensive origin on the spines of the last six thoracic vertebrae, the lumbar vertebrae, the sacrum, and the iliac crest. The lumbodorsal fascia, an aponeurosis, contains the fibers of this origin. The muscle inserts on the proximal portion of the humerus. *Dissect, transect,* and *reflect.*

Deep Thoracic Muscles
Examine the following deeper muscles of the thorax using figures 27.6, 27.7, and 27.8 for reference.

Serratus Ventralis Pull the scapula away from the body wall and rotate it dorsad. Clean off the fascia from the body wall. Note the serrate margin of this muscle due to the origin by digitations from the upper eight or nine ribs.

The cranial portion of this muscle is homologous to the human *levator scapulae,* and the caudal portion is homologous to the *serratus anterior.* *Dissect.*

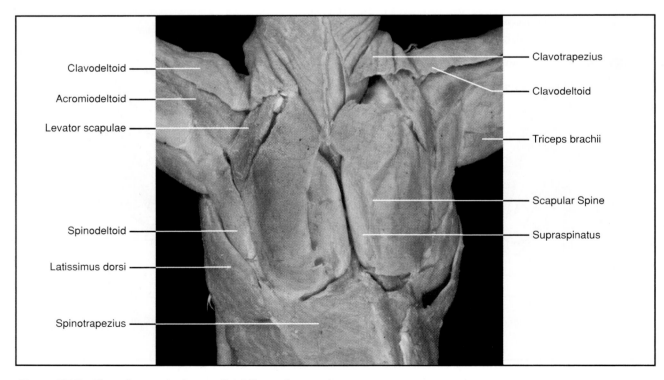

Figure 27.5 Dorsal aspect of superficial thoracic muscles.

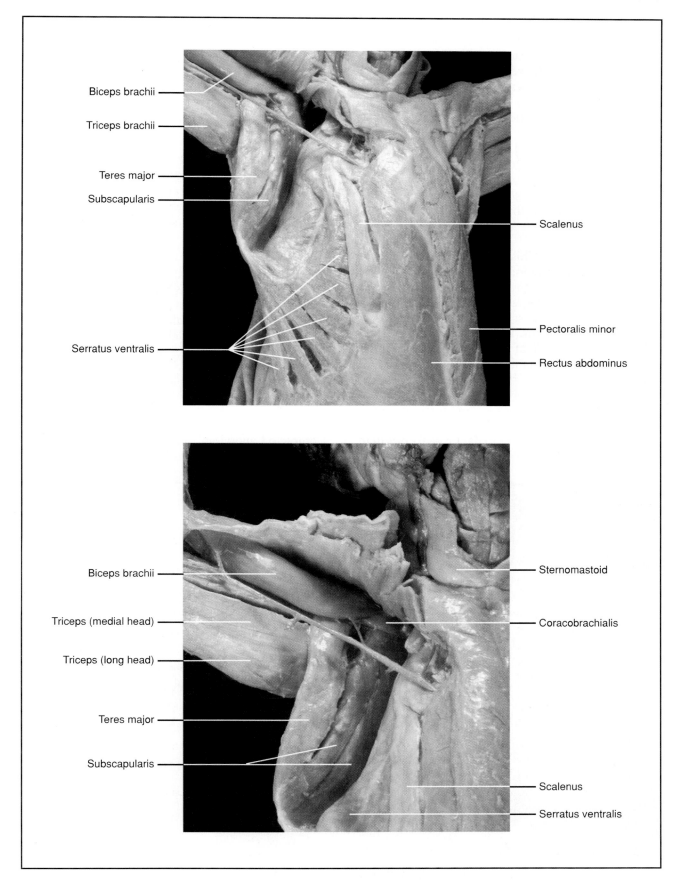

Biceps brachii

Triceps brachii

Teres major

Subscapularis

Scalenus

Serratus ventralis

Pectoralis minor

Rectus abdominus

Biceps brachii

Triceps (medial head)

Triceps (long head)

Teres major

Subscapularis

Sternomastoid

Coracobrachialis

Scalenus

Serratus ventralis

Figure 27.6   Deep muscles of cat thorax and shoulder, ventral aspect.

Rhomboideus This muscle (figure 27.7) lies deep to the acromiotrapezius and spinotrapezius, which should have been transected previously. Pull the two forelegs together on the ventral side. This will abduct the scapulae so that you can see the full extent of the rhomboids.

The rhomboideus extends from the spines of the upper thoracic vertebrae to the vertebral border of the scapula. The cranial (and larger) part of this muscle is homologous to the *rhomboideus minor* muscle in humans, and the caudal (smaller) part is homologous to the *rhomboideus major*. The rhomboideus major is larger than the rhomboideus minor in humans.

Levator Scapulae This is not a separate muscle but a cranial continuation of the serratus ventralis. Its origin is from the transverse processes of the caudal five cervical vertebrae, and it inserts on the vertebral border of the scapula ventral to the insertion of the rhomboideus. In the human, the insertion is comparable to that of the cat, but the origin is roughly comparable to the origin of the cat's levator scapulae ventralis.

Supraspinatus With the acromiotrapezius fully reflected, this muscle can be seen filling the supraspinous fossa of the scapula. From its origin in the supraspinous fossa it passes to the greater tubercle of the humerus.

Infraspinatus Reflect the transected spinodeltoid. The infraspinatus can now be observed extending from its origin in the infraspinous fossa of the scapula to the greater tubercle of the humerus.

Teres Major This thick band of muscle is located caudad to the infraspinatus muscle and covers the axillary border of the scapula. It extends forward to insert on the proximal end of the humerus. *Dissect.* Reflect the latissimus dorsi to expose the full extent of this muscle.

Teres Minor This small muscle is positioned between the infraspinatus and the long head of the biceps brachii. Since it is not labeled in any of the illustrations in this manual, you will have to look for it by reflecting the spinodeltoid. The muscle extends from the axillary border of the scapula to the humerus.

Laboratory Report

Complete Laboratory Report 27 for this exercise.

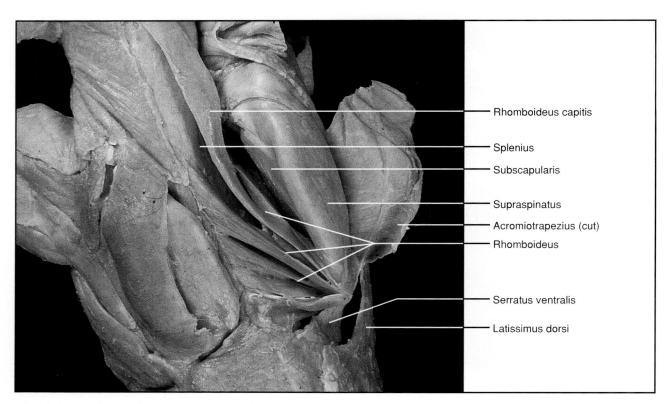

Rhomboideus capitis

Splenius

Subscapularis

Supraspinatus

Acromiotrapezius (cut)

Rhomboideus

Serratus ventralis

Latissimus dorsi

Figure 27.7 Deep muscles of cat's shoulder and back, dorsal aspect.

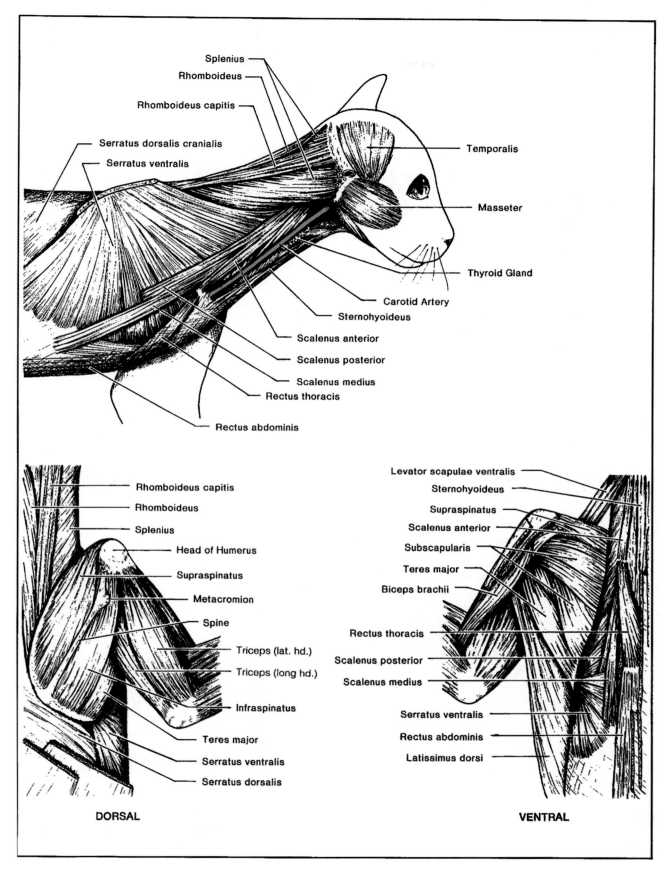

Figure 27.8 Deep trunk, shoulder, and neck muscles.

28

Upper Extremity Muscles

Muscles that move the upper arm are primarily muscles of the shoulder and trunk, which were studied in the last exercise. In this exercise the muscles of the brachium that control movements of the forearm will be studied. One muscle, however, that was not mentioned in the last exercise that does move the upper arm is the *coracobrachialis*. It is a muscle of the upper arm and is illustrated in figure 28.1.

Coracobrachialis This muscle covers a portion of the upper medial surface of the humerus. It takes its *origin* on the apex of the coracoid process of the scapula. Its *insertion* is in the middle of the medial surface of the humerus.

> *Action:* Carries the arm forward in flexion and adducts the arm.

Forearm Movements

The principal movers of the forearm are the *biceps brachii, brachialis, brachioradialis,* and *triceps brachii*. These four muscles are also illustrated in figure 28.1. Note that incorporated into all the names of these upper arm muscles is the Latin term *brachium,* for the upper arm.

Biceps Brachii This is the large muscle on the anterior surface of the upper arm that bulges when the forearm is flexed. Its *origin* consists of two tendinous heads: a medial tendon, which is attached to the coracoid process, and a lateral tendon, which fits into the intertubercular groove on the humerus. The latter tendon is attached to the supraglenoid tubercle of the scapula. At the lower end of the muscle the two heads unite to form a single tendinous *insertion* on the radial tuberosity of the radius.

> *Action:* Flexion of the forearm; also, moves the radius outward to supinate the hand.

Brachialis Immediately under the biceps brachii on the distal anterior portion of the humerus lies the brachialis. Its origin occupies the lower half of the humerus. Its insertion is attached to the front surface of the coronoid process of the ulna.

> *Action:* Flexion of the forearm.

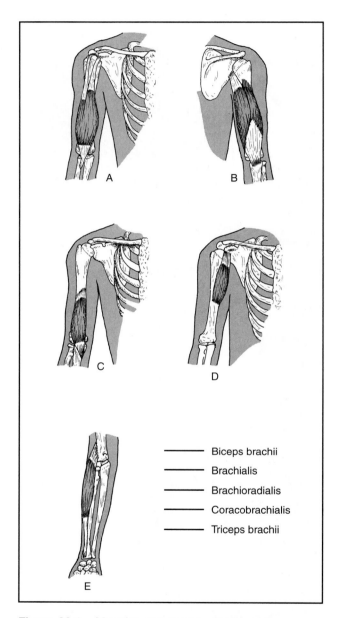

Figure 28.1   Muscles attached to the brachium.

Biceps brachii
Brachialis
Brachioradialis
Coracobrachialis
Triceps brachii

Brachioradialis This muscle is the most superficial muscle on the lateral (radial) side of the forearm. It originates above the lateral epicondyle of the humerus and inserts on the lateral surface of the radius slightly above the styloid process.

> *Action:* Flexion of the forearm.

Triceps Brachii The entire back surface of the brachium is covered by this muscle. It has three heads of *origin:* a long head arises from the scapula, a lateral head from the posterior surface of the humerus, and a medial head from the surface below the radial groove. The tendinous *insertion* of the muscle is attached to the olecranon process of the ulna.

Action: Extension of the forearm; antagonist of the brachialis.

Assignment:

Identify the muscles in figure 28.1 by placing the correct letter in front of each name in the legend.

Hand Movements

Muscles of the arm that cause hand movements are illustrated in figures 28.2 and 28.3. All illustrations are of the right arm; thus, if the thumb points to the left side of the page the anterior aspect is being viewed; when pointing to the right, the posterior view is observed. Keeping this in mind will facilitate understanding the descriptions that follow.

Supination and Pronation

Two muscles can cause supination: the biceps brachii and the supinator.

The **supinator** is a short muscle near the elbow that *arises* from the lateral epicondyle of the humerus and the ridge of the ulna. It curves around the upper portion of the radius and *inserts* on the lateral edge of the radial tuberosity and the oblique line of the radius.

Pronation is achieved by two pronator muscles shown in illustration B, figure 28.2. The upper muscle is the **pronator teres,** which *arises* on the medial epicondyle of the humerus and inserts on the upper lateral surface of the radius. The lower muscle, which is located in the wrist region, is the **pronator quadratus.** It *originates* on the distal portion of the ulna and *inserts* on the distal lateral portion of the radius.

Flexion of the Hand

Two muscles that flex the hand are shown in illustration C, figure 28.2. The muscle that extends diagonally across the forearm is the **flexor carpi radialis.** It *arises* on the medial epicondyle of the humerus and *inserts* on the proximal portions of the second and third metacarpals. The other flexor of the hand is the **flexor carpi ulnaris.** Its *origin* is on the medial epicondyle of the humerus and the posterior surface (olecranon process) of the ulna. Its *insertion* consists of a tendon that attaches to the base of the fifth metacarpal.

Action: Both muscles flex the hand; the radialis muscle causes abduction and the ulnaris muscle causes adduction.

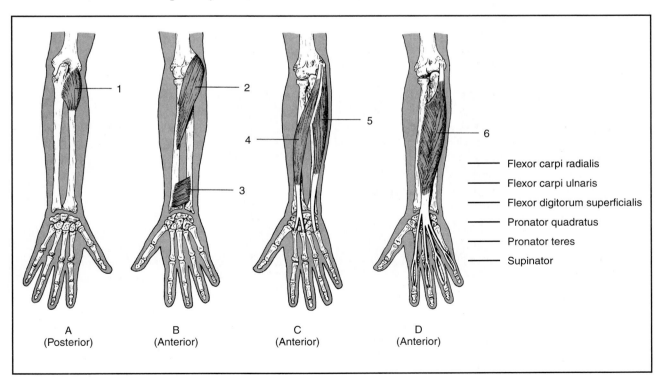

A
(Posterior)

B
(Anterior)

C
(Anterior)

D
(Anterior)

——— Flexor carpi radialis

——— Flexor carpi ulnaris

——— Flexor digitorum superficialis

——— Pronator quadratus

——— Pronator teres

——— Supinator

Figure 28.2 Forearm muscles (right arm shown).

Flexion of Fingers of the Hand

Three muscles flex the fingers. The **flexor digitorum superficialis** in illustration D, figure 28.2 flexes all fingers except the thumb. It *arises* on the humerus, ulna, and radius. Its *insertion* consists of tendons that are attached to the middle phalanges of the second, third, fourth, and fifth fingers.

The second flexor of the fingers is the **flexor digitorum profundus.** It is the larger muscle in illustration A, figure 28.3. It lies directly under the flexor digitorum superficialis. It *originates* on the ulna and the interosseous membrane between the radius and ulna. It *inserts* with four tendons on the distal phalanges of the second, third, fourth, and fifth fingers. This muscle flexes the distal portions of the fingers.

The third flexor is the **flexor pollicis longus** (Latin: *pollex,* thumb). It is the smaller muscle shown in illustration A, figure 28.3. Note that it *arises* on the radius, ulna, and interosseous membrane between these two bones. Its *insertion* consists of a tendon that is anchored to the distal phalanx of the thumb. It flexes only the thumb.

Extension of Wrist and Hand

Three muscles extend the hand at the wrist. Illustrations B and D, figure 28.3, reveal them.

The **extensor carpi radialis longus** and **brevis** are shown in illustration B. The brevis muscle is medial to the longus. Both muscles *originate* on the humerus, with the longus taking a more proximal position. The longus *inserts* on the second metacarpal; the brevis on the middle metacarpal.

The **extensor carpi ulnaris** is the third muscle involved in extending the hand. It is the muscle on the medial edge of the arm in illustration D. It *arises* on the lateral epicondyle of the humerus and part of the ulna. It *inserts* on the fifth metacarpal.

Extension and Abduction of Fingers

Three muscles cause extension of the fingers; one abducts the thumb. The **extensor digitorum communis** lies alongside the extensor carpi ulnaris; it is shown in illustration D, figure 28.3. It *arises* from the lateral epicondyle of the humerus and inserts on the distal phalanges of fingers two through five. It extends all fingers except the thumb.

The **extensor pollicis longus, extensor pollicis brevis,** and **abductor pollicis** move the thumb. The longus muscle is the lower one in illustration C. It *arises* on both the ulna and radius. It *inserts* on the distal phalanx of the thumb. The brevis muscle, which lies superior to it, is not shown in illustration C. It inserts on the proximal phalanx of the thumb and assists the longus in extending the thumb.

The abductor pollicis is the superior muscle shown in illustration C. It takes its *origin* on the interosseous membrane and it *inserts* on the lateral portion of the first metacarpal and trapezium. Its only action is to abduct the thumb.

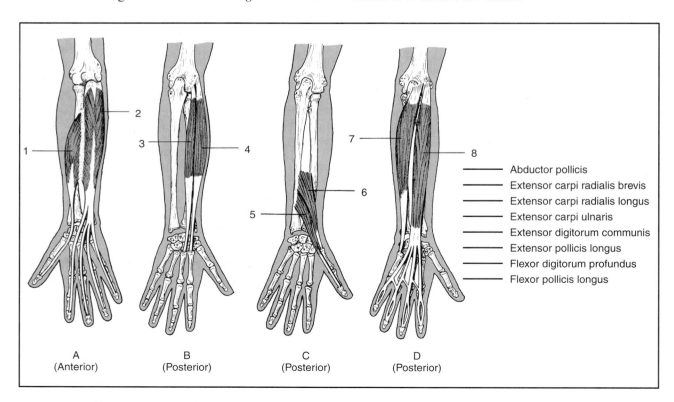

Abductor pollicis
Extensor carpi radialis brevis
Extensor carpi radialis longus
Extensor carpi ulnaris
Extensor digitorum communis
Extensor pollicis longus
Flexor digitorum profundus
Flexor pollicis longus

A (Anterior)   B (Posterior)   C (Posterior)   D (Posterior)

Figure 28.3 Forearm muscles (right arm shown).

Assignment:
Label figures 28.2 and 28.3, and complete the Laboratory Report for this exercise.

Cat Dissection

Forelimb

Muscles of the Brachium

Six homologs of the human brachium are seen in the cat. Except for the anconeus, all the human muscles were identified on pages 142 and 143.

One muscle of the cat's brachium, the *epitrochlearis,* has no homolog in the human and should be removed before studying the six homologs. It is a flat, thin, superficial muscle on the medial side of the brachium and can be seen in figure 28.5. It arises from the latissimus dorsi and is continuous, distally, with the fascia of the forearm. After removing this muscle, proceed to identify and free up with a probe the six homologs in the following order.

Triceps Brachii   This muscle has three heads of origin, which merge to insert by a common tendon on the olecranon process of the ulna. The long head, located on the caudal surface of the arm, is the largest. The lateral head covers much of the lateral surface of the arm. *Dissect.* To reveal the medial head, which is between the long and lateral heads, *transect* and *reflect* the lateral head (see figure 28.6). Only the long head arises from the scapula.

Biceps Brachii   In the cat this muscle lies deep in the pectoral muscles, just medial to their humoral insertions. This muscle is, therefore, located on the

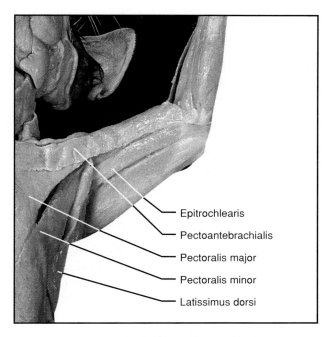

Figure 28.5   Superficial forelimb muscles of the cat.

ventromedial surface of the arm (see figure 28.4). It inserts on the radius. *Dissect.*

Brachialis   This muscle is located on the ventrolateral surface of the arm, craniad to the lateral head of the triceps brachii. Refer to figure 28.6.

The insertion of the pectoralis major separates the biceps brachii and brachialis. Since this muscle arises from the lateral surface of the humerus, you will not be able to run a probe underneath it from its origin to its insertion on the ulna.

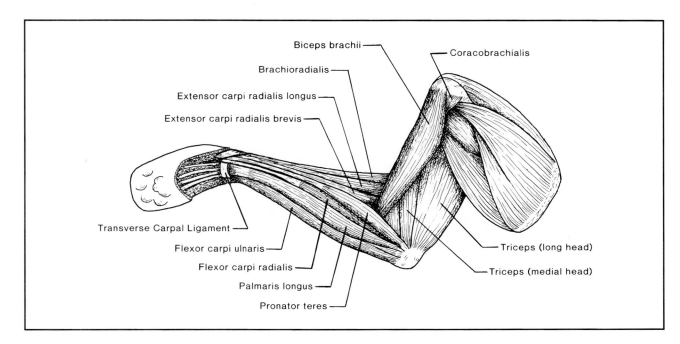

Figure 28.4   Medial aspect of cat forelimb muscles.

Brachioradialis In the cat this muscle (figures 28.4 and 28.8) is a narrow ribbon, which may have been removed with the superficial fascia. It arises on the lateral side of the humerus and passes along the radial side of the forearm to insert on the styloid process of the radius. *Dissect.*

Coracobrachialis Use figures 28.4 and 28.7 to locate this small muscle. Note that it is on the medial side of the scapula caudad of the proximal (origin) end of the biceps brachii. This muscle is almost impossible to locate if the pectoralis major and minor muscles have not been previously transected and reflected.

Anconeus This small muscle is shown in figure 28.6. It consists of short superficial fibers that originate near the lateral epicondyle of the humerus, and it inserts near the olecranon. It assists the triceps in both the cat and human to extend the forearm.

Muscles of the Antebrachium
Eight muscles of the cat's forearm will be identified. As on the brachium homologs, free up each muscle with a probe. They are as follows:

Pronator Teres This muscle extends from the medial epicondyle of the humerus to the middle third of the radius. See figure 28.4. *Dissect.*

Palmaris Longus Note in figure 28.4 that this muscle is a large, flat, broad muscle near the center of the medial surface of the forearm. It arises from the medial epicondyle of the humerus and inserts by

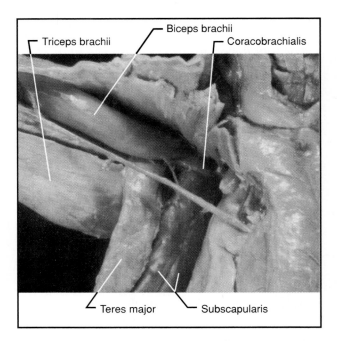

Figure 28.7 **Brachium and shoulder muscles of cat.**

means of four tendons on the digits. Note that at its distal end the four insertion tendons pass under a *transverse carpal ligament*. The palmaris longus is not found in about 10% of humans. *Dissect.*

Flexor Carpi Radialis This long, spindle-shaped muscle lies between the palmaris longus and the pronator teres. It arises from the medial epicondyle of the humerus and inserts at the base of the second and third metacarpals.

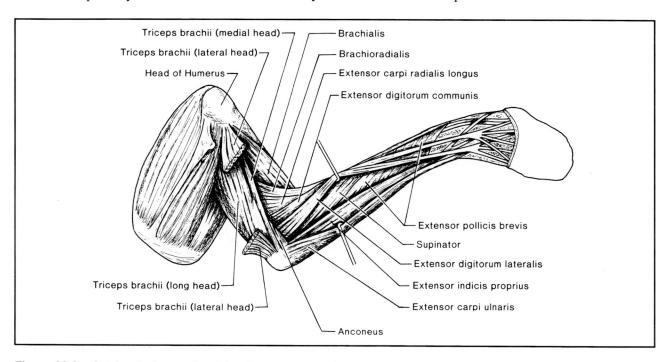

Figure 28.6 **Cat forelimb muscles, lateral aspect.**

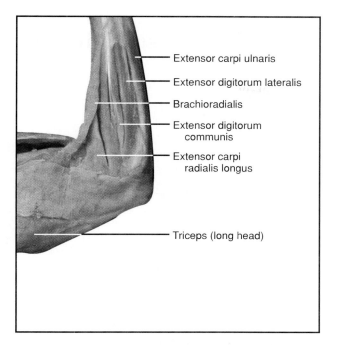

Figure 28.8 Superficial muscles of cat's forearm, lateral aspect.

Flexor Carpi Ulnaris Figure 28.4 shows the location of this muscle. What appears to be two muscles is actually two heads of origin: one originates from the medial epicondyle of the humerus and the other from the olecranon. About the middle of the ulna, the two heads join to form a single band of muscle fibers that inserts on the pisiform and hamate bones. *Dissect.*

Extensor Carpi Ulnaris This muscle, shown in figures 28.6 and 28.8, arises from the lateral epicondyle of the humerus and inserts on the base of the fifth metacarpal. *Dissect.*

Extensor Digitorum Communis Use figure 28.6 to locate this muscle. It arises from the lateral epicondyle of the humerus and extends to the wrist, where it divides into four tendons that insert on the second phalanges of digits 2–5. Note that these four tendons lie superficial to the three tendons of the extensor digitorum lateralis. *Dissect.*

Extensor Carpi Radialis Longus Use figure 28.8 to identify this muscle, which lies adjacent to the brachioradialis. Its origin is on the lateral supracondylar ridge of the humerus, and it inserts at the base of the second metacarpal. *Dissect.*

Supinator This muscle can be exposed by reflecting the extensor digitorum lateralis as shown in figure 28.6. This muscle originates from ligaments on the elbow. It inserts on the proximal end of the radius and acts as a lateral rotator of the radius (supination).

Complete exposure of the supinator would require transection and reflection of the overlying muscles. To avoid irreversible mutilation of your cat, do not do this.

Laboratory Report

Complete Laboratory Report 28 for this exercise.

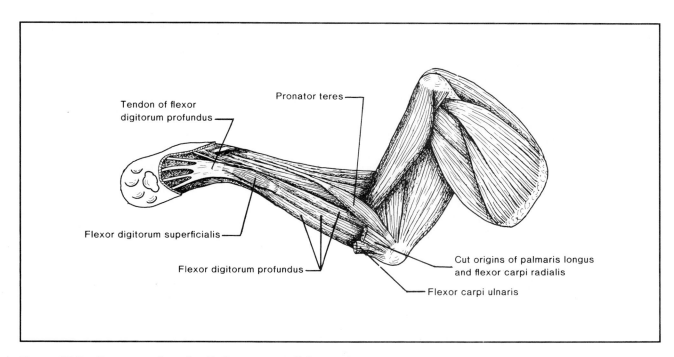

Figure 28.9 Deep muscles of cat's forearm, medial aspect.

29 Abdominal, Pelvic, and Intercostal Muscles

The muscles of the abdominal wall and pelvis will be studied in this exercise. Cat dissection will be utilized for this muscle study.

Abdominal Muscles

The abdominal wall consists of four pairs of thin muscles. The human illustration at the bottom of figure 29.1 has portions of the right abdominal wall removed to reveal the nature of the laminations. Identify the following muscles in this illustration.

Obliquus Externus *(External Oblique)* This muscle is the most superficial layer of the abdominal wall. It is ensheathed by the **aponeurosis of the obliquus externus,** which terminates at the linea alba and inguinal ligament.

    The muscle takes its *origin* on the external surfaces of the lower eight ribs. Although it appears to insert on the linea semilunaris, its actual *insertion* is the **linea alba** (white line), where fibers of the left and right aponeuroses interlace on the midline of the abdomen.

    The lower border of each aponeurosis forms the **inguinal ligament,** which extends from the anterior spine of the ilium to the pubic tubercle. Label 5, figure 29.3 is of this ligament.

Obliquus Internus *(Internal Oblique)* This muscle lies immediately under the external oblique, i.e., between the external oblique and the transversus abdominis.

    Its *origin* is on the lateral half of the inguinal ligament, the anterior two-thirds of the iliac crest, and the thoracolumbar fascia. Its *insertion* is on the costal cartilages of the lower three ribs, the linea alba, and the crest of the pubis.

Transversus Abdominis This is the innermost muscle of the abdominal wall. It *arises* on the inguinal ligament, the iliac crest, the costal cartilages of the lower six ribs, and the thoracolumbar fascia. It *inserts* on the linea alba and the crest of the pubis.

Rectus Abdominis The right rectus abdominis is the long, narrow, segmented muscle running from the rib cage to the pubic bone. It is enclosed in a

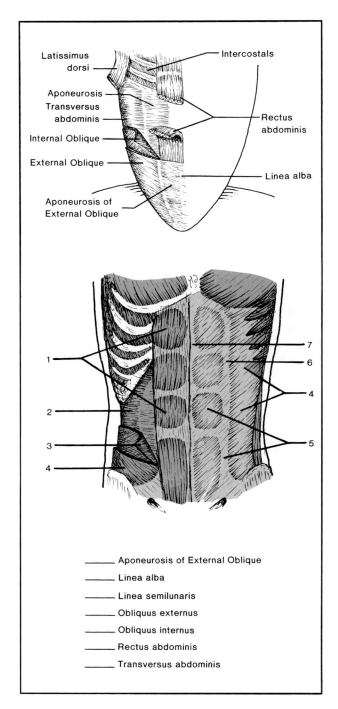

Figure 29.1   Abdominal muscles, cat and human.

_____ Aponeurosis of External Oblique

_____ Linea alba

_____ Linea semilunaris

_____ Obliquus externus

_____ Obliquus internus

_____ Rectus abdominis

_____ Transversus abdominis

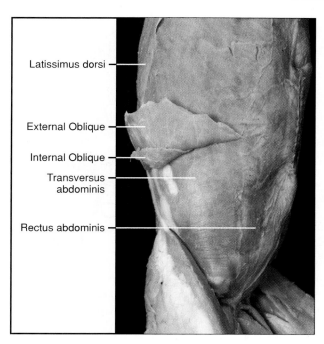

Figure 29.2 Abdominal muscles of the cat.

fibrous sheath formed by the aponeuroses of the above three muscles.

Its *origin* is on the pubic bone. Its *insertion* is on the cartilages of the fifth, sixth, and seventh ribs. The linea alba lies between the pair of muscles. Contraction of the rectus abdominis muscles aids in flexion of the spine in the lumbar region.

Collective Action

The above four abdominal muscles keep the abdominal organs compressed and assist in maintaining intraabdominal pressure.

They act as antagonists to the diaphragm. When the latter contracts, they relax. When the diaphragm relaxes, they contract to effect expiration of air from the lungs. They also assist in defecation, urination, vomiting, and parturition (childbirth delivery). Flexion of the body at the lumbar region is also achieved by these muscles.

Assignment:
Label figure 29.1.

Cat Dissection
Abdominal Muscles

As shown in the upper illustration of figure 29.1 and figure 29.2, the abdominal wall of the cat has the same four muscles as seen in humans. As you separate each muscle layer, pay particular attention to the direction of the fibers. Body wall strength is greatly enhanced by the different directions of the fibers in each layer.

External Oblique This outer muscle of the body wall arises from the external surface of the lower nine ribs. Note how the cranial portion of this muscle interdigitates with the origin of the serratus ventralis. Insertion is on the iliac crest and by an aponeurosis that fuses with that of the opposite side to form the *linea alba.*

Loosen the cranial border of the muscle with a blunt probe and slide the probe underneath the muscle in a caudal direction. *Transect* alongside of the probe and *reflect.* (Remember, when you transect a muscle, your cut should be through the midbelly half way between the origin and insertion.)

Internal Oblique To expose this muscle it will be necessary to make a shallow incision on the right side through the external oblique from the ribs to the pelvis. *Reflect* the external oblique. Since the external oblique is very thin, take care not to cut too deep. Note that the fibers of this muscle layer course at approximately right angles to those of the external oblique. Note, also, how the ends of the muscle fibers are continuous with the aponeurosis.

Transversus Abdominis This innermost muscle layer of the ventrolateral abdominal wall is very difficult to separate from the internal oblique; therefore, only a small "window" should be cut to expose the transversus abdominis.

Try to slide the tip of a blunt probe either underneath a small area of the aponeurosis or under the muscle fibers of the internal oblique. Make a cut to create a small opening and note the general transverse direction of the fibers of this muscle.

Rectus Abdominis This muscle extends from the pubis to the sternum on each side of the linea alba. It extends farther craniad in the cat than in the human. The cranial two-thirds of the aponeurosis of this muscle splits at the lateral border of the rectus abdominis, creating a dorsal leaf that passes dorsal, and a ventral leaf that passes ventral to the rectus abdominis. The dorsal and ventral leaves unite at the medial border of the rectus abdominis to help form the linea alba. Because the rectus abdominis is "wrapped" in the aponeurosis in this manner, it is difficult to pass a probe under this muscle from its origin to the insertion.

Intercostal Muscles
Although the human intercostal muscles were studied in Exercise 27, page 134, the same muscles in the cat were not studied at that time. Using figure 29.1, examine them on your specimen.

These muscles of respiration in the cat are essentially the same as in humans except that there are twelve sets in the cat and eleven sets in humans. Proceed to dissect these muscles as follows:

External Intercostals Reflect the external oblique to expose the lateral aspect of the rib cage. The external intercostal muscles can be seen occupying the intercostal spaces. Note that the direction of the fibers is forward and downward from the caudal border of one rib to the cranial border of the next rib caudad. Note, also, that the external intercostals do not reach the sternum and that the internal intercostals can be seen in the interval.

Internal Intercostals These muscles are immediately deep to the external intercostals. Cut a small window through an external intercostal between two ribs to expose one of these muscles. Note that the fibers course at approximately right angles to the fibers of the external muscle as they pass from the cranial border of the rib to the caudal border of the next rib craniad.

Pelvic Muscles

The principal muscles of the pelvic region are shown in figure 29.3. They are the quadratus lumborum, psoas major, and iliacus. Muscles in this region of the cat will be dissected in the next exercise.

Quadratus Lumborum The muscle that extends from the iliac crest of the os coxa to the lowest rib in figure 29.3 is this muscle. Only the left muscle in figure 29.3 is completely exposed.

Its *origin* is on the iliac crest and the iliolumbar ligament that borders the iliac crest. Its *insertion* is on the lower border of the last rib and the apices of the transverse processes of the upper four lumbar vertebrae.

Occasionally, there is a second portion of this muscle positioned just in front of it. This part takes its *origin* on the upper borders of the transverse processes of the lower three or four lumbar vertebrae and *inserts* on the lower margin of the last rib. It does not extend down to the iliac crest.

Taken as a whole, when both portions of this muscle are present, the transverse processes of some of the lumbar vertebrae act as both origin and insertion—a truly unique situation.

Acting together, the right and left quadratus lumborums extend the spine at the lumbar verte-

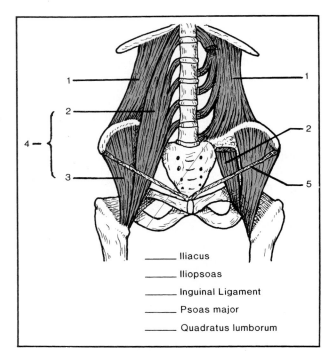

Figure 29.3 Pelvic muscles.

_____ Iliacus

_____ Iliopsoas

_____ Inguinal Ligament

_____ Psoas major

_____ Quadratus lumborum

brae. Lateral flexion or abduction results when one acts independently of the other.

Psoas Major This muscle is the long one shown in figure 29.3 on the right side of the body. A portion of the left psoas major has been cut away to reveal the full extent of the quadratus lumborum.

The psoas major *arises* from the sides of the bodies and transverse processes of the lumbar vertebrae. It *inserts* with the iliacus on the lesser trochanter of the femur.

The two psoas majors work synergistically with the rectus abdominis muscles to flex the lumbar region of the vertebral column.

Iliacus This muscle extends from the iliac crest to the proximal end of the femur. Its *origin* is the whole iliac fossa. Its *insertion* is the lesser trochanter of the femur. The psoas major and iliacus are jointly referred to as the *iliopsoas* because of their intimate relationship at their insertion.

The iliacus works synergistically with the psoas major to flex the femur on the trunk.

Laboratory Report

After labeling figure 29.3, answer the questions on combined Laboratory Report 29,30 that pertain to the muscles studied in this exercise.

Lower Extremity Muscles

Twenty-seven muscles of the leg will be studied in this exercise. As in the case of the upper extremity muscles, they are grouped according to type of movement.

Thigh Movements

Seven muscles that move the femur are shown in figure 30.1. All originate on a part of the pelvis.

Gluteus Maximus This muscle of the buttock region is covered by a deep fascia, the *fascia lata,* that completely invests the thigh muscles. Emerging downward from the fascia lata is a broad tendon, the **iliotibial tract,** which is attached to the tibia.

This muscle *arises* on the ilium, sacrum, and coccyx; it *inserts* on the iliotibial tract and the posterior part of the femur.

Action: Extension and outward rotation of the femur.

Gluteus Medius This muscle lies immediately under the gluteus maximus, covering a good portion of the ilium. Its *origin* is on the ilium and its *insertion* is on the lateral part of the greater trochanter.

Action: Abduction and medial rotation of the femur.

Gluteus Minimus This is the smallest of the three gluteal muscles and is located immediately under the gluteus medius. It, too, *arises* on the ilium; it *inserts* on the anterior border of the greater trochanter.

Action: Abduction, inward rotation, and slight flexing of the femur.

Piriformis This small muscle takes its *origin* on the anterior surface of the sacrum and *inserts* on the upper border of the greater trochanter.

Action: Outward rotation, some abduction, and extension of the femur.

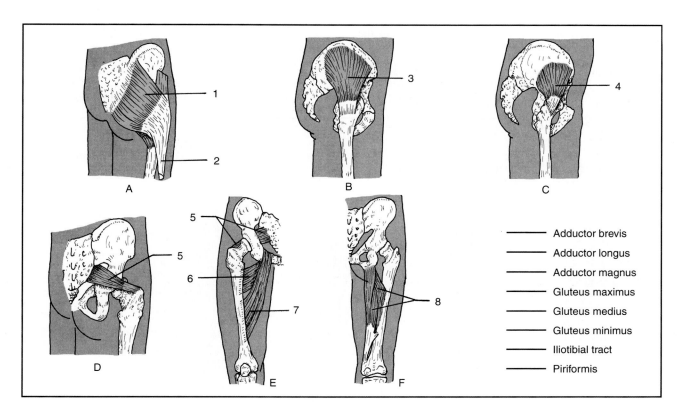

Figure 30.1 Muscles that move the femur.

Adductor Muscles   Three adductors are shown in illustrations E and F of figure 30.1. The **adductor magnus** (illustration F) is the largest one. It *arises* on the inferior portions of the pubis and ischium and it *inserts* on the linea aspera of the femur.

The **adductor longus** (illustration E) *originates* on the front of the pubis and *inserts* on the linea aspera. The **adductor brevis** (illustration E) *arises* on the posterior side of the pubis and *inserts* on the femur above the longus muscle.

Action: In addition to adduction, these muscles flex and rotate the femur medially.

Assignment:
Label figure 30.1.

Thigh and Lower Leg Movements

The muscles illustrated in figure 30.2 are primarily concerned with flexion and extension of the lower part of the leg.

Hamstrings   The three muscles shown in illustration A, figure 30.2, are collectively known as the "hamstrings." They are as follows:

Biceps femoris This muscle occupies the most lateral position of the three hamstrings. It has two heads: one long, and the other short. The long head, which obscures the short one, *arises* on the ischial tuberosity. The short head *originates* on the linea aspera of the femur. *Insertion* of the muscle is on the head of the fibula and the lateral condyle of the tibia.

Semitendinosus Medial to the biceps femoris is this muscle, which *arises* on the ischial tuberosity and *inserts* on the upper end of the shaft of the tibia.

Semimembranosus This muscle occupies the most medial position of the three hamstrings. Its origin consists of a thick semimembranous tendon attached to the ischial tuberosity. Its insertion is primarily on the posterior medial part of the medial condyle of the tibia.

Collective Action: All of these muscles flex the calf upon the thigh. They also extend and rotate the thigh. Rotation by the biceps is outward; the other two muscles cause inward rotation.

Quadriceps Femoris   The large muscle that covers the anterior portion of the thigh is this muscle. It consists of four parts, which are shown in illustrations C and D, figure 30.2. Note that all of them unite to form a common tendon that passes over the patella to *insert* on the tibia.

Rectus femoris This portion of the quadriceps occupies a superficial central position. It *arises* by two tendons: one from the anterior inferior iliac spine and the other from a groove just above the acetabulum. The lower portion of the muscle is a broad aponeurosis that terminates in the tendon of insertion.

Vastus lateralis This lateral portion of the quadriceps is the largest of the four muscles. It *arises* from the lateral lip of the linea aspera.

Vastus medialis The most medial portion of the quadriceps is this muscle. It *arises* from the linea aspera.

Vastus intermedius Since this muscle is deep to the above three, it is shown separately in illustration D. Its *origin* is on the front and lateral surfaces of the femur.

Collective Action: The entire quadriceps femoris extends (straightens) the leg. Flexion of the thigh, however, is achieved by the rectus femoris.

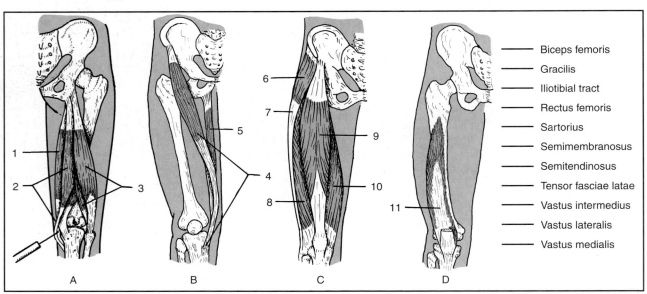

- Biceps femoris
- Gracilis
- Iliotibial tract
- Rectus femoris
- Sartorius
- Semimembranosus
- Semitendinosus
- Tensor fasciae latae
- Vastus intermedius
- Vastus lateralis
- Vastus medialis

Figure 30.2   Muscles that move the tibia and fibula.

Sartorius This is the longest muscle shown in illustration B. It arises from the anterior superior spine of the ilium and inserts on the medial surface of the tibia.

Action: Flexes the calf on the thigh, the thigh upon the pelvis, and rotates the leg laterally.

Tensor Fasciae Latae This muscle is shown in illustration C, figure 30.2. It *arises* from the anterior outer lip of the iliac crest, the anterior superior spine, and from the deep surface of the fascia lata. It is *inserted* between the two layers of the iliotibial tract at about the junction of the middle and upper thirds of the thigh.

Action: Flexes the thigh and rotates it slightly medially.

Gracilis This muscle is located on the medial surface of the thigh. It *arises* from the lower margin of the pubic bone and *inserts* on the medial surface of the tibia near the insertion of the sartorius.

Action: Adducts, flexes, and rotates the thigh medially.

Assignments:
Label figure 30.2.

Lower Leg and Foot Movements

Figures 30.3 and 30.4 illustrate most of the muscles of the lower leg that cause flexion and foot movements.

Triceps Surae The large superficial muscle that covers the calf of the leg is the triceps surae. It consists of two parts that are shown in illustrations A and B, figure 30.3. They are united by a common **tendon of Achilles** that *inserts* on the calcaneous of the foot. **(1) Gastrocnemius:** This outer portion of the triceps surae has two heads that *arise* from the posterior surfaces of the medial and lateral condyles of the femur. It can flex the calf on the thigh as well as cause plantar flexion. **(2) Soleus:** This muscle *arises* on the heads of the fibula and tibia. Plantar flexion is its only action.

Tibialis Anterior The anterior lateral portion of the tibia is covered by this muscle. It *arises* from the lateral condyle and upper two-thirds of the tibia. Its distal end is shaped into a long tendon that passes over the tarsus and *inserts* on the inferior surface of the first cuneiform and first metatarsal bones.

Action: Dorsiflexion and inversion of the foot.

Tibialis Posterior This muscle lies on the posterior surfaces of the tibia and fibula. It *arises* from both these bones and the interosseous membrane. It *inserts* on the inferior surfaces of the navicular, cuneiforms, cuboid, and second through fourth metatarsals.

Action: Plantar flexion and inversion of the foot; assists in maintenance of the longitudinal and transverse arches of the foot.

The Peroneus Muscles (Latin: *peroneus,* fibula) Three peroneus muscles are shown in illustrations A and B, figure 30.4. Note that all three of them *originate* on the fibula. The **peroneus longus** is the

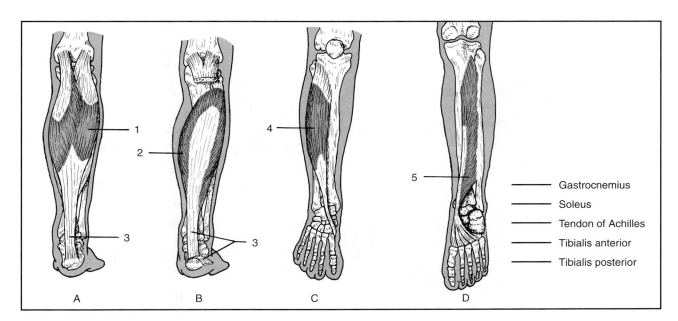

Figure 30.3 **Lower leg muscles.**

153

longest one and takes its *origin* on the head and upper two-thirds of the fibula. The **peroneus tertius** is the smallest one, and the **peroneus brevis** is the one of in-between size. Note that the longus *inserts* on the first metatarsal and second cuneiform bones. The brevis and tertius muscles *insert* at different points on the fifth metatarsal bone.

> *Action:* The longus and brevis muscles cause plantar flexion and eversion of the foot. The peroneus tertius causes dorsiflexion and eversion of the foot.

Flexor Muscles   Two flexor muscles of the foot are shown in illustration C, figure 30.4. The **flexor hallucis longus** (Latin: *hallux,* big toe) has its *origin* on the fibula and intermuscular septa. Its long distal tendon *inserts* on the distal phalanx of the great toe. It flexes the great toe.

The **flexor digitorum longus** is the longer muscle shown in illustration C that *originates* on the tibia and the fascia that covers the tibialis posterior. Its distal end divides into four tendons that *insert* on the bases of the distal phalanges of the second, third, fourth, and fifth toes. It flexes the distal phalanges of the four smaller toes.

Extensor Muscles   Two extensors of the foot are shown in illustration D, figure 30.4. The **extensor digitorum longus** is the longer one that takes its *origin* on the lateral condyle of the tibia, part of the fibula, and part of the interosseous membrane. Its tendon of insertion divides into four parts that *insert* on the superior surfaces of the second and third phalanges of the four smaller toes. It extends the prox-

imal phalanges of the four smaller toes. It also flexes and inverts the foot.

The **extensor hallucis longus** is the other extensor in illustration D. It takes its *origin* on the fibula and interosseous membrane. Its distal tendon *inserts* at the base of the distal phalanx of the great toe. This muscle extends the proximal phalanx of the great toe and aids in dorsiflexion of the foot.

Assignment:
Label figures 30.3 and 30.4.

Cat Dissection

Hip Muscles

Remove the superficial fat and fascia that encase the muscles of the hip and thigh so that muscle fiber direction can be determined. *Take care not to remove the fascia lata and iliotibial tract.* The latter is a white, tendinous band on the lateral side of the thigh. Although the sartorius muscle is usually considered a muscle of the thigh, it will be dissected here first to expose the full extent of the tensor fasciae latae.

Sartorius   This broad straplike muscle occupies the cranial half of the medial side of the thigh. Refer to figures 30.5 and 30.8 for reference. *Dissect, transect,* and *reflect.*

Tensor Fasciae Latae   This fan-shaped muscle is located on the anterolateral side of the hip region somewhat ventral to the other muscles in the group.

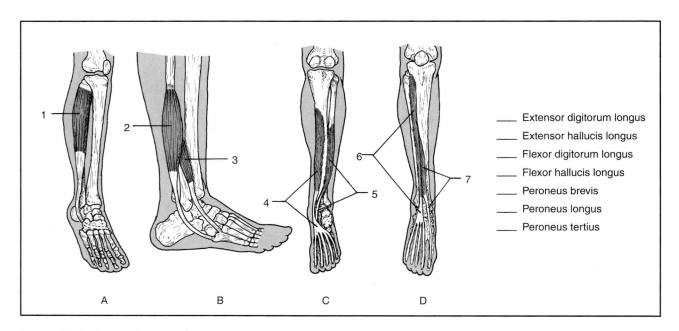

_____ Extensor digitorum longus
_____ Extensor hallucis longus
_____ Flexor digitorum longus
_____ Flexor hallucis longus
_____ Peroneus brevis
_____ Peroneus longus
_____ Peroneus tertius

Figure 30.4   Lower leg muscles.

Refer to figures 30.5 and 30.6. It arises from the lateral ilium and *inserts* in the iliotibial tract portion of the fascia lata. The fascia lata is a tough layer of connective tissue that can be seen covering the vastus lateralis muscle.

Slide a blunt probe underneath the fascia lata to separate it from the vastus lateralis and *transect* the fascia lata so that you can completely *reflect* the tensor fasciae latae. (The tensor muscle usually fuses somewhat with the gluteus maximus. This connection should be cut.)

Gluteus Medius   With the tensor fasciae latae fully reflected, you can see the full extent of the gluteus medius. This thick muscle arises from the lateral surface of the ilium and inserts on the greater trochanter. In the cat this muscle is much larger than the gluteus maximus and will be found craniad of the latter.

Gluteus Maximus   This small hip muscle lies just caudad to the gluteus medius. It arises from the last sacral and first caudal vertebrae and inserts on the greater trochanter; therefore, the origin and insertion in the cat are far less extensive than in the human. *Dissect.*

Caudofemoralis   This small muscle lies just caudad to the gluteus maximus. See figure 30.6. It is difficult to see because it is "hiding" below the cranial border of the biceps femoris. It arises on the

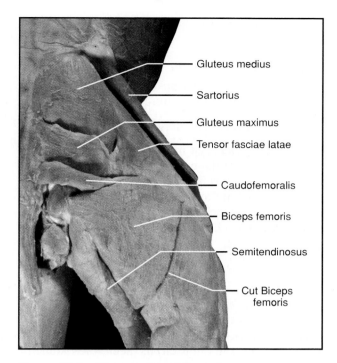

Figure 30.6   Hip muscles of the cat.

caudal vertebrae and inserts by a very thin tendon on the patella.

This muscle is not present in humans; however, in the cat the caudofemoralis and the gluteus maximus together are more comparable to the human

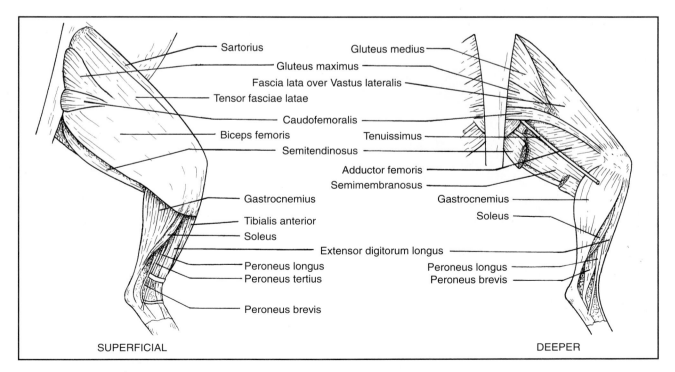

Figure 30.5   Lateral aspect of hindlimb.

gluteus maximus than is the cat's gluteus maximus alone. *Dissect.*

Thigh Muscles

Thirteen muscles make up the musculature of the cat's thigh. Identify them in the following order:

Sartorius   Previously dissected (see p. 154).

Gracilis   A wide, flat muscle covering most of the caudal portion of the medial side of the thigh. Refer to figures 30.7 and 30.9. At the insertion the fibers end in a very thin flat tendon, part of which becomes continuous with the fascia covering the distal portion of the leg. *Dissect, transect,* and *reflect.*

Hamstrings   The cat has the same group of three muscles that are designated as "hamstrings" in humans.

1. **Semimembranosus** This is a large, thick muscle seen just beneath the gracilis; therefore, the latter must be reflected to see its full extent. Dorsal to this muscle is seen the adductor femoris and ventral to it is the semitendinosus.

    Dissect by separating this muscle from the adductor femoris and semitendinosus. Whereas the human homolog inserts only on the tibia, this muscle in the cat inserts on both the femur and tibia.

2. **Semitendinosus** This large band of muscle is on the ventral border of the thigh between the semimembranosus and the biceps femoris. It inserts on the tibia. *Dissect.*

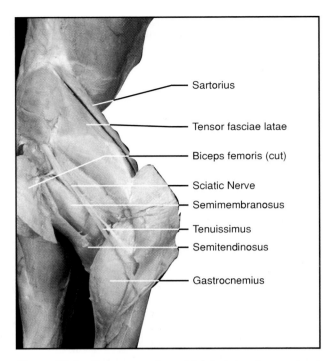

Sartorius

Tensor fasciae latae

Biceps femoris (cut)

Sciatic Nerve

Semimembranosus

Tenuissimus

Semitendinosus

Gastrocnemius

Figure 30.8   **Deep muscles of thigh, lateral aspect.**

3. **Biceps femoris** This large broad muscle covers most of the lateral surface of the thigh. *Dissect, transect,* and *reflect.* When the muscle is transected, be careful not to cut the caudofemoralis and sciatic nerve. The nerve, shown in figure 30.8, appears as a white cord just beneath the biceps femoris.

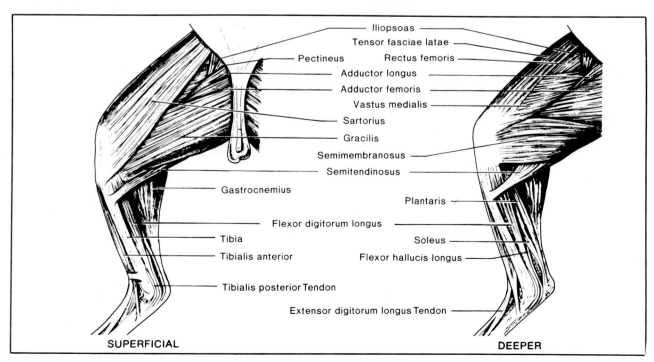

Iliopsoas

Tensor fasciae latae

Pectineus

Rectus femoris

Adductor longus

Adductor femoris

Vastus medialis

Sartorius

Gracilis

Semimembranosus

Semitendinosus

Gastrocnemius

Plantaris

Flexor digitorum longus

Tibia

Soleus

Tibialis anterior

Flexor hallucis longus

Tibialis posterior Tendon

Extensor digitorum longus Tendon

SUPERFICIAL

DEEPER

Figure 30.7   **Medial aspect of hindlimb.**

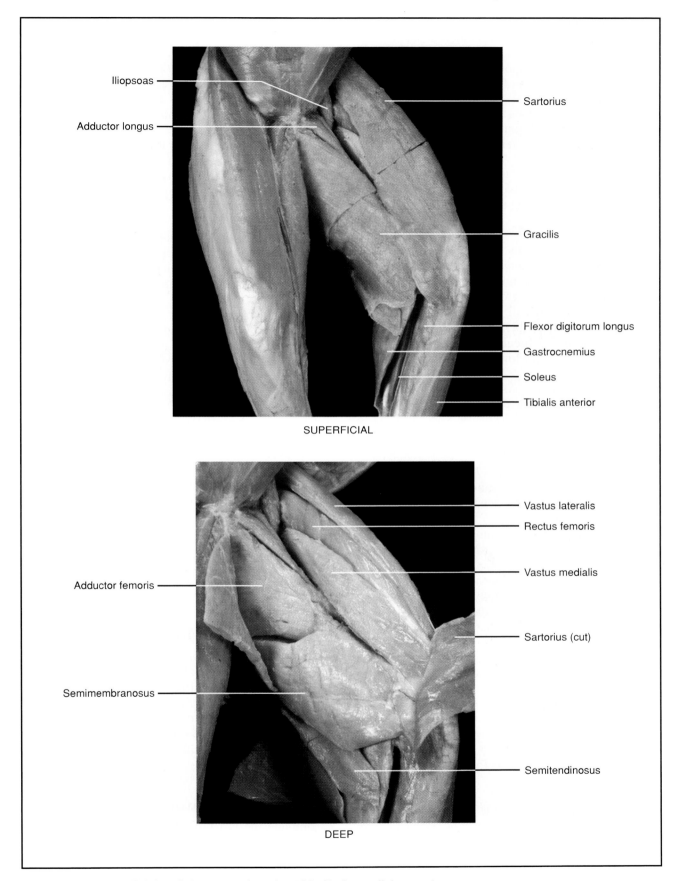

Iliopsoas
Adductor longus
Sartorius
Gracilis
Flexor digitorum longus
Gastrocnemius
Soleus
Tibialis anterior

SUPERFICIAL

Adductor femoris
Vastus lateralis
Rectus femoris
Vastus medialis
Sartorius (cut)
Semimembranosus
Semitendinosus

DEEP

Figure 30.9 Superficial and deep muscles of cat hindlimb, medial aspect.

Adductor Femoris This muscle lies deep to the gracilis and dorsal to the semimembranosus. It originates on the hipbone and inserts on the femur. The adductor femoris corresponds to the adductor magnus and adductor brevis of the human. *Dissect.*

Adductor Longus This thin muscle lies along the cranial border of the adductor femoris and extends from the hipbone to the femur.

Pectineus A very small triangular muscle (see figure 30.7) that lies cranial to the adductor longus and extends from the hipbone to the proximal end of the femur.

Iliopsoas As in humans, this name is applied to a pair of muscles, the *iliacus* and the *psoas major.* Because they have a common insertion on the lesser trochanter, they are grouped together. Only the distal (insertion) end will be observed at this time. Note that the muscle fibers of the iliopsoas are oriented almost perpendicular to the fibers of the pectineus and adductor longus.

Quadriceps Femoris As in humans, the cat has a quadriceps femoris that consists of four muscles. This great extensor muscle of the knee joint has a common insertion by a large tendon that extends to the patella, attaches around the patella, and passes to the tibial tuberosity to insert.

1. **Vastus lateralis** Large muscle deep to the tensor fasciae latae that occupies the craniolateral surface of the thigh.
2. **Vastus medialis** Large muscle deep to the sartorius that occupies the craniomedial surface of the thigh.
3. **Rectus femoris** A cigar-shaped muscle bordered laterally and medially by the vastus medialis. *Dissect, transect,* and *reflect.*
4. **Vastus intermedius** A flat muscle deep to the rectus femoris and attached to the anterior surface of the femur. The rectus femoris must be reflected to see the full extent of this muscle. Try to separate the vastus intermedius from the vastus lateralis and vastus medialis.

Leg and Foot Muscles

Five muscles of the leg and foot of the cat will be identified here.

Gastrocnemius This muscle on the dorsal side of the leg has two heads of origin: a lateral head from the lateral epicondyle of the femur and a medial head from the medial epicondyle of the femur. The muscle inserts by way of the calcaneous (Achilles) tendon onto the calcaneus.

In the cat a large *plantaris* (figure 30.7) muscle can be seen between the heads of the gastrocnemius. In humans the plantaris is a small fusiform muscle of little importance. *Dissect.*

Soleus Deep to the lateral head of the gastrocnemius lies the soleus muscle. Note that, like in humans, this muscle inserts with the gastrocnemius on the calcaneous bone by way of the calcaneous tendon.

Tibialis Anterior This tapered band of muscle is situated on the anterolateral (ventrolateral) aspect of the tibia. Follow the tendon of insertion of this muscle as it crosses the ankle obliquely to reach the medial surface of the foot.

Extensor Digitorum Longus This muscle is covered throughout most of its length by the tibialis anterior. Separate these two muscles and follow the tendon of insertion as it crosses the ankle ventrally and divides into four tendons that are distributed to the digits. *Dissect.*

Peroneus Muscles As do humans, the cat has three peroneus muscles: *peroneus longus, peroneus brevis,* and *peroneus tertius.* These muscles are found on the lateral side of the leg between the extensor digitorum longus and the soleus.

The positions of the tendons of insertion in relation to the lateral malleolus in the cat differ from those in humans; also, the insertion points are somewhat different. Therefore, it is best to identify these muscles simply as a "group" and not try to distinguish which muscle is which.

Surface Muscles Review

Figure 30.10 has been included in this exercise to summarize our study of the muscles of the body as a whole. Only the surface muscles are shown. To determine your present understanding of the human musculature, attempt to label these diagrams first *by not referring back to previous illustrations.* This type of self-testing will determine what additional study is needed.

Laboratory Report

Complete the last portion of combined Laboratory Report 29,30.

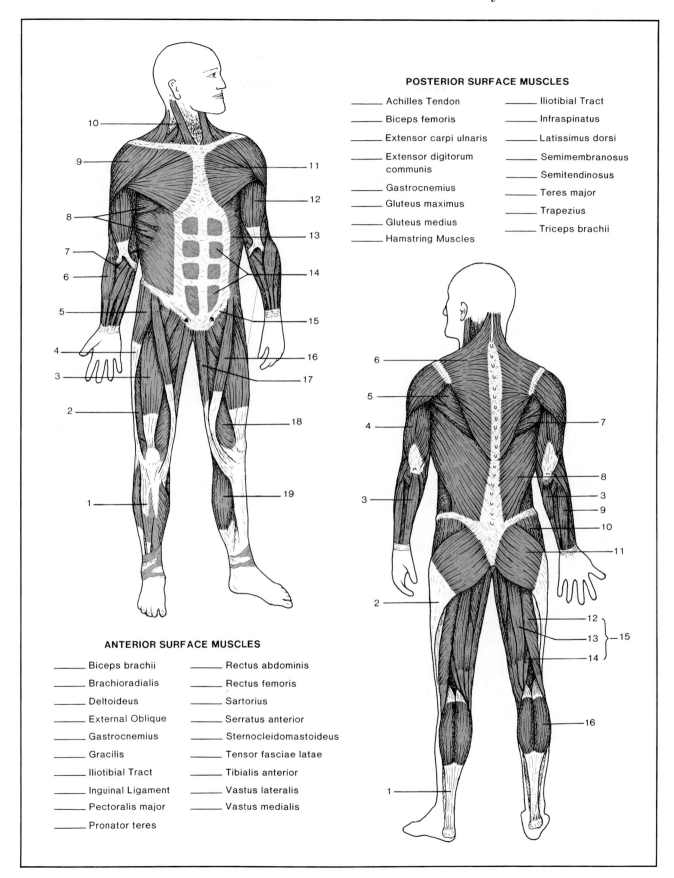

POSTERIOR SURFACE MUSCLES

_____ Achilles Tendon	_____ Iliotibial Tract
_____ Biceps femoris	_____ Infraspinatus
_____ Extensor carpi ulnaris	_____ Latissimus dorsi
_____ Extensor digitorum communis	_____ Semimembranosus
_____ Gastrocnemius	_____ Semitendinosus
_____ Gluteus maximus	_____ Teres major
_____ Gluteus medius	_____ Trapezius
_____ Hamstring Muscles	_____ Triceps brachii

ANTERIOR SURFACE MUSCLES

_____ Biceps brachii	_____ Rectus abdominis
_____ Brachioradialis	_____ Rectus femoris
_____ Deltoideus	_____ Sartorius
_____ External Oblique	_____ Serratus anterior
_____ Gastrocnemius	_____ Sternocleidomastoideus
_____ Gracilis	_____ Tensor fasciae latae
_____ Iliotibial Tract	_____ Tibialis anterior
_____ Inguinal Ligament	_____ Vastus lateralis
_____ Pectoralis major	_____ Vastus medialis
_____ Pronator teres	

Figure 30.10 Major surface muscles of the body.

31

The Spinal Cord, Spinal Nerves, and Reflex Arcs

Automatic stereotyped responses to various types of stimuli enable animals to adjust quickly to adverse environmental changes. These automatic responses, generated by the nervous system, are called **reflexes.**

Reflexes that result in automatic regulation of body function can be either somatic or visceral. **Somatic reflexes** involve skeletal muscle responses, and **visceral reflexes** involve the adjustments of smooth muscle, cardiac muscle, and glands to stimuli.

The neural pathway utilized in performing a reflex is called a **reflex arc.** This pathway involves receptors, spinal nerves, the spinal cord, and effectors. It is the purpose of this exercise to study each of the components that are involved in both types of reflexes.

The Spinal Cord

The spinal cord is a downward extension of the medulla oblongata of the brain. Figure 31.1 reveals its gross structure, as seen posteriorly. To expose it, the posterior portions of the vertebrae and sacrum have been removed.

The spinal cord starts at the upper border of the atlas and terminates as the **conus medullaris** at the lower border of the first lumbar vertebra.

In fetal life the spinal cord occupies the entire length of the vertebral canal (spinal cavity), but as the vertebral column elongates during growth, the spinal cord fails to lengthen with it; thus, the vertebral canal extends downward beyond the end of the spinal cord.

Extending downward from the conus medullaris is an aggregate of fibers called the **cauda equina** (horse's tail), which fills the lower vertebral canal. The innermost fiber of the cauda equina, which is located on the median line, is called the **filum terminale internus.** This delicate prolongation of the conus medullaris becomes the **filum terminale externa** after it passes through the **dural sac** (dura mater), which lines the vertebral canal. The dural sac, incidentally, is continuous with the dura mater that surrounds the brain. Only a portion of it (label 25) is shown in figure 31.1. The remainder of it has been left out of this illustration for clarity.

The Spinal Nerves

The spinal nerves emerge from the spinal cord in pairs through the intervertebral foramina on each side of the spinal cavity. There are eight cervical, twelve thoracic, five lumbar, and five sacral pairs. These thirty pairs of nerves, plus one pair of coccygeal nerves, make a total of thirty-one pairs of spinal nerves.

Reference to the sectional view of the spinal cord in figure 31.1 reveals that each spinal nerve is connected to the spinal cord by structures called **anterior** and **posterior roots.** Note that the posterior root (label 21) has an enlargement, the **spinal ganglion,** which contains cell bodies of sensory neurons. Neuronal fibers from these roots pass through the outer **white matter** of the spinal cord into the inner **gray matter** where neuronal connections (*synapses*) are made.

Cervical Nerves The first cervical nerve (C_1) emerges from the spinal cord in the space between the base of the skull and the atlas. The eighth cervical nerve (C_8) emerges between the seventh cervical and first thoracic vertebrae. Note that the first four cervical nerves are united in the neck region to form a network called the **cervical plexus.** The remaining cervical nerves (C_5, C_6, C_7, and C_8) and the first thoracic nerve unite to form another network called the **brachial plexus** in the shoulder region.

Thoracic (*Intercostal*) **Nerves** The first thoracic nerve (T_1) emerges through the intervertebral foramen between the first and second thoracic vertebrae. The twelfth thoracic nerve (T_{12}) emerges through the foramen between the twelfth thoracic and first lumbar vertebrae. Each of these nerves lies adjacent to the lower margin of the rib above it.

Lumbar Nerves The first lumbar nerve (L_1) emerges from the intervertebral foramen between the first and second lumbar vertebrae. Close to where the nerve emerges from the spinal cavity, it divides to form an upper **iliohypogastric** nerve and a lower **ilioinguinal** nerve.

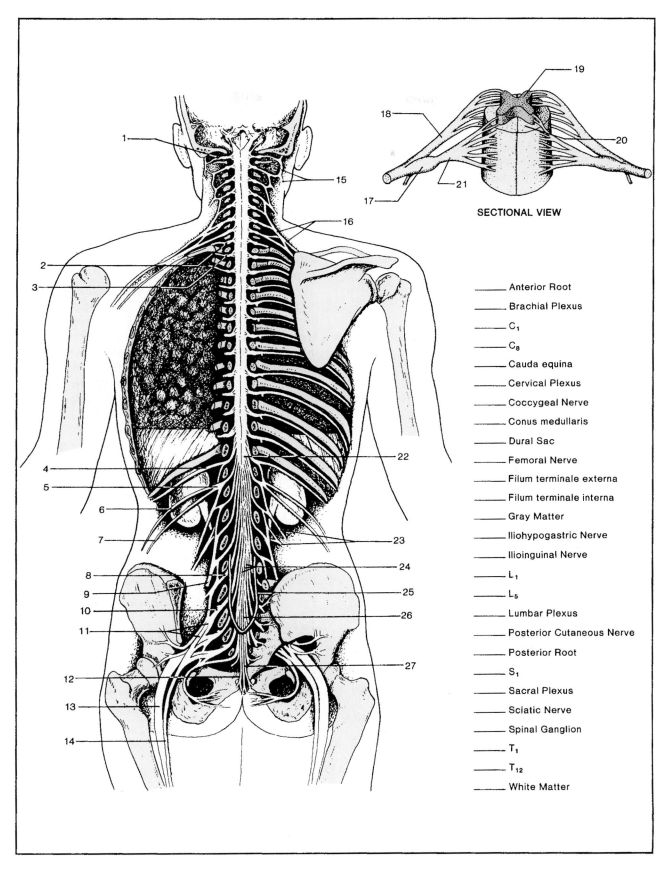

SECTIONAL VIEW

_____ Anterior Root
_____ Brachial Plexus
_____ C_1
_____ C_8
_____ Cauda equina
_____ Cervical Plexus
_____ Coccygeal Nerve
_____ Conus medullaris
_____ Dural Sac
_____ Femoral Nerve
_____ Filum terminale externa
_____ Filum terminale interna
_____ Gray Matter
_____ Iliohypogastric Nerve
_____ Ilioinguinal Nerve
_____ L_1
_____ L_5
_____ Lumbar Plexus
_____ Posterior Cutaneous Nerve
_____ Posterior Root
_____ S_1
_____ Sacral Plexus
_____ Sciatic Nerve
_____ Spinal Ganglion
_____ T_1
_____ T_{12}
_____ White Matter

Figure 31.1 The spinal cord and spinal nerves.

Note in figure 31.1 that L_1, L_2, L_3, and most of L_4 are united to form the **lumbar plexus** (label 23). The largest trunk emanating from this plexus is the **femoral** nerve, which innervates part of the leg. In reality, the femoral is formed by the union of L_2, L_3, and L_4.

Sacral and Coccygeal Nerves The remaining nerves of the cauda equina include five pairs of sacral and one pair of coccygeal nerves. A **sacral plexus** (label 11) is formed by the union of roots of L_4, L_5, and the first three sacral nerves (S_1, S_2, and S_3). The largest nerve that emerges from this plexus is the **sciatic** nerve, which passes down into the leg. The smaller nerve that parallels the sciatic is the **posterior cutaneous** nerve of the leg. The **coccygeal** nerve (label 12) contains nerve fibers from the fourth and fifth sacral nerves (S_4 and S_5).

Assignment:
Label figure 31.1.

Spinal Cord Sectional Details

Injury to the spinal cord is medically grave in that once the spinal cord is traumatized or severed, the damage may be permanent. Fortunately, the bony protection of the spinal cavity and the tissues around the spinal cord afford considerable protection. The presence of various kinds of connective tissue, meninges, and cerebrospinal fluid act as a complex shock absorber to protect the vulnerable cord. The cross-sectional view shown in figure 31.2 illustrates how the cord and spinal nerves are surrounded by these protective layers.

Cord Structure Before examining the surrounding tissues, identify the internal details of the spinal cord in figure 31.2. The most noticeable characteristic of the spinal cord is the pattern of the **gray matter,** which has some semblance to the configuration of the outstretched wings of a swallow-tailed butterfly. Surrounding the darker material is the **white matter,** which consists primarily of myelinated nerve fibers.

Along its median line the spinal cord is divided by a septum and a fissure. Note that the dorsal portion of the cord is bisected by a **posterior median septum** of neuroglial tissue that extends deep into the white matter. On the ventral surface of the cord is an **anterior median fissure.** In the gray matter on the median line lies a tiny **central canal,** evidence of the tubular nature of the spinal cord. This canal is continuous with the ventricles of the brain and extends into the filum terminale. It is lined with ciliated ependymal cells.

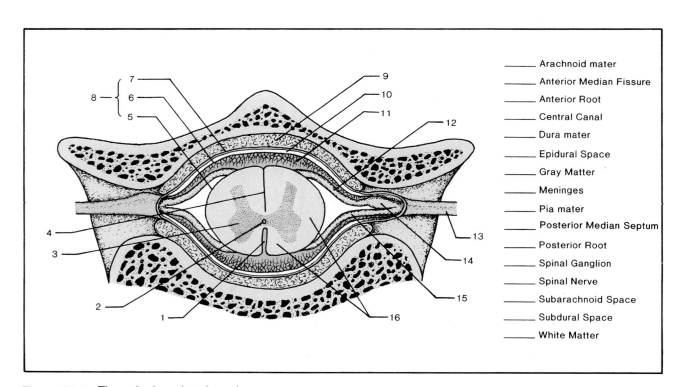

_____ Arachnoid mater
_____ Anterior Median Fissure
_____ Anterior Root
_____ Central Canal
_____ Dura mater
_____ Epidural Space
_____ Gray Matter
_____ Meninges
_____ Pia mater
_____ Posterior Median Septum
_____ Posterior Root
_____ Spinal Ganglion
_____ Spinal Nerve
_____ Subarachnoid Space
_____ Subdural Space
_____ White Matter

Figure 31.2 The spinal cord and meninges.

162

Spinal Nerves Note how the spinal nerves (label 13, figure 31.2) emerge from between the vertebrae on each side of the spinal column. Identify the **posterior root** with its **spinal ganglion** and the **anterior root.**

The Meninges Surrounding the spinal cord (and brain) are three meninges (*meninx*, singular). The outermost membrane is the **dura mater** (label 7), which also forms a covering for the spinal nerves. A cutaway portion of the dura on each side in figure 31.2 reveals the inner spinal nerve roots. The dura mater is the toughest of the three meninges, consisting of fibrous connective tissue. Between the dura and the bony vertebrae is an **epidural space,** which contains loose (areolar) connective tissue and blood vessels.

The innermost meninx is the **pia mater.** It is a thin delicate membrane that is actually the outer surface of the spinal cord. Between the pia mater and dura mater is the third meninx, the **arachnoid mater.** Note that the other surface of the arachnoid mater lies adjacent to the dura mater. The space between the dura and arachnoid mater is actually a potential cavity that is called the **subdural space.** The inner surface of the arachnoid mater has a delicate fibrous texture that forms a netlike support around the spinal cord. The space between the arachnoid mater and the pia mater is the **subarachnoid space.** This space is filled with cerebrospinal fluid that provides a protective fluid cushion around the spinal cord.

Assignment:
Label figure 31.2.

The Somatic Reflex Arc

A brief mention of the somatic reflex arc was made in Exercise 19. In that exercise our prime concern was with the structure of neuronal elements that make up the circuit. Figure 31.3 is very similar to figure 19.1, page 89, although figure 31.3 is simpler.

The principal components of this type of reflex arc are (1) a **receptor,** such as a neuromuscular spindle or a cutaneous end-organ that receives the stimulus; (2) a **sensory** (afferent) **neuron,** which carries impulses through a peripheral nerve and posterior root to the spinal cord; (3) an **interneuron** (association neuron), which forms synaptic connections between the sensory and motor neuron in the gray matter of the spinal cord; (4) a **motor** (efferent) **neuron,** which carries nerve impulses from the central nervous system through the anterior root to the effector via a peripheral nerve; and (5) an **effector** (muscle tissue), which responds to the stimulus by contracting.

Reflex arcs may have more than one interneuron or may be completely lacking in interneurons. If no interneuron is present, as is true of the knee-jerk reflex, the reflex arc is said to be *monosynaptic.* When one or more interneurons exist, the reflex arc is classified as being *polysynaptic.*

The Visceral Reflex Arc

Muscular and glandular responses of the viscera to internal environmental changes are controlled by the **autonomic nervous system.** The reflexes of this system follow a different pathway in the nervous system. Figure 31.4 illustrates the components of a visceral reflex arc.

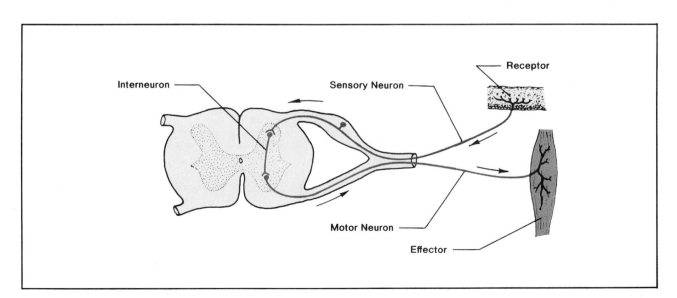

Figure 31.3 The somatic reflex arc.

The principal anatomical difference between somatic and visceral reflex arcs is that visceral reflex arcs have two efferent neurons instead of only one. These two efferent neurons make synaptic connection with each other outside of the central nervous system in an autonomic ganglion.

Two types of autonomic ganglia are shown in figure 31.4: vertebral and collateral. The **vertebral autonomic ganglia** are united to form a chain that lies along the vertebral column. The **collateral ganglia** are located farther away from the central nervous system.

A visceral reflex originates with a stimulus acting on a receptor in the viscera. Impulses pass along the dendrite of a **visceral afferent neuron.** Note that the cell bodies of these neurons are located in the **spinal ganglion.** The axon of the visceral afferent neuron forms a synapse with the **preganglionic efferent neuron** in the spinal cord. Impulses in this neuron are conveyed to the **postganglionic efferent neuron** at synaptic connections in either type of autonomic ganglion (note the short portions of cutoff postganglionic efferent neurons in the vertebral autonomic ganglia). From these ganglia the postganglionic efferent neuron carries the impulses to the organ that is innervated.

In this brief mention of the autonomic nervous system, the student is reminded that the system consists of two parts: the sympathetic and parasympathetic divisions. The *sympathetic (thoracolumbar) division* includes the spinal nerves of the thoracic and lumbar regions. The *parasympathetic (craniosacral) division* incorporates the cranial and sacral nerves. Most viscera of the body are innervated by nerves from both of these divisions. This form of double innervation enables an organ to be stimulated by one system and inhibited by the other.

Assignment:
Label figure 31.4.

Laboratory Report

Complete Laboratory Report 31 for this exercise.

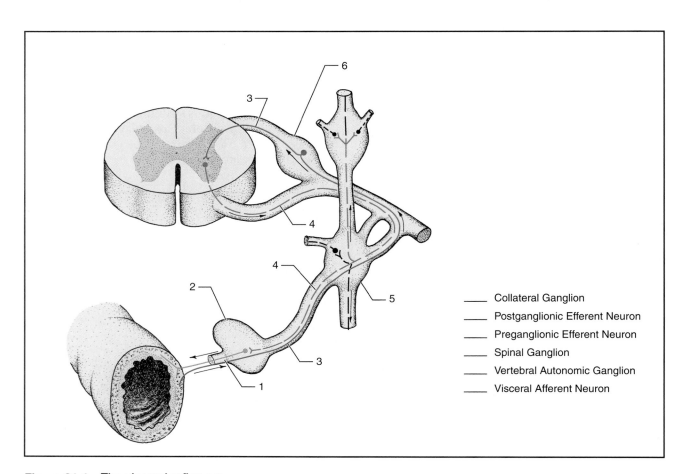

_____ Collateral Ganglion
_____ Postganglionic Efferent Neuron
_____ Preganglionic Efferent Neuron
_____ Spinal Ganglion
_____ Vertebral Autonomic Ganglion
_____ Visceral Afferent Neuron

Figure 31.4   The visceral reflex arc.

Somatic Reflexes

We learned in Exercise 31 that somatic reflexes are automatic responses that occur in skeletal muscles when the proper stimuli are applied to appropriate receptors. In this laboratory period we will study these somatic reflexes, first in the frog, then in humans.

Frog Experiments

Somatic reflexes, whether in amphibians or humans, have five features worthy of study: (1) function, (2) speed of reaction, (3) radiation, (4) inhibition, and (5) synaptic fatigue. Since the frog is a good laboratory specimen to use for studying reflexes, we will use a pithed frog to explore these various facets of muscular response.

Frog Preparation

"Single pith" a frog to destroy its brain, using the following procedure:

Materials:
   small frog
   sharp dissecting needle

   1. Rinse the frog under cool tap water and grip it in your left hand as shown in illustration A, figure 32.1.

2. Use your left index finger to force the frog's nose downward so that the head makes a sharp angle with the trunk. Now, locate the transverse groove between the skull and the vertebral column by pressing down on the skin with your fingernail to mark its location.
3. Push a sharp dissecting needle into the crevice, forcing it forward into the center of the skull through the foramen magnum at the base of the skull. Twist the needle from side to side to destroy the brain. This process of "single pithing" produces what is referred to as a *spinal frog.*
4. After five or six minutes, test the effectiveness of the pithing by touching the cornea of each eye with a dissecting needle. If the lower eyelid on each eye does not rise to cover the eyeball, the pithing is complete.

   This test is valid *only* after approximately five minutes to allow recovery from a temporary state of neural (spinal) shock to the entire spinal cord.
5. Determine whether spinal shock has ended by pinching one of the toes with forceps. If a withdrawal reflex occurs, spinal shock has ended, and your specimen is ready to use.

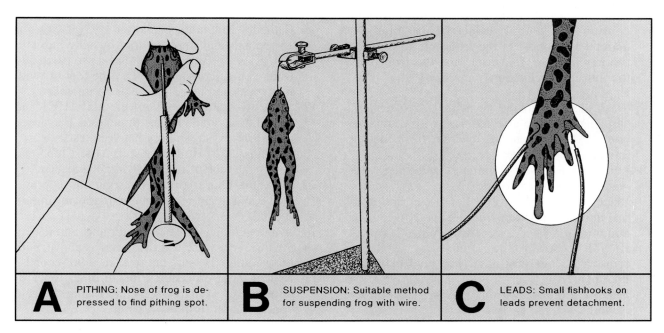

A PITHING: Nose of frog is depressed to find pithing spot.

B SUSPENSION: Suitable method for suspending frog with wire.

C LEADS: Small fishhooks on leads prevent detachment.

Figure 32.1   Frog pithing and setups.

Functional Nature of Reflexes

Muscle tone, movement, and coordinated action in a spinal frog would indicate that these physiological activities occur independently of cognition and are the result of reflex activity. Test for these phenomena as follows:

Materials:

    spinal frog
    ring stand and clamp
    galvanized iron wire, pliers, forceps
    filter paper squares (3 mm)
    28% acetic acid
    10% $NaHCO_3$ in squeeze bottle
    squeeze bottle of tap water
    beaker (250 ml size)

1. With the frog in a squatting position, gently draw out one of the hind legs. Is any pull exerted by the animal? When the foot is released, does it return to its previous position?
2. Squeeze the muscles of the hind legs. Do they feel flaccid or firm? Does muscle tone seem to exist?
3. Can you induce the animal to jump by prodding its posterior?
4. Place the animal in a sink basin filled with water. Does it attempt to swim? Does it float or sink? Report all answers to above questions on the Laboratory Report.
5. Suspend the frog with a wire hook through the lower jaw as shown in illustration B, figure 32.1.
6. With forceps, dip a small square of filter paper in acetic acid, shake off excess, and apply to the ventral surface of the thigh for about 10 seconds. Is the frog able to remove the irritant? Do you think that the frog feels the burning sensation caused by the acid? Explain.
7. Wash the acid off the leg with 10% $NaHCO_3$, spraying it on the affected area with a squeeze bottle. Allow the liquid to drain from the leg into an empty beaker.
8. Let stand for 1 or 2 minutes and rinse with water from squeeze bottle.
9. Place another piece of filter paper and acid on another portion of the body (leg or abdomen). Observe reaction. What happens if you restrain the limb that attempts removal? Does any other adaptive behavior take place?
10. Record all observations on the Laboratory Report.

Reaction Time of Reflexes

To determine whether the reaction time to a stimulus is influenced by the strength of the stimulus, proceed as follows:

Materials:

    7 test tubes (20 mm dia) in test tube rack
    3 beakers (150 ml size)
    1 graduate (10 ml size)
    1 squeeze bottle of tap water
    1 squeeze bottle of 1% HCl
    1 squeeze bottle of 1% $NaHCO_3$
    Kimwipes

1. Dispense 30 ml of 1% $NaHCO_3$ into a beaker.
2. Label seven test tubes: 1.0, .5, .4, .3, .2, .1, and .05%.
3. Measure out 10 ml of 1% HCl from squeeze bottle and pour it into the first tube.
4. Into the other six tubes, measure out the correct amounts of 1% HCl and tap water to make 10 ml each of the correct percentages of HCl. (Examples: 5 ml of 1% HCl and 5 ml of water in the .5% tube; 4 ml of 1% HCl and 6 ml of water in the .4% tube, etc.)
5. Pour the contents from the .05% tube into a clean beaker and allow the long toe of one of the hind legs to be immersed in it for 90 seconds. Be sure to prevent other parts of the foot from touching the sides of the beaker.
    If the foot is withdrawn, record on the Laboratory Report the number of seconds that elapsed from the time of immersion to withdrawal.
    If there is no withdrawal, remove the beaker of .05% HCl at 90 seconds.
6. Bathe the foot in the beaker of 1% $NaHCO_3$ for 15 seconds and rinse the foot by spraying with water. Use the third unused beaker to catch any water sprayed on the foot. Dry the foot off with Kimwipes and let rest for 2 or 3 minutes.
7. Make two more tests with the .05% HCl and record the average for the three tests. Be sure to neutralize, wash, and dry between tests.
8. Pour the .05% HCl back into its test tube and empty the tube of .1% HCl into the beaker.
9. Place the long toe into this solution, as previously, and again record the withdrawal time in seconds.
10. After neutralization, bathing, drying, and resting the foot, repeat the test two times as above, with this solution. Be sure not to exceed 90 seconds.
11. Repeat these procedures for all other tubes of diluted HCl, progressively increasing the acid concentration from low to high.

12. Plot the reaction times against the concentration on the graph provided on the Laboratory Report.

Reflex Radiation

If the foot of a spinal frog is subjected to increasing intensity of electrical stimulation, a phenomenon called *reflex radiation* occurs. To demonstrate it, one applies a tetanic stimulus, first at a low level, and then gradually increases it. Proceed as follows to observe what takes place.

Materials:
  electronic stimulator
  two-wire leads with small fishhooks soldered
    to the leads

1. Connect the jacks of the two-wire leads to the output posts of the stimulator. (These leads have very small fishhooks soldered to one end to enable easy attachment to the skin. Handle them gingerly. They are sharp!)
2. Attach the hooks to the skin of the left foot as shown in illustration C, figure 32.1.
3. Set the stimulator to produce a minimum tetanic stimulus as follows: Voltage at 0.1 volt, Duration at 50 msec, Frequency at 50 pps, Stimulus switch at REGULAR, Polarity at NORMAL, and Output switch on BIphasic.
4. Press the Mode switch to REPEAT and look for a response. If no response, sequentially increase the voltage by doubling each time (i.e., .2, .4, .8, etc.) until a response occurs.

    Describe the response on the Laboratory Report.
5. Increase the voltage in steps of 2 or 3 volts at a time and note any changes that occur in the reflex pattern.

Reflex Inhibition

Just as it is possible to stifle a sneeze or prevent a knee jerk, it should be possible to override the reflex to acid with electrical stimulation. With the stimulator still hooked up to the left foot from the previous experiment, proceed as follows:

1. Lower the toes of the right foot into a beaker of HC1 of a concentration that produced a moderately fast reflex response.
2. As the right foot is lowered into the acid, administer a moderate tetanizing stimulus to the left foot.
3. Adjust the voltage until the foot in the acid is inhibited.
4. Record on the Laboratory Report the acid concentration and voltage that inhibited the reflex action.

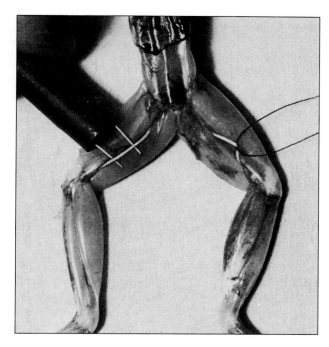

Figure 32.2 Setup for determining when synaptic fatigue occurs. Note string on nerve to prevent trauma when nerve is manipulated.

Synaptic Fatigue

If your specimen is still responding well to stimuli, it may be used for this last test. If it does not prove viable, replace it with a freshly pithed specimen for this experiment.

1. Remove the skin from both thighs (see figure 32.2).
2. Stimulate the sciatic nerve of the *left leg* and note that muscle contraction occurs in both legs.
3. Continue this stimulation until the muscles in the *right leg* fail to respond.
4. After 30 seconds rest, stimulate the left leg again to make sure that *no right leg muscle contraction occurs.*
5. Now, stimulate the sciatic nerve of the right leg instead of the left one.

    Does this cause muscles of the right leg to respond? Where did fatigue take place when the left sciatic nerve was overstimulated?

Reflexes in Medical Diagnosis

Reflex testing is a standard, useful clinical procedure employed by physicians in search of neurological pathology. Damage to intervertebral disks, tumors, polyneuritis, apoplexy, and many other conditions can be better understood with the aid of reflex studies. The diagnostic reflexes studied here are the ones most frequently employed by physicians.

Interpretation of reflex responses is often subjective and requires considerable experience on the part of the diagnostician. Our purpose in performing the tests here, obviously, is not to diagnose, but rather to test, observe, and understand why certain tests are performed.

Two Types of Reflexes

Clinically, reflexes are categorized as being either deep or superficial. The **deep reflexes** include all reflexes that are elicited by a sharp tap on an appropriate tendon or muscle. They are also called *jerk, stretch,* or *myotatic reflexes.* The receptors (spindles) for these reflexes are located in the muscle, not in the tendon. When the tendon is tapped, the muscle is stretched; stretching the muscle, in turn, activates the muscle spindle, which triggers the reflex response.

Superficial reflexes are withdrawal reflexes elicited by noxious or tactile stimulation. They are also called *cutaneous reflexes.* Instead of percussion initiating these reflexes, the skin is stroked or scratched to induce response.

Reflex Aberrations

Abnormal responses to stimuli may be diminished (hyporeflexia), exaggerated (hyperreflexia), or pathological. **Hyporeflexia** may be due to malnutrition, neuronal lesions, aging, deliberate relaxation, etc. **Hyperreflexia** is often accompanied by marked muscle tone due to the loss of inhibitory control by the motor cortex. Among other things, it may also be caused by strychnine poisoning. **Pathological reflexes** are reflex responses that occur in one or more muscles other than the muscle where the stimulus originates.

These three deviations are what a physician looks for in performing these tests. The following scale is used for calculating each response:

$++++$ very brisk, hyperactive; often indicative of pathology; may be associated with clonus (spasms)

  $+++$ brisker than average, may or may not be indicative of pathology

   $++$ average; normal

    $+$ somewhat diminished

    0 no response

Test Procedure

Perform the seven reflex tests shown in figure 32.3 according to the procedures outlined below.

Materials:
reflex hammer

Biceps Reflex   This reflex is a deep reflex that causes flexion of the arm. It is elicited by holding the subject's elbow with the thumb pressed over the tendon of the biceps brachii, as shown in illustration A, figure 32.3. To produce the desired response, strike a sharp blow to the first digit of the thumb with the reflex hammer.

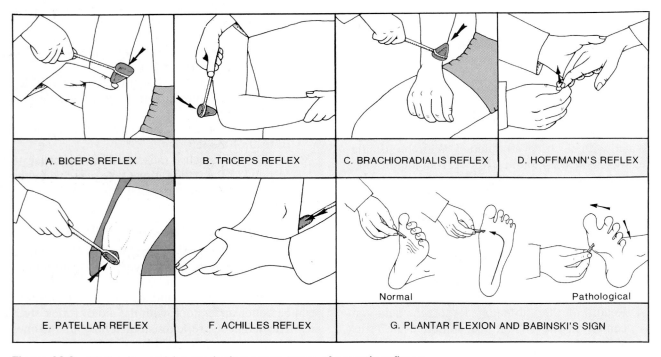

Figure 32.3   Methods used for producing seven types of somatic reflexes.

If the reflex is absent or diminished (0 or +), apply reinforcement by asking the subject to clench his/her teeth or squeeze his/her thigh with the other hand. Test both arms and record the degree of response on the Laboratory Report.

This reflex functions through C_5 and C_6 spinal nerves.

Triceps Reflex This deep reflex causes extension of the arm in normal individuals. To demonstrate this reflex, flex the arm at the elbow, holding the wrist as shown in illustration B, figure 32.3, with the palm facing the body.

Strike the triceps brachii tendon above the elbow with the pointed end of the reflex hammer. Use the same reinforcement techniques as above, if necessary. Test both arms and record the degree of response on the Laboratory Report.

This reflex functions through C_7 and C_8 spinal nerves.

Brachioradialis Reflex In normal individuals this reflex manifests itself in flexion and pronation of the forearm and flexion of the fingers.

To demonstrate this reflex, direct the subject to rest the hand on the thigh in the position shown in illustration C, figure 32.3. Strike the forearm with the wide end of the reflex hammer about one inch above the end of the radius. The illustration reveals the approximate spot. Use reinforcement techniques, if necessary. The tendon that is struck here is for the brachioradialis muscle of the arm.

This reflex functions through the same spinal nerves (C_5 and C_6) as the biceps brachii reflex.

Hoffmann's Reflex This reflex is an example of a deep reflex in which the response does not occur only in the muscle that is stretched. If the response to the stimulus here is broad, affecting several fingers other than the finger stimulated, there is indication of pyramidal tract (spinal cord) damage.

To induce this reflex, flick the terminal phalanx of the index finger upward, as shown in illustration D, figure 32.3.

Pathology is indicated if the thumb is adducted and flexed and the other fingers exhibit twitchlike flexion. Such a broad field of reflex action that results from a brief stretch of the flexor muscles of the index finger indicates deep reflex hyperactivity and pathological characteristics.

Test both hands of the subject and record results on the Laboratory Report.

Patellar Reflex This monosynaptic reflex is variously referred to as the *knee reflex, knee jerk,* or *quadriceps reflex.*

To perform this test, the subject should be seated on the edge of a table with the leg suspended and somewhat flexed over the edge. To elicit the typical response, strike the patellar tendon just below the kneecap. The correct spot is shown in illustration E, figure 32.3.

If the response is negative, utilize the **Jendrassic maneuver** for facilitation. This technique involves the subject locking the fingers of both hands in front of the body and pulling each hand against each other isometrically.

A phenomenon called **clonus** is sometimes seen when inducing this reflex. Clonus is characterized by a succession of jerklike contractions (spasms) that follow the normal response and persist for a period of time. This condition is a manifestation of hyperreflexia and indicates damage within the central nervous system.

Test both legs, recording the results on the Laboratory Report. This reflex functions through L_2, L_3, and L_4 spinal nerves.

Achilles Reflex This reflex is also referred to as the *ankle jerk.* It is characterized by plantar flexion when the Achilles tendon is struck a sharp blow. Stretching this tendon affects the muscle spindles in the triceps surae, causing it to contract.

To perform this test, grip the foot with the left hand, as shown in illustration F, figure 32.3, forcing it upward somewhat. With the subject relaxed, strike the Achilles tendon as shown. Hyporeflexia here is often associated with hypothyroidism.

This reflex functions through S_1 and S_2 spinal nerves.

Plantar Flexion *(Babinski's sign)* This reflex is the only one in this series that is of a superficial nature. Note in illustration G, figure 32.3, that the normal reaction to stroking the sole of the foot in an adult is plantar flexion. If dorsiflexion occurs, starting in the great toe and spreading to the other toes (Babinski's sign), it may be assumed that there is myelin damage to fibers in the pyramidal tracts. Incidentally, dorsiflexion is normal in infants, especially if they are asleep. Babinski's sign disappears in infants once myelinization of nerve fibers is complete. This reflex functions through S_1 and S_2 spinal nerves.

Perform this test using a hard object such as a key. Follow the pattern shown in the middle illustration.

Laboratory Report

Complete Laboratory Report 32 for this exercise.

33

The Patellar Reflex:
Using the Intelitool Flexicomp™

In Exercise 32 we learned that physicians use reflex responses to detect neural pathology. In general, the judgment of the existence or absence of pathology is based on past experience and is necessarily quite subjective. It is possible, however, with the proper software and hardware to quantify certain reflex responses to confirm suspected damage to the nervous system.

The Intelitool Flexicomp™ system will be used here to remove some of the subjectivity in observing the patellar reflex. Since this same hardware can be used to analyze the biceps and triceps reflexes, you may wish to test them also. With the setup illustrated in figure 33.1, you will be able to accomplish the following:

- Quantify the stimulus response in terms of degree of deflection.
- Quantify time intervals, such as the latent period and time to maximal response.
- Demonstrate the effect of neural facilitation on the involuntary reflex response.

- Contrast the involuntary reflex with a voluntary response involving cognition.

It is assumed that you have a basic knowledge of the computer platform that you will be using. The platforms for which Flexicomp is available are:

- IBM or compatible PC (MS-DOS 3.3 or higher)
- Macintosh (System 6.0.8 or higher)
- Apple II (ProDOS)

All equipment will be set up prior to your entering the laboratory. The software will be loaded and ready to use. Figure 33.1 illustrates the setup. You will note as you progress with the experiment that separate instructions are provided to accommodate the software differences between the three platforms.

You will be performing these various experiments with a laboratory partner. One individual will be the subject, and the other will operate the computer and apply the mallet to the subject's knee. Proceed as follows to perform the various tests.

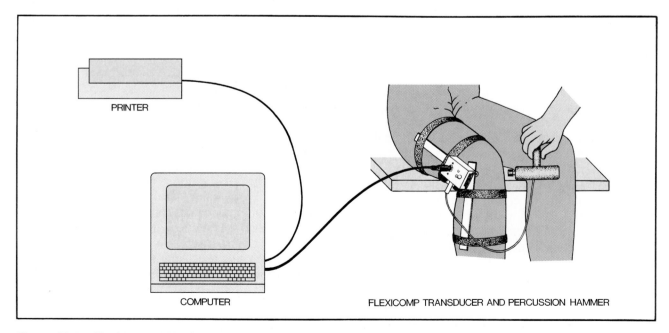

Figure 33.1 Flexicomp setup for monitoring the patellar reflex.

Materials:

computer, monitor, and printer
displacement transducer with Velcro straps
percussion hammer (computer interfaced)
transducer cable (game port connector)
Flexicomp program diskette
data disk (initialized)
calibration template
Flexicomp Lab Manual
Flexicomp User Manual

1. Calibrate the Transducer.

Before hooking up the subject, calibrate the Flexicomp transducer, following the instructions on the screen. It will be calibrated first at 90° and then at 150°. Adjustments are made by turning the aluminum knob on the transducer.

Important: The thumbscrew locking tab in the angle of the transducer **must be loosened** at this point. Failure to do so may cause damage to the transducer. The thumbscrew should always be kept loose whenever data acquisition is not in progress, and the transducer should never be forced if it has hit an internal stop. Be sure to loosen the thumbscrew first!

If calibration fails for any reason during the experiment, the software will detect the flaw and force you to redo the calibration procedure.

2. Attach Flexicomp to Subject's Leg.

With the subject seated on the edge of a table, as shown in figure 33.1, strap the transducer to the right thigh and lower leg.

Note that the transducer box is facing outward and that the hinge of the transducer is aligned with the knee joint. Try to minimize any twisting forces on the transducer.

3. Center the Trace on the Screen.

MS-DOS: Select "Experiment Menu" from the FLEXICOMP MENU. Then select "Real Time Experiment Mode." The data acquisition screen will appear.

With the **thumbscrew loose** press the button on the end of the mallet. As the trace moves across the screen, turn the adjustment knob until the trace is just below the center of the screen (it may take several tries to get it right).

When the trace has been vertically positioned, tighten the thumbscrew locking tab. Press the C key to clear the computer's memory and then Y to confirm. You are now ready to proceed to step 4.

Macintosh: Select **New** from the **File** menu to open a data acquisition window if one is not already

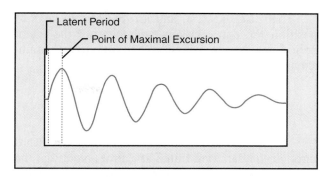

Figure 33.2   Response to stimulus application.

open. With the **thumbscrew loose** press the button on the end of the mallet.

As the trace moves across the screen, turn the adjustment knob until the trace is just below the center of the screen (it may take several tries to get it right).

When the trace has been vertically positioned, tighten the thumbscrew locking tab. Press the C key to clear the computer's memory and then Y to confirm. You are now ready to proceed to step 4.

Apple II: Select "Experiment Mode" from the FLEXICOMP MAIN MENU, and then choose "Real Time Mode." With the **thumbscrew loose** press the button on the end of the mallet.

As the trace moves across the screen, turn the adjustment knob until the trace is just below the center of the screen (it may take several tries to get it right).

When the trace has been vertically positioned, tighten the thumbscrew locking tab. Press the C key to clear the computer's memory and then Y to confirm. You are now ready to proceed to step 4.

4. Begin Data Collection.

Strike the subject's patellar ligament (just below the kneecap) hard enough to elicit a good response. The plot will automatically stop when it reaches the edge of the screen.

Repeat the application of the mallet several times. If your technique is good you should get similar results each time.

5. Review the Data.

MS-DOS: Press ESC to return to the main FLEXICOMP MENU and select "Review/Analyze Menu." Choose "Displacement/Time Analysis" and press the S key to stop the trace.

Then use the Z, J, and arrow (cursor) keys to position the data markers and measure the latent period, maximum excursion, and time to maximum excursion. Record the values you obtain:

Latent Period _____

Maximum Excursion _____

Time to Max. Excursion _____

Use the F or B key to move forward or backward through the data one frame at a time. Note that the left edge of the screen represents the instant the stimulus was applied.

Macintosh: Select **Displacement** from the **Analysis** menu. Click and drag in the plot area to position the data markers and measure the latent period, maximum excursion, and time to maximum excursion. Record the values.

Latent Period _____

Maximum Excursion _____

Time to Max. Excursion _____

To select a specific event, click on the event number at the top of the window. Note that the left edge of the screen represents the instant the stimulus was applied.

Apple II: Press ESC to return to the FLEXICOMP MAIN MENU and select "Review/ Analyze." Choose "Review-Time/Displacement Analysis" and press the S key to stop the trace.

Then use the D, J, and arrow (cursor) keys to position the data markers and measure the latent period, maximum excursion, and time to maximum excursion. Record the values here:

Latent Period _____

Maximum Excursion _____

Time to Max. Excursion _____

Use the F or B key to move forward or backward through the data one frame at a time. Note that the left edge of the screen represents the instant the stimulus was applied.

6. Save the Data to Disk.

MS-DOS: Press ESC to return to the main FLEXICOMP MENU and select "Data File/Disk Commands." Then choose SAVE File and give the file an appropriate name and press the Return (Enter) key.

Macintosh: Select **Save** from the **File** menu and name the file appropriately.

Apple II: From the MAIN MENU select "Save Current Data" and then select "Catalog/Save Data File." On single floppy drive systems, you will be prompted to insert a data disk. Give the file an appropriate name and press the Return key.

7. Demonstrate Facilitation Effects.

With the subject clasping his or her hands in front and **pulling hard,** repeat steps 3 through 6.

Obtain several frames of data in this manner. Note any differences in the measurements between this and the first set of data. What accounts for the differences caused by facilitation?

8. Evaluate Voluntary Reflex.

With the subject facing away from you, repeat steps 3 through 6, as you randomly tap the hammer gently on the table. The subject should respond as soon as he or she hears the tap by gently kicking.

Compare latent periods for this set of data with that of the involuntary knee-jerk responses. Explain the differences you find.

9. Optional Experiments.

If you wish to continue with additional reflex experiments, consult the two Intelitool manuals. The following brief descriptions describe the general procedures:

Biceps Jerk With the subject's arm relaxed on the desk, gently press his or her biceps tendon with your thumb. The biceps tendon can be felt in the center of the hollow of the elbow. Strike your thumb with the hammer to elicit the reflex response.

Triceps Jerk Have the subject lie down on the table with the elbow bent so that his or her arm lies loosely across the abdomen. Support the arm being tested at the forearm balance point (near the elbow) with your supporting hand balled up into a fist. Strike the triceps tendon about 2 inches above the elbow.

Ankle Jerk The subject should kneel on a chair, with back to the examiner and feet projecting over the edge of the chair (shoes and socks off).

Strike the Achilles tendon in its center (at the level of the ankle) and observe the plantar extension of the foot.

Laboratory Report

Since there is no Laboratory Report for this experiment, your instructor will indicate how to write up this experiment.

Brain Anatomy:
External

This study of the human brain will be done in conjunction with dissection of the sheep brain. The ample size and availability of sheep brains make them preferable to brains of most other animals; also, the anatomical similarities that exist between human and sheep brains are considerably greater than their differences.

Preserved human brains will also be available for study in this laboratory period. Dissection, however, will be confined to the sheep brain. A discussion of the meninges will precede the sheep brain dissection.

The Meninges

It was observed in Exercise 31 that the spinal cord, as well as the brain, is surrounded by three membranes called the *meninges.* Due to the importance of these protective membranes it will be necessary to reemphasize certain anatomical characteristics about them.

Dura Mater The outermost meninx is the *dura mater,* which is a thick tough membrane made up of fibrous connective tissue. The lateral view of the opened skull in figure 34.1 reveals the dura mater lifted away from the exposed brain. Note that on the median line of the skull it forms a large **sagittal sinus** (label 6) that collects blood from the surface of the brain. Observe, also, that many cerebral veins empty into this sinus.

Extending downward between the two halves of the cerebrum is an extension of the dura mater, the **falx cerebri.** This structure shows up only in the frontal section of figure 34.1. At its anterior inferior end, the falx cerebri is attached to the crista galli of the ethmoid bone.

Arachnoid Mater Inferior to the dura mater lies the second meninx, the *arachnoid mater* (label 8, figure 34.1). Between this delicate netlike membrane and the surface of the brain is the **subarachnoid space,** which contains cerebrospinal fluid. Note that the arachnoid mater has small projections of tissue, the **arachnoid granulations,** that extend up into the interior of the sagittal sinus. These granulations allow cerebrospinal fluid to diffuse from the subarachnoid space into the venous blood of the sagittal sinus.

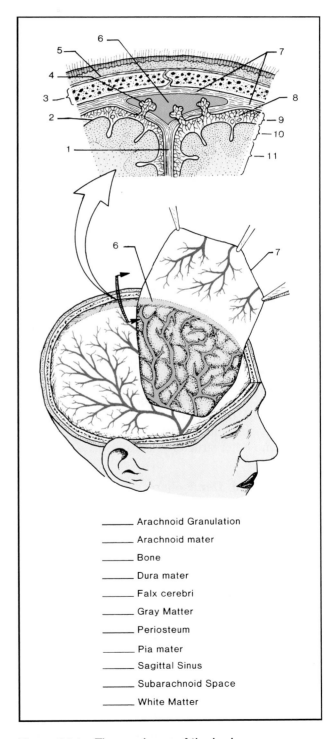

_____ Arachnoid Granulation

_____ Arachnoid mater

_____ Bone

_____ Dura mater

_____ Falx cerebri

_____ Gray Matter

_____ Periosteum

_____ Pia mater

_____ Sagittal Sinus

_____ Subarachnoid Space

_____ White Matter

Figure 34.1 The meninges of the brain.

Pia Mater The surface of the brain is covered by the third meninx, the *pia mater.* This membrane is very thin. Note that the surface of the brain has grooves, or **sulci,** that increase the surface of the **gray matter** (label 10). As in the case of the spinal cord, the gray matter is darker than the inner **white matter** because it contains cell bodies of neurons. The white matter consists of neuron fibers that extend from the gray matter to other parts of the brain and spinal cord.

Assignment:
Label figure 34.1.

Sheep Brain Dissection

As you dissect the sheep brain and identify the various structures on it, compare the sheep brain with the human brain. Note that the illustrations of comparable views of both brain types are on the same page or opposite pages. Once you have located a structure on the sheep brain, try to find a similar structure on the human brain. Performing the dissection in this way will help you to learn human brain anatomy.

Materials:
> preserved sheep brains
> preserved human brains
> human brain models
> dissecting instruments
> dissecting trays

Place a sheep brain on a dissecting tray and, by comparing it with figures 34.2 and 34.4, identify the **cerebrum, cerebellum, pons Varolii, midbrain,** and **medulla oblongata.** Study each of these five major divisions to identify the fissures, grooves, ridges, and other structures as follows:

The Cerebral Hemispheres In both the sheep and human, the cerebrum is the predominant portion of the brain. Note that when observed from the dorsal or ventral aspect, the cerebrum is divided along the median line to form two *cerebral hemispheres* by the **longitudinal cerebral fissure.**

Observe that the surface of the cerebrum is covered by ridges and furrows of variable depth. The deeper furrows are called **fissures,** and the shallow ones, **sulci.** The ridges or convolutions are called **gyri.** The frontal section of figure 34.1 reveals how the infolding of the cerebral surface greatly increases the amount of gray matter of the brain.

The principal fissures of the human cerebrum are the **central sulcus** (fissure of Rolando), **lateral cerebral fissure** (fissure of Sylvius), and the **parieto-occipital** fissure. In figure 34.2 the central sulcus is label 14; the lateral cerebral fissure is label 12. The greater portion of the parieto-occipital fissure is seen on the medial surface of the cerebrum (label 1, figure 35.2 in the next exercise). If the lateral cerebral fissure is retracted, as shown in the lower illustration of figure 34.2, an area called the **insula** (island of Reil) is exposed.

Each half of the human cerebrum is divided into four lobes. The **frontal lobe** is the most anterior portion. Its posterior margin falls on the central sulcus and its inferior border consists of the lateral cerebral fissure. The **occipital lobe** is the most posterior lobe of the brain. The anterior margin of this lobe falls on the parieto-occipital fissure. Since most of this fissure is seen on the medial face of the cerebrum, an imaginary dotted line in figure 34.2 reveals its position. The **temporal lobe** lies inferior to the lateral cerebral fissure and extends back to the parieto-occipital fissure. An imaginary horizontal dotted line provides a border between the temporal and **parietal lobes.** Labels 1, 4, 8, and 10 are for these four lobes.

The Cerebellum Examine the surface of the sheep cerebellum. Note that its surface is furrowed with sulci. How do these sulci differ from those of the cerebrum? Note that the human cerebellum is constricted in the middle to form right and left hemispheres. Can the same be said for the sheep's cerebellum? This part of the brain plays an important role in the maintenance of posture and the coordination of complex muscular movements.

The Brain Stem

The remaining structures to be observed are components of what is called the *brain stem.* This axial portion of the brain includes the midbrain, diencephalon, pons Varolii, and medulla oblongata. It excludes the cerebrum and cerebellum.

Midbrain Since the dorsal portion of the midbrain lies concealed under the cerebellum and cerebrum, it can only be exposed by forcing the cerebellum downward with the thumb as illustrated in figure 34.3. Observe that this maneuver exposes four roundish bodies that constitute the *corpora quadrigemina.* The two larger bodies are the **superior colliculi.** The pair of smaller bodies are the **inferior colliculi.** In some animals, the superior colliculi are called "optic lobes" because of their close association with the optic tracts. In animals the colliculi are important analytical centers concerned with brightness and sound discrimination.

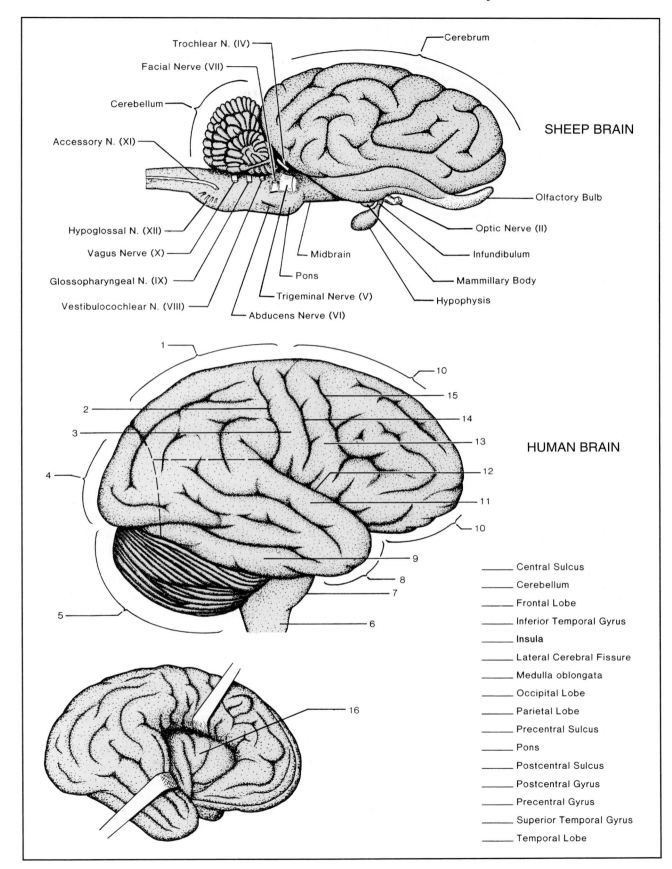

Trochlear N. (IV)

Facial Nerve (VII)

Cerebellum

Accessory N. (XI)

Hypoglossal N. (XII)

Vagus Nerve (X)

Glossopharyngeal N. (IX)

Vestibulocochlear N. (VIII)

Abducens Nerve (VI)

Trigeminal Nerve (V)

Pons

Midbrain

Cerebrum

SHEEP BRAIN

Olfactory Bulb

Optic Nerve (II)

Infundibulum

Mammillary Body

Hypophysis

HUMAN BRAIN

_____ Central Sulcus

_____ Cerebellum

_____ Frontal Lobe

_____ Inferior Temporal Gyrus

_____ Insula

_____ Lateral Cerebral Fissure

_____ Medulla oblongata

_____ Occipital Lobe

_____ Parietal Lobe

_____ Precentral Sulcus

_____ Pons

_____ Postcentral Sulcus

_____ Postcentral Gyrus

_____ Precentral Gyrus

_____ Superior Temporal Gyrus

_____ Temporal Lobe

Figure 34.2   Lateral aspects of sheep and human brains.

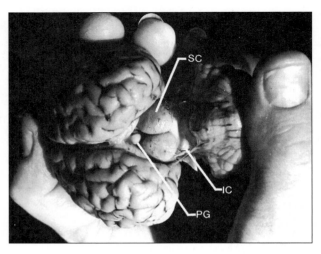

Figure 34.3 Exposing structures of the diencephalon and midbrain: SC—superior colliculus; IC—inferior colliculus, PG—pineal gland.

Diencephalon *(Interbrain)* This part of the brain is difficult to observe because it lies dorsal and anterior to the midbrain and ventral to the cerebrum. The **pineal gland,** shown in figure 34.3 is a part of the diencephalon. The thalamus, hypothalamus, and mammillary bodies are also parts of the diencephalon. More will be said about these parts later.

The pineal gland is thought to be an evolutionary remnant of the third eye that exists in certain reptiles. The gland produces a hormone called *melatonin,* which in lower animals inhibits the estrus cycle. The exact role of melatonin in humans is not completely understood. At present it is marketed as a sleep inducer. It is probably significant that this organ is, histologically, glandular until puberty, and then becomes fibrous after puberty.

Pons Varolii Locate this portion of the sheep's brain by examining its ventral surface and comparing it with the upper illustration in figure 34.4. This part of the brain stem contains fibers that connect parts of the cerebellum and medulla with the cerebrum. It also contains nuclei of the 5th, 6th, 7th, and 8th cranial nerves.

Medulla Oblongata This portion of the brain stem is also known as the *spinal bulb.* It contains centers of gray matter that control the heart, respiration, and vasomotor reactions. The last four cranial nerves emerge from the medulla.

Assignment:
Label figure 34.2.

The Cranial Nerves

There are twelve pairs of *cranial nerves* that emerge from various parts of the brain and pass through foramina of the skull to innervate parts of the head and trunk. Each pair has a name as well as a position number. Although most of these nerves contain both motor and sensory fibers *(mixed nerves),* a few contain only sensory fibers *(sensory nerves).* The sensory fibers have their cell bodies in ganglia outside of the brain; the cell bodies of the motor neurons, on the other hand, are situated within nuclei of the brain.

Compare your sheep brain with figures 34.2 and 34.4 to assist you in identifying each pair of nerves. As you proceed with the following study, avoid damaging the brain tissue with dissecting instruments. Keep in mind, also, that once a nerve is broken off it is difficult to determine where it was. The pairs of cranial nerves are as follows:

I Olfactory Nerve This cranial nerve contains sensory fibers for the sense of smell. Since it extends from the mucous membranes in the nose through the ethmoid bone to the **olfactory bulb,** it cannot be seen on your specimen. In the olfactory bulb synapses are made with fibers that extend inward to olfactory areas in the cerebrum. Note that on the human brain an **olfactory tract** (label 2, figure 34.4) extends between the bulb and the brain.

II Optic Nerve This sensory nerve functions in vision. It contains axon fibers from ganglion cells of the retina of the eye. Part of the fibers from each optic nerve cross over to the other side of the brain as they pass through the **optic chiasma.** From the optic chiasma the fibers pass through the **optic tracts** to the thalamus and finally to the visual areas of the cerebrum. Identify these structures on the sheep brain.

III Oculomotor Nerve This nerve emerges from the midbrain and supplies somatic nerve fibers to the *levator palpebrae* (eyelid muscle) and four of the extrinsic ocular muscles *(superior rectus, medial rectus, inferior rectus,* and *inferior oblique).* The *superior oblique* and *lateral rectus* muscles are innervated by IV and VI, respectively.

The oculomotor nerve also supplies parasympathetic fibers to the *constrictor pupillaris* of the iris and the *ciliaris muscle* of the ciliary body. These fibers constrict the iris and change the lens shape as needed in accommodation (focusing).

If the oculomotor nerves are not visible on your sheep brain, it may be that they are obscured by the hypophysis (pituitary gland). The sheep brain illustrated in figure 34.4 lacks this structure. To remove

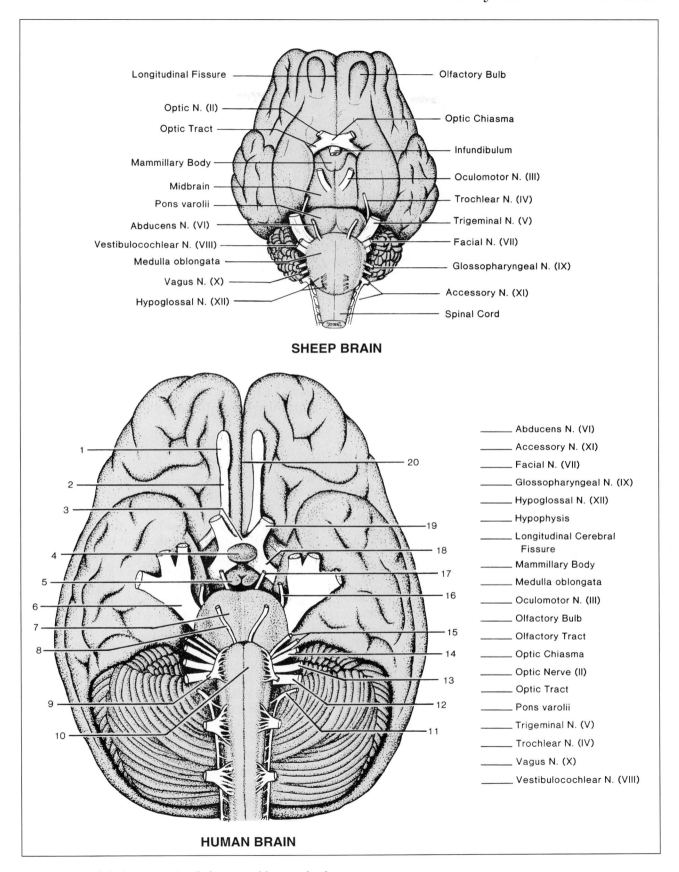

SHEEP BRAIN

Longitudinal Fissure
Optic N. (II)
Optic Tract
Mammillary Body
Midbrain
Pons varolii
Abducens N. (VI)
Vestibulocochlear N. (VIII)
Medulla oblongata
Vagus N. (X)
Hypoglossal N. (XII)

Olfactory Bulb
Optic Chiasma
Infundibulum
Oculomotor N. (III)
Trochlear N. (IV)
Trigeminal N. (V)
Facial N. (VII)
Glossopharyngeal N. (IX)
Accessory N. (XI)
Spinal Cord

HUMAN BRAIN

_____ Abducens N. (VI)
_____ Accessory N. (XI)
_____ Facial N. (VII)
_____ Glossopharyngeal N. (IX)
_____ Hypoglossal N. (XII)
_____ Hypophysis
_____ Longitudinal Cerebral
          Fissure
_____ Mammillary Body
_____ Medulla oblongata
_____ Oculomotor N. (III)
_____ Olfactory Bulb
_____ Olfactory Tract
_____ Optic Chiasma
_____ Optic Nerve (II)
_____ Optic Tract
_____ Pons varolii
_____ Trigeminal N. (V)
_____ Trochlear N. (IV)
_____ Vagus N. (X)
_____ Vestibulocochlear N. (VIII)

Figure 34.4 Inferior aspects of sheep and human brains.

the hypophysis without damaging other structures, gently lift it up with forceps and sever the infundibulum with a scalpel or a pair of scissors.

IV Trochlear Nerve   This nerve provides muscle sense and motor stimulation of the *superior oblique* muscle of the eye. As in the case of the oculomotor, this nerve also emerges from the midbrain. This nerve is very small in diameter and one of the more difficult ones to locate.

V Trigeminal Nerve   Just posterior to the trochlear nerve lies the largest cranial nerve, the trigeminal. Although it is a mixed nerve, its sensory functions are much more extensive than its motor functions. Because of its extensive innervation of the mouth and face, it is described in detail on pages 180, 181, and 182.

VI Abducens Nerve   This small nerve provides innervation of the *lateral rectus* muscle of the eye. It is a mixed nerve in that it provides muscle sense as well as muscular contraction. On the human brain it emerges from the lower part of the pons near the medulla.

VII Facial Nerve   This nerve consists of motor and sensory divisions. Label 15, figure 34.4, reveals that this nerve on the human brain has two branches. It innervates muscles of the face, salivary glands, and taste buds of the anterior two-thirds of the tongue.

VIII Vestibulocochlear *(Auditory)* **Nerve**   This nerve goes to the inner ear. It has two branches: the *vestibular division,* which innervates the semicircular canals to function in equilibrium, and the *cochlear division,* which innervates the cochlea to function in hearing. The vestibulocochlear nerve emerges just posterior to the facial nerve on the human brain.

IX Glossopharyngeal Nerve   This mixed nerve emerges from the medulla posterior to the vestibulocochlear. It functions in reflexes of the heart, taste, and swallowing. Taste buds on the back of the tongue are innervated by some of its fibers. Efferent fibers innervate muscles of the pharynx (swallowing) and the parotid salivary glands (secretion).

X Vagus Nerve   This cranial nerve, which originates on the medulla, exceeds all of the other cranial nerves in its extensive ramifications. In addition to supplying parts of the head and neck with nerves, it has branches that extend down into the chest and abdomen. It is a mixed nerve.

Sensory fibers in this nerve go to the heart, external acoustic meatus, pharynx, larynx, and thoracic and abdominal viscera.

Motor fibers pass to the pharynx, base of the tongue, larynx, and to the autonomic ganglia of the thoracic and abdominal viscera. Many references will be made to this nerve in subsequent discussions of the physiology of the lungs, heart, and digestive organs.

XI Accessory Nerve   Since this nerve emerges from both the brain and spinal cord it is sometimes called the *spinal accessory* nerve. Note that on both the sheep and human brains it parallels the lower part of the medulla and the spinal cord with branches going into both structures. On the human brain it appears to occupy a more posterior position than the twelfth nerve. Afferent and efferent spinal components innervate the *sternocleidomastoid* and *trapezius* muscles. The cranial portion of the nerve innervates the pharynx, upper larynx, uvula, and palate.

XII Hypoglossal Nerve   This nerve emerges from the medulla to innervate several muscles of the tongue. It is a mixed nerve. On the human brain in figure 34.4, it has the appearance of a cut-off tree trunk with roots extending into the medulla.

The Hypothalamus

That portion of the diencephalon that includes the mammillary bodies, infundibulum, and part of the hypophysis is collectively called the *hypothalamus.* The entire diencephalon, including the hyopthalamus, will be studied more in detail in the next exercise, when the internal parts of the brain are revealed by dissection. For now, let's consider the mammillary bodies and hypohysis.

Mammillary Bodies   This part of the hypothalamus lies superior to the hypophysis. Although it appears as a single body on the sheep, it consists of a pair of rounded eminences just posterior to the infundibulum on the human. These bodies receive fibers from the olfactory areas and ascending pathways of the brain. Fibers also exit from these bodies to the thalamus and other brain nuclei.

Hypophysis   This structure, also known as the *pituitary gland,* is attached to the base of the brain by a stalk, the **infundibulum.** It consists of two distinctly different parts: an anterior adenohypophysis and a posterior neurohypophysis. Only the neurohypophysis and infundibulum are considered to be part of the hypothalamus because they both have the same embryological origin as the brain.

The adenohypophysis originates as an outpouching of the ectodermal stomodeum (mouth cavity) and is quite different histologically from the neurohypophysis. A close functional relationship exists between the hypothalamus, hypophysis, and the autonomic nervous system.

Assignment:
Label figure 34.4.

The remainder of this exercise pertains to the labeling of figures 34.5 and 34.6 and does not pertain to sheep brain dissection. To continue the sheep brain dissection, proceed to Exercise 35.

Functional Localization of the Cerebrum

Extensive experimental studies on monkeys, apes, and humans have resulted in considerable knowledge of the functional areas of the cerebrum.

Figure 34.5 shows the locations of some of these centers. Before attempting to identify each of these areas be sure you know the boundaries (sulci and fissures) of the cerebral lobes (figure 34.2).

Somatomotor Area   This area occupies the surface of the precentral gyrus of the frontal lobe. Electrical stimulation of this portion of the cerebral cortex in a conscious human results in movement of specific muscular groups.

Premotor Area   The large area anterior to the somatomotor area is the *premotor area*. It exerts control over the motor area.

Somatosensory Area   This area is located on the postcentral gyrus of the parietal lobe. It functions to localize very precisely those points on the body where

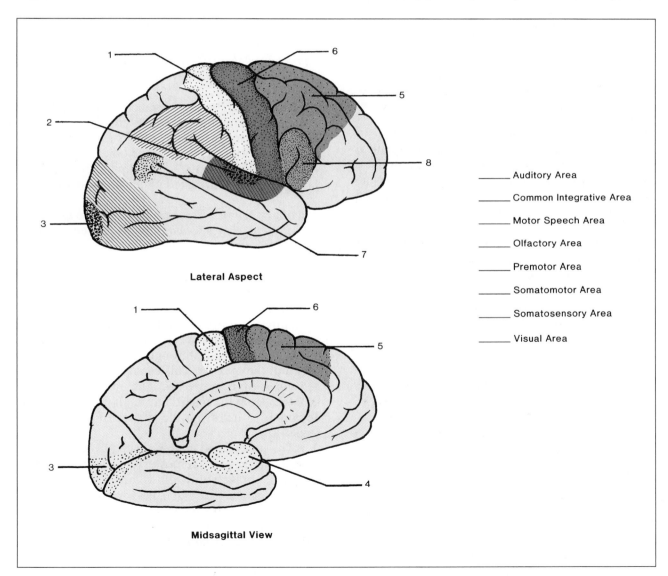

_____ Auditory Area

_____ Common Integrative Area

_____ Motor Speech Area

_____ Olfactory Area

_____ Premotor Area

_____ Somatomotor Area

_____ Somatosensory Area

_____ Visual Area

Lateral Aspect

Midsagittal View

Figure 34.5   Functional areas of the human brain.

sensations of light touch and pressure originate. It also assists in determining organ position. Other sensations, such as aching pain, crude touch, warmth, and cold are localized by the thalamus rather than this area.

Motor Speech Area This area is located in the frontal lobe just above the lateral cerebral fissure, anterior to the somatomotor area. It exerts control over the muscles of the larynx and tongue that produce speech.

Visual Area This area is located on the occipital lobe. Note that only a small portion of it is seen on the lateral aspect; the greater portion of this area is on the medial surface of the cerebrum. On this surface it extends anteriorly to the parieto-occipital fissure, becoming narrower as it approaches the fissure. This area receives impulses from the retina via the thalamus. Destruction of this region causes blindness, although light and dark are still discernible.

Auditory Area This area receives nerve impulses from the cochlea of the inner ear via the thalamus. It is responsible for hearing and speech understanding. It is located on the superior temporal gyrus, which borders on the lateral cerebral fissure.

Olfactory Area This area is located on the medial surface of the temporal lobe. The recognition of various odors occurs here. Tumors in this area cause individuals to experience nonexistent odors of various kinds, both pleasant and unpleasant.

Association Areas Adjacent to the somatosensory, visual, and auditory areas are *association areas* that lend meaning to what is felt, seen, or heard. These areas are lined regions in 34.5.

Common Integrative Area The integration of information from the above three association areas and the olfactory and taste centers is achieved by a small area that is called the *common integrative*, or *gnostic area*. This area is located on the **angular gyrus,** which is positioned approximately midway between the three association areas.

Assignment:
Label figure 34.5.

The Trigeminal Nerve

Although a detailed study of all cranial nerves is precluded in an elementary anatomy course, the trigeminal nerve has been singled out for a thorough study here. This fifth cranial nerve is the largest one that innervates the head and is of particular medical-dental significance.

Figure 34.6 illustrates the distribution of this nerve. You will note that it has branches that supply the teeth, tongue, gums, forehead, eyes, nose, and lips. The following description identifies the various ganglia and branches.

On the left margin of the diagram is the severed end of the nerve at a point where it emerges from the brain. Between this cut end and the three main branches is an enlarged portion, the **Gasserian ganglion.** This ganglion contains the nerve cell bodies of the sensory fibers in the nerve. The three branches that extend from this ganglion are the *ophthalmic, maxillary,* and *mandibular* nerves. Locate the bony portion of the skull in figure 34.6 through which these three branches pass.

Ophthalmic Nerve The ophthalmic nerve is the upper branch that passes out of the cranium through the *superior orbital fissure.* It has three branches that innervate the lacrimal gland, the upper eyelid, and the skin of the nose, forehead, and scalp. One of its branches also supplies the cornea of the eye, the iris, and the ciliary body (a muscle attached to the lens of the eye).

Superior to the lower branch of the ophthalmic is the **ciliary ganglion,** which contains nerve cell bodies that are incorporated into reflexes controlling the ciliary body. Unlike the other two divisions of the trigeminal, the ophthalmic nerve and its branches are not involved in dental anesthesia.

Maxillary Nerve This nerve, also known as the *second division of the trigeminal nerve,* is a sensory nerve that provides innervation to the nose, upper lip, palate, maxillary sinus, and upper teeth. It is the branch that comes off just inferior to the ophthalmic nerve.

Note that the maxillary nerve has two short branches that extend downward to an oval body, the **sphenopalatine ganglion.** The major portion of the maxillary nerve becomes the **infraorbital nerve,** which passes through the *infraorbital nerve canal* of the maxilla. This canal is shown with dotted lines in figure 34.6. The infraorbital nerve emerges from the maxilla through the **infraorbital foramen** to innervate the tissues of the nose and upper lip.

The upper teeth are innervated by three superior alveolar nerves. The **posterior superior alveolar nerve** is a branch of the maxillary nerve that enters the posterior surface of the upper jaw and innervates the molars. The **anterior superior alveolar nerve** is a branch of the infraorbital nerve that supplies the anterior teeth and bicuspids with nerve fibers. The anterior superior alveolar nerve forms a loop with the **middle superior alveolar nerve.** This latter nerve is shown clearly where a portion of the

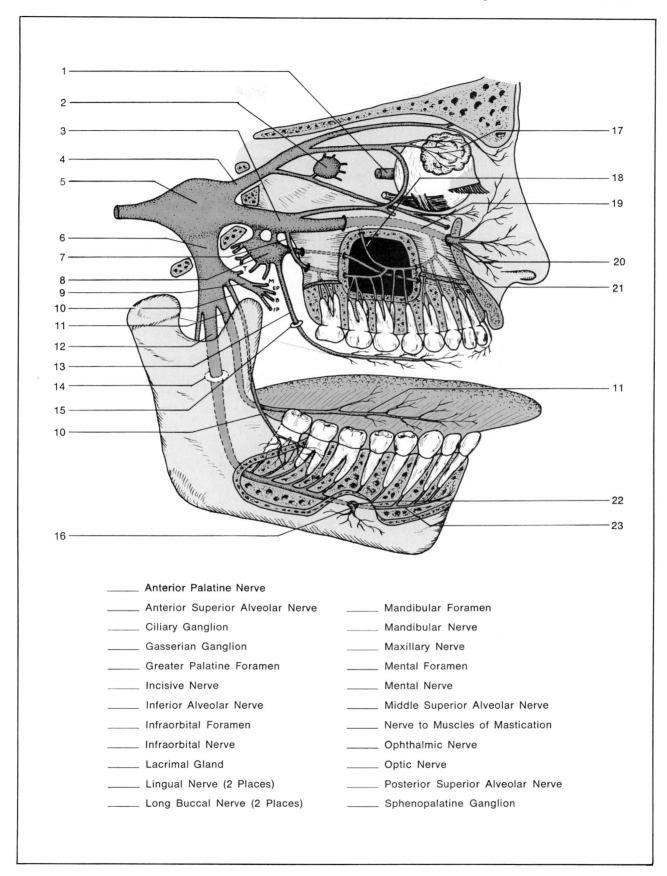

Figure 34.6   The trigeminal nerve.

_____ **Anterior Palatine Nerve**

_____ Anterior Superior Alveolar Nerve   _____ Mandibular Foramen

_____ Ciliary Ganglion   _____ Mandibular Nerve

_____ Gasserian Ganglion   _____ Maxillary Nerve

_____ Greater Palatine Foramen   _____ Mental Foramen

_____ Incisive Nerve   _____ Mental Nerve

_____ Inferior Alveolar Nerve   _____ Middle Superior Alveolar Nerve

_____ Infraorbital Foramen   _____ Nerve to Muscles of Mastication

_____ Infraorbital Nerve   _____ Ophthalmic Nerve

_____ Lacrimal Gland   _____ Optic Nerve

_____ Lingual Nerve (2 Places)   _____ Posterior Superior Alveolar Nerve

_____ Long Buccal Nerve (2 Places)   _____ Sphenopalatine Ganglion

maxilla has been cut away. It contains fibers that pass to the bicuspids and first molar.

To desensitize all of the upper teeth on one side of the maxilla, a dentist can inject anesthetic near either the maxillary nerve or the sphenopalatine ganglion. Desensitization of the maxillary nerve is called a *second division nerve block.* This type of nerve block will affect the palate as well as the teeth because it involves fibers of the **anterior palatine nerve.** The latter nerve extends downward from the sphenopalatine ganglion to the soft tissues of the palate through the *greater palatine foramen.*

Mandibular Nerve The mandibular nerve, or *third division of the trigeminal nerve,* is the most inferior branch of the trigeminal nerve. It innervates the teeth and gums of the mandible, the muscles of mastication, the anterior part of the tongue, the lower part of the face, and some skin areas on the side of the head.

Moving downward from the Gasserian ganglion, the first branch of the mandibular nerve that we encounter is the **nerve to the muscles of mastication.** This nerve has five branches with small identifying letters near their cut ends (**T** for temporalis, **M** for masseter, **EP** for external pterygoid, **B** for buccinator, and **IP** for internal pterygoid). The fibers in these nerves control the motor activities of these five muscles.

The largest branch of the mandibular nerve is the **inferior alveolar nerve,** which enters the ramus of the mandible through the **mandibular foramen** and supplies branches to all the teeth on one side of the mandible. At the mental foramen, the inferior alveolar nerve becomes the **incisive nerve,** which innervates the anterior teeth. Emerging from the mental foramen is the **mental nerve,** which supplies the soft tissues of the chin and lower lip.

Desensitization of the mandibular teeth is generally achieved by injecting anesthetic near the mandibular foramen. This type of injection, called a *lower nerve block,* or *third division nerve block,* is widely used in dentistry because the external and internal alveolar plates of the mandibular alveolar process are too dense to facilitate anesthesia by infiltration of the solid bone. The lower nerve block of one side of the mandible desensitizes all of the teeth on one side except for the anterior teeth. The fact that the anterior teeth receive nerve fibers from both the left and right incisive nerves prevents complete desensitization.

A branch of the mandibular nerve that emerges just above the inferior alveolar nerve and innervates the tongue is the **lingual nerve.** As in the case of the inferior alveolar nerve, it is sensory in function.

The **long buccal nerve** is another sensory branch of the mandibular nerve that innervates the buccal gum tissues of the molars and bicuspids. This nerve arises from a point that is just superior to the lingual nerve.

Assignment:
Label figure 34.6.

Laboratory Report
Complete Laboratory Report 34 for this exercise.

Brain Anatomy:
Internal

35

The nature of the interior of the brain is best seen by studying sections cut longitudinally and frontally through it. In this laboratory period we will study several sectional views to identify the various cavities (ventricles) and deeper portions such as the thalamus, fornix, and hypothalamus. Proceed as follows:

Materials:
> preserved sheep brains
> preserved human brains
> human brain models
> dissecting instruments
> long sharp knife
> dissecting trays

Midsagittal Section

With the preserved sheep brain resting ventral side down on a dissecting tray, place a long sharp meat-cutting knife (butcher knife) in the longitudinal cerebral fissure. Carefully slice through the tissue, cutting the brain into right and left halves on the midline. If only a scalpel is available, attempt to cut the tissue as smoothly as possible. Proceed to identify the structures on the sheep brain and their counterparts on the human brain as described below.

If the brain has been cut exactly on the midline, the only portion of the cerebrum that is cut is the **corpus callosum.** Locate this long band of white matter on the sheep brain by referring to figure 35.1. Also, compare your specimen with the human brain in figure 35.2. The corpus callosum is a commissural tract of fibers that connects the right and left cerebral hemispheres, correlating their functions. Note that this tract is significantly thicker on the human brain than on the sheep brain.

The Diencephalon

This portion of the brain lies between the corpus callosum of the cerebrum and the midbrain. It is also called the *interbrain*. It contains the thalamus, intermediate mass, hypothalamus, fornix, and third ventricle.

Fornix Locate this structure on the sheep brain. This body contains fibers that are an integral part of the *rhinencephalon* (olfactory mechanism) of the brain. Its equivalent on the human brain is label 18, figure 35.2. Note how much larger, proportionately, it is on the sheep brain.

Intermediate Mass This portion of the diencephalon is the only part of the thalamus that can be seen in the midsagittal sections of the sheep and human brains. The **thalamus** is a large ovoid

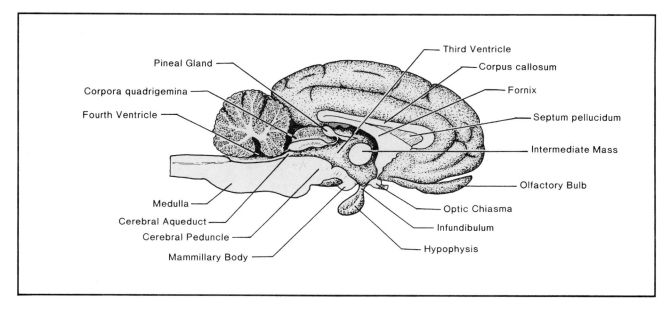

Figure 35.1 Midsagittal section of the sheep brain.

183

structure on the lateral walls of the third ventricle. It is an important relay station in which sensory pathways of the spinal cord and brain form synapses on their way to the cerebral cortex. The intermediate mass contains fibers that pass through the third ventricle, uniting the two sides of the thalamus.

Hypothalamus   This portion of the diencephalon extends about two centimeters up into the brain from the ventral surface. It has many nuclei that control various physiological functions of the body (temperature regulation, hypophysis control, coordination of the autonomic nervous system). The only part of the hypothalamus that is labeled in figure 35.2 is the **mammillary body** (label 7). For a better view of the hypothalamus refer to figure 35.4 (label 11).

Inferior and anterior to the mammillary body in figure 35.2 lie the **infundibulum** and **hypophysis.** Note how the latter lies encased in the bony recess of the *sella turcica.* The **optic chiasma** is the small round body just above the hypophysis, anterior to the infundibulum.

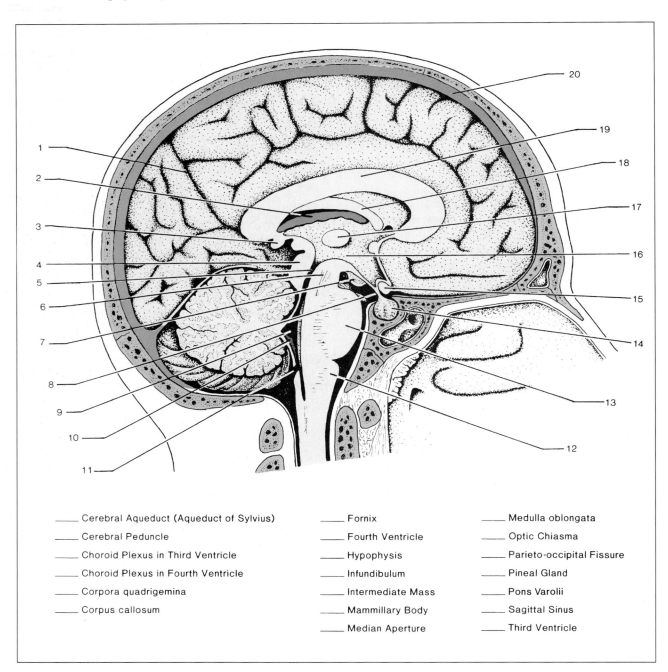

_____ Cerebral Aqueduct (Aqueduct of Sylvius)	_____ Fornix	_____ Medulla oblongata
_____ Cerebral Peduncle	_____ Fourth Ventricle	_____ Optic Chiasma
_____ Choroid Plexus in Third Ventricle	_____ Hypophysis	_____ Parieto-occipital Fissure
_____ Choroid Plexus in Fourth Ventricle	_____ Infundibulum	_____ Pineal Gland
_____ Corpora quadrigemina	_____ Intermediate Mass	_____ Pons Varolii
_____ Corpus callosum	_____ Mammillary Body	_____ Sagittal Sinus
	_____ Median Aperture	_____ Third Ventricle

Figure 35.2 Midsagittal section of the human brain.

The Midbrain and Diencephalon

Locate the **corpora quadrigemina, cerebral peduncles,** and **cerebral aqueduct** on the midbrain. On the diencephalon locate the **pineal gland.**

The *cerebral peduncles* of the midbrain consist of a pair of cylindrical bodies made up largely of ascending and descending fiber tracts that connect the cerebrum with the other three brain divisions.

The *cerebral aqueduct* is a duct that runs longitudinally through the midbrain, connecting the third and fourth ventricles.

The Cerebellum

Examine the cut surface of the cerebellum and note that gray matter exists near the outer surface. The cerebellum receives nerve impulses along cerebellar peduncles from motor and visual centers of the brain, the semicircular canals of the inner ears, and muscles of the body. Nerve impulses pass from this region to all the motor centers of the body wall to maintain posture, equilibrium, and muscle tonus. None of the activities of the cerebellum are of a conscious nature. Unlike the cerebrum, each hemisphere of the cerebellum controls muscles on its own side of the body.

The Pons and Medulla

These two parts make up the greater portion of the brain stem. As stated previously, the pons, medulla, and midbrain make up the entire brain stem. The pons is that portion between the midbrain and the medulla. Note that it consists primarily of white matter (fiber tracts). Locate the pons and medulla on both the sheep and human brains.

Within the brain stem is a complex interlacement of nuclei and white fibers (gray and white matter) that constitutes the *reticular formation.* Nuclei of this formation generate a continuous flow of impulses to other parts of the brain to keep the cerebrum in a state of alert consciousness.

The Ventricles and Cerebrospinal Fluid

The brain has four cavities called *ventricles* that contain cerebrospinal fluid. There are two **lateral ventricles** situated in the lower medial portion of each cerebral hemisphere. Between these two ventricles on the median line is a thin membrane, the **septum pellucidum** (shown in figure 35.1, but not in figure 35.2). Refer to figure 35.4 to better visualize the two lateral ventricles (label 5) and the septum pellucidum.

The other two ventricles of the brain are the **third** and **fourth ventricles.** These ventricles are positioned on the median line and are visible in figures 35.1 and 35.2. Locate them first on the sheep brain and then on figure 35.2. Note that the **intermediate mass of the thalamus** passes through the third ventricle.

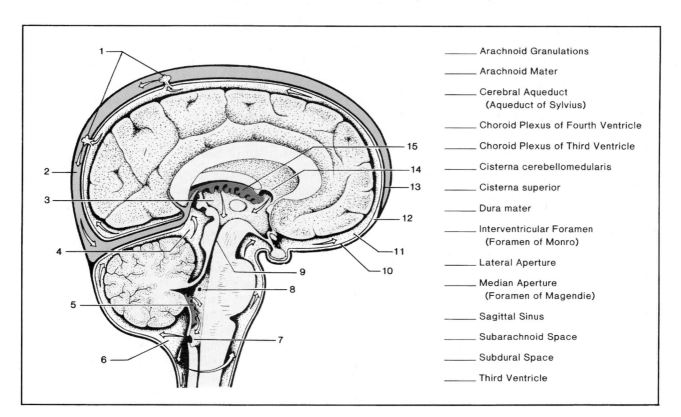

_____ Arachnoid Granulations

_____ Arachnoid Mater

_____ Cerebral Aqueduct
(Aqueduct of Sylvius)

_____ Choroid Plexus of Fourth Ventricle

_____ Choroid Plexus of Third Ventricle

_____ Cisterna cerebellomedularis

_____ Cisterna superior

_____ Dura mater

_____ Interventricular Foramen
(Foramen of Monro)

_____ Lateral Aperture

_____ Median Aperture
(Foramen of Magendie)

_____ Sagittal Sinus

_____ Subarachnoid Space

_____ Subdural Space

_____ Third Ventricle

Figure 35.3   Origin and circulation of cerebrospinal fluid.

Within the third ventricle is seen a **choroid plexus** (label 2, figure 35.2), that secretes cerebrospinal fluid (CSF). Although not shown in any of the human brain illustrations here, there are choroid plexuses in the lateral ventricles also. A choroid plexus in the fourth ventricle is shown in figure 35.2.

Cerebrospinal Fluid Pathway   The CSF is constantly being produced by choroid plexuses of all the ventricles. Figure 35.3 indicates the pathway of the fluid in the brain.

CSF that originates in the lateral ventricles enters the third ventricle through the **interventricular foramen** (foramen of Monro). Note in figure 35.3 that this foramen lies anterior to the fornix and choroid plexus of the third ventricle.

From the third ventricle the CSF passes into the **cerebral aqueduct** (aqueduct of Sylvius). This duct conveys the CSF to the fourth ventricle. The **choroid plexus of the fourth ventricle** lies on the posterior wall of this ventricle.

From the fourth ventricle the CSF exits into the subarachnoid space that surrounds the cerebellum through three apertures: one on the median line and two lateral apertures.

In figure 35.3 the **median aperture** (foramen of Magendie) is the lower opening that has two arrows leading from it. That part of the subarachnoid space that it empties into is the **cisterna cerebellomedullaris.** Since the **lateral apertures** (foramina of Luschka) are located on the lateral walls of the fourth ventricle, only one is shown in figure 35.3.

From the cisterna cerebellomedullaris the CSF moves up into the **cisterna superior** (above the cerebellum) and finally into the subarachnoid space around the cerebral hemispheres. From the subarachnoid space this fluid escapes into the blood of the sagittal sinus through the **arachnoid granulations.** The CSF passes down the subarachnoid space of the spinal cord on the posterior side and up along the anterior surface.

Assignment:
After identifying all the structures of the sheep brain, label figure 35.2.
Label figure 35.3.

Frontal Sections

Illustrations A and B of figure 35.4 reveal frontal sections through two slightly different regions of the sheep brain. Once you study sections such as these, the human brain section in illustration C will be relatively easy to label.

Infundibular Section (Sheep Brain)

Place a whole sheep brain on a dissecting tray with the *dorsal side down.* Cut a frontal section with a long knife by starting at the point of the infundibulum. Cut through perpendicularly to the tray and to the longitudinal axis of the brain. Then, make a second cut parallel to the first cut ¼ inch farther back. The latter cut should be through the center of the hypophysis.

Place this slice, *with the first cut upward,* on a piece of paper toweling for study. Illustration A, figure 35.4, is of this surface. Proceed to identify the following structures. You may need to refer back to figure 35.1.

Note the pattern and thickness of the **gray matter** of the cerebral cortex. Force a probe down into a sulcus. What meninx do you break through as the probe moves inward?

Locate the two triangular **lateral ventricles.** Probe into one of the ventricles to see if you can locate the **choroid plexus.** Identify the **corpus callosum** and **fornix.**

Identify the **third ventricle,** which is situated on the median line under the fornix. Also, identify the **thalamus** and **intermediate mass.** Does the thalamus appear to consist of white or gray matter? What is the function of the white matter between the thalamus and the cerebral cortex?

Hypophyseal Section (Sheep Brain)

Prepare a section of the sheep brain by cutting another ¼ inch slice of the posterior portion of the brain. Place this slice, *anterior face upward,* on paper toweling. This section through the hypophysis should look like illustration B, figure 35.4. Identify the structures in illustration B that are labeled.

Infundibular Section of Human Brain

Now that you have identified the parts of the sheep brain, identify the following structures in illustration C in figure 35.4:

Locate first the two triangular **lateral ventricles** and the **third ventricle.** Note how the **longitudinal cerebral fissure** extends downward to the **corpus callosum,** which forms the upper wall of the lateral ventricles. Although there is no label for the *septum pellucidum,* can you find it? Identify the two masses of gray matter just below the septum pellucidum that constitute the **fornix.**

Identify the areas on each side of the third ventricle that make up the nuclei of the **hypothalamus.** Approximately eight separate masses of gray matter are seen here. What physiological functions are regulated by these nuclei?

Locate the two large areas on the walls of the lateral ventricles that constitute the **thalamus.** Note, also, how these two portions of the thalamus are united by the **intermediate mass** that passes through the third ventricle.

Labels 12, 13, 14, and 15 are masses of gray matter in the cerebral hemispheres that are collectively known as the *basal ganglia.* The largest basal ganglia are the **putamen** and **globus pallidus** (labels 12 and 13). The latter is medial to the putamen. Together the putamen and globus pallidus form a triangular mass, the **lentiform nucleus.** These two ganglia exert a steadying effect on voluntary muscular movements.

The **caudate nuclei** are smaller basal ganglia located in the walls of the lateral ventricles superior to the thalamus. These nuclei have something to do with muscular coordination, since surgically produced lesions in these centers can correct certain kinds of palsy.

Although some anatomists prefer to designate the thalamus and hypothalamus as basal ganglia, functionally, they differ considerably.

Assignment:
Label figure 35.4.

Laboratory Report
Complete Laboratory Report 35 for this exercise.

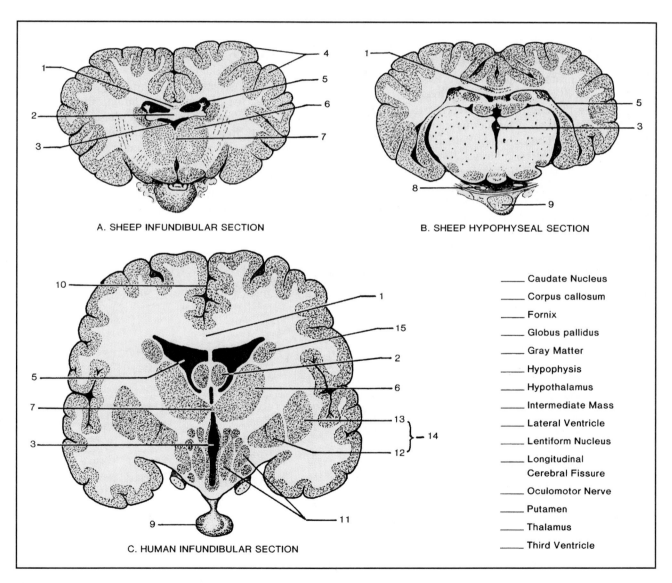

A. SHEEP INFUNDIBULAR SECTION

B. SHEEP HYPOPHYSEAL SECTION

C. HUMAN INFUNDIBULAR SECTION

_____ Caudate Nucleus
_____ Corpus callosum
_____ Fornix
_____ Globus pallidus
_____ Gray Matter
_____ Hypophysis
_____ Hypothalamus
_____ Intermediate Mass
_____ Lateral Ventricle
_____ Lentiform Nucleus
_____ Longitudinal Cerebral Fissure
_____ Oculomotor Nerve
_____ Putamen
_____ Thalamus
_____ Third Ventricle

Figure 35.4 Frontal sections of sheep and human brains.

36

Anatomy of the Eye

In preparation for the study of eye function in the next exercise, our focus here will be on the anatomy of the eye through dissection, ophthalmoscopy, and microscopy. Before performing these studies, however, label figures 36.1, 36.2, and 36.3, which pertain to the lacrimal system, internal anatomy, and the extrinsic muscles.

The Lacrimal Apparatus

Each eye lies protected in a bony recess of the skull and is covered externally by the **eyelids** *(palpebrae)*. Lining the eyelids and covering the exposed surface of the eyeball is a mucous membrane called the **conjunctiva.** The conjunctival surfaces are kept constantly moist by protective secretions of the lacrimal glands, which lie between the upper eyelid and the eyeball.

Figure 36.1 reveals the external anatomy of the right eye, with particular emphasis on the lacrimal apparatus. Note that the lacrimal gland consists of two portions: a large **superior lacrimal gland** and a smaller **inferior lacrimal gland.** Secretions from these glands pass through the conjunctiva of the eyelid via six to twelve small ducts.

The flow of fluid from these ducts moves across the eyeball to the *medial canthus* (angle) of the eye, where the fluid enters two small orifices, the **lacrimal puncta.** These openings lead into the **superior** and **inferior lacrimal ducts;** the superior one is in the upper eyelid and the inferior one is in the lower eyelid. These two ducts empty directly into an enlarged cavity, the **lacrimal sac,** which, in turn, is drained by the **nasolacrimal duct.** The tears finally exit into the nasal cavity through the end of the nasolacrimal duct.

Two other structures, the caruncula and the plica semilunaris, are seen near the puncta in the medial canthus of the eye. The **caruncula** is a small red conical body that consists of a mound of skin containing sebaceous and sweat glands and a few small hairs. It is this structure that produces a whitish secretion that constantly collects in this region. Lateral to the caruncula is a curved fold of conjunctiva, the **plica semilunaris.** This structure contains some smooth muscle fibers; in the cat and many other animals it is more highly developed than in humans and is often referred to as the "third eyelid."

Assignment:
Label figure 36.1.

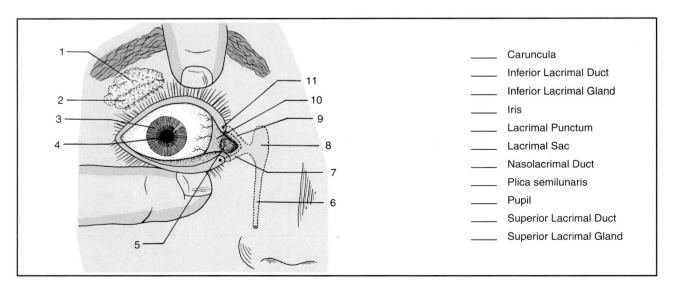

_____ Caruncula

_____ Inferior Lacrimal Duct

_____ Inferior Lacrimal Gland

_____ Iris

_____ Lacrimal Punctum

_____ Lacrimal Sac

_____ Nasolacrimal Duct

_____ Plica semilunaris

_____ Pupil

_____ Superior Lacrimal Duct

_____ Superior Lacrimal Gland

Figure 36.1   **The lacrimal apparatus of the eye.**

Internal Anatomy

Figure 36.2 is a horizontal section of the right eye as seen looking down upon it. Note that the wall of the eyeball consists of three layers: an outer **scleroid coat,** a middle **choroid coat,** and an inner **retina.** The continuity of the surface of the retina is interrupted only by two structures, the optic nerve and the fovea centralis.

The **optic nerve** (label 3) contains fibers leading from the rods and cones of the retina to the brain. That point on the retina where the optic nerve makes its entrance is lacking in rods and cones. The absence of light receptors at this spot makes it insensitive to light; thus, it is called the **blind spot.** It is also known as the **optic disk.**

Note that the optic nerve is wrapped in a sheath, the **dura mater,** which is an extension of the dura mater that surrounds the brain. To the right of the blind spot is a pit, the **fovea centralis,** which is the center of a round yellow spot, the *macula lutea.* The macula, which is only a half a millimeter in diameter, is composed entirely of cones and is that part of the eye where all critical vision occurs.

Light entering the eye is focused on the retina by the **lens,** an elliptical crystalline clear structure suspended in the eye by a **suspensory ligament.** The substance of the lens is a completely transparent protein called *crystallin,* which is formed from cells during embryological development. The cells that produce this protein self-destruct (apoptosis) before birth.

Attached to the suspensory ligament is the **ciliary body,** a circular smooth muscle that can change the shape of the lens to focus on objects at different distances.

In front of the lens is a cavity that contains a watery fluid, the **aqueous humor.** Between the lens and the retina is a larger cavity that contains a more viscous, jellylike substance, the **vitreous body.** Immediately in front of the lens lies the circular colored portion of the eye, the **iris.** It consists of circular and radiating muscle fibers that can change the size of the **pupil** of the eye. The iris regulates the amount of light that enters the eye through the pupil.

Covering the anterior portion of the eye is the **cornea,** a clear, transparent structure that is an extension of the scleroid coat. It acts as a window to the eye, allowing light to enter.

Aqueous humor is constantly being renewed in the eye. The enlarged inset of the lens-iris area in figure 36.2 illustrates where the aqueous humor is

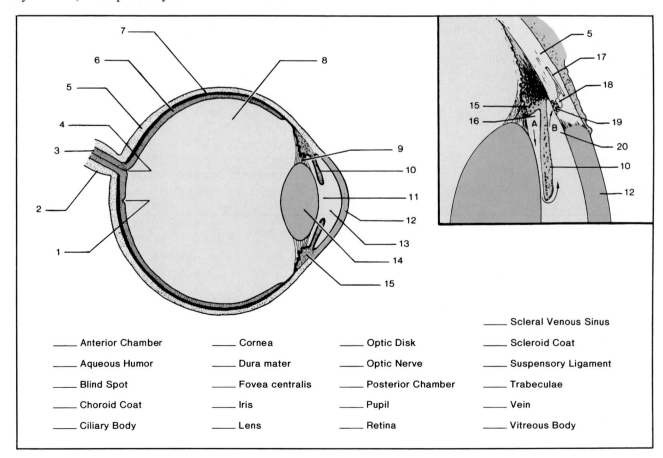

____ Anterior Chamber	____ Cornea	____ Optic Disk	____ Scleral Venous Sinus
____ Aqueous Humor	____ Dura mater	____ Optic Nerve	____ Scleroid Coat
____ Blind Spot	____ Fovea centralis	____ Posterior Chamber	____ Suspensory Ligament
____ Choroid Coat	____ Iris	____ Pupil	____ Trabeculae
____ Ciliary Body	____ Lens	____ Retina	____ Vein
			____ Vitreous Body

Figure 36.2 Internal anatomy of the eye.

produced (point A) and where it is reabsorbed (point B). At point A, which is the **posterior chamber,** aqueous humor is produced by capillaries of the ciliary body. As this fluid is produced, it passes through the pupil into the **anterior chamber** and is reabsorbed into minute spaces called **trabeculae** at the juncture of the cornea and the iris (point B). From the trabeculae the fluid passes to the **scleral venous sinus** (canal of Schlemm), label 18. This venous sinus drains into the venous system, represented as a short **vein** in figure 36.2.

Assignment:
Label figure 36.2.

The Ocular Muscles

The ability of each eye to move in all directions within its orbit is controlled by six *extrinsic ocular muscles.* Movement of the upper eyelid is controlled by a single muscle, the **superior levator palpebrae.**

Figure 36.3, which is a lateral view of the right eye, reveals the insertions of all seven of these muscles. Their origins are on various bony formations at the back of the eye orbit.

The muscle on the side of the eyeball that has a portion of it removed is the **lateral rectus.** Opposing it on the medial side of the eye is the **medial rectus.** The muscle that inserts on top of the eye is the **superior rectus;** its antagonist on the bottom of the eye is the **inferior rectus.** Just above the stub of the lateral rectus is seen the insertion of the **superior oblique,** which passes through a cartilaginous loop, the **trochlea.** Opposing the superior oblique is the **inferior oblique,** of which only a portion of the insertion can be seen.

Assignment:
Label figure 36.3.

Beef Eye Dissection

The nature of the tissues of the eye can best be studied by actual dissection of an eye. A beef eye will be used for this study. Figure 36.4 illustrates various steps that will be followed.

Materials:
beef eye, preferably fresh
dissecting instruments
dissecting tray

1. Examine the external surfaces of the eyeball. Identify the **optic nerve** and look for remnants of **extrinsic muscles** on the sides of the eyeball.
2. Note the shape of the **pupil.** Is it round or elliptical? Identify the **cornea.**
3. Holding the eyeball as shown in illustration 1, figure 36.4, make an incision into the scleroid coat with a sharp scalpel about one-quarter of an inch away from the cornea. Note how difficult it is to penetrate.
4. Insert scissors into the incision and carefully cut all the way around the cornea, *taking care not to squeeze the fluid out of the eye.* Refer to illustration 2, figure 36.4.
5. Gently lift the front portion off the eye and place it on the tray with the inner surface upward. The lens usually remains attached to the vitreous body, as shown in illustration 3. If the eye is preserved (not fresh), the lens may remain in the anterior portion of the eye.
6. Examine the inner surface of the anterior portion and identify the thickened, black circular **ciliary body.** What function does the black pigment perform?
7. Study the **iris** carefully. Can you distinguish **circular** and **radial muscle fibers?**

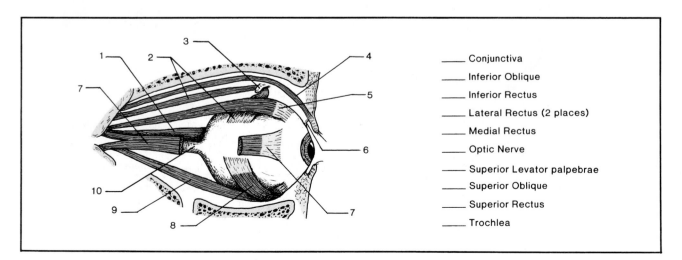

____ Conjunctiva
____ Inferior Oblique
____ Inferior Rectus
____ Lateral Rectus (2 places)
____ Medial Rectus
____ Optic Nerve
_____ Superior Levator palpebrae
____ Superior Oblique
____ Superior Rectus
____ Trochlea

Figure 36.3 The extrinsic muscles of the eye.

8. Is there any **aqueous humor** left between the cornea and iris? Compare its consistency with the **vitreous body.**

9. Separate the lens from the vitreous body by working a dissecting needle around its perimeter as shown in illustration 3. If the lens is from a fresh eye (not preserved), hold the lens up by its edge with a forceps and look through it at some distant object.

   What is unusual about the image? Next, place the lens on printed matter. Are the letters magnified?

   Note: Lenses from preserved specimens are usually not transparent due to coagulation of the crystallin by preservatives.

10. Compare the consistency of the lens at its center and near its circumference by squeezing the lens between your thumb and forefinger. Do you detect any differences?

11. Locate the **retina** at the back of the eye. It is a thin colorless inner coat that separates easily from the pigmented **choroid coat.**

    Now locate the **blind spot** *(optic disk).* This is the area where the optic nerve forms a juncture with the retina.

12. Note the irridescent nature of a portion of the choroid coat. This reflective surface is called the **tapetum lucidum.** It enables the eyes of animals to reflect light at night and appears to enhance night vision by reflecting some light back into the retina.

13. Answer all questions on the Laboratory Report that pertain to this dissection.

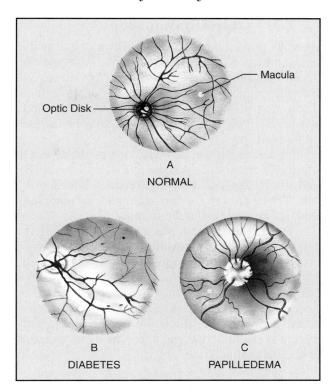

A
NORMAL

B
DIABETES

C
PAPILLEDEMA

Figure 36.5 Normal and abnormal retinas as seen through an ophthalmoscope. In illustration B *(diabetic retinopathy)* the peripheral vessels are generally irregular, fewer in number, and often exhibit small hemorrhages. *Papilledema,* or "choked disk" (illustration C), may be evidence of hypertension, brain tumor, or meningitis. In this case the veins are considerably enlarged and often hemorrhagic.

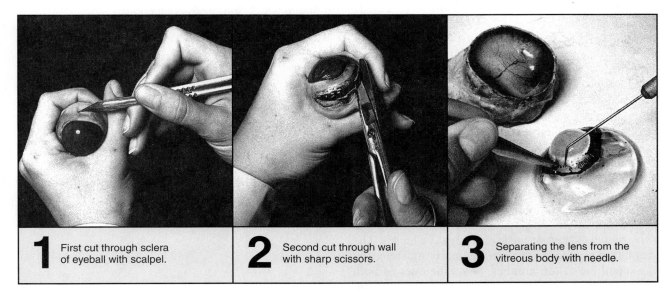

1 First cut through sclera of eyeball with scalpel.

2 Second cut through wall with sharp scissors.

3 Separating the lens from the vitreous body with needle.

Figure 36.4 Three steps in beef eye dissection.

Ophthalmoscopy

Routine physical examinations invariably involve an examination of the **fundus,** or interior of the eye. A careful examination of the fundus provides valid information about the general health of the patient. The retina of the eye is the only part of the body where relatively large blood vessels may be inspected without surgical intervention.

Figure 36.5 illustrates what one might expect to see in a normal eye (A), the eye of a diabetic (B), and the eye of one who exhibits a "choked disk" (C). These are only a few examples of pathology that can be detected in the fundus.

The instrument that one uses for examining the fundus is called an **ophthalmoscope.** In this exercise you will have an opportunity to examine the interior of the eyes of your laboratory partner. Your partner, in turn, will examine your eyes. It will be necessary to make this examination in a darkened room. Before this examination can be made, however, it is essential that you understand the operation of the ophthalmoscope.

The Ophthalmoscope

Figure 36.6 reveals the construction of a Welch Allyn ophthalmoscope. Within its handle are two C-size batteries that power an illuminating bulb that is located in the viewing head. Note that at the top of the handle there is a **rheostat control** for regulating the intensity of the light source. Only after depressing the lock button can this control be rotated.

The head of the instrument has two rotatable notched disks: a large lens selection disk and a smaller aperture selection disk. A viewing aperture is located at the upper end of the head.

While looking through the viewing aperture, the examiner is able to select one of twenty-three small lenses by rotating the **lens selection disk** with the index finger held as shown in figure 36.7. Twelve of the lenses on this disk are positive (convex) and eleven of them are negative (concave). Numbers that appear in the **diopter window** indicate which lens is in place. The positive lenses are represented as black numbers, the negative lens numbers are red. A black "0" in the window indicates that no lens is in place.

If the eyes of both subject and examiner are normal *(emmetropic),* no lens is needed. If the eye of the subject or examiner is farsighted *(hypermetropic),* the positive lenses will be selected. Nearsighted *(myopic)* eyes require the use of negative lenses. The degree of myopia or hypermetropia is indicated by the magnitude of the number. Since the eyes of both the examiner and subject affect the lens selection, the selection becomes entirely empirical.

The **aperture selection disk** enables the examiner to change the character of the light beam that is

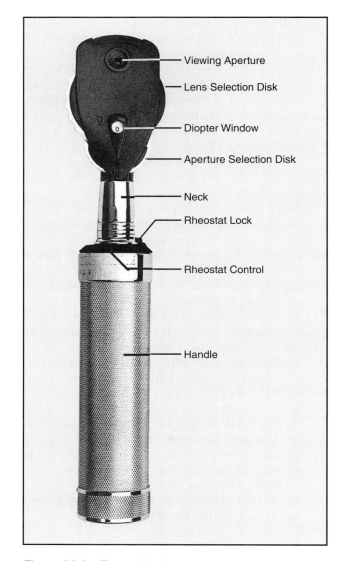

Figure 36.6   The ophthalmoscope.

Labels: Viewing Aperture, Lens Selection Disk, Diopter Window, Aperture Selection Disk, Neck, Rheostat Lock, Rheostat Control, Handle

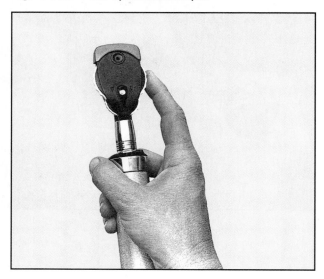

Figure 36.7   Viewing lenses of ophthalmoscope are rotated into position by moving the lens selection disk with the index finger.

projected into the eye. The light may be projected as a grid, straight line, white spot, or green spot. **The green spot is most frequently used** because it is less irritating to the eye of the subject, and it causes the blood vessels to show up more clearly.

Using an ophthalmoscope for the first time seems difficult for some, relatively easy for others, and nearly impossible for a few. If the examiner and subject have normal eyes, usually there is a minimum of difficulty. However, when the examiner's eyes are severely myopic, astigmatic, or hypermetropic, difficulties can occur. Even for examiners with normal vision there is the problem of reflection of light from the retina back into the examiner's eye. The best way to minimize this problem is to **direct the light beam toward the edge of the pupil** rather than through the center.

A precautionary statement must be heeded concerning light exposure. **Avoid subjecting an eye to more than one minute of light exposure.** After one minute exposure, allow several minutes of rest for recovery of the retina.

Procedure

Prior to using the ophthalmoscope in the darkened room, it will be necessary for you to become completely familiar with its mechanics.

Materials:
ophthalmoscope
metric tape or ruler

Desk Study of Ophthalmoscope

Turn on the light source by depressing the lock button and rotating the rheostat control. Observe how the light intensity can be varied from dim to very bright. Rotate the aperture selection disk as you hold the ophthalmoscope about 2 inches away from your desktop. How many different light patterns are there? Select the large white spot for the next phase of this study.

Rotate the lens selection disk until a black "5" appears in the diopter window. While peering through the viewing window at the print on this page, bring the letters into sharp focus. Have your lab partner measure with a metric tape the distance in millimeters between the ophthalmoscope and the printed page when the lettering is in sharp focus. Record this distance on the Laboratory Report. Do the same thing for the 10D, 20D, and 40D lenses, recording all measurements.

Now, set the lens selection disk on "0" and the light source on the green spot. The instrument is properly set now for the beginning of an eye examination. Read over the following procedure *in its entirety* before entering the darkened room. This will prepare you for what you will be doing.

The Examination

Figures 36.8 and 36.9 illustrate how the ophthalmoscope is held when viewing different eyes. Note that the examiner uses the right hand when examining the right eye and the left hand when examining the left eye. Proceed as follows:

1. With the "0" in the diopter window and the light turned on, grasp the ophthalmoscope, as shown in figure 36.8. Note that the index finger rests on the lens selection disk.
2. Place the viewing aperture in front of your right eye and steady the ophthalmoscope by resting

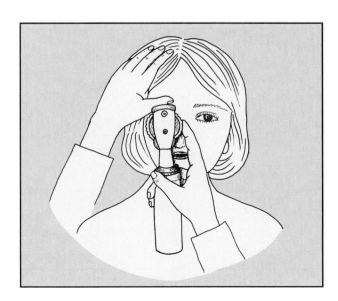

Figure 36.8 When examining the right eye, the ophthalmoscope is held with the right hand.

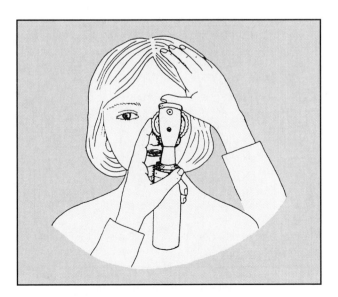

Figure 36.9 When examining the left eye, the ophthalmoscope is held with the left hand.

the top of it against your eyebrow. See figure 36.10.

Start viewing the subject's right eye at a distance of about 12 inches. Keep your subject on your right and instruct your subject to **look straight ahead at a fixed object at eye level.**

3. Direct the beam of light into the pupil and examine the lens and vitreous body. As you look through the pupil, a **red reflex** will be seen. If the image is not sharp, adjust the focus with the lens selection disk.

4. While keeping the pupil in focus, move in to within 2 inches of the subject, as illustrated in figure 36.11. The red reflex should become more pronounced. **Try to direct the light beam toward the edge of the pupil** rather than in the center to minimize reflection from the retina.

5. Search for the **optic disk** and adjust the focus, as necessary, to produce a sharp image. Observe how blood vessels radiate out from the optic disk. Follow one or more of them to the periphery.

Locate the **macula.** It is situated about two disk diameters *laterally* from the optic disk. Note that it lacks blood vessels. It is easiest to observe when the subject is asked to look directly into the light, a position that **must be limited to one second only.**

6. Examine the entire retinal surface. Look for any irregularities, depressions, or protrusions from its surface. It is not unusual to see a roundish elevation, not unlike a pimple on its surface.

To examine the periphery, instruct the subject to (1) **look up** for examination of the superior retina, (2) **look down** for examination of the inferior retina, and (3) **look laterally and medially** for those respective areas.

7. **Caution:** Remember to limit the right eye examination to **one minute!** After this time limit is reached, examine the left eye, using the left hand to hold the ophthalmoscope and reversing the entire procedure.

8. Switch roles with your laboratory partner. When you have examined each other's eyes examine the eyes of other students. Report any unusual conditions to your instructor.

Histological Studies

Most prepared slides available for laboratory study are made from either the rabbit or the monkey. Both animals are suitable. Proceed as follows to identify structures depicted in the photomicrographs of figures HA-16 and HA-17 of the Histology Atlas.

Materials:
prepared slides of:
    monkey eye, x.s. (H10.635, H10.64 or H10.76)
    rabbit eye, sagittal section (H10.61 or 10.62)

The Cornea With the lowest powered objective on your microscope scan a cross section of the eye of a rabbit or monkey until you find the cornea. Refer to illustrations A and B, figure HA-16. Note that the cornea consists of three layers: the epithelium, stroma, and endothelium. For fine detail, as in illustration B, use high-dry.

The **corneal epithelium** is the outer portion that consists of stratified squamous cells. It is, essentially, the conjunctival portion of the cornea. Note that the outermost cells of this layer are flattened squamous cells and that the basal cells are columnar in shape.

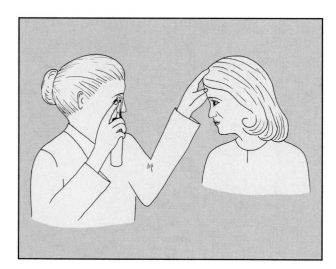

Figure 36.10 When viewing the lens and vitreous body, use this position, relative to the subject.

Figure 36.11 When examining the retinal surfaces move in close as illustrated here.

The **corneal endothelium** is the innermost layer of cells, which is adjacent to the aqueous humor. This thin layer consists of low cuboidal cells.

The **corneal stroma** *(substantia propria)* constitutes nine-tenths of the thickness of the cornea. It consists of collagenous fibrils, fibroblasts, and cementing substance. The fibrils are arranged in lamellae that run parallel to the corneal surface. The components of this stroma are held together by a mucopolysaccharide cement. The chemical structure and fibril arrangement contribute to the transparency of the cornea.

The Lens   Note (illustration C, figure HA-16) that the lens is enclosed in a homogenous elastic **capsule** to which the **suspensory ligament** is attached. The lens is formed from epithelial cells that become elongated to a fibrous shape (lens fibers), losing most of their organelles and nuclei. All that remains in a mature lens fiber are a few microtubules and clumps of free ribosomes. The principal component of the lens is a protein called *crystallin,* which is formed during embryological development.

The Ciliary Muscle   This ring of smooth muscle tissue is a part of the body wall of the eye. Identify the **ciliary processes,** which form ridges on the ciliary muscle. The latter provide an anchor for the suspensory ligaments.

Body Wall of the Eye   Examine a section through the body wall at the back of the eye. Identify the sclera, choroid coat, and retina. Note that the **sclera** consists of closely packed collagenous fibers, elastic fibers, and fibroblasts.

Note, also, the large amount of pigment in the **choroid coat** that is produced by **melanocytes.** This is the layer that supplies nourishment to the retina and scleroid coat. Look for blood vessels.

The **retina** is the inner photosensitive layer of the body wall.

Retinal Layers   The retina is composed of five main classes of neurons: the *photoreceptors* (rods and cones), *bipolar cells, ganglion cells, horizontal cells,* and *amacrine cells.* The first three form a direct pathway from the retina to the brain. The horizontal and amacrine cells form laterally directed pathways that modify and control the message being passed along the direct pathway.

Study a section of the retina and identify the various layers. Refer to figure HA-17 for reference. Any slide of monkey eye section is suitable for retinal study.

At the base of the retina, where it meets the choroid layer, are the **rods** and **cones.** The nuclei for these photoreceptors are located in the **outer nuclear layer.** To stimulate these receptors light must pass through all the other cell layers.

Nuclei of the amacrine and bipolar neurons are located in the **inner nuclear layer.** Dendrites of these association neurons make synaptic connections with axons of the rods and cones in the **outer synaptic layer.**

The nuclei of ganglion cells are located closest to the exposed surface of the retina in a layer designated as the **ganglion cell layer.** The dendrites of ganglion cells make synaptic connections with amacrine and bipolar cells in the **inner synaptic layer.** Nonmyelinated axons of the ganglion cells fill up the **nerve fiber layer;** they converge at the optic disk to form the optic nerve. Some neuroglial cell nuclei can be seen in the nerve fiber layer.

The Fovea Centralis   Study a slide that reveals the structure of the fovea (H10.635). Compare your slide with the photomicrographs in illustrations B and C in figure HA-17.

Laboratory Report

Answer the questions on the first portion of combined Laboratory Report 36,37.

Answers to Histology Self-Quiz No. 1

1. Fibrous connective tissue
2. Fibroblasts
3. Collagenous (white) fibers
4. Squamous
5. Lining of mouth
6. Plain columnar
7. Basal lamina
8. Adipose
9. Food storage, insulation, protection
10. Stratified squamous
11. Stratum basale
12. Dermis
13. Loose (areolar) connective
14. Collagenous (white)
15. Elastic (yellow)
16. Epidermis
17. Hair shaft
18. Hair follicle
19. Internal root sheath
20. External root sheath
21. Hypodermis (subcutaneous)
22. Loose (areolar) connective
23. Mast cell
24. Fibroblasts
25. Transitional epithelial
26. Basal lamina
27. Lamina propria
28. Cuboidal (glandular)
29. Glands
30. Elastic cartilage
31. Reticular connective
32. Liver, lymph glands, etc.
33. Binucleate
34. Transitional
35. Plain columnar
36. Goblet
37. Brush border
38. Fibrocartilage
39. Chondrocytes
40. Collagenous (white)
41. Compact bone
42. Perforating canal
43. Smooth muscle tissue
44. Nuclei
45. Pseudostratified ciliated
46. Cilia
47. Goblet cell
48. Smooth muscle
49. Canaliculi
50. Osteocytes
51. Central canal
52. Cardiac muscle tissue
53. Intercalated disks
54. Heart muscle, walls of vena cava
55. Striated muscle tissue
56. Endomysium
57. Cuboidal epithelial tissue
58. Stratum spinosum
59. Dermis
60. Keratinization
61. Striated muscle
62. Sarcolemma
63. Sarcomere
64. Pacinian corpuscle
65. Sensitive to pressure
66. Bone tissue (membranous)
67. Osteoclasts
68. Bone reabsorption
69. Sweat gland
70. Ducts of sweat gland
71. Dermis
72. External root sheath
73. Hair shaft
74. Bulb
75. Follicular papilla

Answers to Histology Self-Quiz No. 2

1. Cerebellum
2. Molecular layer
3. Granule cell layer
4. Purkinje cells
5. Pyramidal cell
6. Cerebrum
7. Axon
8. Apical dendrite
9. Cochlea of the ear
10. Vestibular membrane
11. Scala vestibuli
12. Spiral ligament
13. Scala tympani
14. Endolymph
15. Tectorial membrane
16. Rod of Corti
17. Outer hair cells
18. Basilar membrane
19. Motor neuron
20. Gray matter of spinal cord
21. Motor neuron
22. Nissl bodies
23. Dendrite
24. Axon
25. Crista amullaris
26. Ampulla of semicircular duct
27. Hair cells
28. Endolymph
29. Sensory neuron
30. Sensory ganglion
31. Satellite cells
32. Lens of the eye
33. Lens capsule
34. Suspensory ligament
35. Ciliary process
36. Nerve fiber layer
37. Inner nuclear layer
38. Outer synaptic layer
39. Rods and cones of the eye
40. Inner nuclear layer
41. Melanocytes
42. Rods and cones
43. Fovea centralis
44. Critical vision
45. Iris of the eye
46. Cornea
47. Corneal epithelium
48. Corneal stroma
49. Anterior chamber/aqueous humor
50. Posterior chamber/vitreous body

Visual Tests

The various tests outlined in this exercise relate to the observation of normal conditions as well as the detection of some of the more common types of abnormalities. By performing these tests you will learn much about the physiology of vision.

Materials:

 penlite (for image suppression)
 12″ ruler or tape measure (for blind spot test)
 3″ × 5″ card and pins (for Scheiner's experiment)
 Snellen eye charts (for visual acuity test)
 laboratory lamp (for reflexes)

Image Suppression (The Purkinje Tree)

When you examined the fundus of the eye with an ophthalmoscope in the last laboratory exercise, you observed that blood vessels and capillaries were very much in evidence. Although these branching vessels lie very close to receptor cells and cast sharp shadows on most regions of the retina, one's vision is not impaired by them. The reason for no visual impairment by these shadows is that the brain has the capacity to suppress the disturbing images.

To illustrate that these blood vessels do cast an image on the retina and that they can be brought into one's field of vision, one need only illuminate the interior of the eye with rays of light that enter the eye at an abnormal angle. If light rays are brought into the eyeball at an oblique angle through the eyelid and sclera instead of directly through the pupil, the brain will not suppress the shadows that are cast by the blood vessels. The branching image that one sees with this type of illumination is called the *Purkinje tree.*

To perform this experiment on your own eye, proceed as follows: Hold a penlite with your right hand against the eyelid of your right eye at an angle of about 45° to the right as shown in figure 37.1. The eyelid may be open or closed. If the eyelid is open, look at a dimly lighted wall. If the eyelid is closed, do not face a brightly lighted window.

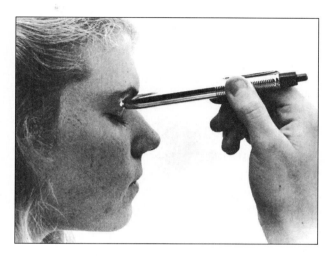

Figure 37.1 Proper position for penlite when observing the Purkinje tree.

While exposing the eye to the light, move the penlite from side to side to produce the Purkinje image. The amount of movement should be very slight, only a few millimeters from side to side. A most striking image can be produced if this experiment is performed in a dark closet. Describe the image on Laboratory Report 36,37.

The Blind Spot

The examination of the fundus in Exercise 35 clearly revealed the nature of the optic disk. It was noted that where the optic nerve enters the eyeball there is an absence of rods and cones, which renders that part of the retina insensitive to light. The illustration in figure 37.2 can be used to detect the presence of this blind spot in each eye.

To test the right eye, close the left eye and stare at the plus sign with the right eye as this page is moved from about 18 inches toward the face. At first, both the plus sign and dot are seen simultaneously. Then, at a certain distance from the eye, the dot will disappear as it comes into focus on the optic disk of the retina.

Perform this test on both of your eyes. To test the left eye, it is necessary to look at the dot instead

Figure 37.2 The blind spot test.

of the plus sign. Have your laboratory partner measure the distance from your eye to the test chart with a ruler or tape measure.

Record the measurements on the chart on Laboratory Report 36,37.

Near Point Determination (Distance Accommodation)

The ability of the lens of the eye to produce a sharp image on the retina is partially a function of its elasticity. When focusing on close objects, the lens must be considerably more spherical, or convex, than when focusing on distant objects. To become more convex, the tension on the lens periphery is relaxed as the ciliary muscle is stimulated.

In an infant the ability of the eye to accommodate to distance is at its maximum. On average, an infant's eye lens has a range of 17 diopters from distance sighting to near objects. As an individual gets older, the lens gradually becomes less elastic, and the degree of accommodation diminishes.

Between the ages of 45 and 50, one's accommodation is reduced to 2 diopters. Beyond 60 years of age, accommodation is nearly nonexistent. This loss of lens elasticity causes a form of far-sightedness, called **presbyopia.** The condition is probably due to denaturation of protein in the lens. It is for this reason that most individuals over 45 years of age find it necessary to acquire bifocal or trifocal glasses.

A measure of the elasticity of the lens in the eye can be made by determining the near point of the eye. The **near point** is the closest distance at which one can see an object in sharp focus. At 20 years of age the near point is approximately 3½″; at 30, 4½″; at 40, 6½″; at 50, 20½″; and at 60 it may be 33″.

To determine the near point in each eye, use the letter "T" at the beginning of this paragraph. Close one eye and move the page up to the eye until the letter becomes blurred; then move it away until you get a clear undistorted image.

Have your laboratory partner measure the distance from your eye to the page with a ruler or tape measure. The closest distance at which the image is clear will be the near point. Test the other eye also, and record your results on the Laboratory Report.

Scheiner's Experiment Another method for determining the near point is with Scheiner's experiment. With this method one peers through two small pinholes in a card at an object such as a common pin. Figure 37.3 illustrates the method.

The pinholes should be in the center of the card and be no more than 2 mm apart, center to center. While peering through the two holes, it will be noted that the holes overlap so that it seems as if there is only a single opening. Holes that are too far apart, however, will appear as two distinct openings with an opaque bridge between them. Perform this experiment as follows:

1. Pierce two small pinholes in the center of a card that are no more than 2 mm apart, center to center.
2. Peer through the holes at a common pin that is held up at **arm's length.** If the two holes merge as a single opening, they are the correct distance apart. **It is important that the pin be viewed through the space where the two luminous circles overlap.**
3. With the pin still at arm's length, focus on a distant object. The pin should now appear double since the eye is focused on infinity.
4. Now focus on the pin and note that a single image is seen.
5. Move the pin slowly toward your eye, keeping it in focus until the image changes from single to double. **Be sure to keep the eye looking through the overlapping area of the two holes.**

Figure 37.3 Scheiner's experiment. An object is viewed through two pinholes in a card.

The distance from the eye at which the image changes from single to double is the **near point.** Have your laboratory partner measure the distance with a ruler or tape measure.

6. Repeat this experiment several times to establish an average near point. How does this measurement compare with the first method used? Record the measurements on Laboratory Report 36,37.

Visual Acuity

A completely normal human eye lacking any form of deviation is able to differentiate, at a distance of 10 meters, two points that are only 1 millimeter apart. Points that are less than a millimeter apart at this distance will be seen as a single spot. The size and proximity of cones in the fovea determine this degree of *visual acuity.*

If pinpoints of light from two different objects strike adjacent cones, only a single image will be seen. This is due to the fact that the brain lacks the ability to differentiate stimuli that it receives from adjacent cones. However, if the images fall on two cones separated by an unexcited cone, the brain recognizes two separate points.

Because the diameter of a cone is approximately two micrometers (1/500 mm), the images on the fovea must be at least 2 μm apart. On the basis of this information it can be calculated that the brain can differentiate two pinpoints that enter the eye at an angle of 26 seconds. This is somewhat less than one-half a degree!

The Snellen eye chart (figure 37.4) has been developed with this mathematics in mind. When you stand at a certain distance from it, usually 20 feet, and are able to read the letters on a line designated to be read at 20 feet, you are said to have 20/20 vision. The ability to read these letters indicates that there are no aberrations of the lens or cornea to interfere with the angle of pinpoints of light reaching the retina of the eye. If you are only able to read the larger letters, such as those that should be read at 200 feet, you are said to have 20/200 vision.

At some place in the laboratory, a Snellen eye chart will be posted on the wall. A 20-foot mark will also be designated on the floor. Working with your laboratory partner, test each other's eyes. Test one eye at a time, with the other eye covered. Record your results on the Laboratory Report.

Test for Astigmatism

If the lens of an eye has an uneven curvature on one of its surfaces, a condition called *astigmatism*

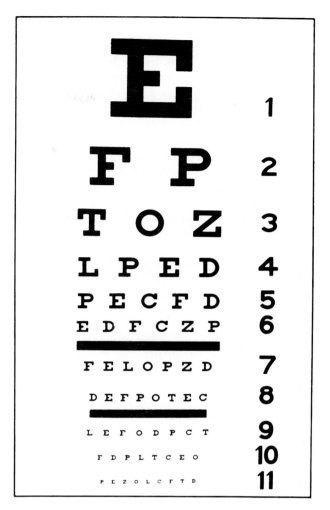

Figure 37.4   The Snellen eye chart.

exists. Figure 37.5 reveals the difference between normal and astigmatic lenses. Note that the astigmatic lens has a greater curvature on its left upper surface than on the lower left quadrant. This type of lens will cause greater bending of light rays as they pass through one axis of the lens than when passing through another axis. The image that one sees with this type of lens will be blurred in one axis and sharp in other axes.

To determine the presence of astigmatism, look at the center of the diagram in figure 37.6 and note if all radiating lines are in focus and have the same intensity of blackness. If all lines are sharp and equally black, no astigmatism exists. Of course, the presence of other refractive abnormalities can make this test impractical.

Look at the wheel-like chart with each eye while closing the other and record your conclusions on the Laboratory Report.

Test for Color Blindness

The perception of color is a sensation achieved partly by the retina and partly by the brain. The receptors of the retina that are sensitive to color are the **cones.** According to the *Young-Helmholtz theory* of color perception, there are three different types of cones, each of which responds maximally to a different color. The three colors that cause maximum response in these cones are **red, blue,** and **green.** The degree of stimulation that each type of cone gets from a particular wavelength of light determines what color is perceived by the brain.

When the retina is exposed to *red* monochromatic light (wavelength of 610 nanometers), the red cones are stimulated at 75%, the green cones at 13%, and the blue cones not at all. The ratio of stimulation for red is thus 75:13:0. When the brain receives this ratio of stimulation from the three types of cones, the interpretation is for red color.

When a monochromatic *blue* light (wavelength of 450 nanometers) strikes the retina, the red cones are not stimulated at all (0), the green cones are stimulated to a value of 14%, and the blue cones to a value of 86%. The ratio here of 0:14:86 is interpreted by the brain as blue.

For *green,* the ratio is 50:85:15. Orange-yellow produces a ratio of 100:50:0. When exposed to white light, which has no specific wavelength since it is a mixture of all colors of the spectrum, the three types of cones are stimulated equally.

Color blindness is a sex-linked hereditary condition that affects 8% of the male population and 0.5% of females. The most common type is red-green color blindness, in which either the red or green cones are lacking. If red cones are lacking, a conditon called **protanopia** exists. Individuals that have this conditon see blue-greens and purplish-tinted reds as gray. A lack of green cones is designated as **deuteranopia.**

Although both protanopes and deuteranopes have difficulty differentiating reds and greens, their visual spectrums differ enough so that they can be diagnosed with color test charts. Illustration 4, figure 37.7, is a test plate for differentiating protanopes and deuteranopes. While a normal person would see the number 96 on this plate, a protanope would see only the number 6 and a deuteranope would see only the number 9.

Other than protanopia and deuteranopia, there are rarer forms of color vision deficiencies, such as blue-color weakness, yellow-color weakness, and total-color blindness. These forms of color vision deficiencies are not as well understood.

Using Ishihara Plates

In our color blindness test here we will use Ishihara's book of plates (*Ishihara's Tests for Colour Blindness,* Concise Edition). Procure one of these books from the supply table.

Working with your laboratory partner, test each other by holding the test plates about 30 inches away from the subject. If sunlight is available, use it.

Start with plate #1 and proceed consecutively through all 14 plates. The subject should respond with an answer for each plate within **3 seconds.** The

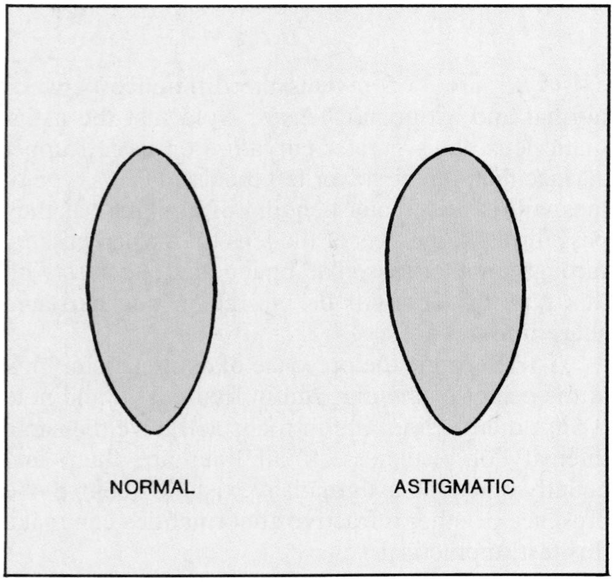

Figure 37.5 Normal and astigmatic lenses.

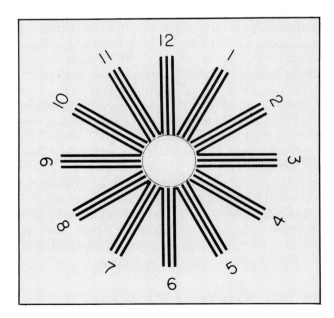

Figure 37.6 Astigmatism test chart.

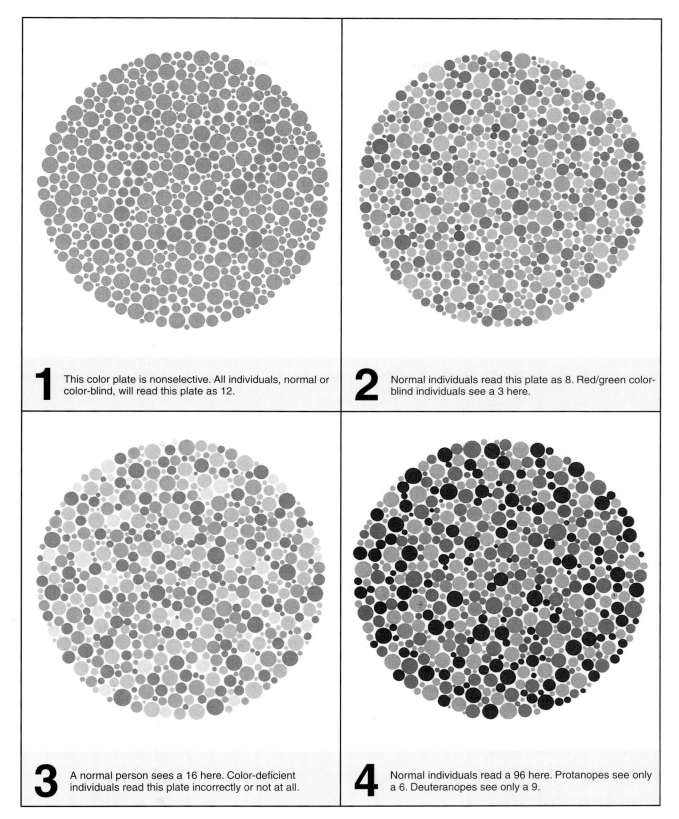

1 This color plate is nonselective. All individuals, normal or color-blind, will read this plate as 12.

2 Normal individuals read this plate as 8. Red/green color-blind individuals see a 3 here.

3 A normal person sees a 16 here. Color-deficient individuals read this plate incorrectly or not at all.

4 Normal individuals read a 96 here. Protanopes see only a 6. Deuteranopes see only a 9.

Figure 37.7 **Ishihara color test plates.** (These patterns have been reproduced from *Ishihara's Tests for Colour Blindness* published by Kanehara & Co., Ltd., Tokyo, Japan. Due to reproduction inaccuracies, these color replicas should not be used for accurate color-blind testing.)

examiner will record the responses on the chart on the Laboratory Report.

After all plates have been observed and recorded, compare the responses with the correct answers and make a statement at the bottom of the chart (conclusion).

Pupillary Reflexes

The tests and experiments performed so far have been concerned primarily with the role of the lens and retina in vision. The following experiments illustrate the roles of the pupil in adjustment of the eye to distance and light intensity.

Accommodation to Distance   Three events take place when the eyes change their focus from a distant object to a close object:

* Eye muscles react to achieve convergence of the eyes.
* The lens becomes more convex.
* A change occurs in the size of the pupil.

Working with your laboratory partner or a selected member of the class, perform the following experiment to observe what happens to the pupils when the eyes focus on a near object after looking at a distant object.

This experiment works best if the subject has **pale blue eyes.** If your laboratory partner does not qualify in this respect, request some other class member to be the subject and allow several other students to observe the results. Have the subject look at the wall on the side of the room opposite to windows. The eyes are now relaxed and focused at infinity.

While closely watching the pupils, place the printed page of your laboratory manual within 6 inches of the subject's face and ask the individual to focus on the print. Incidentally, the light intensity on the printed page and distant wall should be equal.

Do the pupils remain the same size when looking at the printed page? Repeat the experiment several times and record your results on the Laboratory Report.

Accommodation to Light Intensity   Sudden exposure of the retina to a bright light causes immediate reflex contraction of the pupil in direct proportion to the degree of light intensity. The pupil contracts to approximately 1.5 mm when the eye is exposed to intense light, and it enlarges to almost 10 mm in complete darkness. This approximates a total difference in pupillary area of about 40 times. In this reflex, impulses pass from the retina via the optic nerve through two centers in the brain and then return to the sphincter of the iris through the ciliary ganglion.

Perform this simple experiment to learn a little more about the pathway of this pupillary reflex. As in the previous experiment, select a subject with pale blue eyes so that the pupil size is more easily observed.

Have the subject hold this laboratory manual vertically, with the spiral binding close to the forehead and extending downward along the bridge of the nose. Now, position an unlighted laboratory lamp about 6 inches from the right eye.

While watching the pupil of the left eye, turn on the lamp for one second and then turn it off. Make sure that no light spills over from the right side of the book.

Did the unlighted pupil of the left eye react to light that exposed the right eye? What does this reaction tell us with respect to the pathways of the nerve impulses?

Laboratory Report

Complete the last half of combined Laboratory Report 36,37.

The Ear:
Its Role in Hearing

38

Four aspects of the ear will be studied in this exercise: (1) its anatomical components that pertain to hearing, (2) the physiology of hearing, (3) hearing tests, and (4) histological studies. The vestibular apparatus, which functions in equilibrium, will be studied in the next exercise. A brief statement pertaining to the characteristics of sound will precede our study of the ear. The sections on ear anatomy and the physiology of hearing should be completed prior to the hearing tests and microscopy.

Characteristics of Sound

Sound waves moving through the air are propagated in waveforms that possess characteristics relative to the vibratory motion that generates them. The ear is able to distinguish tones that differ in pitch, loudness, and quality. **Pitch** is determined by the frequency of vibration. **Loudness** pertains to the intensity of the vibration. **Quality** relates to the nature of the vibrations as revealed by the wave shape.

Figure 38.1 illustrates several curves depicting both the shapes of sound waves and the characteristics of the vibrations that produce them. The sine curves A and B in illustration I differ in frequency, with B producing the tone of higher pitch. Curves A and C in the middle group are of the same frequency; they differ, however, in amplitude, or loudness. Although the amplitude of A is twice that of C, the loudness of A will be four times that of C. This is because the intensity (I) of sound is directly proportional to the square of the amplitude (a):

$$I = 2\pi^2 V f^2 a^2 d$$

V = velocity of wave propagation

f = frequency

d = density of medium

Waves A and D in illustration III differ in shape, D having some components of higher frequency that are not present in A. These curves represent sounds of different quality.

The ability of the ear to distinguish differences in pitch, amplitude, and quality depend on (1) the conduction and pressure amplification of sound waves through the ossicles of the middle ear, (2) the stimulation of receptor cells in the cochlea, and (3) the conveyance of action potentials in the cochlear nerve to the auditory centers in the brain for interpretation. Interference of any component of this system results in hearing loss. The types of hearing loss will be studied as hearing tests are performed. Let us now explore the anatomy of the ear as it relates to auditory function.

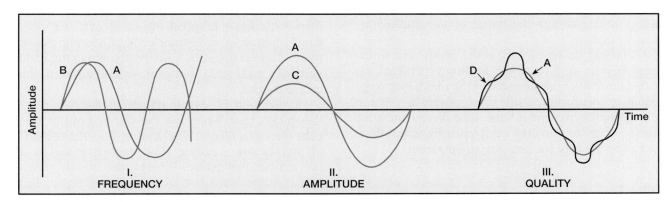

Figure 38.1 Differences in sound waves.

Components of the Ear

Anatomically, the human ear consists of three distinct divisions: the external ear, the middle ear, and the internal ear. Figure 38.2 is a diagrammatic representation of its various components.

The External Ear

The outer, or external, ear consists of two parts: the auricle and external auditory meatus. The **auricle,** or **pinna,** is the outer shell of skin and elastic cartilage that is attached to the side of the head. The **external auditory meatus** is a canal about one inch in length that extends from the auricle into the head through the temporal bone. The skin lining this canal contains some small hairs and modified apocrine sweat glands that produce a wax secretion called *cerumen.* The inner end of this meatus terminates at the **tympanic membrane,** or eardrum. The auricle serves to collect and direct sound waves into the tympanic membrane through the meatus.

The Middle Ear

This middle ear consists of a small cavity in the temporal bone between the tympanic membrane and the inner ear. It contains three small bones *(ossicles)* that are united to form a system of levers. The outermost ossicle, which is attached to the tympanic membrane, is the **malleus,** a hammer- or club-shaped bone. The middle bone, an anvil-shaped structure, is the **incus.** The innermost bone, which fits into the oval window of the inner ear, is stirrup-shaped and called the **stapes.**

It is the role of these ossicles to transfer the forces from the eardrum to the cochlear fluids of the inner ear through the **oval window** *(fenestra vestibuli).* Although most sound waves reach the inner ear via the ossicles *(ossicular conduction),* some sounds, specifically very loud ones, reach the cochlea through the bones of the skull *(bone conduction).*

Leading downward from the middle ear to the nasopharynx is a duct, the **auditory** (Eustachian) **tube,** that allows air pressure in the middle ear to be equalized with the outside atmosphere. A valve at the nasopharynx end of the tube keeps the tube closed. Acts of yawning or swallowing cause it to open temporarily for pressure equalization.

The Internal Ear

The internal ear consists of two labyrinths, the osseous and membranous labyrinths. The **osseous labyrinth** is shown in illustration B. It is the hollowed-out portion of the temporal bone that contains an inner tubular structure of membranous tissue, the **membranous labyrinth.** The entire membranous labyrinth is shown in the next exercise (figure 39.1). Some portions of the membranous labyrinth are shown in cutaway portions of the osseous labyrinth in illustration B, figure 38.2.

Within the membranous labyrinth is a fluid, the **endolymph.** Between the membranous and osseous labyrinths is a different fluid, the **perilymph.** These fluids act as conduction media for the forces involved in hearing and maintaining equilibrium.

The osseous labyrinth consists of three semicircular canals, the vestibule, and the cochlea. The **vestibule** is that portion that has the oval window on its side into which the stapes fits. The three **semicircular canals** branch off the vestibule to one side, and the **cochlea,** which is shaped like a snail's shell, emerges from the other side.

On the osseous labyrinth are two nerves, which are branches of the eighth cranial (vestibulocochlear) nerve. The **vestibular nerve** is the upper nerve branch that passes from the sensory areas of the semicircular ducts, saccule, and utricle. The other branch is the **cochlear nerve,** that emerges from the cochlea.

The cochlea consists of a coiled, bony tube with three chambers extending along its full length. Illustration C in figure 38.2 reveals a cross section of the cochlear tube. The upper chamber, or **scala vestibuli** (label 16), is so named because it is continuous with the vestibule. The lower larger chamber is the **scala tympani.** The **round window** *(fenestra cochlea),* label 1, is a membrane-covered opening that is on the osseous wall of the scala tympani. Between these two chambers is a triangular cross section that represents the **cochlear duct.** This duct is bounded on its upper surface by the **vestibular membrane** and on its lower surface by the **basilar membrane.** On the upper surface of the basilar membrane lies the **spiral organ** *(organ of Corti),* which contains the receptor cells of hearing. All of these chambers contain fluid: perilymph in the scala vestibuli and scala tympani; endolymph in the cochlear duct.

Illustration D reveals the detailed structure of the spiral organ. Note how the hair tips of the **hair cells** are embedded in a gelatinous flap, the **tectorial membrane.** Leading from each hair cell is a nerve fiber that passes through the basilar membrane and becomes part of the cochlear nerve. The upper margins of the hair cells are held in place by the **reticular lamina.** Between the reticular lamina and the basilar membrane are reinforcing structures

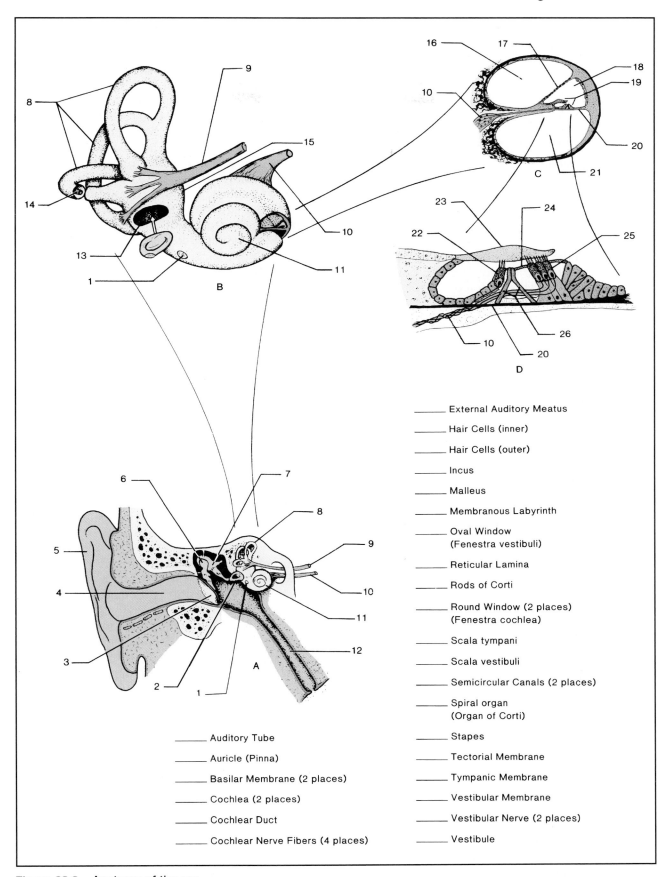

Figure 38.2   Anatomy of the ear.

_____ External Auditory Meatus

_____ Hair Cells (inner)

_____ Hair Cells (outer)

_____ Incus

_____ Malleus

_____ Membranous Labyrinth

_____ Oval Window
(Fenestra vestibuli)

_____ Reticular Lamina

_____ Rods of Corti

_____ Round Window (2 places)
(Fenestra cochlea)

_____ Scala tympani

_____ Scala vestibuli

_____ Semicircular Canals (2 places)

_____ Spiral organ
(Organ of Corti)

_____ Stapes

_____ Tectorial Membrane

_____ Tympanic Membrane

_____ Vestibular Membrane

_____ Vestibular Nerve (2 places)

_____ Vestibule

_____ Auditory Tube

_____ Auricle (Pinna)

_____ Basilar Membrane (2 places)

_____ Cochlea (2 places)

_____ Cochlear Duct

_____ Cochlear Nerve Fibers (4 places)

called the **rods of Corti.** The significance of all these structures must be taken into account in any plausible theory of hearing.

Assignment:
Label figure 38.2.

The Physiology of Hearing

As stated previously, hearing occurs when action potentials are received by the auditory centers of the brain. These action potentials, which pass along the cochlear nerve fibers of the eighth cranial nerve, are initiated by the hair cells in the spiral organ. Activation of these sensory hair cells depends on forces within the cochlear fluids, basilar membrane structure, secondary energy transfer, and endo-cochlear potential. A discussion of the roles of each of these factors follows.

Role of Cochlear Fluids   In figure 38.3 the cochlea has been uncoiled to reveal the relationship of various chambers to each other. The perilymph here is colored yellow, and the endolymph in the cochlear duct is colored green.

When the stapes moves in and out of the oval window, sound wave energy moves through the perilymph of the scala vestibuli into the scala tympani through a small opening, the helicotrema. Since the perilymph is incompressible, in this enclosure, the elasticity of the round window membrane accommodates the pressure changes by moving in and out in synchrony with the movements of the stapes.

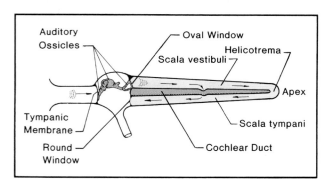

Figure 38.3 Pathway of sound wave transmission in the ear.

The energy of sound waves moving through the cochlea reach the round window in two ways: (1) directly into the scala tympani through the helicotrema and (2) through the flexible cochlear duct to the scala tympani. Arrows in figure 38.3 illustrate both routes of propagation.

Basilar Membrane Structure   The first structure of the spiral organ that reacts to vibrations of different sound frequencies is the basilar membrane. Within this membrane are approximately 20,000 fibers that project from the bony center of the cochlea toward the outer wall. The fibers closest to the stapes are short (0.04 mm) and stiff; they vibrate when stimulated by vibrations caused by high-frequency sounds. The fibers at the end of the cochlea near the helicotrema are long (0.5 mm) and more limber, and vibrate in harmony with low-frequency sounds. These fibers are elastic reedlike structures that are not fixed at their distal ends. Because they are free at one end, they are able to vibrate like reeds of a harmonica to specific frequencies.

Secondary Energy Transfer   Once fibers at a particular point on the basilar membrane are set into motion by a specific sound frequency, the rods of Corti transfer the energy from the basilar membrane to the reticular lamina, which supports the upper portion of the hair cells (see illustration D, figure 38.2).

This secondary transfer of energy causes all these components to move as a unit, with the end result that the hairs of the hair cells are bent and stressed. The back-and-forth bending of the hairs, in turn, causes alternate changes in the electrical potential across the hair cell membrane. This alternating potential, known as the *receptor potential of the hair cell,* stimulates the nerve endings that are at the base of each hair cell, and an action potential in the cochlear nerve fibers is initiated.

Endocochlear Potential   The sensitivity of the hair cells to depolarization is greatly enhanced by the chemical composition differences between the perilymph and endolymph. Perilymph has a high sodium-to-potassium ratio; endolymph, conversely, has a high potassium-to-sodium ratio.

These ionic differences result in an *endocochlear potential* of 80 millivolts. It is significant that the tops of the hair cells project through the reticular lamina into the endolymph of the scala media, but the lower portions of these cells lie bathed in perilymph. It is believed that this potential difference of 80 mV between the top and bottom of each hair cell greatly sensitizes the cell to slight movements of the hairs.

Frequency Range   The variability in length and stiffness of the fibers within the basilar membrane enables the ear to differentiate sound waves as low as 30 cycles per second near the apex of the cochlea and as high as 20,000 cps near the base of the cochlea. In between these two extremes, at various spots on the basilar membrane, the membrane responds to the in-between frequencies. This method of pitch localization by the basilar membrane is called the *place principle.*

Loudness Determination As noted in figure 38.1, the loudness of a sound is expressed in the amplitude of the sine wave. Increased loudness of sounds in the cochlear fluids causes an increase in the amplitude of vibration of the basilar membrane.

With an increase in basilar membrane vibration, more and more hair cells become stimulated, causing *spatial summation of impulses;* that is, more nerve fibers are carrying more impulses, and this is interpreted by the brain as increased loudness.

Assignment:
Answer the questions on Laboratory Report 38 that pertain to the physiology of hearing.

Hearing Tests

Although deafness may have many different origins, there are essentially two principal kinds of deafness: nerve and conduction deafness. If the cochlea or cochlear nerve is damaged, the condition is referred to as **nerve deafness.** Damage to the eardrum or ossicles, on the other hand, will result in **conduction deafness.** While conduction deafness can usually be remedied with surgery or hearing aids, nerve deafness cannot be corrected.

The three kinds of hearing tests outlined here are ones that are most frequently used. Each type of test serves a specific function. Perform those tests for which equipment is available.

Watch Tick Method

This is one of the simplest (and oldest) methods used for determining hearing acuity. Its only instrument is a spring-wound pocket watch that produces an audible ticking sound. The inability of a patient to hear a ticking watch at prescribed distances from the ear can alert the examining physician to a hearing problem. When used, it is used only for screening purposes. Further tests using tuning forks or the audiometer can then be brought into play.

Materials:
spring-wound pocket watch
cotton earplugs
meterstick or measuring tape

1. Seat the subject comfortably in a chair and plug one ear with cotton.
2. Instruct the subject to use the index finger to signal when sound is first heard as the watch approaches or when sound disappears as the watch is moved away from the ear.
3. With the subject looking straight ahead, place the watch at a position that is out of hearing range of the ear, usually about 3 feet. Keep the face of the watch parallel to the side of the head.
4. Move the watch toward the ear at **a speed of about 1/2 inch per second.** This is a very slow movement. The subject should signal when the **first tick** is heard.
5. Measure the distance with the meterstick or tape measure, record the information on the Laboratory Report, and repeat the test two or three times. On repeat tests change the approaches to the ear to prevent the factor of subject anticipation.
6. Reverse the procedure by starting close to the ear and moving away from it. In this test, the subject signals the **last tick** that is heard.
7. Transfer the cotton plug to the other ear and follow the same procedure again.

Tuning Fork Methods

The distinction between nerve and conduction deafness can be readily made with two tuning fork methods: the *Rinne* and *Weber* tests. In one test the base of the tuning fork is applied to the mastoid process behind the ear; in the other test the fork is placed on the forehead. In both tests the tuning fork is set in vibration by bouncing the fork off the heel of the hand. These tests may be performed by yourself or with the aid of your laboratory partner.

Materials:
tuning fork
cotton earplugs

The Rinne Test This test, which utilizes the mastoid process to differentiate the two types of deafness, should be performed as follows:

1. Plug one ear with cotton.
2. Set a tuning fork in motion by striking it on the heal of the hand, as shown in illustration 1, figure 38.4.
3. Place the tuning fork about 3 to 6 inches away from the unplugged ear with the tine facing the ear, as shown in illustration 2, figure 38.4. If there is a minimum of hearing loss, the vibrating sound will be heard immediately. If no sound is heard go to step 4.

As the sound intensity diminishes, a point will be reached at which the sound finally disappears. As soon as the sound disappears, place the stem of the tuning fork against the mastoid process as shown in illustration 3, figure 38.4. If the sound vibrations reappear while in contact with the mastoid process, **conduction deafness is present.**

4. If considerable hearing loss is present and the tuning fork vibrations cannot be easily heard with the tuning fork near the ear, place the stem of the vibrating fork against the mastoid process and note if the sound can be heard through bone conduction. If the sound is loud through the bone, **conduction deafness is present.** If no sound is heard while in contact with the skull, **nerve deafness is present.**

5. Repeat the test on the other ear.

6. Record results on the Laboratory Report.

The Weber Test   When the stem of a vibrating tuning fork is placed on the forehead of a person with normal hearing, the sound of the fork is heard at equal intensity in both ears. If an individual with **conduction deafness** is tested in this manner, **the sound will be heard louder in the deaf ear than in the normal ear.** The reason is that the deaf ear, which is normally not activated by sound waves through the air, is more acutely attuned to sound waves being conducted to the cochlea through bone. On the other hand, if this test is performed on a person who has one normal ear and **one ear with nerve deafness,** the **sound will be more intense in the normal ear** than in the deaf one. Place a vibrating tuning fork on your forehead and compare the sounds heard. Record results on the Laboratory Report.

Audiometry

The audiometer is an instrument used for determining hearing losses in the audible range of normal speech. It measures the ability of the ear to hear sounds in the *frequency range of 125 to 8000 cps.* Although the hearing range of the human ear may be as broad as 30 to 20,000 cps, it is the 125–8000 cps range that is important in hearing the spoken word.

This instrument also measures the *level of intensity* of sounds in the audible range that one can hear. The sensation of loudness experienced by the ear is not related in a simple way to the intensity of sound that strikes the tympanic membrane. While the range of sensitivity between a faint whisper and the loudest noise is one trillion times, the ear interprets this great difference as approximately a 10,000-fold change. What happens here is that the scale of intensity is compressed by the action of the tympanic membrane, ossicles, and cochlear duct, enabling the ear to have a much broader sensitivity range.

Because of this extreme range in sound intensities, it is necessary to express the intensity in terms of the logarithms of their actual intensities. The basic unit is called a **bel** (after Alexander Bell). Sounds that differ by one bel in intensity differ by ten times. A **decibel** is one-tenth of a bel. It is in decibel units that hearing is measured, mainly because this is the smallest unit that the human ear is able to differentiate.

The audiometer is, essentially, an electronic oscillator with earphones. As shown in figure 38.5, it has one control that regulates the frequency and another control that regulates the intensity, or

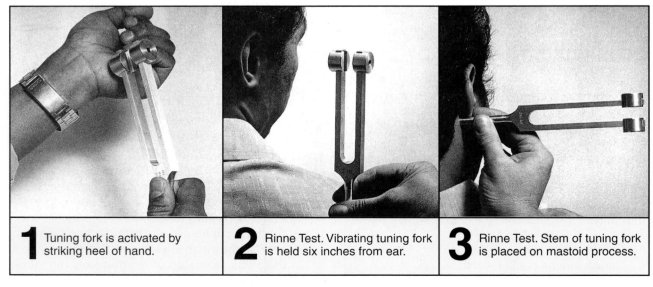

| **1** Tuning fork is activated by striking heel of hand. | **2** Rinne Test. Vibrating tuning fork is held six inches from ear. | **3** Rinne Test. Stem of tuning fork is placed on mastoid process. |

Figure 38.4   Tuning fork manipulations in hearing tests.

Figure 38.5   **The audiometer.**

loudness. Two separate switches, or tone controls, near the bottom are used to release the tone into the earphones. The left switch (blue) is for the left earphone and the right (red) one is for the right earphone.

The Hearing Level Control is calibrated so that the zero intensity level of sound at each frequency is the loudness that can be barely heard by a normal person. The numbers on this control represent decibels. If it is necessary to adjust this control to 30 at 125 cps frequency, it means that the patient has a hearing loss of 30 decibels.

In this experiment, work with your laboratory partner to plot an audiogram on the Laboratory Report for each ear. Proceed as follows:

Materials:
    audiometer
    red and blue pencils

1. Place the headset securely on the subject's head so that the ear cushions make good contact over the ears and are not obstructed by clothing, earrings, etc.
   Important: Make certain that the earphone with the red cord is placed over the right ear.
2. To enable the subject to become familiar with the tone, set the Frequency control on 1000 cps and the Hearing Level Control at 50 dB.
   Now, depress the right (red) Tone Control so that the subject can hear the tone in the right ear.

3. Rotate the Frequency control throughout all frequencies at this 50 db level to let the subject hear all frequencies prior to testing.
   Depress the Tone Control *only* when the dial is set on each frequency.
4. Return the Frequency Control to 1000 cps and, while depressing the red Tone Control, rotate the Hearing Level Control counterclockwise slowly until you reach the point where the subject is just barely able to recognize the tone.
   If, for instance, the subject can hear the tone at 30 decibels, but not hear the tone at 25 decibels, the subject's threshold is established at 30 decibels for that frequency setting. Have the subject signal with a hand signal when the tone is heard.
5. Record this measurement on the audiogram chart of the Laboratory Report by marking a red "O" on the vertical line for 1000 cps, where it intersects the line for the decibel value. (In the above example, it would be where the 30 decibels line intersects the 1000 cps line.)
6. Now rotate the Frequency Control to 2000 cps following the above procedures. After testing at this frequency, test the right ear at the remaining higher frequencies up to 8000 cps.
7. Finally, finish the test on the right ear by testing at frequencies of 250 and 500 cps.
8. Test the left ear, following steps 2 through 7 above. Use the blue Tone Control switch for testing the left ear.
   Record each threshold level with an "X" on the appropriate frequency line with a blue pencil.
9. Connect the points on the chart with appropriately colored lines to produce a hearing graph for the ear.

Histological Study

Examine slides H10.53 and H10.54 to study the parts of the cochlea and crista ampullaris shown in figure HA-18 of the Histology Atlas. Both of these slides are made from guinea pig tissue. In particular, attempt to identify all the structures of the spiral organ that are shown in illustration C. Note in illustration B that the cupula is obscured by the collapsed wall of the ampulla.

Laboratory Report

Complete Laboratory Report 38 for this exercise.

39

The Ear:
Its Role in Equilibrium

The semicircular ducts, saccule, and utricle function in maintaining equilibrium. These three components are combined into a unit called the *vestibular apparatus.*

In addition to the vestibular apparatus, three other factors play important roles in maintaining equilibrium. They are (1) visual recognition of the horizontal position, (2) proprioceptor sensations in joint capsules, and (3) cutaneous sensations through certain exteroceptors. Sensations from these four sources are continuously subconsciously synthesized at the cortical level to provide the correct reflex responses necessary to maintain equilibrium.

In this exercise we will evaluate the relative significance of visual, proprioceptive, and vestibular sensations in postural control. Some clinical tests for evaluating static and dynamic equilibrium mechanisms will also be explored.

Two types of equilibrium are identifiable: static and dynamic. **Static equilibrium** pertains to the effect of gravity on receptors. **Dynamic equilibrium** compensates for angular movements of the body in different directions. The physiology of each type follows.

Static Equilibrium

The sensory receptors of the vestibular apparatus that respond to gravitational forces are located in the saccule and utricle. Note in figure 39.1 that the **saccule** connects directly to the cochlear duct, and the **utricle** lies at the base of the semicircular ducts. These two structures, being a part of the entire membranous labyrinth, contain **endolymph.**

On the inner walls of the saccule and utricle lie two sensory structures, the **maculae.** Each macula consists of **hair cells,** a **gelatinous matrix,** and calcium carbonate crystals called **otoconia.** The otoconia are embedded in the gelatinous matrix. Fibers of the vestibular nerve carry impulses from the hair cells to the brain.

When the head is tilted slightly in any direction, the otoconia are pulled by gravitational forces, causing the cilia of the hair cells to be bent by the action of the gelatinous matrix against them. Bending of the cilia in one direction causes impulse traffic in the vestibular nerve fibers to increase markedly; bending in another direction decreases the impulse traffic to a point of no reaction at all.

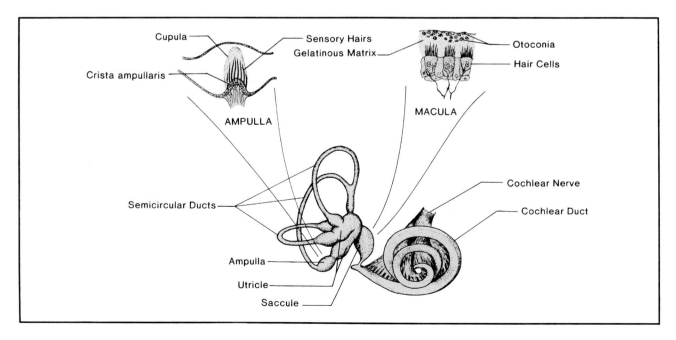

Figure 39.1 The membranous labyrinth.

This system is so sensitive to malequilibrium that as little as 1/2° shifting of the head in any direction from vertical is detectable.

Simply tilting the head, however, does not bring about a sense of malequilibrium. Fortunately, proprioceptors in the joints of the neck transmit inhibitory signals to the brain stem that neutralize the effects of the vestibular receptors. It is only when the whole body becomes disoriented that the hair-cell initiated impulses are not inhibited.

Balancing Test One of the simplest tests for determining the integrity of this static equilibrium mechanism is to have the subject stand perfectly still with eyes closed. If there is any damage to this system, the subject will waiver and tend to fall. Individuals with long-term damage, however, are often able to stand fairly well due to well-developed proprioceptive mechanisms. Work with your partner to test each other's sense of balance with this test.

Dynamic Equilibrium

The receptor hair cells for dynamic equilibrium are contained in the **ampullae** of the **semicircular ducts.** These hair cells are arranged in a crest, the **crista ampullaris,** within each ampulla. The hair tufts, in turn, are embedded in a gelatinous mass to form a structure called the **cupula.** Locate all these structures in figure 39.1.

The three semicircular ducts are arranged at right angles to each other in three different planes. When the head moves in any direction, the semicircular duct in the plane of directional movement moves relative to the endolymph within it; in other words, *the fluid is stationary as the duct moves.*

This movement causes the cupula to act as a trapdoor, moving in either direction because of the force of the endolymph. The hair cells in this structure act much like the hair cells of the macula, in that bending in one direction produces increased action potentials, and bending in the other direction is inhibitory. Nerve impulses along the vestibular nerve reflexively excite the appropriate muscles to maintain equilibrium.

Nystagmus

The principal function of the semicircular ducts is to maintain equilibrium *in the initial stages* of angular or rotational movement. As soon as one begins to change direction, the semicircular ducts trigger reflexes to the proper muscles in anticipation of malequilibrium. A reflex movement of the eyes, called *nystagmus,* occurs during this movement to produce fixed momentary images on the retina.

Nystagmus is characterized by rapid and slow jerky movements of the eyes as the body is rotated. It is caused by reflexes transmitted through the vestibular nuclei, the cerebellum, and the medial longitudinal fasciculus to the ocular nuclei. *Its function is to permit fixed images rather than blurred ones during head movements.*

If the head is slowly rotated to the right, the eyes will, at first, move slowly to the left. This reflex eye movement is the **slow component** of nystagmus. As soon as the eyes have moved as far to the left as they possibly can, they quickly shift to the right, producing what is called the **fast component** of nystagmus. The first, slow component is initiated by the vestibular apparatus; the second, fast component involves the brain stem.

Group Demonstration

Proceed as follows to demonstrate nystagmus. Work in teams of five students, using one member as a subject. The subject, seated in a swivel chair, will be rotated by the other four members. By using several individuals to anchor and rotate the chair, the subject will be provided with maximum safety. Before beginning the test, **the chair's back should be securely tightened to prevent backward tilting.**

Materials:
    swivel-type armchair

1. Have the subject sit in the chair, cross-legged and gripping the arms of the chair.
2. The other four members of the team will form a circle around the chair, each member placing one foot against the base of the chair to prevent chair base movement.
3. Revolve the subject in a **clockwise** direction and observe the eye movement as the subject gazes straight ahead during the rotation. Make ten revolutions at a rate of one revolution per second.
4. Allow the subject to remain seated for one or two minutes after the test to regain stability.
5. Repeat the procedure with another subject, but rotate the individual in a **counterclockwise** direction.
6. With the remaining three team members, follow the same procedure, but have the subjects keep their eyes closed during rotation. Immediately upon stopping the chair, have the subjects open their eyes for observation.
7. Record all observations on the Laboratory Report.

Ice Water Test (Caloric Stimulation Test)

Caution: This test must not be performed here without special permission. It may induce nausea and vomiting!

A simple test that may be used to determine whether the vestibular apparatus of one ear is functioning properly is to irrigate the external ear canal with a small amount of ice water. If the vestibular apparatus is intact and functional, the subject will experience a discomforting sensation of rotation, and *nystagmus will immediately be initiated.* If the subject experiences no discomfort or nystagmus, it may be assumed that damage to the vestibular apparatus or the vestibular nerve has occurred.

The physiological explanation of this reaction is that the cold water increases the density of the endolymph in a portion of the semicircular canals. With some of the endolymph heavier than other endolymph, convection currents are created in the semicircular ducts that cause a sense of malequilibrium.

This test is often used to test for vestibular nerve damage resulting from overdosage of certain antibiotics such as streptomycin.

Proprioceptive Influences

To observe the relative roles of proprioceptors, vision, and vestibular reflexes in equilibrium, perform the following experiments. Retain the same five-member teams, as previously.

Materials:
    swivel-type armchair
    pencil
    blindfold

At-Rest Reactions

Direct the subject to sit in the swivel chair and perform the following simple acts to establish norms for comparison.

1. **With eyes closed,** have the subject place the heel of the right foot on the toes of the left foot.
2. **With eyes closed,** have the subject bring the index finger of the right hand to the tip of the nose from an extended arm position.
   Question: How do proprioceptors in the appendages function to accomplish these two

feats? Record your conclusions on the Laboratory Report.

3. **With eyes open,** have the subject raise his or her hand from the right knee vertically and forward to point his or her finger at the eraser of a pencil held about 2 feet directly in front of the subject. Have the subject repeat this pointing six times, once per second on command of the person holding the pencil. After each pointing, the hand is returned to rest on the right knee.
4. **With eyes closed,** have the subject repeat step 3. How does the subject approach the eraser with eyes closed? Report the results on the Laboratory Report.

Effects of Rotation

The role of the eyes in kinesthetic efficiency will now be determined in tests similar to those above. Proceed as follows:

1. Direct the subject to sit cross-legged with eyes open. Revolve the chair at one revolution per second for ten full turns. Halt the chair in exactly the same starting position.
2. Ask the subject to point to the pencil eraser held in the same position as in the previous test. As before, the subject points to the eraser one time per second for a total of six times.
   After each pointing, the hand is returned to rest on the right knee. Record observations on the Laboratory Report.
3. Blindfold the subject, orient the chair to the same starting position, and by feel, show the subject where the pencil eraser is located.
4. Rotate the subject for ten full turns, stopping the chair at the same starting point.
5. As soon as the chair has come to rest, ask the subject to point to the spot where he or she thinks the eraser is, **but don't let the subject touch the eraser.**
6. Repeat the pointing a total of six times. Remember, the pencil eraser must be held in the same spot as previously and must not be touched.
7. Repeat the entire procedure with another subject, but reverse the direction of rotation.

Laboratory Report

Complete Laboratory Report 39 for this exercise.

HISTOLOGY ATLAS

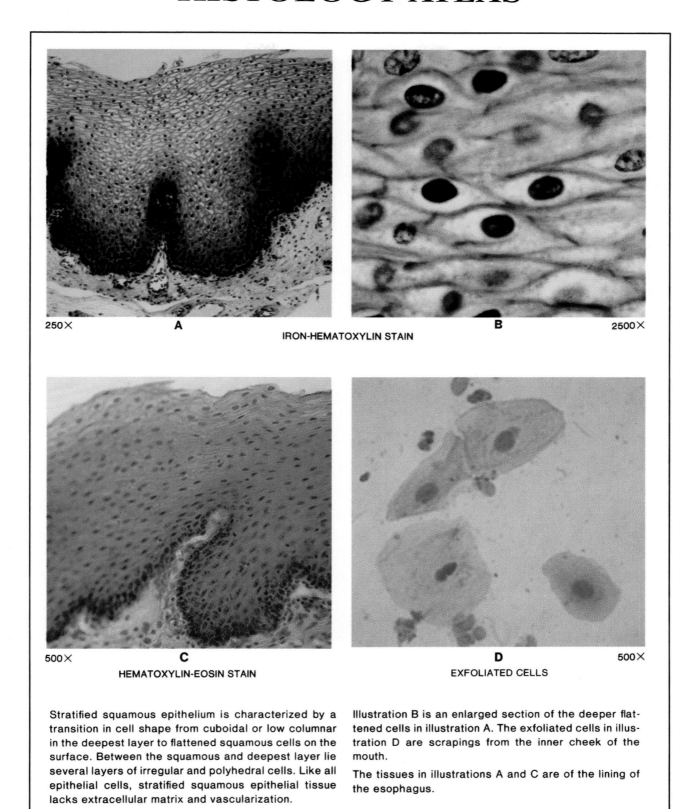

250×   **A**   **B**   2500×

IRON-HEMATOXYLIN STAIN

500×   **C**   **D**   500×

HEMATOXYLIN-EOSIN STAIN   **EXFOLIATED CELLS**

Stratified squamous epithelium is characterized by a transition in cell shape from cuboidal or low columnar in the deepest layer to flattened squamous cells on the surface. Between the squamous and deepest layer lie several layers of irregular and polyhedral cells. Like all epithelial cells, stratified squamous epithelial tissue lacks extracellular matrix and vascularization.

Illustration B is an enlarged section of the deeper flattened cells in illustration A. The exfoliated cells in illustration D are scrapings from the inner cheek of the mouth.

The tissues in illustrations A and C are of the lining of the esophagus.

Figure HA-1   Stratified squamous epithelium.

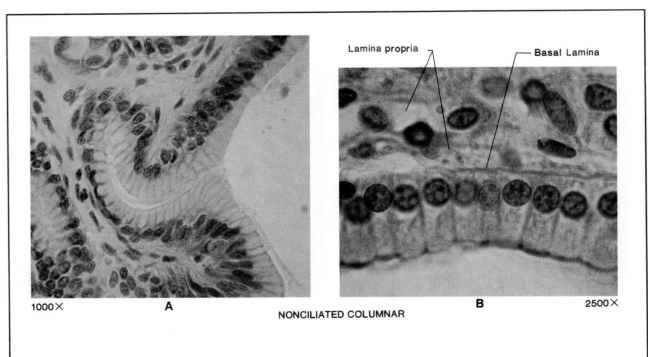

1000×   **A**

NONCILIATED COLUMNAR   2500×

Lamina propria   Basal Lamina

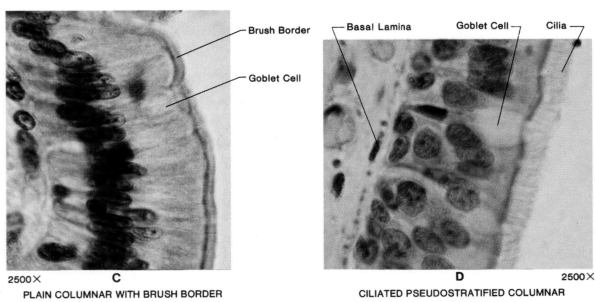

2500×   **C**

PLAIN COLUMNAR WITH BRUSH BORDER

Brush Border

Goblet Cell

Basal Lamina   Goblet Cell   Cilia

D   2500×

CILIATED PSEUDOSTRATIFIED COLUMNAR

Columnar epithelia of the digestive and respiratory tracts exhibit large numbers of goblet cells that produce mucus. Although only illustrations C and D have them labeled, they can also be seen in illustration A.

The distinct differences between cilia and a brush border are seen in illustrations C and D. Whereas cilia form from centrioles, brush borders are modified microvilli. Another modification of microvilli are cilialike structures called stereocilia. These nonmotile organelles are seen on columnar cells that line the vas deferens (see figure HA-38).

Note the distinct line of demarcation that constitutes the basal lamina in illustrations B and D. This thin layer between the epithelial cells and the lamina propria consists of a colloidal complex of protein, polysaccharide, and reticular fibers.

The lamina propria, to which all epithelial tissues are connected, consists of connective tissue, vascular and lymphatic channels, lymphocytes, plasma cells, eosinophils, and mast cells.

Figure HA-2   Columnar epithelium.

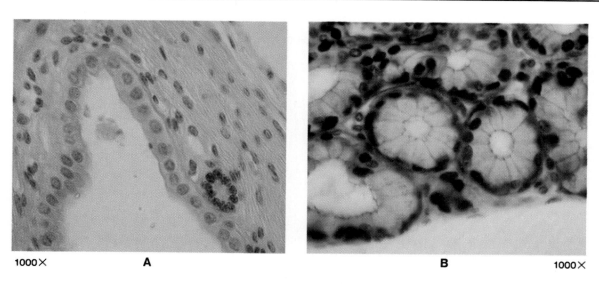

1000×   **A**   **B**   1000×

CUBOIDAL EPITHELIUM

Lamina propria   Basal Lamina   Binucleate Cells

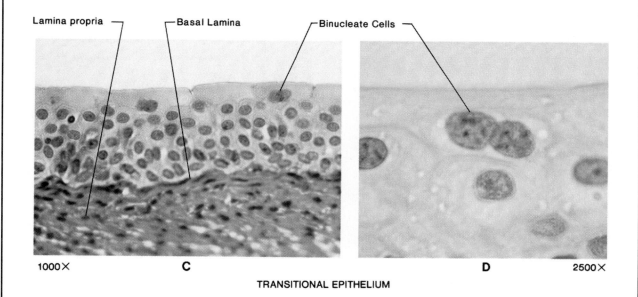

1000×   **C**   **D**   2500×

TRANSITIONAL EPITHELIUM

Although cuboidal cells are usually thought of as having a squarish appearance, as in illustration A, they often take on a pyramidal structure when observed surrounding the lumen of a duct or a small gland as in illustration B. Cuboidal epithelia may serve both absorptive and secretory functions, as in the case of tubules in the kidney.

Transitional epithelium is a stratified epithelium whose surface cells do not fall into squamous, cuboidal, or columnar categories. Note that the surface cells are dome-shaped and often binucleate, whereas the basal cells are more like stratified columnar cells. Between the surface cells and basal cells can be seen layers of pear-shaped cells in a loose configuration. This type of tissue is seen in the wall of the urinary bladder, the urethra, and certain places in the kidneys. The loose nature of the cells makes them desirable in places where organ distention demands elasticity of tissue.

Figure HA-3   Cuboidal and transitional epithelium.

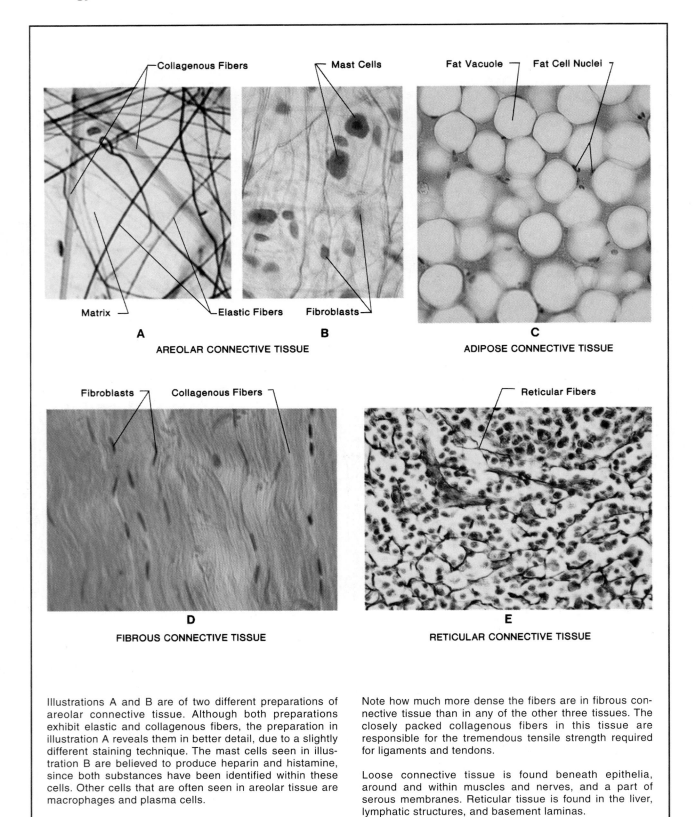

A AREOLAR CONNECTIVE TISSUE

B

C ADIPOSE CONNECTIVE TISSUE

D FIBROUS CONNECTIVE TISSUE

E RETICULAR CONNECTIVE TISSUE

Illustrations A and B are of two different preparations of areolar connective tissue. Although both preparations exhibit elastic and collagenous fibers, the preparation in illustration A reveals them in better detail, due to a slightly different staining technique. The mast cells seen in illustration B are believed to produce heparin and histamine, since both substances have been identified within these cells. Other cells that are often seen in areolar tissue are macrophages and plasma cells.

Note how much more dense the fibers are in fibrous connective tissue than in any of the other three tissues. The closely packed collagenous fibers in this tissue are responsible for the tremendous tensile strength required for ligaments and tendons.

Loose connective tissue is found beneath epithelia, around and within muscles and nerves, and a part of serous membranes. Reticular tissue is found in the liver, lymphatic structures, and basement laminas.

Figure HA-4   Connective tissues (1000×).

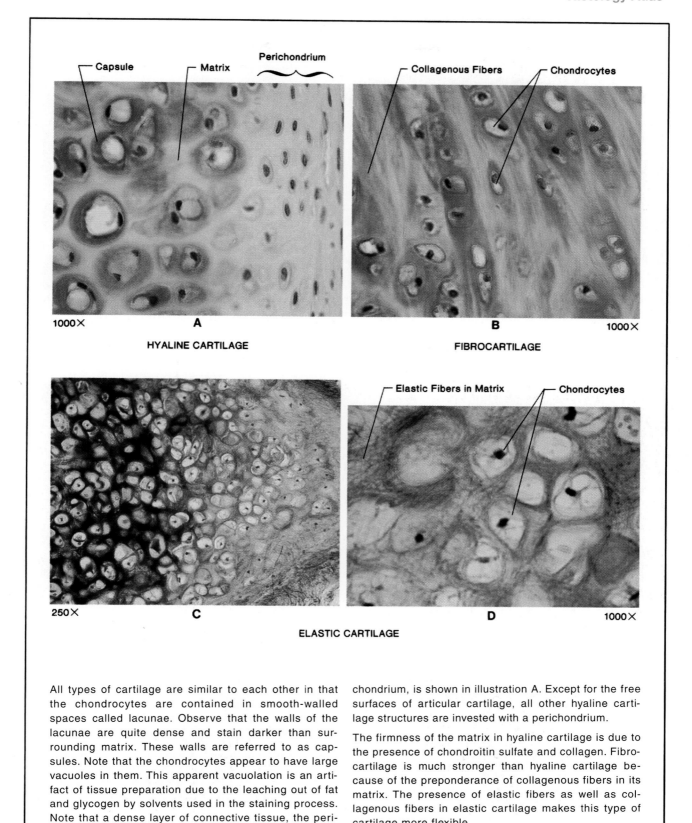

Figure HA-5   Types of cartilage.

All types of cartilage are similar to each other in that the chondrocytes are contained in smooth-walled spaces called lacunae. Observe that the walls of the lacunae are quite dense and stain darker than surrounding matrix. These walls are referred to as capsules. Note that the chondrocytes appear to have large vacuoles in them. This apparent vacuolation is an artifact of tissue preparation due to the leaching out of fat and glycogen by solvents used in the staining process. Note that a dense layer of connective tissue, the peri-chondrium, is shown in illustration A. Except for the free surfaces of articular cartilage, all other hyaline cartilage structures are invested with a perichondrium.

The firmness of the matrix in hyaline cartilage is due to the presence of chondroitin sulfate and collagen. Fibrocartilage is much stronger than hyaline cartilage because of the preponderance of collagenous fibers in its matrix. The presence of elastic fibers as well as collagenous fibers in elastic cartilage makes this type of cartilage more flexible.

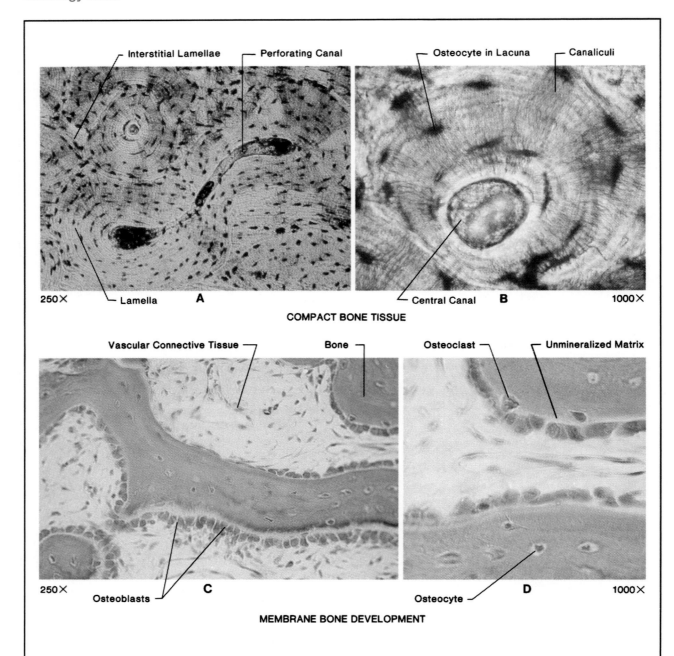

COMPACT BONE TISSUE

MEMBRANE BONE DEVELOPMENT

Bone formation has two origins: (1) from cartilage and (2) from osteogenic mesenchymal connective tissue. Compact bone of the long bones forms from cartilage; bones of the skull, on the other hand, develop as shown in illustrations C and D. The latter are often referred to as "membrane bones."

Note the loci of the different types of bone cells in the above illustrations. Illustrations C and D reveal that osteoblasts in membranous bone formation are seen on the leading edge of the forming bone. The first stage is the secretion of matrix by the osteoblasts. Reticular fibers are then added to the matrix by surrounding mesenchymal cells. Finally, mineralization occurs due to osteoblastic activity.

Osteoclasts are larger multinucleated cells that are involved in shaping bone structure by bone resorption. These cells are surrounded by a clear area (Howship's lacuna), evidence of mineral resorption. The osteoclast seen in illustration D is in an early stage of development; thus, Howship's lacuna is not very large. Once a bone cell becomes completely surrounded by bone, it is referred to as an osteocyte. Nourishment in mature compact bone reaches osteocytes through canaliculi.

Figure HA-6   Bone histology.

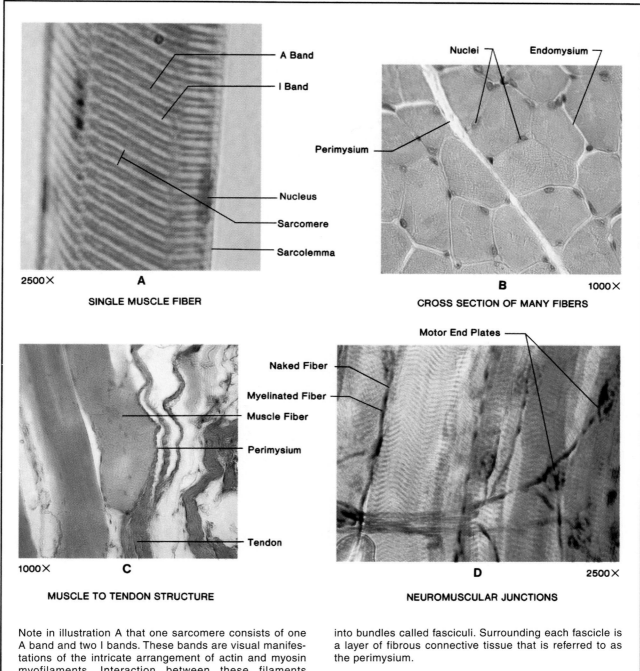

A Band
I Band
Nucleus
Sarcomere
Sarcolemma
2500×   **A**
SINGLE MUSCLE FIBER

Nuclei   **Endomysium**
Perimysium
B   1000×
CROSS SECTION OF MANY FIBERS

Naked Fiber
Myelinated Fiber
Muscle Fiber
Perimysium
Tendon
1000×   **C**
MUSCLE TO TENDON STRUCTURE

Motor End Plates
D   2500×
NEUROMUSCULAR JUNCTIONS

Note in illustration A that one sarcomere consists of one A band and two I bands. These bands are visual manifestations of the intricate arrangement of actin and myosin myofilaments. Interaction between these filaments causes muscle contraction.

A significant characteristic of skeletal muscle cells is that they are multinucleated, or syncytial. Note that the nuclei are elongated and situated near the sarcolemma of the cell. The peripheral location of the nuclei shows up well in illustration B.

Illustration B reveals that muscle fibers are separated from each other by endomysium, and grouped together into bundles called fasciculi. Surrounding each fascicle is a layer of fibrous connective tissue that is referred to as the perimysium.

Illustration C illustrates how the fibrous connective tissue of the endomysium, perimysium, and epimysium is continuous with the tendons which attach muscles to bone. The connective tissue of tendons, in turn, is continuous with the periosteum of bone.

Note in illustration D how the myelinated motor nerve fibers lose their myelin sheaths and become "naked" where they join the motor end plates.

Figure HA-7   Skeletal muscle microstructure.

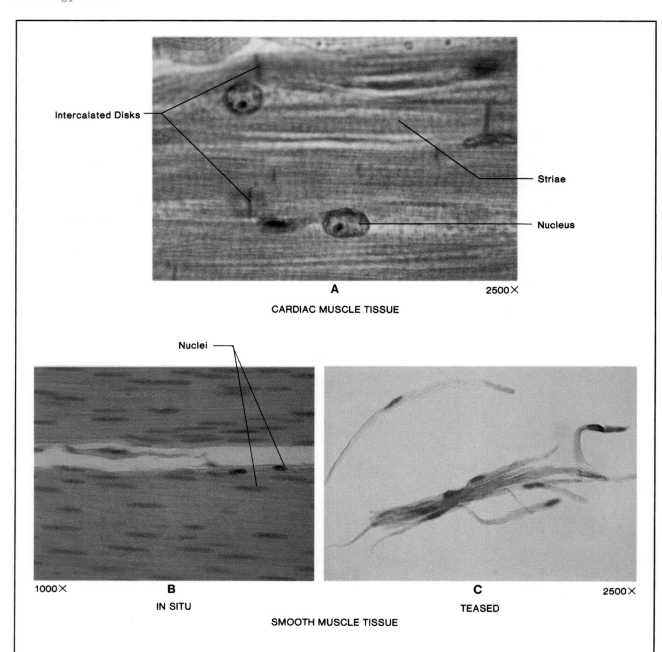

Intercalated Disks

Striae

Nucleus

A 2500×

CARDIAC MUSCLE TISSUE

Nuclei

1000× **B**

IN SITU

C 2500×

TEASED

SMOOTH MUSCLE TISSUE

The distinct contrast of the striae in the cardiac muscle preparation in illustration A is evidence that these fibers are in a state of complete relaxation; fibers prepared from tissue in contraction generally do not reveal striae as distinctly. Although this tissue may appear to be syncytial, it is not. Cardiac fibers have only one nucleus per fiber, with intercalated disks forming the limits for each unit.

The smooth muscle tissue seen in illustration B is of a portion of the intestinal wall. Due to the closely packed nature of the cells in such structures it is very difficult to see individual cells. Illustration C reveals how cells of this type of tissue appear when separated with a probe prior to staining. Each cell has a single nucleus and lacks the cross-striations seen in skeletal and cardiac muscle tissue.

Figure HA-8 Cardiac and smooth muscle tissue.

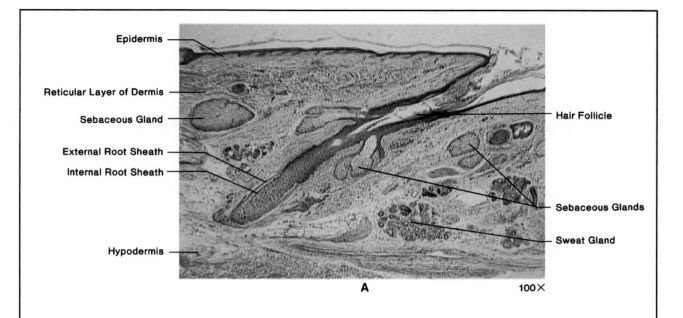

Epidermis

Reticular Layer of Dermis

Sebaceous Gland

External Root Sheath

Internal Root Sheath

Hypodermis

Hair Follicle

Sebaceous Glands

Sweat Gland

A   100×

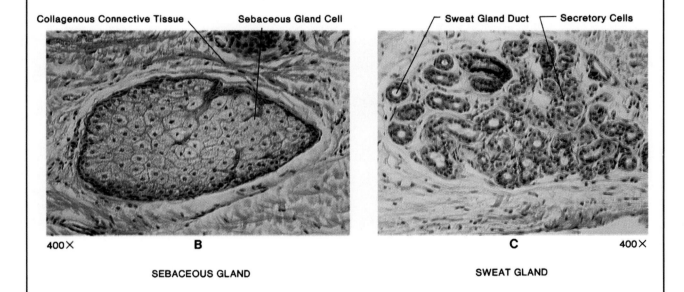

Collagenous Connective Tissue   Sebaceous Gland Cell

Sweat Gland Duct   Secretory Cells

400×   B

C   400×

SEBACEOUS GLAND

SWEAT GLAND

The relationship of the sebaceous and sweat glands to other structures of the skin is illustrated on this page. The low magnification of illustration A reveals that the epidermis is a very thin layer as compared to the dermis. Except for the palms and soles of the feet, the epidermis is usually only 0.1 mm thick. The dermis or corium, which underlies the epidermis, varies from 0.3 to 4.0 mm thick in different parts of the body. Note that both the sebaceous and sweat glands are located in the reticular layer of the dermis.

Observe that the sweat glands are tubulo-alveolar structures lined with cuboidal or columnar epithelium. The sebaceous glands, on the other hand, consist of rounded masses of cells, cuboidal at the periphery, and polygonal in the center. The secretion of sebaceous glands is formed by the breaking down of the central cells into an oily complex which is forced out into the space between the follicle and hair shaft. Although not shown in illustration A, the sweat glands open out directly onto the surface of the skin.

Figure HA-9   Scalp histology.

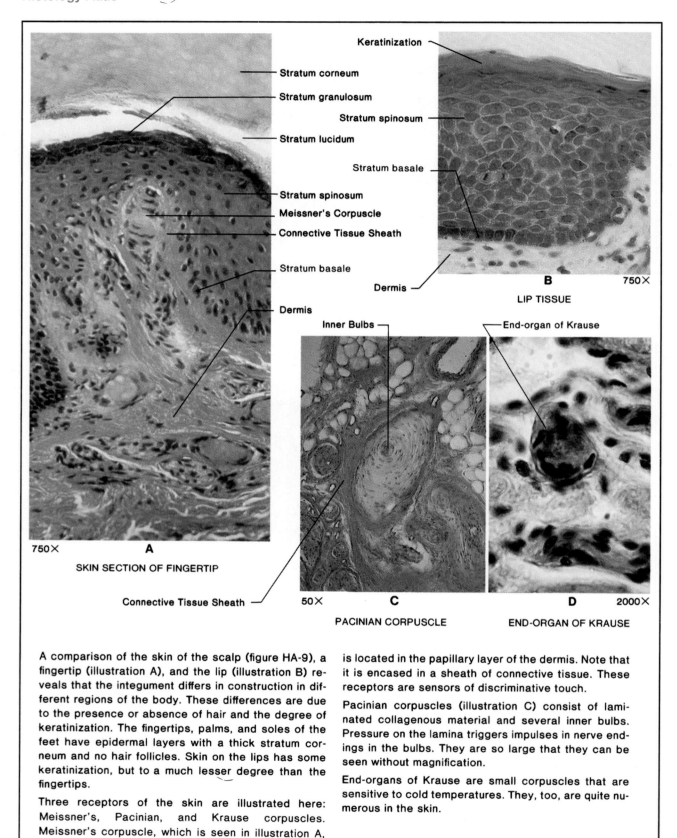

Stratum corneum

Stratum granulosum

Stratum lucidum

Stratum spinosum

Meissner's Corpuscle

Connective Tissue Sheath

Stratum basale

Dermis

750× **A**

SKIN SECTION OF FINGERTIP

Connective Tissue Sheath

Keratinization

Stratum spinosum

Stratum basale

Dermis

B 750×

LIP TISSUE

Inner Bulbs

End-organ of Krause

50× **C**

PACINIAN CORPUSCLE

D 2000×

END-ORGAN OF KRAUSE

A comparison of the skin of the scalp (figure HA-9), a fingertip (illustration A), and the lip (illustration B) reveals that the integument differs in construction in different regions of the body. These differences are due to the presence or absence of hair and the degree of keratinization. The fingertips, palms, and soles of the feet have epidermal layers with a thick stratum corneum and no hair follicles. Skin on the lips has some keratinization, but to a much lesser degree than the fingertips.

Three receptors of the skin are illustrated here: Meissner's, Pacinian, and Krause corpuscles. Meissner's corpuscle, which is seen in illustration A,

is located in the papillary layer of the dermis. Note that it is encased in a sheath of connective tissue. These receptors are sensors of discriminative touch.

Pacinian corpuscles (illustration C) consist of laminated collagenous material and several inner bulbs. Pressure on the lamina triggers impulses in nerve endings in the bulbs. They are so large that they can be seen without magnification.

End-organs of Krause are small corpuscles that are sensitive to cold temperatures. They, too, are quite numerous in the skin.

Figure HA-10   Skin structure.

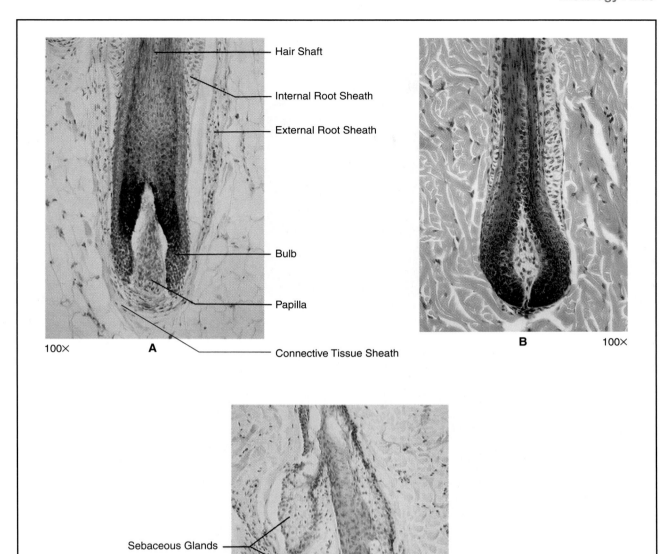

Hair Shaft

Internal Root Sheath

External Root Sheath

Bulb

Papilla

100×   **A**

Connective Tissue Sheath

B   100×

Sebaceous Glands

Arrector pili Muscle

C   100×

Hair consists of horny threads that are derived from the epidermis. Its shaft is enveloped by a tube, the follicle. Illustration A reveals that there are three components to a hair follicle: an internal root sheath, an external root sheath, and a connective tissue sheath. The two inner sheaths are derived from the epidermis; the outer connective tissue sheath is formed from the dermis.

Note that the separation of these two sheaths is less distinct in illustration B.
Illustration C reveals the relationship of the arrector pili muscle to the sebacous glands. These small muscles, which consist of smooth muscle tissue, play a role in forcing sebum out into the hair follicle. The result of their action produces what we call "goose bumps" on the skin.

Figure HA-11   Hair structure.

Nail Plate

Nail Bed

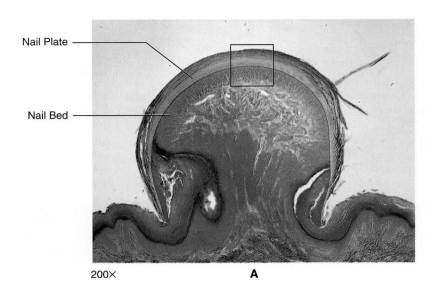

200×   **A**

Nail Plate

Nail Bed

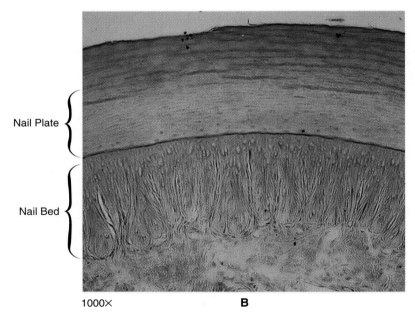

1000×   **B**

Fingernails and toenails are modified structures of the epidermis. Illustration A is a section through the fingernail of an infant. Note that the body of the nail consists of several layers of flattened clear cells that take the place of the stratum corneum of the skin. These cells are much harder than the stratum corneum layers and have shrunken nuclei.

Although the nail bed (shown in illustration B) is modified epidermis, it lacks the stratum granulosum and stratum lucidum that one would see in a typical section of the skin. Only the deeper epidermal layers are present.

Figure HA-12   **Fingernail structure.**

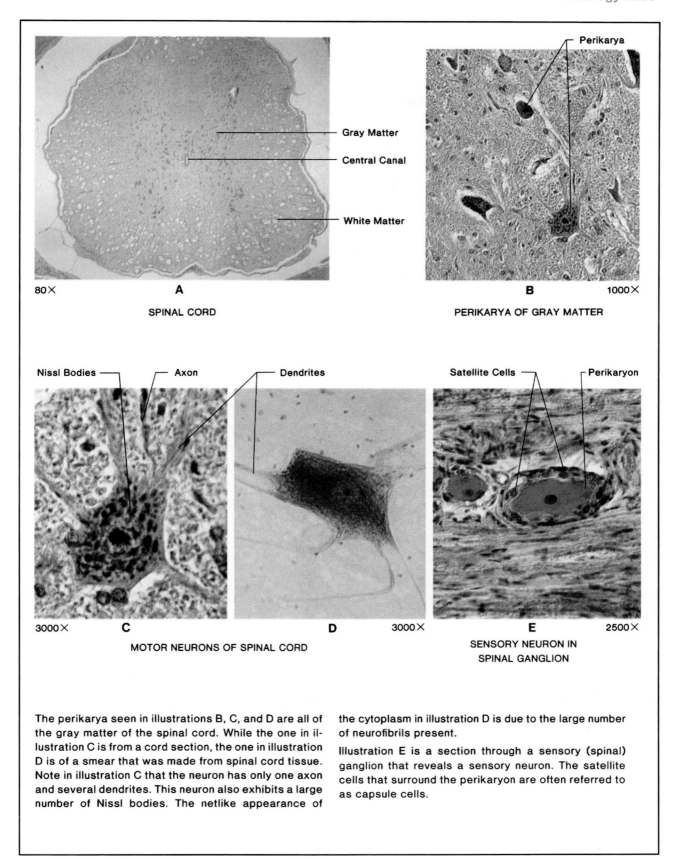

Gray Matter

Central Canal

White Matter

80×   **A**

SPINAL CORD

Perikarya

B   1000×

PERIKARYA OF GRAY MATTER

Nissl Bodies   Axon   Dendrites   Satellite Cells   Perikaryon

3000×   **C**   **D**   3000×   **E**   2500×

MOTOR NEURONS OF SPINAL CORD   SENSORY NEURON IN SPINAL GANGLION

The perikarya seen in illustrations B, C, and D are all of the gray matter of the spinal cord. While the one in illustration C is from a cord section, the one in illustration D is of a smear that was made from spinal cord tissue. Note in illustration C that the neuron has only one axon and several dendrites. This neuron also exhibits a large number of Nissl bodies. The netlike appearance of the cytoplasm in illustration D is due to the large number of neurofibrils present.

Illustration E is a section through a sensory (spinal) ganglion that reveals a sensory neuron. The satellite cells that surround the perikaryon are often referred to as capsule cells.

Figure HA-13   Neurons of the spinal cord and spinal ganglia.

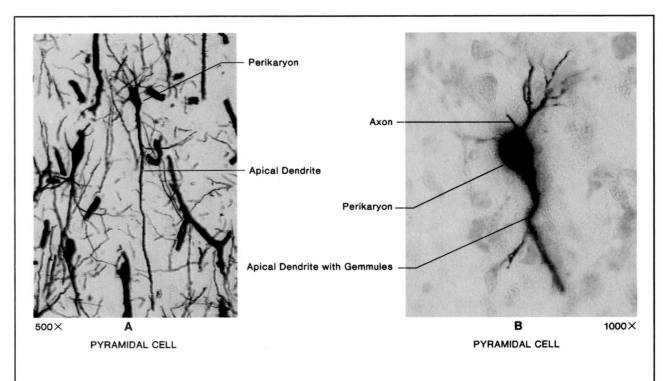

500×   **A**

PYRAMIDAL CELL

                    Perikaryon

                    Apical Dendrite

Apical Dendrite with Gemmules

Axon ——

Perikaryon ——

B   1000×

PYRAMIDAL CELL

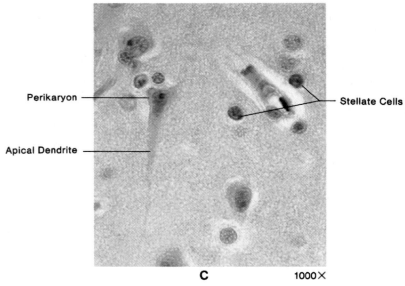

Perikaryon ——

Apical Dendrite ——

—— Stellate Cells

C   1000×

PYRAMIDAL AND STELLATE CELLS

The principal nerve cells of the cerebrum are pyramidal cells. Note in each of the above photomicrographs that a long apical dendrite extends from the perikaryon. The apical dendrite is always oriented toward the surface of the cerebrum. In illustration B there is evidence of "gemmules" on the surface of the dendrite. Gemmules are small processes that greatly increase the surface area of dendrites, allowing large neurons to receive as many as 100,000 separate axon terminals or synapses.

Note the difference in size between the axon and dendrites in illustration B. The axon is much smaller in diameter and has a smoother surface due to the absence of gemmules.

The stellate cells shown in illustration C are association neurons that provide connections between pyramidal cells of the cerebrum. These interneurons are also present in the cerebellum, where they provide linkage between the Purkinje cells.

Figure HA-14   Neurons of the cerebrum.

HA-14

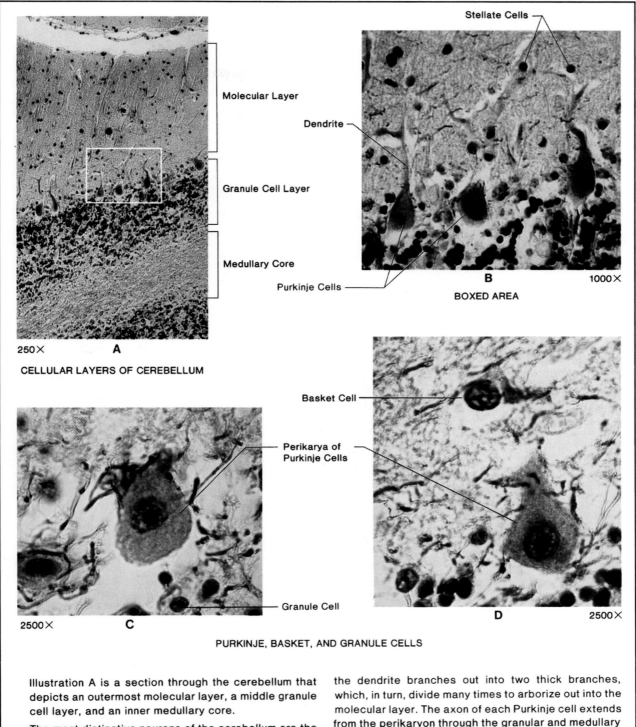

Molecular Layer

Granule Cell Layer

Medullary Core

250× **A**

CELLULAR LAYERS OF CEREBELLUM

Stellate Cells

Dendrite

Purkinje Cells

B 1000×

BOXED AREA

Basket Cell

Perikarya of
Purkinje Cells

Granule Cell

2500× **C**

D 2500×

PURKINJE, BASKET, AND GRANULE CELLS

Illustration A is a section through the cerebellum that depicts an outermost molecular layer, a middle granule cell layer, and an inner medullary core.

The most distinctive neurons of the cerebellum are the Purkinje cells. Note in illustration A that these neurons form a layer deep within the molecular layer near the edge of the granule cell layer. Each flask-shaped cell has a thick dendrite that is directed toward the cerebellar cortex. A short distance out from the perikaryon the dendrite branches out into two thick branches, which, in turn, divide many times to arborize out into the molecular layer. The axon of each Purkinje cell extends from the perikaryon through the granular and medullary layers, and finally, through collaterals, reenters the molecular layer to contact other Purkinje cells.

Stellate, basket, and granule cells are all multipolar association neurons. Note the proximity of the perikarya of the basket cells to the Purkinje cells.

Figure HA-15   Neurons of the cerebellum.

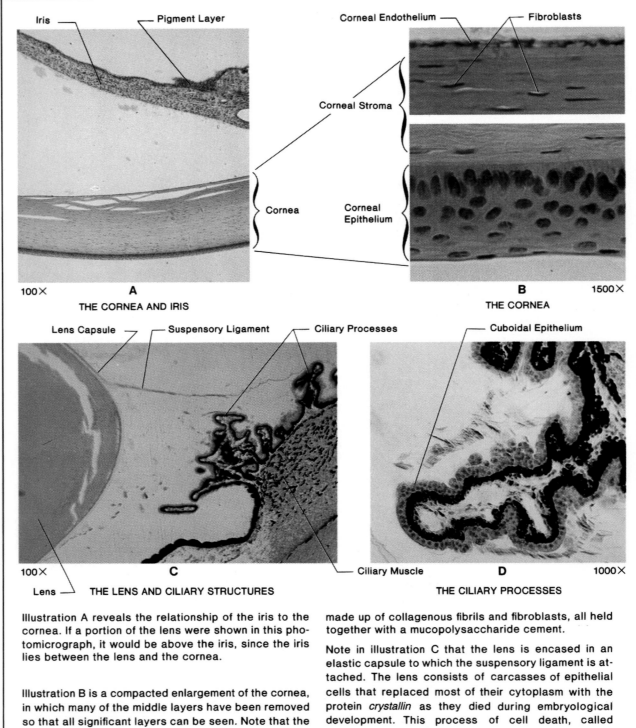

Figure HA-16   The cornea, lens, iris, and ciliary structures.

Illustration A reveals the relationship of the iris to the cornea. If a portion of the lens were shown in this photomicrograph, it would be above the iris, since the iris lies between the lens and the cornea.

Illustration B is a compacted enlargement of the cornea, in which many of the middle layers have been removed so that all significant layers can be seen. Note that the outer corneal epithelium consists of stratified squamous epithelium, and that the inner corneal endothelium consists of a single layer of low cuboidal cells. Between these two layers is the corneal stroma (*substantia propria*), which makes up nine-tenths of the thickness of the cornea. This transparent stroma is made up of collagenous fibrils and fibroblasts, all held together with a mucopolysaccharide cement.

Note in illustration C that the lens is encased in an elastic capsule to which the suspensory ligament is attached. The lens consists of carcasses of epithelial cells that replaced most of their cytoplasm with the protein *crystallin* as they died during embryological development. This process of cell death, called *apoptosis,* is a common phenomenon in cellular physiology (consider keratinization). Note, also, that the ciliary processes shown in illustration D are covered with cuboidal epithelial cells. It is these cells that produce the aqueous humor that fills the space between the lens and the cornea.

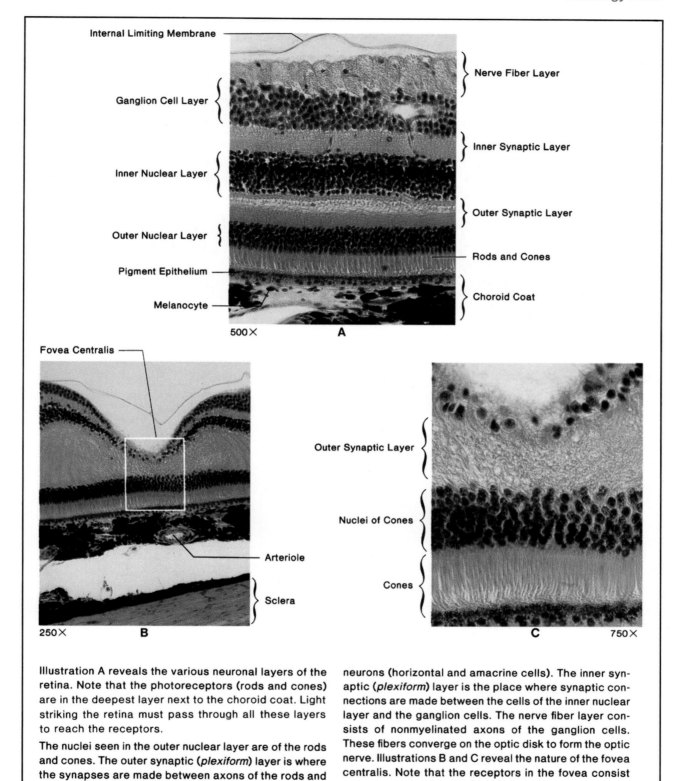

The following labels appear on the illustrations:

A (500×)
- Internal Limiting Membrane
- Ganglion Cell Layer
- Inner Nuclear Layer
- Outer Nuclear Layer
- Pigment Epithelium
- Melanocyte
- Nerve Fiber Layer
- Inner Synaptic Layer
- Outer Synaptic Layer
- Rods and Cones
- Choroid Coat

B (250×)
- Fovea Centralis
- Arteriole
- Sclera

C (750×)
- Outer Synaptic Layer
- Nuclei of Cones
- Cones

Illustration A reveals the various neuronal layers of the retina. Note that the photoreceptors (rods and cones) are in the deepest layer next to the choroid coat. Light striking the retina must pass through all these layers to reach the receptors.

The nuclei seen in the outer nuclear layer are of the rods and cones. The outer synaptic (*plexiform*) layer is where the synapses are made between axons of the rods and cones and the dendrites of bipolar cells and the processes of horizontal cells. The inner nuclear layer contains nuclei of bipolar neurons and association neurons (horizontal and amacrine cells). The inner synaptic (*plexiform*) layer is the place where synaptic connections are made between the cells of the inner nuclear layer and the ganglion cells. The nerve fiber layer consists of nonmyelinated axons of the ganglion cells. These fibers converge on the optic disk to form the optic nerve. Illustrations B and C reveal the nature of the fovea centralis. Note that the receptors in the fovea consist only of cones and that the inner synaptic and nerve fiber layers are lacking.

Figure HA-17   The retina of the eye.

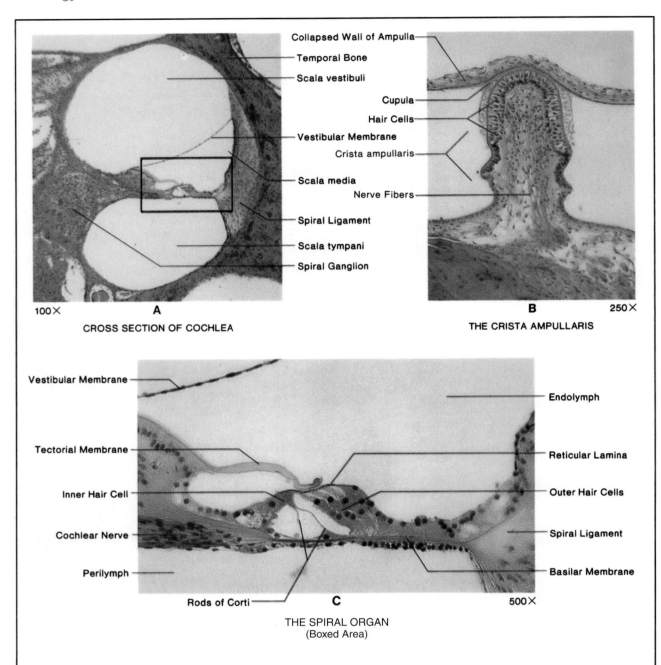

100× **A**
CROSS SECTION OF COCHLEA

- Collapsed Wall of Ampulla
- Temporal Bone
- Scala vestibuli
- Cupula
- Hair Cells
- Vestibular Membrane
- Crista ampullaris
- Scala media
- Nerve Fibers
- Spiral Ligament
- Scala tympani
- Spiral Ganglion

B 250×
THE CRISTA AMPULLARIS

- Vestibular Membrane
- Endolymph
- Tectorial Membrane
- Reticular Lamina
- Inner Hair Cell
- Outer Hair Cells
- Cochlear Nerve
- Spiral Ligament
- Perilymph
- Basilar Membrane
- Rods of Corti
- **C** 500×

THE SPIRAL ORGAN
(Boxed Area)

The cochlea of the ear lies within the temporal bone. Along its outer wall is seen a spiral ligament (illustration A) that is formed from periosteum of the bone. Note that fibers of the spiral ligament are continuous with the basilar membrane. Observe that the cochlea has three chambers: scala vestibuli, scala tympani, and scala media. The cochlear duct is the flexible portion of the cochlea that is bounded on one side by the vestibular membrane and on the other side by the basilar membrane. Endolymph is present in the cochlear duct; perilymph fills the scala vestibuli and scala tympani.

The spiral organ, (or organ of Corti) shown in illustration C is an enlargement of the boxed area in illustration A. Note how the gelatinous flap of the tectorial membrane lies in contact with hairs that emerge from the hair cells. Discrimination of sound frequencies results from nerve impulses sent to the brain via the cochlear nerve in conjunction with vibrations in the endolymph and basilar membrane.

The obliteration of the cupula in illustration B occurred during slide preparation. Other structures in the crista ampullaris are normal.

Figure HA-18   The cochlea and crista ampullaris.

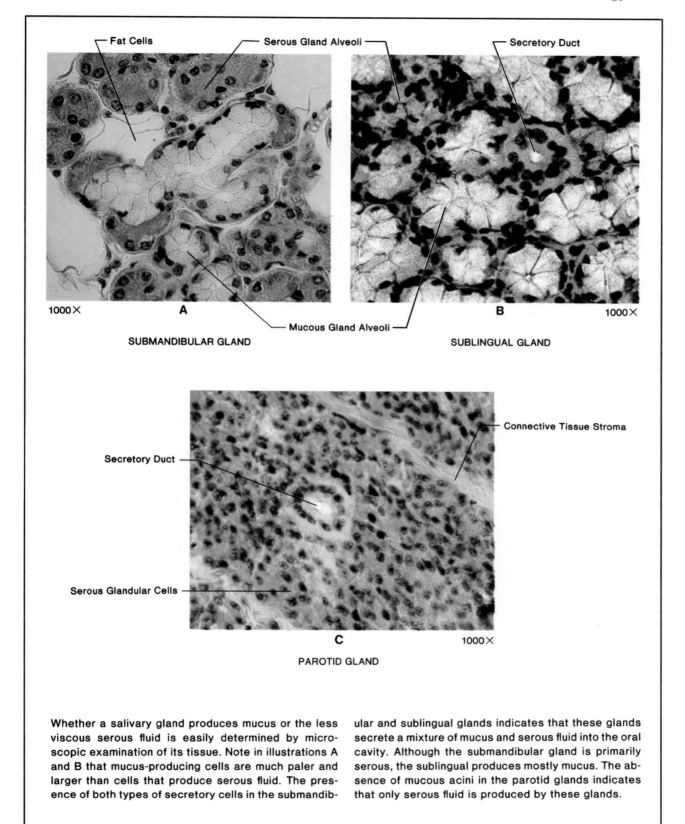

Fat Cells — Serous Gland Alveoli — Secretory Duct

1000× A

SUBMANDIBULAR GLAND

— Mucous Gland Alveoli —

B 1000×

SUBLINGUAL GLAND

Connective Tissue Stroma

Secretory Duct

Serous Glandular Cells

C 1000×

PAROTID GLAND

Whether a salivary gland produces mucus or the less viscous serous fluid is easily determined by microscopic examination of its tissue. Note in illustrations A and B that mucus-producing cells are much paler and larger than cells that produce serous fluid. The presence of both types of secretory cells in the submandibular and sublingual glands indicates that these glands secrete a mixture of mucus and serous fluid into the oral cavity. Although the submandibular gland is primarily serous, the sublingual produces mostly mucus. The absence of mucous acini in the parotid glands indicates that only serous fluid is produced by these glands.

Figure HA-19   Histology of major salivary glands.

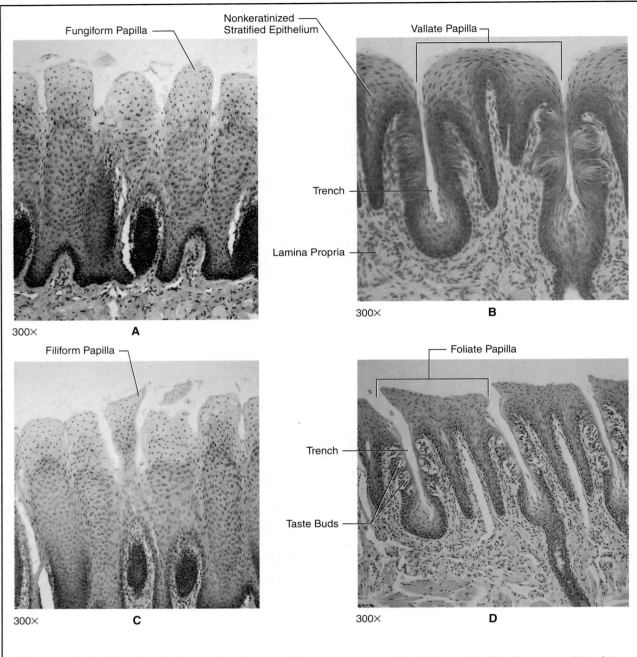

The four principal papillae that cover the dorsum of the tongue are the fungiform, filiform, vallate, and foliate papillae. All of them are projections of the dermis of the skin.

The fungiform papillae (illustration A) have rounded eminences and are deep red. Although they are found primarily on the apex and sides of the tongue, they are scattered irregularly over the dorsum.

Filiform papillae (illustration C) have a threadlike (filiform) appearance. At the back of the tongue they are arranged in lines parallel with the two rows of vallate papillae. On the apex of the tongue their arrangement is in lines extending from one side of the tongue to the other.

Vallate papillae (illustration B) are large projections located at the back of the tongue. They vary from eight to twelve in number. Note that each papilla is covered with stratified squamous epithelium and is surrounded by a deep trench. In the walls of these trenches are located hundreds of taste receptors (taste buds).

Foliate papillae (illustration D) are ridges seen on the sides of the tongue. They resemble vallate papillae in that they have taste buds on their trench walls.

Figure HA-20   Papillae of the tongue.

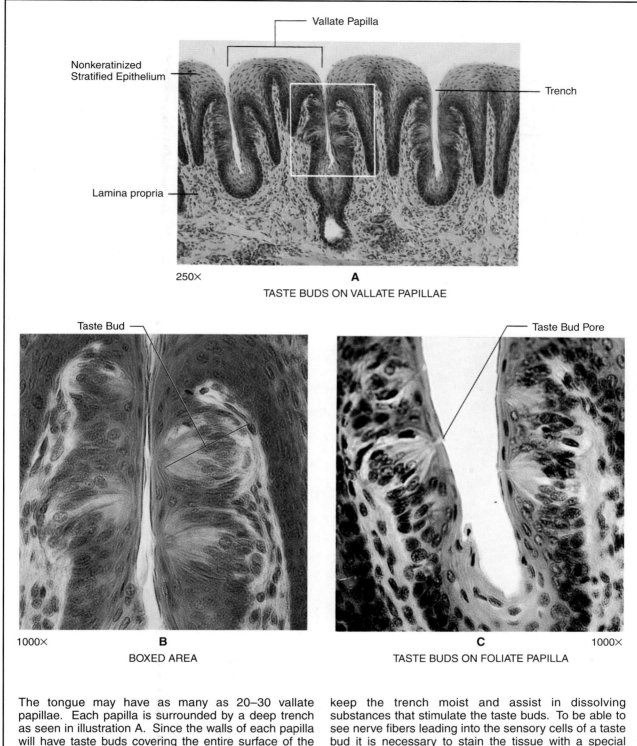

250×

A

TASTE BUDS ON VALLATE PAPILLAE

1000× **B** BOXED AREA

C 1000× TASTE BUDS ON FOLIATE PAPILLA

The tongue may have as many as 20–30 vallate papillae. Each papilla is surrounded by a deep trench as seen in illustration A. Since the walls of each papilla will have taste buds covering the entire surface of the trench, it is conceivable that hundreds of taste receptors are present on these structures.

Note that each taste bud has a pore that opens out into the trench. At the base of each trench are duct openings from serous glands (glands of von Ebner) that keep the trench moist and assist in dissolving substances that stimulate the taste buds. To be able to see nerve fibers leading into the sensory cells of a taste bud it is necessary to stain the tissue with a special silver dye. No such dye was used on these slide preparations.

Foliate papillae (illustration C) are located on the sides of the tongue. The superior clarity of the taste buds in this photomicrograph is due to superb tissue preparation.

Figure HA-21 **Taste buds.**

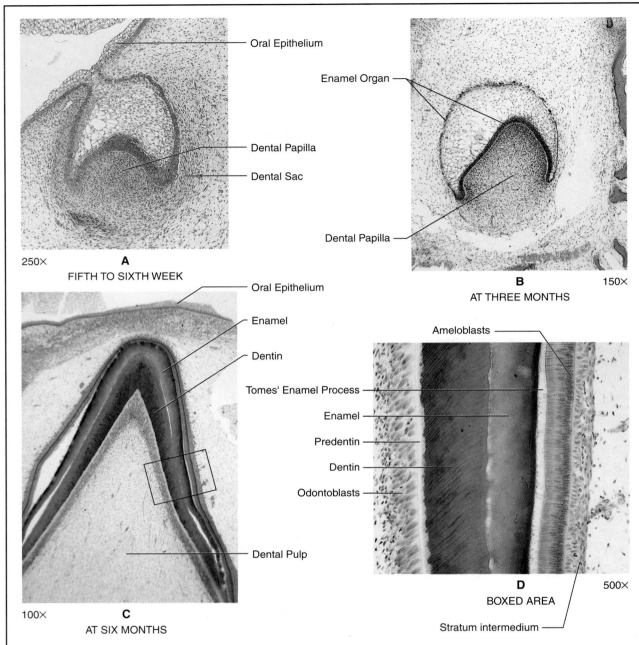

FIFTH TO SIXTH WEEK — A — 250×

Oral Epithelium
Dental Papilla
Dental Sac

AT THREE MONTHS — B — 150×

Enamel Organ
Dental Papilla

AT SIX MONTHS — C — 100×

Oral Epithelium
Enamel
Dentin
Dental Pulp

BOXED AREA — D — 500×

Ameloblasts
Tomes' Enamel Process
Enamel
Predentin
Dentin
Odontoblasts
Stratum intermedium

In each dental arch of the embryo ten tooth buds form by proliferation of cells in the oral epithelium. At first the buds are solid and rounded, but as the cells multiply they push inward (invaginate) as shown in illustration A. The oral epithelium (ectoderm) forms a two-layered structure called the enamel organ. Mesenchymal cells (mesoderm) in the distal portion of the bud push in to form the dental papilla. Illustration C reveals that by the fifth or sixth month the enamel organ produces the enamel of the tooth, and the cells of the dental papilla differentiate to form dentin.

Dentin is laid down just before the appearance of enamel. It is produced by a layer of elongated cells called odontoblasts that form from mesenchymal cells of the dental papilla. Note in illustration D that uncalcified dentin, or predentin, is formed first. Predentin is quickly converted to mature dentin, which consists of collagenous fibers and a calcified component called apatite.

Enamel is produced by cells called ameloblasts. Note in illustration D that these cells produce a clear layer called Tomes' enamel process, which becomes converted to mature enamel by calcification with mineral salts. As new enamel forms, the ameloblasts move outward away from the dentin and odontoblasts. When the tooth erupts, the ameloblasts are sloughed off.

Figure HA-22 Embryological stages in tooth development.

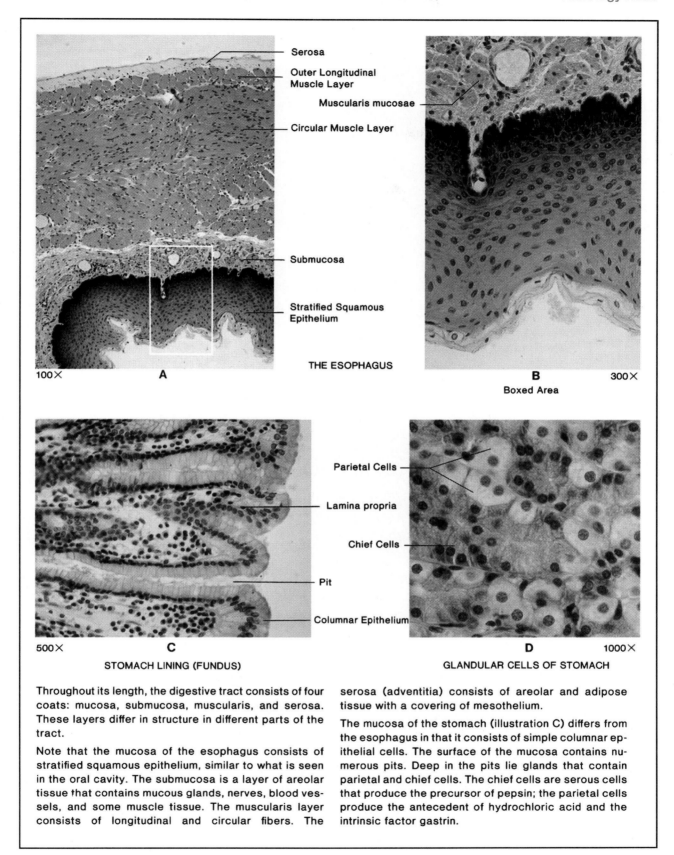

Serosa

Outer Longitudinal
Muscle Layer

Muscularis mucosae

Circular Muscle Layer

Submucosa

Stratified Squamous
Epithelium

THE ESOPHAGUS

100×   **A**

B   300×

Boxed Area

Parietal Cells

Lamina propria

Chief Cells

Pit

Columnar Epithelium

500×   **C**

D   1000×

STOMACH LINING (FUNDUS)

GLANDULAR CELLS OF STOMACH

Throughout its length, the digestive tract consists of four coats: mucosa, submucosa, muscularis, and serosa. These layers differ in structure in different parts of the tract.

Note that the mucosa of the esophagus consists of stratified squamous epithelium, similar to what is seen in the oral cavity. The submucosa is a layer of areolar tissue that contains mucous glands, nerves, blood vessels, and some muscle tissue. The muscularis layer consists of longitudinal and circular fibers. The

serosa (adventitia) consists of areolar and adipose tissue with a covering of mesothelium.

The mucosa of the stomach (illustration C) differs from the esophagus in that it consists of simple columnar epithelial cells. The surface of the mucosa contains numerous pits. Deep in the pits lie glands that contain parietal and chief cells. The chief cells are serous cells that produce the precursor of pepsin; the parietal cells produce the antecedent of hydrochloric acid and the intrinsic factor gastrin.

Figure HA-23   The esophagus and stomach.

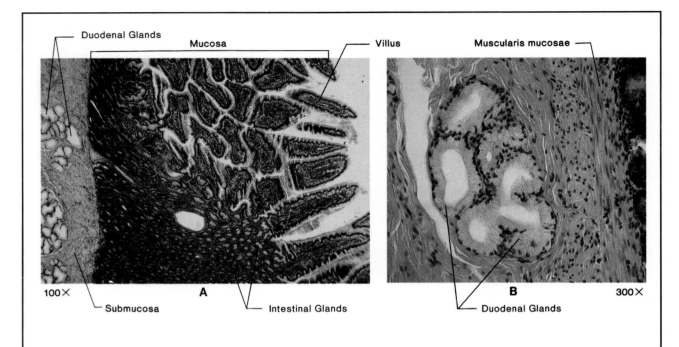

Duodenal Glands

Mucosa

Villus

Muscularis mucosae

100×   A

Submucosa

Intestinal Glands

B   300×

Duodenal Glands

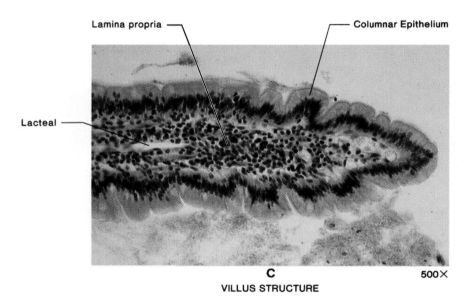

Lamina propria

Columnar Epithelium

Lacteal

C   500×

VILLUS STRUCTURE

The mucosa of the duodenum consists of the epithelium and the lamina propria. Within the lamina propria are numerous intestinal glands (crypts of Lieberkühn). The entire mucosal surface is covered with millions of small fingerlike projections called villi.

Where the lamina propria meets the muscularis mucosae, the mucosa ends. Note the large number of mucus-secreting duodenal (Brunner's) glands (illustration B) located in the submucosa.

Illustration C reveals the structure of a single villus. The entire epithelium consists of simple columnar cells interspersed with goblet cells. The core of the villus is the lamina propria, which contains loose connective tissue, smooth muscle fibers, blood vessels, and a lymphatic vessel, the lacteal. The muscularis layer (not shown in illustration C) contains an inner circular layer of smooth muscle tissue and an outer longitudinal layer of smooth fibers. The outer surface of the duodenum is covered by the serosa.

Figure HA-24   The duodenum.

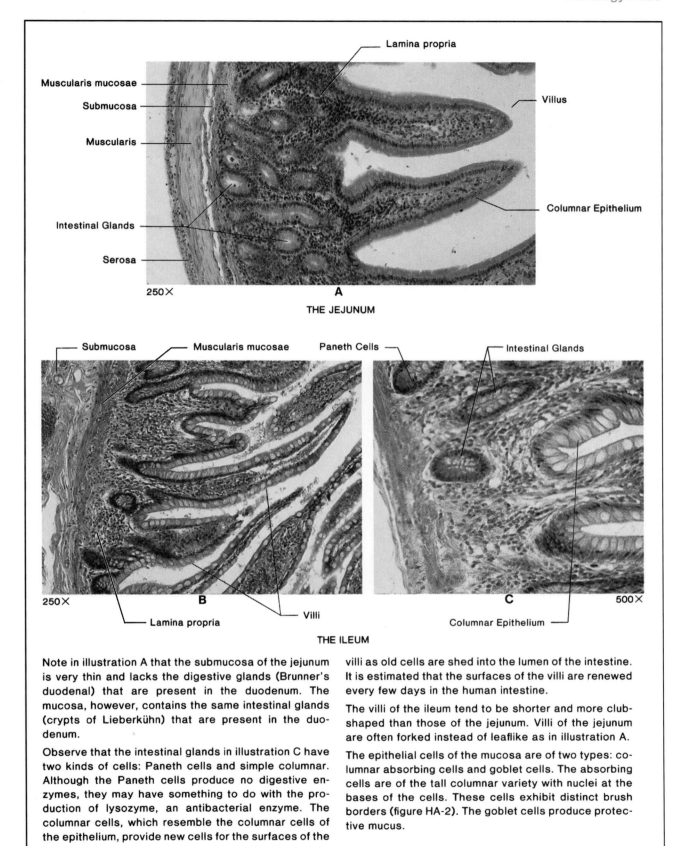

Lamina propria

Muscularis mucosae

Submucosa

Muscularis

Intestinal Glands

Serosa

Villus

Columnar Epithelium

250×

A

THE JEJUNUM

Submucosa   Muscularis mucosae   Paneth Cells   Intestinal Glands

250×

B

Villi

Lamina propria

C

500×

Columnar Epithelium

THE ILEUM

Note in illustration A that the submucosa of the jejunum is very thin and lacks the digestive glands (Brunner's duodenal) that are present in the duodenum. The mucosa, however, contains the same intestinal glands (crypts of Lieberkühn) that are present in the duodenum.

Observe that the intestinal glands in illustration C have two kinds of cells: Paneth cells and simple columnar. Although the Paneth cells produce no digestive enzymes, they may have something to do with the production of lysozyme, an antibacterial enzyme. The columnar cells, which resemble the columnar cells of the epithelium, provide new cells for the surfaces of the villi as old cells are shed into the lumen of the intestine. It is estimated that the surfaces of the villi are renewed every few days in the human intestine.

The villi of the ileum tend to be shorter and more club-shaped than those of the jejunum. Villi of the jejunum are often forked instead of leaflike as in illustration A.

The epithelial cells of the mucosa are of two types: columnar absorbing cells and goblet cells. The absorbing cells are of the tall columnar variety with nuclei at the bases of the cells. These cells exhibit distinct brush borders (figure HA-2). The goblet cells produce protective mucus.

Figure HA-25   The jejunum and ileum.

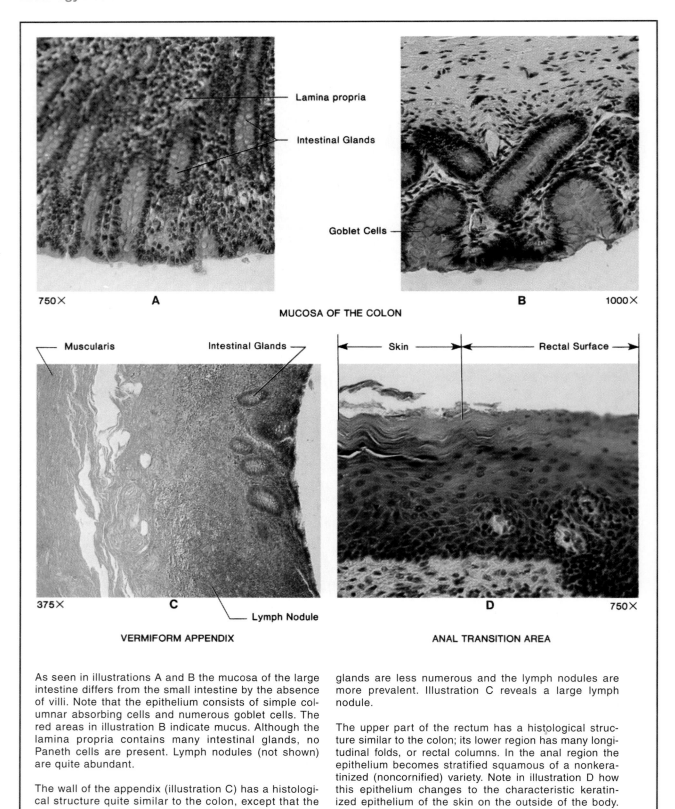

As seen in illustrations A and B the mucosa of the large intestine differs from the small intestine by the absence of villi. Note that the epithelium consists of simple columnar absorbing cells and numerous goblet cells. The red areas in illustration B indicate mucus. Although the lamina propria contains many intestinal glands, no Paneth cells are present. Lymph nodules (not shown) are quite abundant.

The wall of the appendix (illustration C) has a histological structure quite similar to the colon, except that the glands are less numerous and the lymph nodules are more prevalent. Illustration C reveals a large lymph nodule.

The upper part of the rectum has a histological structure similar to the colon; its lower region has many longitudinal folds, or rectal columns. In the anal region the epithelium becomes stratified squamous of a nonkeratinized (noncornified) variety. Note in illustration D how this epithelium changes to the characteristic keratinized epithelium of the skin on the outside of the body.

Figure HA-26   The lower digestive tract.

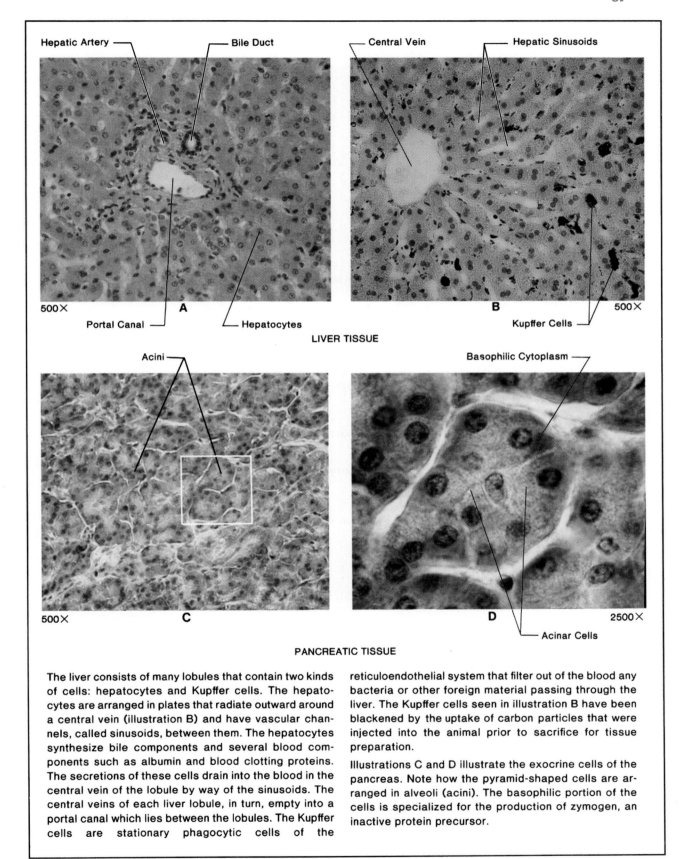

Hepatic Artery — | — Bile Duct

Portal Canal — | — Hepatocytes

500× **A**

LIVER TISSUE

— Central Vein | — Hepatic Sinusoids

B 500×

Kupffer Cells —

Acini —

500× **C**

Basophilic Cytoplasm —

D 2500×

— Acinar Cells

PANCREATIC TISSUE

The liver consists of many lobules that contain two kinds of cells: hepatocytes and Kupffer cells. The hepatocytes are arranged in plates that radiate outward around a central vein (illustration B) and have vascular channels, called sinusoids, between them. The hepatocytes synthesize bile components and several blood components such as albumin and blood clotting proteins. The secretions of these cells drain into the blood in the central vein of the lobule by way of the sinusoids. The central veins of each liver lobule, in turn, empty into a portal canal which lies between the lobules. The Kupffer cells are stationary phagocytic cells of the reticuloendothelial system that filter out of the blood any bacteria or other foreign material passing through the liver. The Kupffer cells seen in illustration B have been blackened by the uptake of carbon particles that were injected into the animal prior to sacrifice for tissue preparation.

Illustrations C and D illustrate the exocrine cells of the pancreas. Note how the pyramid-shaped cells are arranged in alveoli (acini). The basophilic portion of the cells is specialized for the production of zymogen, an inactive protein precursor.

Figure HA-27   Liver and pancreas histology.

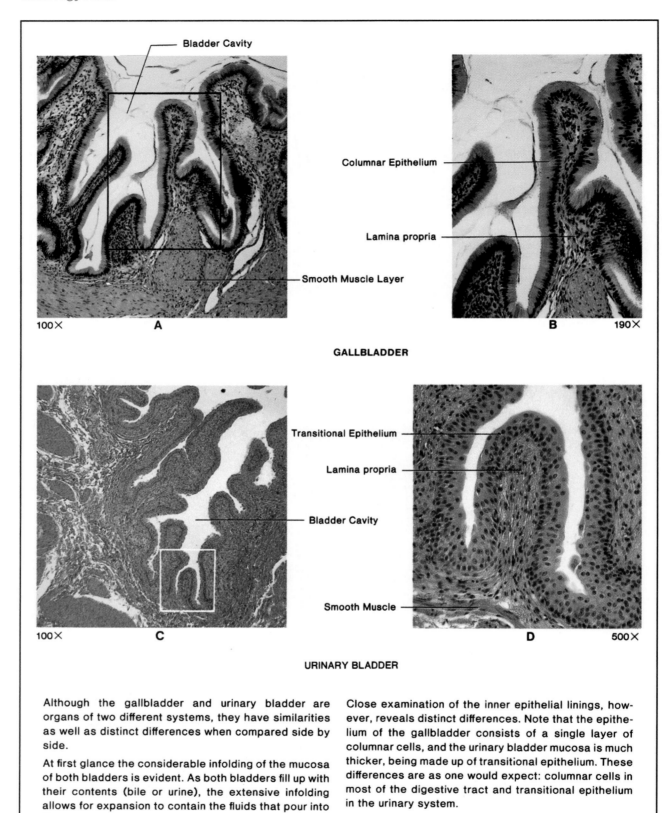

Figure HA-28   The gallbladder and urinary bladder.

Although the gallbladder and urinary bladder are organs of two different systems, they have similarities as well as distinct differences when compared side by side.

At first glance the considerable infolding of the mucosa of both bladders is evident. As both bladders fill up with their contents (bile or urine), the extensive infolding allows for expansion to contain the fluids that pour into them.

Close examination of the inner epithelial linings, however, reveals distinct differences. Note that the epithelium of the gallbladder consists of a single layer of columnar cells, and the urinary bladder mucosa is much thicker, being made up of transitional epithelium. These differences are as one would expect: columnar cells in most of the digestive tract and transitional epithelium in the urinary system.

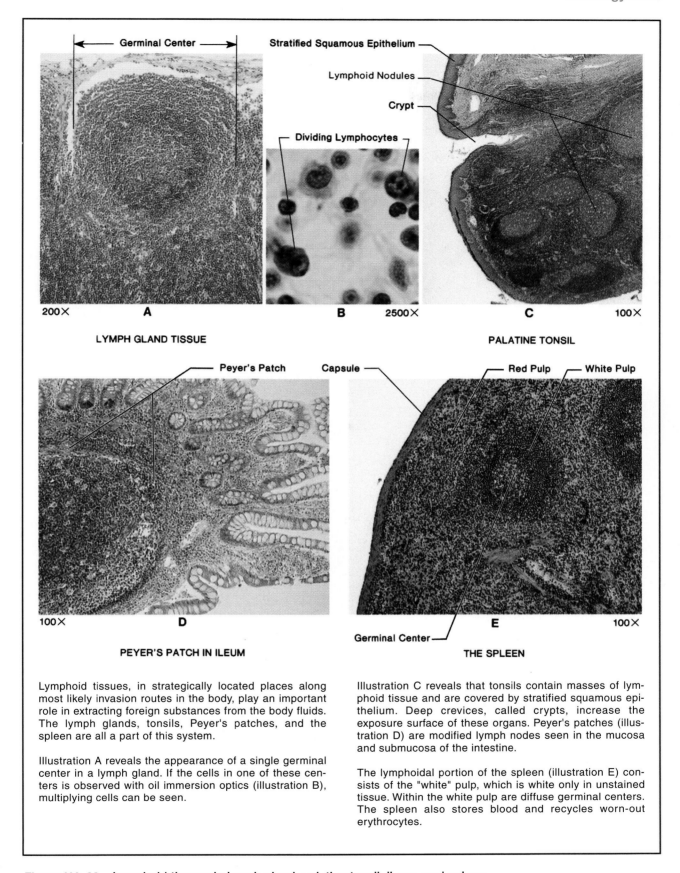

Figure HA-29   Lymphoid tissues in lymph gland, palatine tonsil, ileum, and spleen.

Lymphoid tissues, in strategically located places along most likely invasion routes in the body, play an important role in extracting foreign substances from the body fluids. The lymph glands, tonsils, Peyer's patches, and the spleen are all a part of this system.

Illustration A reveals the appearance of a single germinal center in a lymph gland. If the cells in one of these centers is observed with oil immersion optics (illustration B), multiplying cells can be seen.

Illustration C reveals that tonsils contain masses of lymphoid tissue and are covered by stratified squamous epithelium. Deep crevices, called crypts, increase the exposure surface of these organs. Peyer's patches (illustration D) are modified lymph nodes seen in the mucosa and submucosa of the intestine.

The lymphoidal portion of the spleen (illustration E) consists of the "white" pulp, which is white only in unstained tissue. Within the white pulp are diffuse germinal centers. The spleen also stores blood and recycles worn-out erythrocytes.

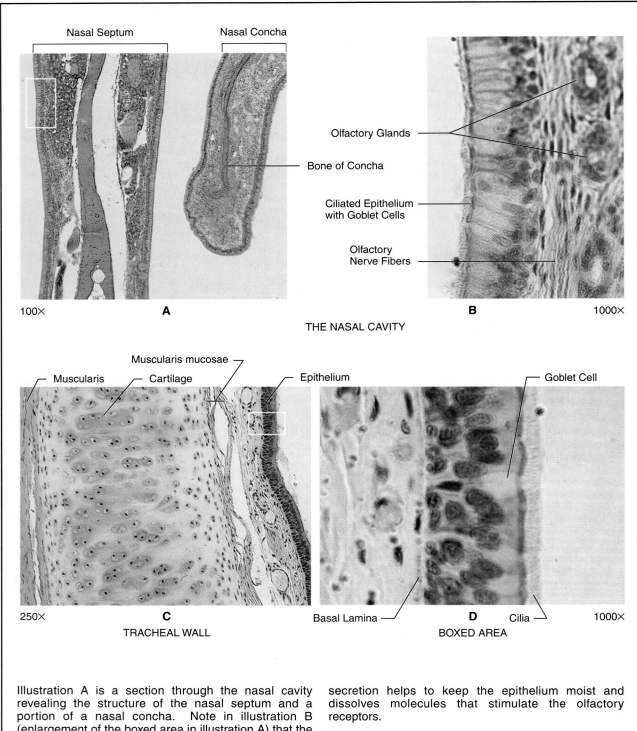

Nasal Septum

Nasal Concha

Olfactory Glands

Bone of Concha

Ciliated Epithelium
with Goblet Cells

Olfactory
Nerve Fibers

100×   A

B   1000×

THE NASAL CAVITY

Muscularis mucosae

Muscularis   Cartilage   Epithelium

Goblet Cell

250×   C   Basal Lamina   D   Cilia   1000×

TRACHEAL WALL   BOXED AREA

Illustration A is a section through the nasal cavity revealing the structure of the nasal septum and a portion of a nasal concha. Note in illustration B (enlargement of the boxed area in illustration A) that the nasal epithelium consists of ciliated columnar cells. Interspersed between these columnar cells are olfactory receptors that are not readily visible here to the absence of silver dye staining. The olfactory (Bowman's) glands in the lamina propria produce a secretion made up of mucus and serous fluid. This secretion helps to keep the epithelium moist and dissolves molecules that stimulate the olfactory receptors.

Illustration C reveals the structure of the tracheal wall. Tracheal rigidity is provided by hyaline cartilaginous rings. The epithelium consists of pseudostratified ciliated epithelium as shown in illustration D which is a view of the boxed section in illustration C. Note the presence of goblet cells that produce mucus.

Figure HA-30   Histology of nasal passages and the trachea.

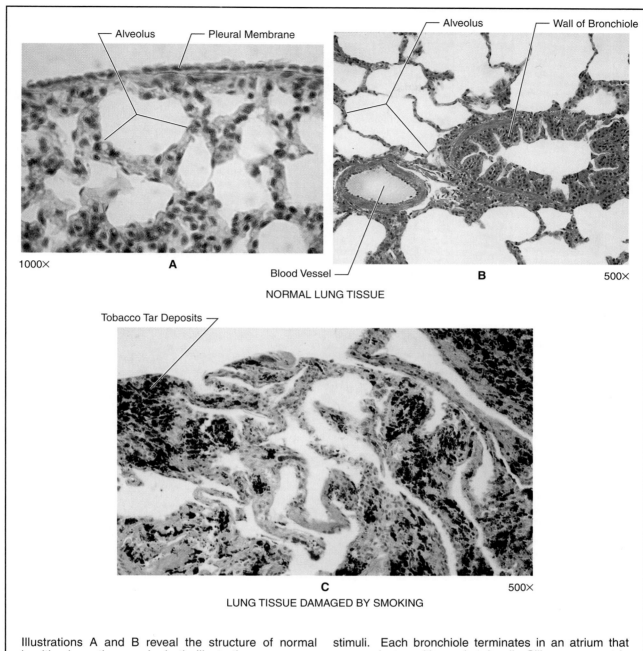

NORMAL LUNG TISSUE

LUNG TISSUE DAMAGED BY SMOKING

Illustrations A and B reveal the structure of normal healthy lung tissue. In both illustrations are seen numerous sacs called alveoli, which provide extensive surface area in each lung for the exchange of oxygen, carbon dioxide and water between the air and blood. Note that the surface of the lung in illustration A is covered with a thin serous membrane, the pulmonary pleura.

Note, also, the considerable thickness of the wall of the bronchiole in illustration B. The wall consists of infoldings of low columnar epithelial tissue surrounded by a very thin layer of smooth muscle tissue. The presence of muscle tissue in the walls of bronchioles allows them to expand and contract due to various stimuli. Each bronchiole terminates in an atrium that communicates with nearby alveoli. Cilia are present in the proximal portions of bronchioles and absent distally.

Illustration C reveals what happens to lung tissue that has been subjected to years of cigarette smoking. Note, first of all, the loss of alveoli. Smoking causes the thin walls of the alveoli to break down, reducing the total absorptive area of the lung. The end result of this deterioration is shortness of breath (emphysema). The second assault on lung tissue is the deposition of particulate material that cannot be disposed of in the tissue. These tar deposits are carcinogenic and cause lung cancer.

Figure HA-31   Lung tissue (healthy and smoke damaged).

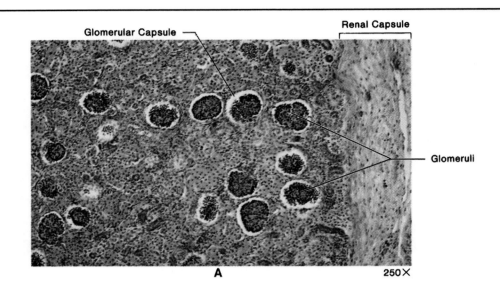

Glomerular Capsule

Renal Capsule

Glomeruli

A 250×

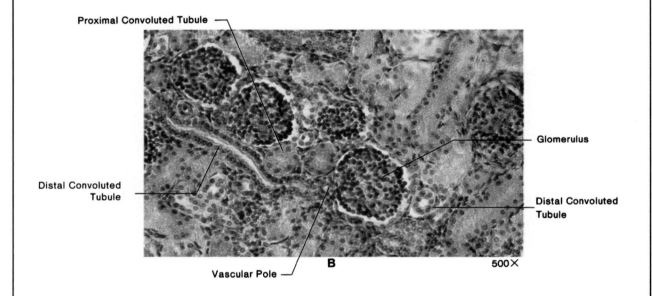

Proximal Convoluted Tubule

Distal Convoluted Tubule

Glomerulus

Distal Convoluted Tubule

Vascular Pole

B 500×

Illustration A reveals the structure of the outermost portion of the cortex of the kidney. The entire organ is enclosed by a thick fibrous capsule.

Blood enters each glomerulus through its vascular pole (shown in illustration B). High vascular pressure in the glomeruli results in the production of a glomerular filtrate consisting of water, glucose, amino acids, and other substances. This glomerular filtrate is collected by the glomerular capsule and passes down the length of the nephron collecting tubule, being altered in composition as it approaches the calyx of the kidney. Note that the proximal convoluted tubules can be differentiated from the distal convoluted tubules by the size of the cuboidal cells that make up their walls: large in the proximal tubule and small in the distal tubule. Eighty percent of water absorption occurs in the proximal convoluted tubules.

Figure HA-32   Histology of the cortex of the kidney.

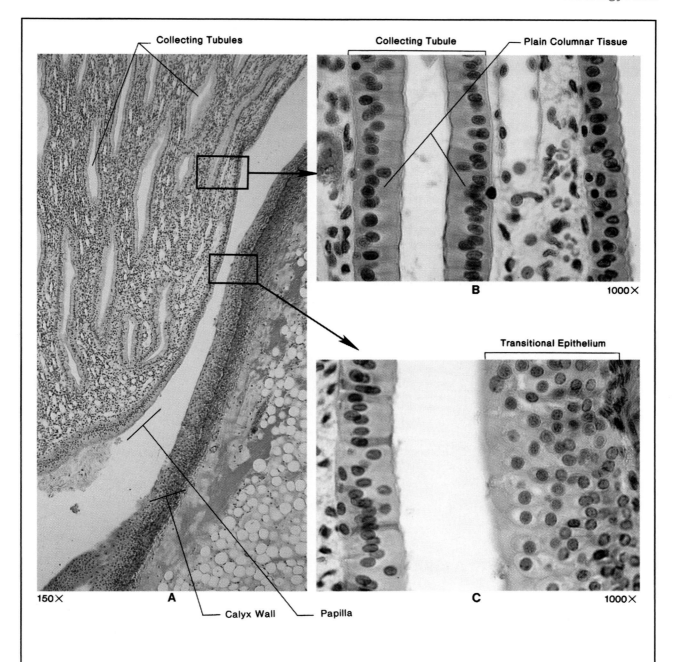

Illustration A portrays a portion of the tip of a pyramid and the wall of the calyx. Illustrations B and C reveal the histological nature of these two structures. The collecting tubules, which collect urine from several nephrons, are lined with distinct low columnar epithelium. Examination of these cells with oil immersion optics will reveal the existence of a distinct brush border. Both absorption and secretion take place through these columnar cells.

Note in illustration C that the wall of the calyx is lined with transitional epithelium. This same type of tissue is seen lining the ureters, urethra, and the urinary bladder (See figures HA28 and HA37.)

Figure HA-33   **Histology of the papilla and calyx of the kidney.**

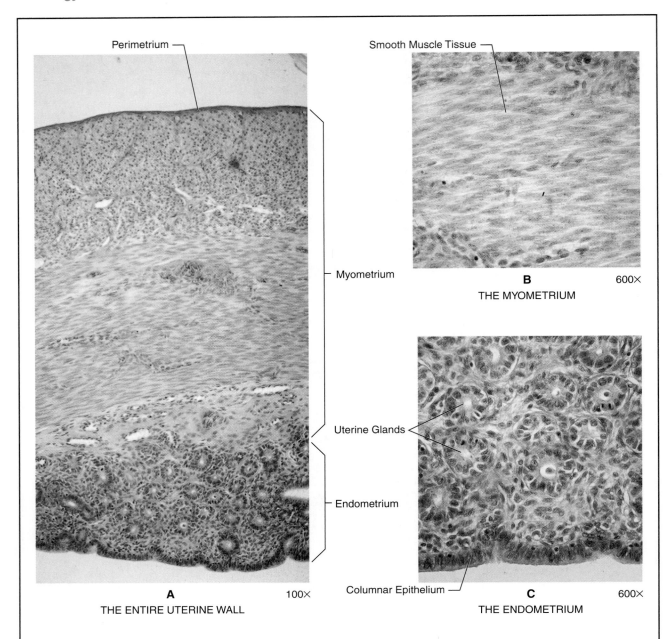

Perimetrium

Smooth Muscle Tissue

Myometrium

B   600×
THE MYOMETRIUM

Uterine Glands

Endometrium

Columnar Epithelium

A   100×
THE ENTIRE UTERINE WALL

C   600×
THE ENDOMETRIUM

The wall of the uterus consists of three layers: (1) the endometrium, which corresponds to the mucosa and submucosa; (2) the myometrium, or muscularis; and (3) the perimetrium, a typical serous membrane.

The myometrium, which forms three-fourths of the uterine wall, consists of three layers of smooth muscle fibers: an inner longitudinal layer, a middle circular layer, and an outer oblique layer. Some glands extend into it from the endometrium.

The endometrial surface cells (illustration C) are columnar and partially ciliated (no cilia are shown in illustration C). During the proliferative stage (immediately after menstruation) the endometrium undergoes regeneration of the columnar cells and increased growth of the mucosal glands. Vascularity of the tissue also becomes more pronounced. The proliferative stage terminates at the 13th or 14th day of the menstrual cycle.

Figure HA-34   **The uterus.**

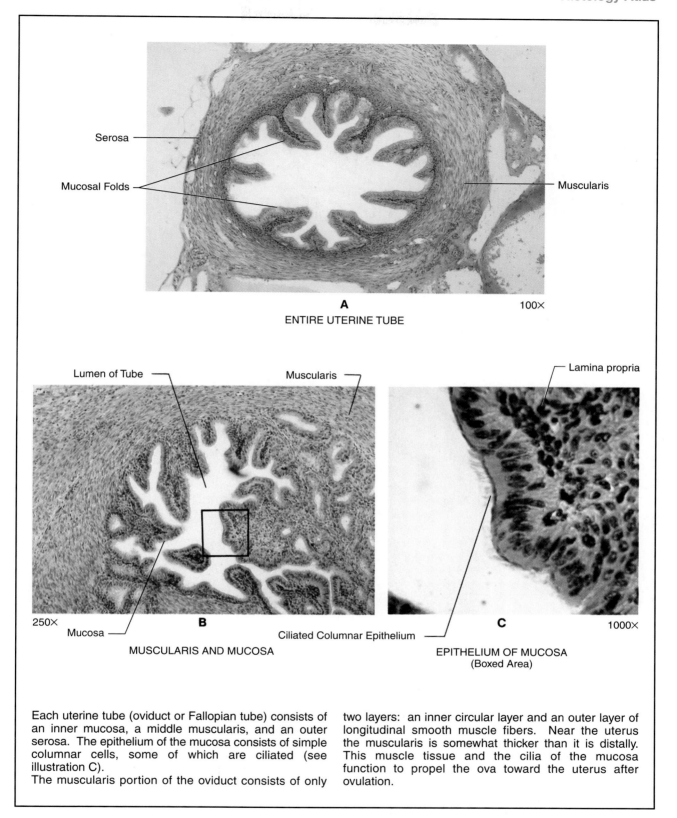

Serosa

Mucosal Folds

Muscularis

A
100×
ENTIRE UTERINE TUBE

Lumen of Tube

Muscularis

Lamina propria

250×

Mucosa

B

Ciliated Columnar Epithelium

MUSCULARIS AND MUCOSA

C
1000×

EPITHELIUM OF MUCOSA
(Boxed Area)

Each uterine tube (oviduct or Fallopian tube) consists of an inner mucosa, a middle muscularis, and an outer serosa. The epithelium of the mucosa consists of simple columnar cells, some of which are ciliated (see illustration C).

The muscularis portion of the oviduct consists of only two layers: an inner circular layer and an outer layer of longitudinal smooth muscle fibers. Near the uterus the muscularis is somewhat thicker than it is distally. This muscle tissue and the cilia of the mucosa function to propel the ova toward the uterus after ovulation.

Figure HA-35   The uterine tube.

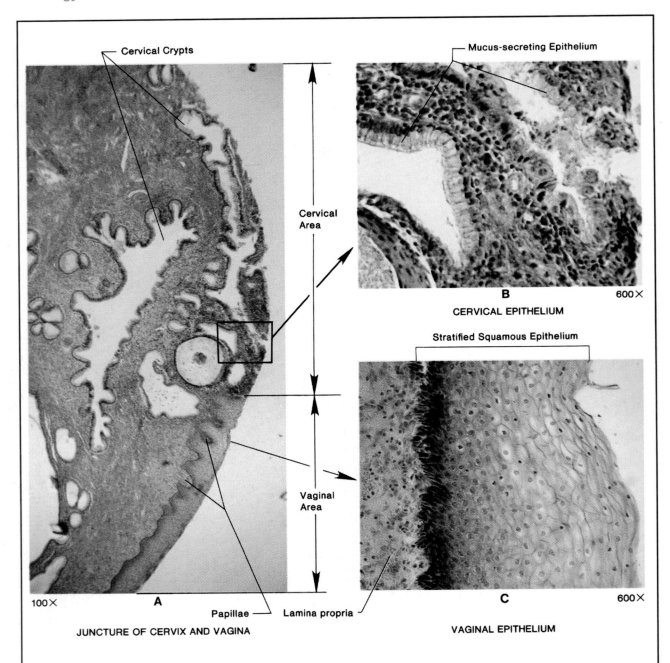

JUNCTURE OF CERVIX AND VAGINA

CERVICAL EPITHELIUM

VAGINAL EPITHELIUM

Illustration A shows the contrasting histological differences that exist between the epithelia of the vagina and cervix. Note in illustration C that the vaginal epithelium consists of nonkeratinized stratified squamous epithelium. Examination of the lamina propria of the vaginal mucosa with high-dry or oil immersion optics will reveal the presence of large numbers of lymphocytes. Note in illustration A that the epithelium has a large number of papillae.

The most distinguishing characteristic of the cervical portion of the uterus is the presence of approximately one hundred mucus-secreting cervical crypts. As indicated in illustration B, the epithelium of these crypts consists, primarily, of plain columnar cells; some ciliated cells are also present, however. Approximately 20 to 60 mg of mucus is produced daily. During ovulation mucus production increases to over 700 mg daily. Mucus plays an important role in fertility since it is the first secretion met by the sperm entering the female tract. Occasionally the exit of a crypt will become occluded, causing a cyst to form. The large round structure near the box in illustration A is such a structure.

Figure HA-36   The cervix and vagina.

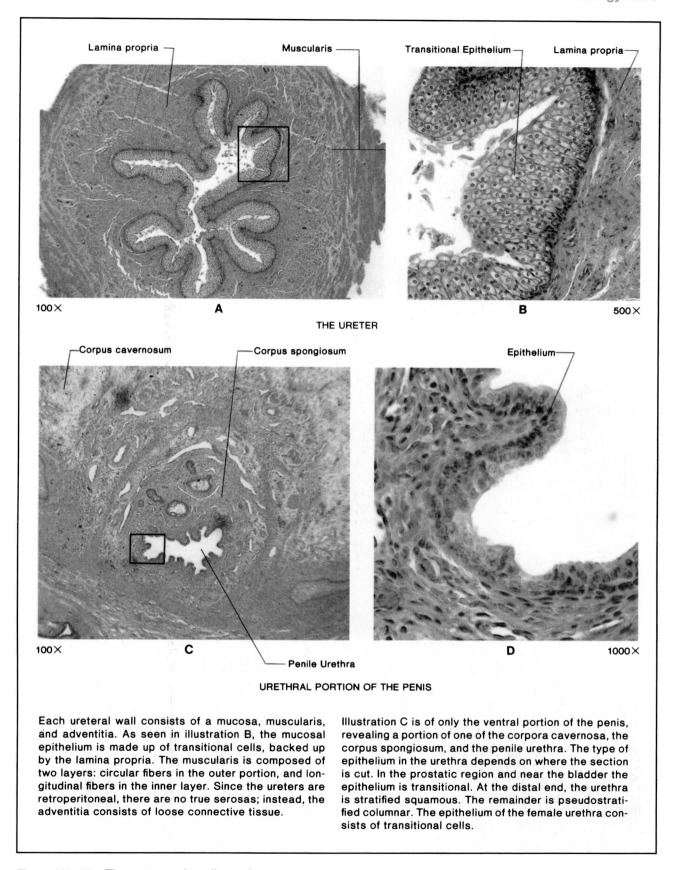

Lamina propria — Muscularis — Transitional Epithelium — Lamina propria —

100×   A   B   500×

THE URETER

Corpus cavernosum — Corpus spongiosum — Epithelium —

100×   C   D   1000×

— Penile Urethra

URETHRAL PORTION OF THE PENIS

Each ureteral wall consists of a mucosa, muscularis, and adventitia. As seen in illustration B, the mucosal epithelium is made up of transitional cells, backed up by the lamina propria. The muscularis is composed of two layers: circular fibers in the outer portion, and longitudinal fibers in the inner layer. Since the ureters are retroperitoneal, there are no true serosas; instead, the adventitia consists of loose connective tissue.

Illustration C is of only the ventral portion of the penis, revealing a portion of one of the corpora cavernosa, the corpus spongiosum, and the penile urethra. The type of epithelium in the urethra depends on where the section is cut. In the prostatic region and near the bladder the epithelium is transitional. At the distal end, the urethra is stratified squamous. The remainder is pseudostratified columnar. The epithelium of the female urethra consists of transitional cells.

Figure HA-37   The ureter and penile urethra.

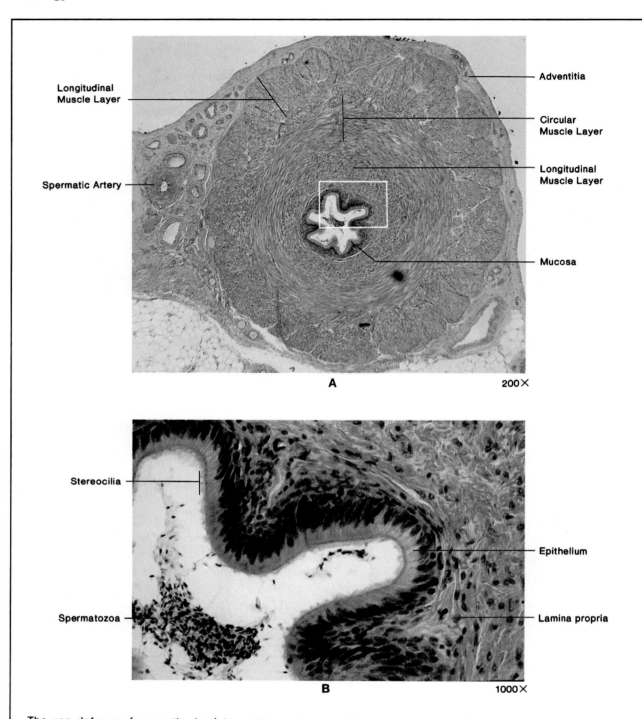

Longitudinal Muscle Layer

Spermatic Artery

Adventitia

Circular Muscle Layer

Longitudinal Muscle Layer

Mucosa

A   200×

Stereocilia

Spermatozoa

Epithelium

Lamina propria

B   1000×

The vas deferens (spermatic duct) is a thick-walled muscular tube that transports spermatozoa from the epididymis to the ejaculatory duct, where fluid from the seminal vesicles joins the spermatozoa. The wall has three distinct areas: an outer adventitia, a middle muscularis (three-layered), and an inner mucosa. Note that the muscularis is very thick, with two layers of longitudinal fibers and a middle layer of circular fibers. Peri-staltic waves produced by these muscles help to propel the spermatozoa during ejaculation.

The mucosa, consisting of the lamina propria and epithelium, is unique in that the columnar cells that make up the epithelium have stereocilia. These structures are nonmotile enlarged microvilli that cover the exposed surfaces of the cells.

Figure HA-38   The vas deferens.

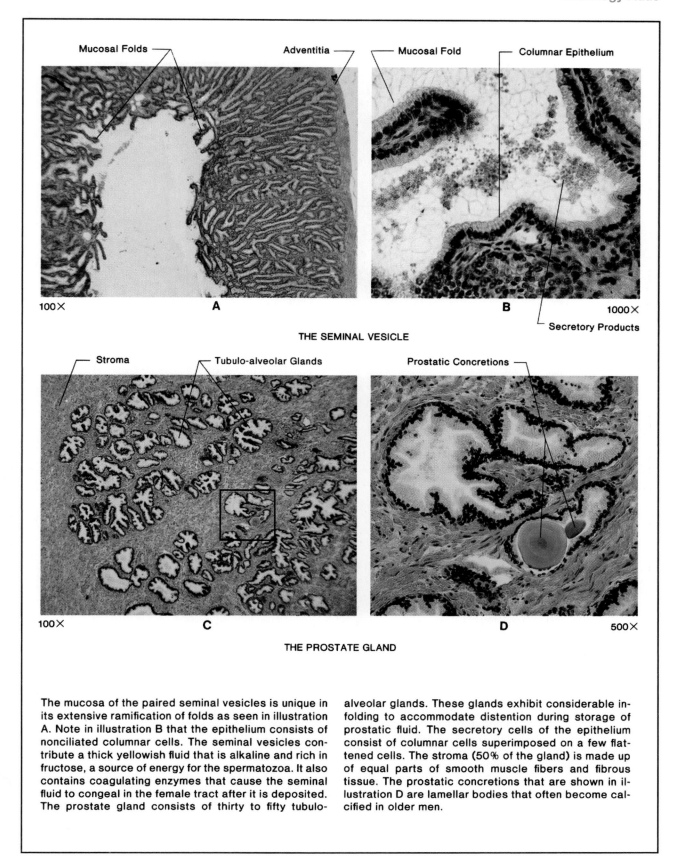

Mucosal Folds **Adventitia** **Mucosal Fold** **Columnar Epithelium**

100× **A**

1000× **B**

Secretory Products

THE SEMINAL VESICLE

Stroma **Tubulo-alveolar Glands** **Prostatic Concretions**

100× **C**

D 500×

THE PROSTATE GLAND

The mucosa of the paired seminal vesicles is unique in its extensive ramification of folds as seen in illustration A. Note in illustration B that the epithelium consists of nonciliated columnar cells. The seminal vesicles contribute a thick yellowish fluid that is alkaline and rich in fructose, a source of energy for the spermatozoa. It also contains coagulating enzymes that cause the seminal fluid to congeal in the female tract after it is deposited. The prostate gland consists of thirty to fifty tubulo-alveolar glands. These glands exhibit considerable infolding to accommodate distention during storage of prostatic fluid. The secretory cells of the epithelium consist of columnar cells superimposed on a few flattened cells. The stroma (50% of the gland) is made up of equal parts of smooth muscle fibers and fibrous tissue. The prostatic concretions that are shown in illustration D are lamellar bodies that often become calcified in older men.

Figure HA-39 Seminal vesicle and prostate gland.

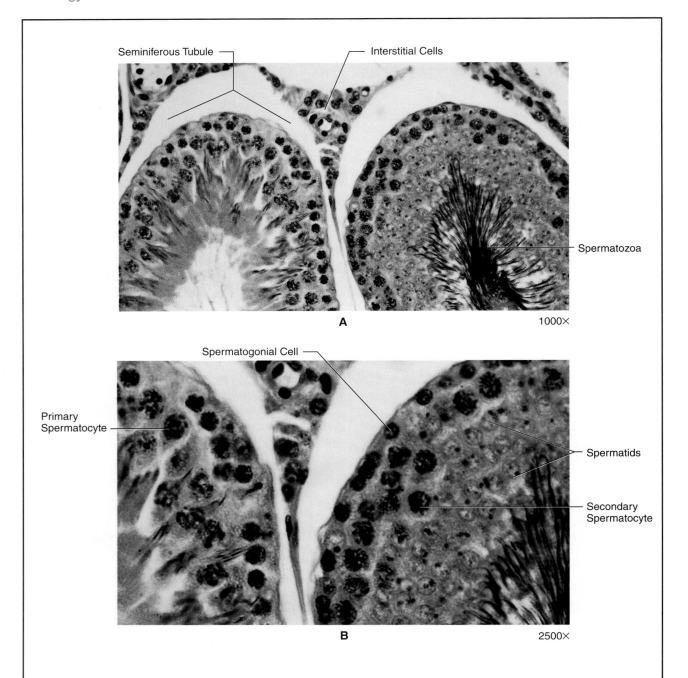

Seminiferous Tubule

Interstitial Cells

Spermatozoa

A

1000×

Spermatogonial Cell

Primary
Spermatocyte

Spermatids

Secondary
Spermatocyte

B

2500×

The production of spermatozoa, known as spermato-genesis, occurs in the seminiferous tubules of the testes. Illustration A reveals portions of two of these structures. External to the tubules is seen a tri-angular structure that is a group of interstitial cells. These cells are the endocrine cells that secrete testosterone into the circulatory system.

All spermatozoa originate from the primordial germ cells called spermatogonia. Note in illustration B that they are located at the periphery of the seminiferous tubule. These cells have a diploid complement of 46 chromosomes. The first step in spermatogenesis takes place in these cells with the union (synapsis) of the homologous chromosomes to form tetrads. Once synapsis has occurred the new cell is called a primary spermatocyte. Note the proximity of the primary spermatocytes to the spermatogonial cells in illustration B.

The next step is for each primary spermatocyte to divide, meiotically, to produce two secondary sper-matocytes that have a haploid number (23) of chromosomes. Note in illustration B that the secon-dary spermatocytes are closer to the lumen of the sem-iniferous tubule. Finally, the two secondary spermato-cytes divide meiotically, again, to produce a total of four spermatids, which contain 23 chromo-somes each. The spermatids continue to develop into mature spermatozoa.

Figure HA-40   Spermatogenesis.

Exercises 40 through 45 pertain to a series of blood tests that are performed routinely for basic blood testing. When collected prior to surgery, this information can reveal certain types of problems that might occur in the operating room. In diagnostic studies, the presence of infection, and even the nature of an infection, may be revealed by some of those tests.

Of the many kinds of blood tests that are performed in laboratories today, we have selected only seven of the simpler, more common ones. If all of these tests are performed, considerable information can be made available for diagnostic purposes.

A convenient way to utilize this group of tests is for the student to perform each blood test on his or her own blood, recording all the information on tables 1 and 2 of Laboratory Report 40–45. Normal values for each test are given in table 2. To perform these tests, however, students will have to work in pairs. A significant feature of these tests is that no venipuncture is required; instead, only a drop of blood from a punctured finger is necessary for each test.

Precautions

There are two hazards that must be avoided in doing blood tests: (1) self-infection from environmental contaminants, and (2) transmission of infections (i.e., hepatitis, HIV, etc.) from person to person. All the tests in this manual can be performed safely if the following safeguards are rigidly observed:

- Sponge down your tabletop at the beginning of the period with an appropriate disinfectant.
- Wash your hands with soap and water before and after doing blood tests.
- If you are perforating another person's finger for a blood sample, avoid contact with the blood. *Wearing surgical rubber gloves is the best method to avoid contact.*
- Use disposable lancets only one time. Deposit used lancets into a receptacle that contains disinfectant. **Never** toss used lancets into the wastebasket!
- Before perforating the finger for a blood sample, disinfect the skin with alcohol.
- At the end of the laboratory period wash all equipment with soap and water. All pipettes must be cleaned with solutions drawn from the pipette cleaning kits.
- Scrub down the tabletop at the end of the period and wash your hands again with soap and water. A final rinse of the hands with a disinfectant will provide additional protection.

40

A White Blood Cell Study:
The Differential WBC Count

The blood is an opaque, rather viscous fluid that is bright red when oxygenated and dark red when depleted of oxygen. Its specific gravity is normally around 1.06, and its pH is slightly alkaline. When centrifuged, blood becomes separated into a dark red portion made up of **formed elements** and a clear straw-colored portion called **plasma.** The formed element content of blood is around 47% in men and 42% in women.

In this exercise we will study the leukocytic portion of the formed elements in blood. This study may be made from a prepared stained microscope slide or from a slide made from your own blood. Figures 40.1 and 40.2 will be used to identify the various kinds of leukocytes. If you are going to prepare a slide from your own blood, *be sure to review the sanitary precautions that are provided on page 253.*

The formed elements of blood consist of erythrocytes (red blood cells), leukocytes (white blood cells), and blood platelets. As illustrated in figure 40.1, the *erythrocytes* are nonnucleated cells with depressed centers that contain hemoglobin. They are the most numerous elements in blood—around 4.5 to 5.5 million cells per cubic millimeter.

The *leukocytes* are of two types: granulocytes and agranulocytes. The *granulocytes* are so named because they have conspicuous granules in their cytoplasm. **Neutrophils, eosinophils,** and **basophils** fall into this category. The *agranulocytes* lack pronounced granules in the cytoplasm; the **monocytes** and **lymphocytes** are of this type.

The *blood platelets* are noncellular elements in the blood that assist in the blood clotting process.

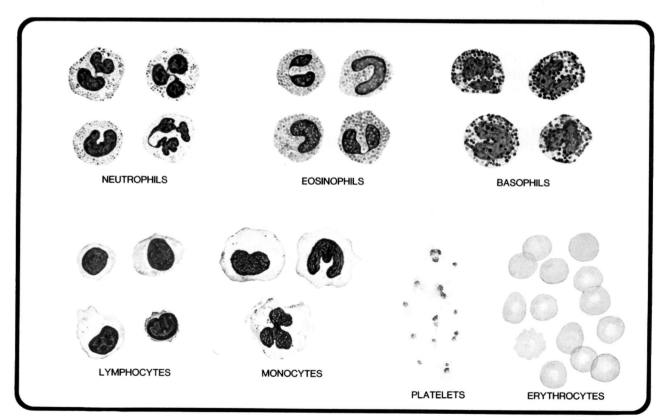

NEUTROPHILS   EOSINOPHILS   BASOPHILS

LYMPHOCYTES   MONOCYTES   PLATELETS   ERYTHROCYTES

K.P. Talaro

Figure 40.1   Formed elements of blood.

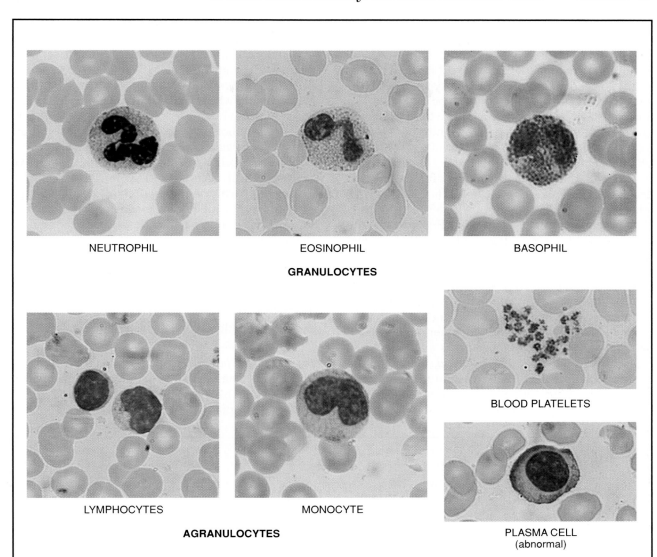

NEUTROPHIL   EOSINOPHIL   BASOPHIL

GRANULOCYTES

LYMPHOCYTES   MONOCYTE

AGRANULOCYTES

BLOOD PLATELETS

PLASMA CELL
(abnormal)

Neutrophils differ from the other two granulocytes in having smaller and paler granules in the cytoplasm. Nuclei of these leukocytes are characteristically lobulated with thin bridges between the lobules (see figure 40.1). During infections, when many of these are being produced in the bone marrow, the nuclei are horseshoe-shaped; these juvenile cells are often called "stab cells."

Eosinophils, generally, have bilobed nuclei as above; however, juveniles have nuclei that are horseshoe-shaped (figure 40.1). Note that the eosinophilic granules in the cytoplasm are large and pronounced.

Basophils, which are few in number, are distinguished from the other granulocytes by the presence of dark basophilic granules in the cytoplasm. The nuclei of these cells are large and varied in shape. The fact that heparin has been identified in these cells indicates that they probably secrete this substance into the blood.

Two sizes of lymphocytes—large and small—are shown above and in figure 40.1. The small ones, which are more abundant than the large ones, have large, dense nuclei surrounded by a thin layer of basophilic cytoplasm. The large lymphocytes have indented nuclei and more cytoplasm than small lymphocytes.

Monocytes are the largest cells found in normal blood. The nucleus may be ovoid, indented, or horseshoe-shaped, as illustrated in figure 40.1. Monocytes have considerably more cytoplasm than lymphocytes and are voracious phagocytes.

Blood platelets, which form from megakaryocytes in the bone marrow, function in the formation of blood clots.

Plasma cells are only rarely seen in the blood. The photomicrograph above was made from the blood of a patient with plasma cell leukemia. Although they resemble lymphocytes, the plasma cells have more cytoplasm that is very basophilic.

Figure 40.2   Photomicrographs of formed elements in the blood (5000×).

In this study of the formed elements we will attempt to determine the percentage of each type of leukocyte on a slide. By scanning an entire slide and counting the various types, you will have an opportunity to encounter most, if not all, types. The erythrocytes and blood platelets will be ignored. The main purpose of this exercise will be to learn how to distinguish the various types of white blood cells, not to become proficient in performing differential WBC counts.

The normal percentage ranges for the various leukocytes are as follows: neutrophils 50–70%, lymphocytes 20–30%, monocytes 2–6%, eosinophils 1–5%, and basophils 0.5–1%. Because of the extremely low percentages of eosinophils and basophils, it is necessary to examine at least 100 leukocytes to increase the possibility of encountering one or two of them on a slide.

Deviations from the above percentages may indicate serious pathological conditions. High neutrophil counts, or *neutrophilia,* often signal localized infections such as appendicitis or abscesses in some part of the body. *Neutropenia,* a condition in which there is a marked decrease in the number of neutrophils, occurs in typhoid fever, undulant fever, and influenza. *Eosinophilia* (high eosinophil count)

may indicate allergic conditions or invasions by parasitic roundworms such as *Trichinella spiralis,* the pork worm. Counts of eosinophils may rise to as much as 50% of total leukocytes in cases of trichinosis. High lymphocyte counts, or *lymphocytosis,* are present in whooping cough and some viral infections. Increased numbers of monocytes, or *monocytosis,* may indicate the presence of the Epstein-Barr virus, which causes infectious mononucleosis.

A white cell count that determines the relative percentages of each type of leukocyte is called a **differential white blood cell count.** It is this type of count that will be performed here in this laboratory period. If a prepared slide is to be used, ignore the instructions below under the heading of "Preparation of a Slide" and proceed to the heading "Performing the Cell Count".

CAUTION: See page 253.

Materials:
   prepared blood slide stained with Wright's or
      Giemsa's stains (H2.851 or H2.852)

for staining a blood smear:
   2 or 3 clean microscope slides (with polished
      edges)
   sterile disposable lancets

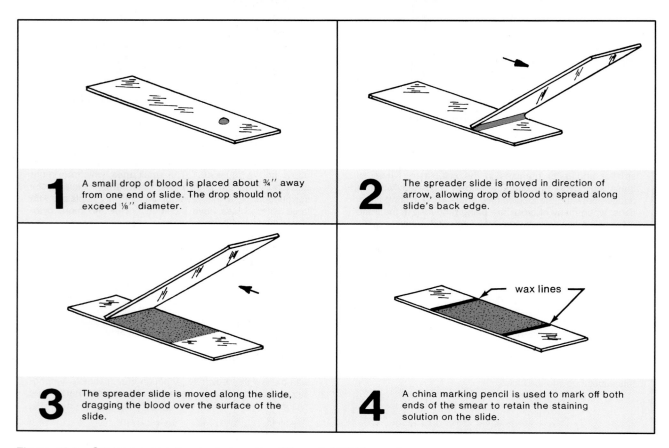

1 A small drop of blood is placed about ¾″ away from one end of slide. The drop should not exceed ⅛″ diameter.

2 The spreader slide is moved in direction of arrow, allowing drop of blood to spread along slide's back edge.

3 The spreader slide is moved along the slide, dragging the blood over the surface of the slide.

4 A china marking pencil is used to mark off both ends of the smear to retain the staining solution on the slide.

wax lines

Figure 40.3   Smear preparation technique for differential WBC count.

sterile absorbent cotton and 70% alcohol (or disposable alcohol swabs)
Wright's stain
distilled water in dropping bottle
wax pencil
bibulous paper
immersion oil

Preparation of a Slide

Figure 40.3 illustrates the procedure that will be used to make a stained slide of a blood smear. The most difficult step in making such a slide is getting an ideal spread of the blood, which is thick at one end and thin at the other end. If done properly, the smear will have a gradient of cellular density that will make it possible to choose an area that is ideal for study. The angle at which the spreading slide is held will determine the thickness of the smear. It may be necessary for you to make more than one smear to get one that is suitable.

1. Clean three or four slides with soap and water. Handle them with care to avoid getting their flat surfaces soiled. Although only two slides may be needed, it is often necessary to repeat the spreading process; thus, the extra slides.
2. Scrub the middle finger with 70% alcohol and perforate it with a lancet.
3. Put a drop of blood on the slide about 3/4″ from one end and spread with another slide in the manner illustrated in figure 40.3. *Note that the blood is dragged over the slide, not pushed.*

   Do not pull the slide over the smear a second time. *If you don't get an even smear the first time, repeat the process on a fresh clean slide.* To get a smear that will be the proper thickness, hold the spreading slide at an angle somewhat greater than 45°.
4. Draw a line on each side of the smear with a wax pencil to confine the stain, which is to be applied to the slide. (Note: This step is helpful for beginners and usually omitted by professionals.)
5. Cover the film with Wright's stain, counting the drops as you add them. Stain for **4 minutes** and then add the same number of drops of distilled water to the stain and let stand for another

10 minutes. Blow gently on the mixture every few minutes to keep the solutions mixed.
6. Gently wash off the slide under running water for **30 seconds** and shake off the excess. Blot dry with bibulous paper.

Performing the Cell Count

Whether you are using a prepared slide or one that you have just stained, the counting procedure is essentially the same. Ideally, one should use an oil immersion lens for this procedure; high-dry optics can be used, but they are not as reliable for best results. Proceed as follows:

1. Scan the slide with the low-power objective to find an area where cell distribution is best. A good area is one in which the cells are not jammed together or scattered too far apart.
2. Once an ideal area has been selected, place a drop of immersion oil on the slide near one edge and lower the oil immersion objective into the oil. If you are using a slide you have just prepared you needn't worry about oil being placed directly on the stained smear.

   If the high-dry objective is to be used, omit placing oil on the slide.
3. Systematically scan the slide, following the pathway indicated in figure 40.4. As each leukocyte is encountered identify it, using figures 40.1 and 40.2 for reference.

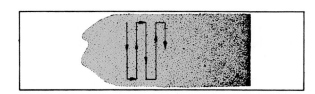

Figure 40.4   Examination path for a differential count.

4. Tabulate your count on the sheet for Laboratory Report 40–45 according to the instructions there. It is best to remove your Lab Report sheet from the back of the manual for this identification and tabulation procedure.

Histology Self-Quiz No. 2

If at this point you have completed all the histology assignments in Exercises 19, 36, and 38, you should be ready to test your knowledge of the histology of these areas by taking self-quiz no. 2 on page 412. After recording your answers on a sheet of paper, check your answers against the key that you will find on page 196.

41

Total Blood Cell Counts

The number of leukocytes and erythrocytes present in a cubic millimeter of blood is routinely determined in diagnostic blood analyses. While the differential WBC count performed in the last exercise is very useful, it alone cannot give us a complete picture of the health of the patient. One must also know the total number of white and red blood cells per unit of volume (a cubic millimeter).

In this exercise we will have an opportunity to determine accurate blood cell counts using a special cell-counting slide called a **hemacytometer.** Although using a device such as this for counting blood cells is very satisfactory, the procedure is tedious and has been supplanted in the modern medical laboratory by automated electronic counting devices that are much more rapid. Since no automated equipment of this type is available here, we will use the hemacytometer.

Total WBC Count

The number of white blood cells in a cubic millimeter of normal blood will range from 5000 to 9000 cells. The exact number of cells will vary with the time of day, exercise, and other factors. If a patient has a high WBC count, such as 17,000 leukocytes per cubic millimeter, and an abnormally high neutrophil count, as determined by a differential count, one can safely assume that an infection is present somewhere in the body.

To determine the number of leukocytes in a cubic millimeter of blood, one must dilute the blood and count the WBCs on a hemacytometer, using the low-power objective of a microscope. Figures 41.1 and 41.2 illustrate the various steps in diluting the blood and charging a hemacytometer for this count.

Note in figure 41.1 that blood is drawn up into a special pipette and then diluted in the pipette with a weak acid solution. After shaking the pipette to mix the acid and blood, a small amount of the diluted blood is allowed to flow under the cover glass of the hemacytometer as shown in illustration 1, figure 41.2.

When examined under the low-power objective, the diluted blood will reveal only white blood cells since all red blood cells have been destroyed by the acid in the dilution fluid. Illustration 2, figure 41.2, reveals the five areas on the hemacytometer that are tabulated for a WBC total count. Mathematical calculations are used to determine the number of leukocytes per cubic millimeter. Proceed as follows:

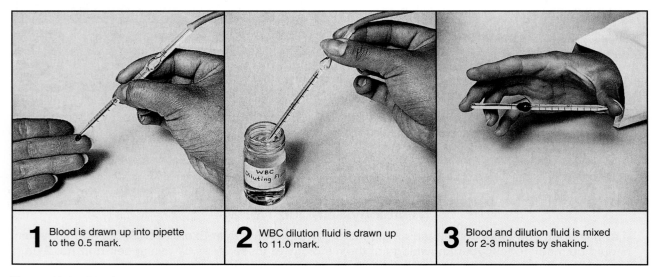

1 Blood is drawn up into pipette to the 0.5 mark.

2 WBC dilution fluid is drawn up to 11.0 mark.

3 Blood and dilution fluid is mixed for 2-3 minutes by shaking.

Figure 41.1 Dilution and mixing procedures for WBC blood count.

Preparation of Hemacytometer

Working with your laboratory partner, assist each other to prepare a "charged" hemacytometer as follows:

Materials:

> hemacytometer and cover glass
> WBC diluting pipette and rubber tubing
> WBC diluting fluid
> mechanical hand counter
> pipette cleaning solutions
> cotton, alcohol, lancets, clean cloth

1. Wash the hemacytometer and cover glass with soap and water, rinse well, and dry with a clean cloth or Kimwipes.
2. Produce a free flow of blood, wipe away the first drop, and draw the blood up into the diluting pipette to the 0.5 mark. See illustration 1, figure 41.1. Any slight excess over the 0.5 mark can be withdrawn by placing the pipette tip on blotting paper.

    If the blood goes substantially past the 0.5 mark, discharge the blood from the pipette and flush it out with acid, water, alcohol, and acetone (in that order). See illustration 3, figure 41.2.

    The ideal way is to draw up the blood *exactly* to the mark on the first attempt.
3. As shown in illustration 2, figure 41.1, draw the WBC diluting fluid up into the pipette until it reaches the 11.0 mark.
4. Place your thumb over the tip of the pipette, slip off the tubing, and place your third finger over the other end (illustration 3, figure 41.1).

5. Mix the blood and diluting fluid in the pipette for **2–3 minutes** by holding it as shown in illustration 3, figure 41.1. The pipette should be held parallel to the tabletop and moved through a 90° arc, with the wrist held rigidly.
6. Discharge one-third of the bulb fluid from the pipette by allowing it to drop onto a piece of paper toweling.
7. While holding the pipette as shown in illustration 1, figure 41.2, deposit a **tiny drop** on the sloping polished surface of the counting chamber next to the edge of the cover glass. **Do not let the tip of the pipette touch the polished surface for more than an instant.** If it is there too long, the chamber will overfill.

    A properly filled chamber will have diluted blood filling only the space between the cover glass and counting chamber. No fluid should run down into the moat.
8. Charge the other side if the first side was overfilled.

Performing the Count

Place the hemacytometer on the microscope stage and bring the grid lines into focus under the **low-power** (10×) objective. Use the coarse adjustment knob and reduce the lighting somewhat to make both the cells and lines visible.

Locate one of the "W" (white) sections shown in illustration 2, figure 41.2. Since the diluting fluid contains acid, all erythrocytes have been destroyed (hemolyzed); only the leukocytes will show up as very small dots.

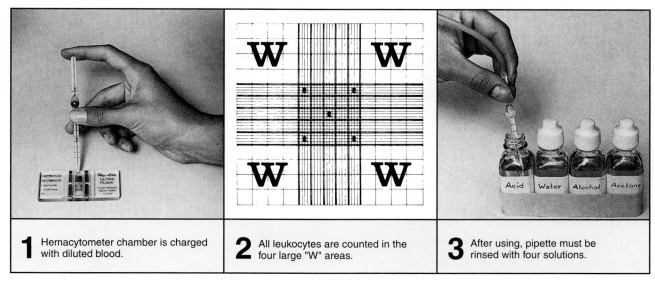

1 Hemacytometer chamber is charged with diluted blood.

2 All leukocytes are counted in the four large "W" areas.

3 After using, pipette must be rinsed with four solutions.

Figure 41.2　Charging the hemacytometer, counting areas, and pipette cleaning.

Do the cells seem to be evenly distributed? If not, charge the other half of the counting chamber after further mixing. If the other chamber had been previously charged unsuccessfully by overflooding, wash off the hemacytometer and cover glass, shake the pipette for 2–3 minutes, and recharge it.

Count all the cells in the four "W" areas, using a mechanical hand counter. To avoid overcounting of cells at the boundaries, **count the cells that touch the lines on the left and top sides only.** Cells that touch the boundary lines on the right and bottom sides should be ignored. This applies to the boundaries of each entire "W" area.

Discharge the contents of the pipette and rinse it out sequentially with acid, water, alcohol, and acetone (illustration 3, figure 41.2).

Calculations

To determine the number of leukocytes per cubic millimeter, multiply the total number of cells counted in the four "W" areas by 50. The factor of 50 is the product of the volume correction factor and dilution factor, or

$$2.5 \times 20 = 50$$

The volume correction factor of 2.5 is arrived at in this way: Each "W" area is exactly one square millimeter by 0.1 mm deep. Therefore, the volume of each "W" is 0.1 mm^3. Since four "W" sections are counted, the total amount of diluted blood that is examined is 0.4 mm^3. And since we are concerned with the number of cells in one cubic millimeter instead of 0.4 mm^3, we must multiply our count by 2.5, derived by dividing 1.0 by 0.4.

Record your results on the Laboratory Report.

Red Blood Cell Count

Normal RBC counts for adult males average around 5,400,000 ($\pm$600,000) cells per cubic millimeter. The normal average for women is 4,600,000 ($\pm$500,000) per cubic millimeter. The difference in sexes does not exist before puberty. At high altitudes, higher values will be normal for all individuals.

Low and high RBC counts result respectively in anemia or polycythemia. Although low RBC counts

do result in anemia, individuals can be anemic and still have normal RBC counts. If the cells are small, or if they lack sufficient hemoglobin, anemia will result. Thus, **anemia** may be defined, simply, as a condition in which the oxygen-carrying capacity of the blood is reduced.

Polycythemia, a condition characterized by above-normal RBC counts, may be due to living at high altitudes *(physiological polycythemia)* or red marrow malignancy *(polycythemia vera).* In physiological polycythemia, counts may run as high as 8,000,000/mm^3. In polycythemia vera, counts of 10–11 million are not uncommon.

Although the general procedures for the RBC count are essentially the same as for counting white blood cells, the diluting fluid, pipette, and mathematics are different. The RBC pipette differs from the WBC pipette in that the RBC pipette has 101 scribed above the bulb instead of 11. The RBC diluting fluid may be one of several isotonic solutions such as physiological saline or Hayem's solution. To perform a red blood count, follow this procedure:

Materials:
  RBC diluting pipette
  rubber tubing and mouthpiece
  RBC diluting fluid
  other supplies used for WBC count

1. Draw the blood up to the 0.5 mark and the diluting fluid up to the 101 mark.
2. Mix the blood and diluting fluid in the same manner as performed for WBC counts.
3. Count all the cells in the five "R" areas (see illustration 2, figure 41.2). Use the **high-power objective** (40 or 50$\times$) and observe the same rules for "line counts" as applied to WBC counts.
4. Multiply the total count of the five areas by 10,000 (dilution factor is 200, volume correction factor is 50).
5. Clean the pipette with acid, water, alcohol, and acetone at the end of the period.

Laboratory Report

Record your results in section B and table 2 (section E) of Laboratory Report 40–45.

Hemoglobin Percentage Measurement

42

Since hemoglobin is the essential oxygen-carrying ingredient of erythrocytes, its quantity in the blood determines whether or not a person is anemic. We learned in Exercise 41 that one may have a normal RBC count and still be anemic if the erythrocytes are smaller than normal. Such a condition, which is called *microcytic anemia,* results in one having an insufficient amount of hemoglobin. In other cases, a normal count in which the erythrocytes have a low hemoglobin percentage *(hypochromic cells)* may also result in anemia. Thus, it is apparent that the amount of hemoglobin present in a unit volume of blood is the critical factor in determining whether an individual is anemic.

Determination of the hemoglobin content of blood can be made by various methods. One of the oldest methods, and incidentally a very inaccurate one, is the Tallquist scale. In this technique a piece of blotting paper that has been saturated with a drop of blood is compared with a color chart to determine the percentage of hemoglobin. Very few, if any, physicians utilize this method today. A rapid, accurate method is to insert a cuvette of blood into a photocolorimeter that is calibrated for blood samples. Such a method,

however, requires a rather expensive piece of electronic equipment. It also requires a considerable quantity of blood.

A relatively inexpensive instrument for hemoglobin determinations is the *hemoglobinometer.* The *Hb-Meter,* manufactured by American Optical, is such a device. Figures 42.1 through 42.4 illustrate the procedure to follow in using the Hb-Meter. It is this piece of equipment that we will use in this laboratory period to determine the hemoglobin content of a blood sample.

The hemoglobin content of blood is expressed in terms of grams per 100 ml of blood. Three different standards are used depending on the community where the test is performed. For our purposes we will use the 15.6 gm/100 ml as standard.

For adult males, the normal on this scale is 14.9 ± 1.5 gm/100 ml; for adult females, the normal on this scale is 13.7 ± 1.5; for children at birth, 21.5 ± 3; for children at 4 years, 13 ± 1.5 gm/100 ml. A conversion scale exists on the side of the A/O Hb-Meter that allows one to determine hemoglobin percentages from grams Hb/100 ml. Proceed as follows to determine your own hemoglobin percentage:

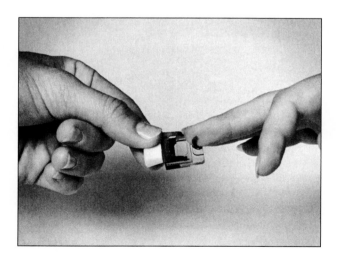

Figure 42.1 A fresh drop of blood is added to the moat plate of the blood chamber assembly. The blood should flow freely without excessive finger pressure.

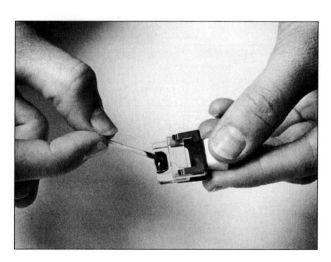

Figure 42.2 Blood sample is hemolyzed on moat plate with a wooden hemolysis applicator. Thirty-five to forty-five seconds are required for complete hemolysis.

Materials:
   American Optical Hb-Meter
   hemolysis applicators
   lancets
   cotton
   alcohol

1. Disassemble the blood chamber by pulling the two pieces of glass from the metal clip. Note that one piece of glass has an H-shaped moat cut into it. This piece will receive the blood. The other piece of glass has two flat surfaces and serves as a cover plate.
2. Clean both pieces of glass with alcohol and Kimwipes. Handle by edges to keep clean.
3. Reassemble the glass plates in the clip so that the grooves on the moat plate face the cover plate. The moat plate should be inserted only halfway to provide an exposed surface to receive the drop of blood. See figures 42.1 and 42.2.
4. Disinfect and puncture the finger with a disposable lancet.
5. Place a drop of blood on the exposed surface of the moat plate, as shown in figure 42.1.
6. Hemolyze the blood on the plate by mixing the blood with the pointed end of a hemolysis appli-cator as shown in figure 42.2. It will take 30–45 seconds for all the red blood cells to rupture. Complete hemolysis has occurred when the blood loses its cloudy appearance and becomes a transparent red liquid.
7. Push the moat plate in flush with the cover plate and insert the sample into the side of the instrument, as shown in figure 42.3.
8. Place the eyepiece to your eye with the left hand in such a manner that the left thumb rests on the light switch button on the bottom of the Hb-Meter.
9. While pressing the light button with the left thumb, move the slide button on the side of the instrument back and forth with the right index finger until the two halves of the split field match. The index mark on the slide knob indicates the grams of hemoglobin per 100 ml of blood. Read the percent hemoglobin on the 15.6 scale.

Laboratory Report

Record your results on table 2 in section E of Laboratory Report 40–45.

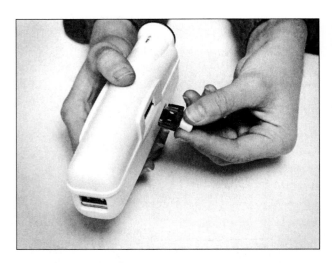

Figure 42.3 Charged blood chamber is inserted into slot of Hb-Meter. Before insertion, the unit should be tested to make certain the batteries are active.

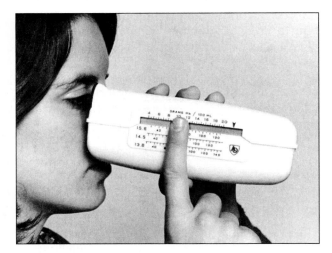
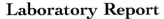

Figure 42.4 Blood sample is analyzed by moving slide button with right index finger. When the two colors match in density, the gm/100 ml is read on the scale.

Packed Red Cell Volume

43

In Exercises 41 and 42 we employed two different tests for determining the presence or absence of anemia: (1) the RBC count, and (2) the use of a hemoglobinometer. We have seen that anemia may be due to a dilution of the number of erythrocytes or to a deficiency of hemoglobin. It was observed that the most significant factor is the amount of hemoglobin that is present in a given volume of blood. A third method that can be used in anemia diagnosis is to determine the volume percentage of blood that is occupied by the red blood cells that have been packed by centrifugation. This method is called the **VPRC,** or volume of packed red cells.

The VPRC is determined by centrifuging a blood sample in a special centrifuge tube called a *hematocrit* or in a special size of heparinized capillary tubing. The centrifuge must be specifically designed to accommodate the hematocrit or capillary tubing.

In this exercise, we will utilize a micro method for determining the VPRC. As illustrated in figure 43.1, only a drop of blood is needed to perform the test. Capillary tubing, which has been heparinized, is used to collect the blood sample. It is centrifuged at high speed for only 4 minutes. (Macro methods that utilize hematocrits usually centrifuge the blood for 30 minutes at a much slower speed.) After centrifugation, the percentage of total volume occupied by the packed red

blood cells is read on a special tube reader (figure 43.6) or directly on the head of the centrifuge, depending on the type of centrifuge.

An interesting relationship exists between grams of hemoglobin per 100 ml and the VPRC: the VPRC is usually three times the gm Hb/100 ml. In men, the normal VPRC range is between 40% and 54%, with 47% as average. In women, the normal range is between 37% and 47%, with 42% as average. The principal advantages of this method over total RBC counts and Hb-Meter usage are: (1) simplicity, (2) speed, and (3) high degree of accuracy. Although this micro method is not as accurate as macro methods, it still functions in an accuracy range of ±2% for most blood samples. Proceed as follows to determine the VPRC of your own blood.

Caution: See page 253.

Materials:
    lancets
    cotton
    alcohol
    hematocrit centrifuge
    tube reader
    heparinized capillary tubes
    Clay-Adams Seal-Ease

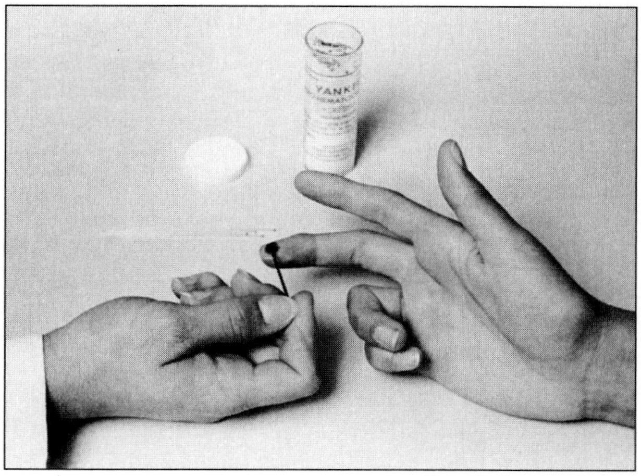

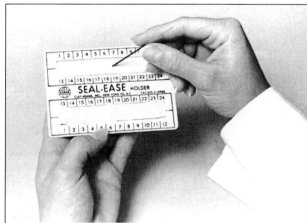

Figure 43.1 Blood is drawn up into a heparinized capillary tube for VPRC determination.

Figure 43.2 After filling with blood, the end of the capillary tube is sealed with clay.

263

1. Produce a free flow of blood on a finger. Wipe away the first drop of blood.
2. Place the red marked end of the capillary tube into the drop of blood and allow the blood to be drawn up about two-thirds of the way into the tube. Holding the open end of the tube downward from the blood source will cause the tube to fill more rapidly.
3. Seal the blood end of the tube with Seal-Ease. See figure 43.2.
4. Place the tube into the centrifuge with the sealed end against the ring of rubber at the circumference. Load the centrifuge with an even number of tubes (2, 4, 6, etc.) to properly balance the head.
5. Secure the inside cover with a wrench as illustrated in figure 43.4.

6. Lower the outside cover and securely latch it.
7. Turn on the centrifuge, setting the timer for 4 minutes.
8. Remove the centrifuged sample from the centrifuge and determine the VPRC by placing the tube in the mechanical tube reader. Instructions for reading are on the instrument.
   Note: On some centrifuges the tubes can be read directly on the head without using a tube reader.

Laboratory Report

Record your results on table 2 in section E of Laboratory Report 40–45.

Figure 43.3 The capillary tubes are placed in the centrifuge with the sealed end toward the perimeter.

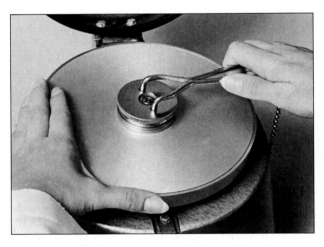

Figure 43.4 The inner safety lid is tightened in place with a lock wrench.

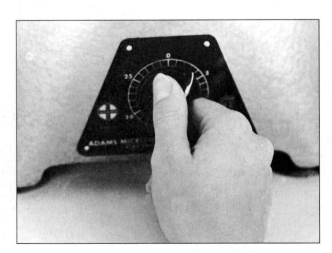

Figure 43.5 The timer on the centrifuge is set for 4 minutes by turning the dial clockwise.

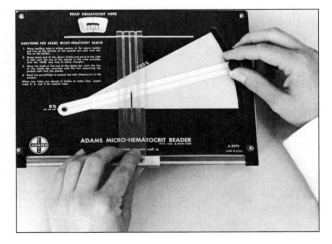

Figure 43.6 The capillary tube is placed in the tube reader to determine the VPRC.

Coagulation Time

44

The coagulation of blood is a complex phenomenon involving over thirty substances. The majority of these substances inhibit coagulation and are called *anticoagulants;* the remainder, which promote coagulation, are designated as *procoagulants.* Whether or not the blood will coagulate depends on which group predominates in a given situation. Normally, the anticoagulants predominate, but when a vessel is ruptured, the procoagulants in the affected area assume control, causing a clot *(fibrin)* to form in a relatively short period of time.

Of the various methods that have been devised to determine the rate of blood coagulation, the one outlined here is very easy to perform and very reliable. Figures 44.1 through 44.3 illustrate the general procedure. Proceed as follows:

Materials:
lancets, cotton, alcohol
capillary tubes (0.5 mm diameter)
3-cornered file
nonheparinized capillary tubes (0.5 mm diameter)

1. Puncture the finger to expose a free flow of blood. **Record the time.**
2. Place one end of the capillary tube into the drop of blood. Hold the tube so that the other end is **lower** than the drop of blood so that the force of gravity will aid the capillary action.
3. At **one-minute intervals,** break off small portions of the tubing by scratching the glass first with a file.
   Important: Separate the broken ends slowly and gently while looking for coagulation. Coagulation has occured when threads of fibrin span the gap between the broken ends as shown in figure 44.3.
4. **Record the time** as that from which the blood first appeared on that finger to the formation of fibrin.

Laboratory Report

Record your results on table 2, section E, of Laboratory Report 40–45.

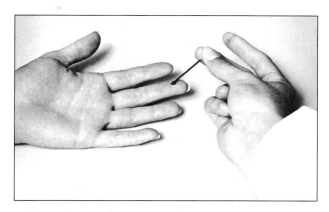

Figure 44.1 Blood is drawn up into a nonheparinized capillary tube (0.5 mm diameter).

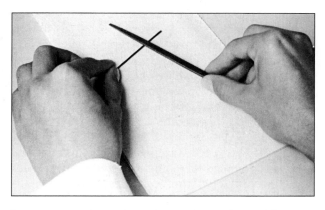

Figure 44.2 At 1-minute intervals, sections of the tube are filed and broken off for the test.

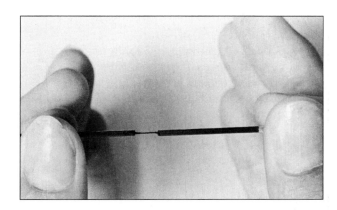

Figure 44.3 When the blood has coagulated, strands of fibrin will extend between broken ends of tube.

45

Blood Typing

The surfaces of red blood cells possess various antigenic molecules that have been designated by letters such as A, B, C, D, E, c, d, e, M, N, etc. Although the exact chemical composition of all these substances is undetermined, some of them appear to be carbohydrate residues (oligosaccharides).

The presence or absence of these various substances determines the type of blood possessed by an individual. Since the chemical makeup of cells is genetic, an individual's blood type is the same in old age as it is at birth. It never changes.

The only factors that we are concerned with here in this exercise are A, B, and D (Rh) antigens since they are most commonly involved in transfusion reactions.

To determine an individual's blood type, drops of blood-typing sera are added to suspensions of red blood cells to detect the presence of **agglutination** (clumping) of the cells. ABO typing may be performed at room temperature with saline suspensions of red blood cells as shown in figure 45.1. Rh typing for the D factor, on the other hand, requires higher temperatures (around 50° C) and whole blood instead of diluted blood. For ABO typing the diluted blood procedure is preferable. For convenience, however, the warming box method may be used for combined ABO and Rh typing. Two procedures are provided here. Your instructor will indicate which method will be used.

ABO Blood Typing

(Saline Dilution Method)

Materials:
> small vial (10 mm dia × 50 mm long)
> disposable lancets (*B-D Microlance,*
>   *Sera-sharp,* etc.)
> 70% alcohol and cotton
> wax pencil and microscope slides
> typing sera (anti-A and anti-B)
> applicators or toothpicks
> saline solution (0.85%)
> 1 ml pipettes

1. Mark a slide down the middle with a marking pencil, dividing the slide into two halves (see figure 45.1). Write ANTI-A on the left side and ANTI-B on the right side.
2. Pipette approximately 1 ml of saline solution into a small vial or test tube.
3. Scrub the middle finger with a piece of cotton saturated with 70% alcohol and pierce it with a sterile disposable lancet.
4. Allow 2 or 3 drops of blood to mix with the saline by holding the finger over the end of the vial and washing it with the saline by inverting the vial several times.
5. Place a drop of this red cell suspension on each side of the slide.
6. Add a drop of anti-A serum to the left side of the slide and a drop of anti-B serum to the right side. *Do not contaminate the tips of the serum pipettes with the material on the slide.*
7. After mixing each side of the slide with separate applicators or toothpicks, look for agglutination. The slide should be held about 6 inches above an illuminated white background and rocked gently for 2 or 3 minutes.
8. Record your results on Laboratory Report as of 3 minutes.

Combined ABO and Rh Typing

(Warming Box Method)

As stated above, Rh typing must be performed with heat on blood that has not been diluted with saline. A warming box such as the one in figure 45.2 is essential in this procedure. In performing this test, two factors are of considerable importance: first, only a small amount of blood must be used (a drop of about 3 mm diameter on the slide), and second, proper agitation must be executed. The agglutination that occurs in this antibody-antigen reaction results in finer clumps; therefore, closer examination is also essential. If the agitation is not properly performed, agglutination may not be as apparent as it should be.

In this combined method, we will use whole blood for the ABO typing also. Although this method works out satisfactorily as a classroom demonstration for the ABO groups, it is not as

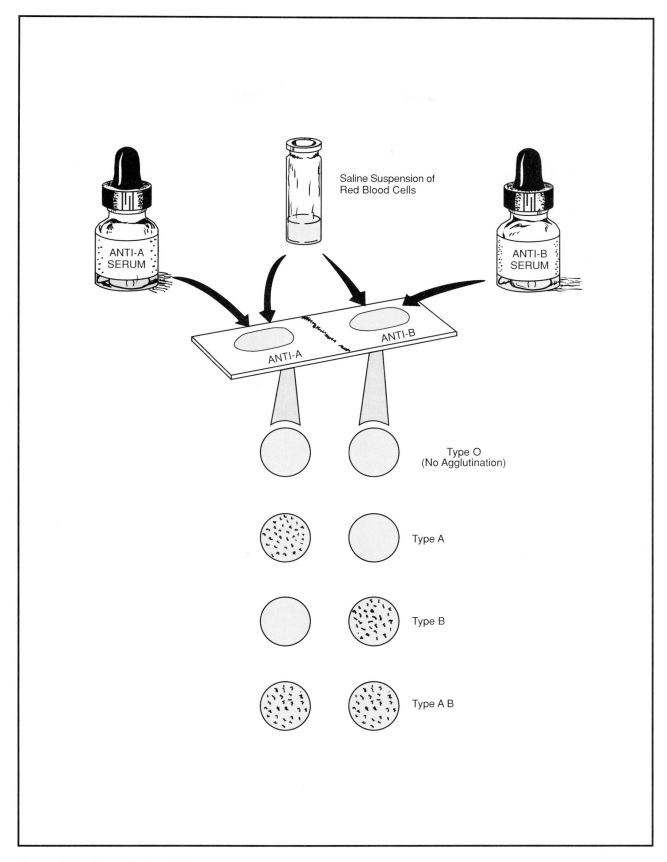

Figure 45.1   Blood typing ABO groups.

reliable as the other method in which saline and room temperature are used. *For clinical applications, whole undiluted blood, with heat, is not recommended.*

Materials:
slide warming box with a special marked slide
anti-A, anti-B, and anti-D typing sera
applicators or toothpicks
70% alcohol
cotton
disposable sterile lancets (*B-D Microlance, Sera-sharp, etc.*)

1. Scrub the middle finger with a piece of cotton saturated with 70% alcohol and pierce it with a sterile disposable lancet. Place a small drop in each of these squares on the marked slide on the warming box. To get the proper proportion of serum to blood, do not use a drop that is larger than 3 mm diameter on the slide.

2. Add a drop of anti-D serum to the blood in the anti-D square, mix with a toothpick, and note the time. **Only two minutes should be allowed for agglutination.**

3. Add a drop of anti-B serum to the anti-B square and a drop of anti-A to the anti-A square. Mix the sera and blood in both squares with *separate fresh* toothpicks.

4. Agitate the mixtures on the slide by slowly rocking the box back and forth on its pivot. At the end of two minutes, examine the anti-D square carefully for agglutination. If no agglutination is apparent, consider the blood to be Rh negative. By this time, the ABO type can also be determined.

Laboratory Report

Record your results in section C of Laboratory Report 40–45, and answer all the questions.

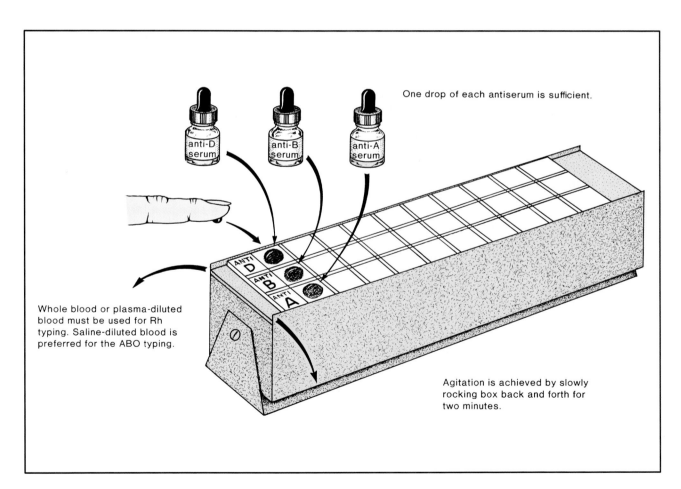

One drop of each antiserum is sufficient.

anti-D serum

anti-B serum

anti-A serum

ANTI D

ANTI B

ANTI A

Whole blood or plasma-diluted blood must be used for Rh typing. Saline-diluted blood is preferred for the ABO typing.

Agitation is achieved by slowly rocking box back and forth for two minutes.

Figure 45.2   Blood typing with a warming box.

46

Anatomy of the Heart

In this study of the anatomy of the human heart we will use a sheep heart for dissection and comparison. Before beginning the dissection, however, study figures 46.1 and 46.2.

Internal Anatomy

Figure 46.1 reveals the internal anatomy of the human heart. Red-colored vessels carry oxygenated blood and blue ones carry deoxygenated blood.

Chambers of the Heart

Note that the heart has four chambers: two small upper atria and two larger ventricles. The atria receive all the blood that enters the heart and the ventricles pump it out. Note that the **right atrium** in the frontal section is on the left side of the illustration and the **left atrium** is on the right. Observe, also, that although the **right ventricle** is somewhat larger than the left ventricle, the **left ventricle** has a thicker wall. Separating the two ventricles is a partition, the **ventricular septum.**

Wall of the Heart

The wall of the heart consists of three layers: the myocardium, the endocardium, and the epicardium. The **myocardium** is the muscular portion of the wall that is composed of cardiac muscle tissue. Note in the enlarged section of the wall that the myocardium makes up the bulk of the wall thickness.

Lining the inner surface of the heart is the **endocardium.** It is a thin serous membrane that is continuous with the endothelial lining of the arteries and veins. An infection of this membrane is called *endocarditis.*

Attached to the outer surface of the heart is another serous membrane, the **epicardium,** or *visceral pericardium.* The production of serous fluid by this membrane on the outer surface of the heart enables the heart to move freely within the pericardial sac. Note that the **pericardial sac,** or *parietal pericardium,* consists of two layers: an inner serous layer and an outer fibrous layer. Between

the heart and the parietal pericardium is the **pericardial cavity** (label 19). Its dimension has been exaggerated here.

Vessels of the Heart

The major vessels of the heart are the two venae cavae, the pulmonary trunk, the four pulmonary veins, and the aorta. The **superior vena cava** is the upper blue vessel on the right atrium; it conveys deoxygenated blood to the right atrium from the head and arms. The **inferior vena cava** is the lower blue vessel that empties deoxygenated blood from the trunk and legs into the right atrium. From the right atrium blood passes to the right ventricle, where it leaves the heart through the **pulmonary trunk.** This large vessel branches into the **right** and **left pulmonary arteries,** which carry blood to the lungs.

Blood is drained from the lungs by means of the four **pulmonary veins** (four small red vessels), which carry it to the left atrium. From the left atrium the blood passes into the left ventricle, where it exits the heart through the *aorta.* Although the initial emergence of the aorta is obscured, it can be seen as a large curved red vessel, the **aortic arch,** that has three smaller arteries emerging from its superior surface. The aorta carries oxygenated blood to the entire systemic circulatory system.

Note what appears to be a short vessel between the pulmonary trunk and the aortic arch. It is called the **ligamentum arteriosum.** During prenatal life this ligament is a functional blood vessel, the *ductus arteriosus,* that allows blood to pass from the pulmonary trunk to the aorta.

Valves of the Heart

The heart has two atrioventricular and two semilunar valves. Between the right atrium and the right ventricle is the **tricuspid valve.** Between the left atrium and the left ventricle is the **bicuspid,** or *mitral,* **valve.** The differences between these two valves are seen in the superior sectional view of figure 46.1. Note that the tricuspid valve has three flaps or cusps, and the bicuspid valve has only two

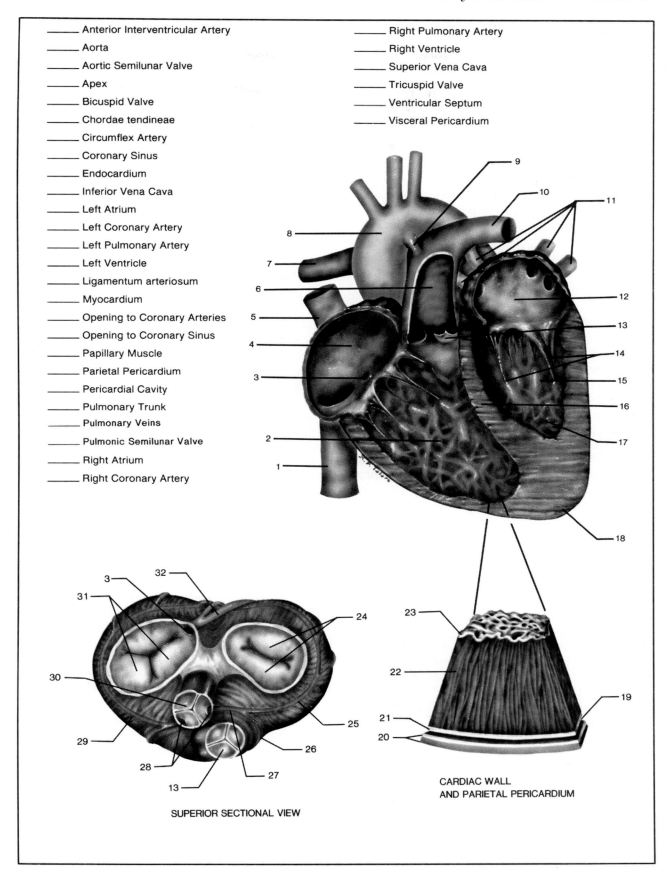

_____ Anterior Interventricular Artery
_____ Aorta
_____ Aortic Semilunar Valve
_____ Apex
_____ Bicuspid Valve
_____ Chordae tendineae
_____ Circumflex Artery
_____ Coronary Sinus
_____ Endocardium
_____ Inferior Vena Cava
_____ Left Atrium
_____ Left Coronary Artery
_____ Left Pulmonary Artery
_____ Left Ventricle
_____ Ligamentum arteriosum
_____ Myocardium
_____ Opening to Coronary Arteries
_____ Opening to Coronary Sinus
_____ Papillary Muscle
_____ Parietal Pericardium
_____ Pericardial Cavity
_____ Pulmonary Trunk
_____ Pulmonary Veins
_____ Pulmonic Semilunar Valve
_____ Right Atrium
_____ Right Coronary Artery

_____ Right Pulmonary Artery
_____ Right Ventricle
_____ Superior Vena Cava
_____ Tricuspid Valve
_____ Ventricular Septum
_____ Visceral Pericardium

SUPERIOR SECTIONAL VIEW

CARDIAC WALL
AND PARIETAL PERICARDIUM

Figure 46.1   Internal anatomy of the heart.

cusps. Observe, also, that the edges of the cusps have fine cords, the **chordae tendineae,** which are anchored to **papillary muscles** on the wall of the heart (see frontal section). These cords and muscles prevent the cusps from being forced up into the atria during systole (ventricular contraction).

The other two valves are the **pulmonic semilunar** and **aortic semilunar valves.** They are located at the bases of the pulmonary trunk and the aorta. They prevent blood in those vessels from flowing back into the heart during diastole (relaxation phase). Note in the superior sectional view that each of these valves has three small cusps.

Coronary Circulation

The vessels that supply the heart muscle with blood comprise the *coronary circulatory system.* Oxygenated blood in the aorta passes into right and left coronary arteries through two small openings at the point just superior to the aortic semilunar valve. These **openings to the coronary arteries** are seen in the wall of the aorta in the superior sectional view in figure 46.1. One opening leads into a **left coronary artery** (label 27) and the other opening leads into the **right coronary artery.**

The left coronary artery has two branches: a **circumflex artery** that passes around the heart in the left atrioventricular sulcus and an **anterior interventricular artery** (label 26) that passes down the interventricular sulcus on the anterior surface of the heart. The right coronary artery lies in the right atrioventricular sulcus and has branches that supply the posterior and anterior surfaces of the ventricular muscle.

Once the coronary blood has been relieved of its oxygen and nutrients in the heart muscle, it is picked up by the various veins that parallel the arteries and is emptied into a large vein, the **coronary sinus** (label 32). Blood in the coronary sinus is emptied into the right atrium through a portal, the **opening to the coronary sinus** (label 3).

To identify the remainder of the vessels of the coronary system, proceed to study the external features of the heart in figure 46.2.

Assignment:
Label figure 46.1

External Anatomy

The study of the external anatomy of the heart is relatively simple once the internal anatomy is understood. Figure 46.2 reveals two external views of the heart.

Anterior Aspect

Note that the heart lies within the **mediastinum** at a slight tilt to the left so that the *apex,* or tip of the heart, is somewhat on the left side. The obscured appearance of the coronary vessels in this view is due to the fact that the **parietal pericardium** lies intact over the organ. The clarity of the structures on the posterior view is due to the fact that this pericardium has been removed.

Differentiating the right ventricle from the left ventricle is best done by locating the *interventricular sulcus* first. This sulcus is the slight depression over the ventricular septum that contains the **anterior interventricular artery** and the **great cardiac vein.** The yellow material seen in the sulcus consists of fat deposits. The left ventricle is on the left side of the interventricular sulcus, and the right ventricle is to the right of it.

Three portions of the aorta are labeled on the anterior aspect: a short **ascending aorta** that emerges from the left ventricle, a curving **aortic arch,** and a portion of the **descending aorta** that passes down through the mediastinum behind the heart.

Locate the **right** and **left pulmonary arteries** that emerge from behind the aortic arch. The pulmonary trunk that gives rise to these arteries is obscured by the aorta. Also, identify the **ligamentum arteriosum.** The coronary vessels in the atrioventricular sulcus are the **right coronary artery** and the **small cardiac vein.**

Posterior Aspect

The posterior view of the heart in figure 46.2 reveals more clearly the position of the four pulmonary veins and additional coronary vessels. The two **right pulmonary veins** are seen near the superior vena cava. The **two left pulmonary veins** are situated to the left of the right pulmonary veins. The appearance of a single blue vessel lying below the aorta is actually the site of the division of the pulmonary trunk into the **right** and **left pulmonary arteries.**

The following coronary arteries are seen on this side of the heart: the **right coronary artery** (label 4), which lies in the right atrioventricular sulcus, the **posterior descending right coronary artery,** which is a downward extension of the right coronary artery, the **posterior interventricular artery** (label 20), and the **circumflex artery** (label 23).

The major coronary veins on this side are the **middle cardiac vein,** which parallels the posterior descending right coronary artery, the **left posterior**

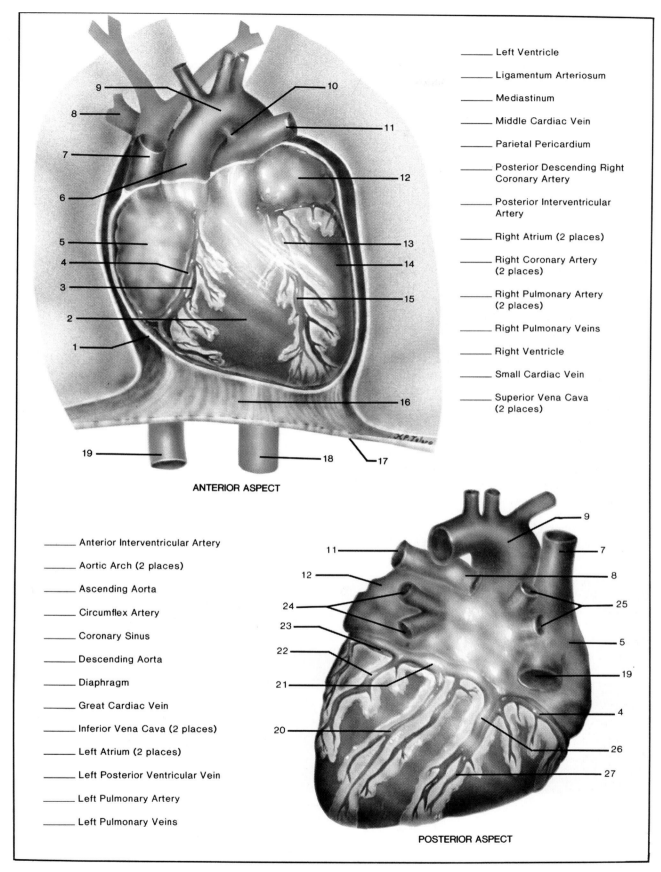

Left Ventricle

Ligamentum Arteriosum

Mediastinum

Middle Cardiac Vein

Parietal Pericardium

Posterior Descending Right
Coronary Artery

Posterior Interventricular
Artery

Right Atrium (2 places)

Right Coronary Artery
(2 places)

Right Pulmonary Artery
(2 places)

Right Pulmonary Veins

Right Ventricle

Small Cardiac Vein

Superior Vena Cava
(2 places)

ANTERIOR ASPECT

Anterior Interventricular Artery

Aortic Arch (2 places)

Ascending Aorta

Circumflex Artery

Coronary Sinus

Descending Aorta

Diaphragm

Great Cardiac Vein

Inferior Vena Cava (2 places)

Left Atrium (2 places)

Left Posterior Ventricular Vein

Left Pulmonary Artery

Left Pulmonary Veins

POSTERIOR ASPECT

Figure 46.2   External anatomy of the heart.

ventricular vein (label 22), and the **posterior interventricular vein** (not labeled), which lies in the interventricular sulcus with the similarly named artery. All these coronary veins empty into the large **coronary sinus,** which lies in the atrioventricular sinus.

Assignment:
Label figure 46.2.

Sheep Heart Dissection

While dissecting the sheep heart, attempt to identify as many structures as possible that are shown in figures 46.1 and 46.2.

Materials:
> sheep heart, fresh or preserved
> dissecting instruments and tray

1. Rinse the heart with cold water to remove excess preservative or blood. Allow water to flow through the large vessels to irrigate any blood clots out of its chambers.
2. Look for evidence of the **parietal pericardium.** This fibroserous membrane is usually absent from laboratory specimens, but there may be remnants of it attached to the large blood vessels of the heart.
3. Attempt to isolate the **visceral pericardium** *(epicardium)* from the outer surface of the heart. Since it consists of only one layer of squamous cells, it is very thin. With a sharp scalpel try to peel a small portion of it away from the myocardium.
4. Identify the **right** and **left ventricles** by squeezing the walls of the heart as shown in illustra-

tion 1, figure 46.3. The right ventricle will have the thinner wall.
5. Identify the **anterior interventricular sulcus,** which lies between the two ventricles. It is usually covered with fat tissue. Carefully trim away the fat in this sulcus to expose the **anterior interventricular artery** and the **great cardiac vein.**
6. Locate the **right** and **left atria** of the heart. The right atrium is seen on the left side of illustration 1, figure 46.3.
7. Identify the **aorta,** which is the large vessel just to the right of the right atrium as seen in illustration 1, figure 46.3. Carefully peel away some of the fat around the aorta to expose the **ligamentum arteriosum.**
8. Locate the **pulmonary trunk,** which is the large vessel between the aorta and the left atrium as seen when looking at the anterior side of the heart. If the vessel is of sufficient length, trace it to where it divides into the **right** and **left pulmonary arteries.**
9. Examine the posterior surface of the heart. It should appear as in illustration 2, figure 46.3. Note that only the right atrium and two ventricles can be seen on this side. The left atrium is obscured by fat from this aspect. Look for the four thin-walled **pulmonary veins** that are embedded in the fat. Probe into these vessels and you will see that they lead into the left atrium.
10. Locate the **superior vena cava,** which is attached to the upper part of the right atrium. Insert one blade of your dissecting scissors into this vessel (illustration 3, figure 46.3), and cut

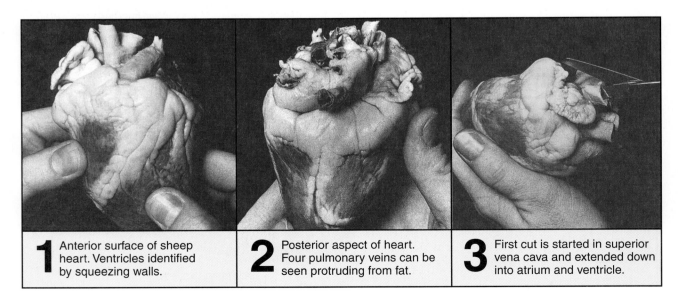

1 Anterior surface of sheep heart. Ventricles identified by squeezing walls.

2 Posterior aspect of heart. Four pulmonary veins can be seen protruding from fat.

3 First cut is started in superior vena cava and extended down into atrium and ventricle.

Figure 46.3 **Sheep heart dissection.**

through it into the atrium to expose the **tricuspid valve** between the right atrium and right ventricle. Don't cut into the ventricle at this time.

11. Fill the right ventricle with water, pouring it in through the tricuspid valve. Gently squeeze the walls of the ventricle to note the closing action of the valve's cusps.

12. Drain the water from the heart and continue the cut with scissors from the right atrium through the tricuspid valve down to the apex of the heart.

13. Open the heart and flush it again with cold water. Examine the interior. The open heart should look like illustration 1, figure 46.4.

14. Examine the interior wall of the right atrium. This inner surface has ridges, giving it a comb-like appearance; thus, it is called the **pectinate muscle** (*pecten*, comb).

15. Look for the **opening to the coronary sinus.** The white arrow in illustration 1, figure 46.4 shows its location. It is through this opening that blood of the coronary circulation is returned to the venous circulation.

16. Insert a probe under the cusps of the tricuspid valve. Are three flaps evident?

17. Locate the **papillary muscles** and **chordae tendineae** in the right ventricle. How many papillary muscles are there?

18. Look for the **moderator band,** a reinforcement cord that extends from the ventricular septum to the ventricular wall. Its presence prevents excessive stress from occurring in the myocardium of the right ventricle.

19. With scissors, cut the right ventricular wall up along its lower margin parallel to the anterior interventricular sulcus to the pulmonary trunk.

Continue the cut through the exit of the right ventricle into the **pulmonary trunk.** Spread the cut surfaces of this new incision to expose the **pulmonic semilunar valve.** Wash the area with cold water to dispel blood clots.

20. Insert one blade of your scissors into the left atrium, as shown in illustration 2, figure 46.4. Cut through the atrium to the left ventricle. Also, cut from the left ventricle into the aorta, slitting this vessel longitudinally.

21. Examine the **bicuspid** (mitral) **valve,** which lies between the left atrium and left ventricle. With a probe, reveal the two cusps.

22. Examine the pouches of the **aortic semilunar valve.** Compare the number of pouches of this valve with the pulmonic valve.

23. Look for the two **openings to the coronary arteries,** which are in the walls of the aorta just above the aortic semilunar valve. Force a blunt probe into each hole. Note how they lead into the two coronary arteries.

Laboratory Report

Complete Laboratory Report 46 for this exercise.

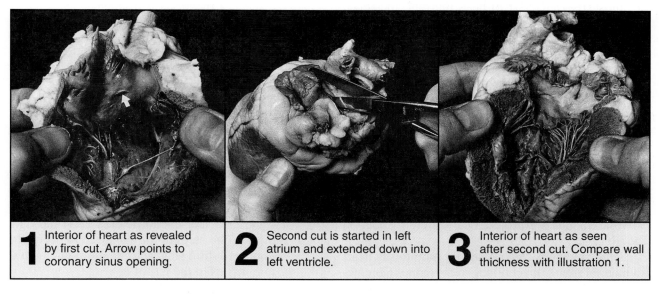

1 Interior of heart as revealed by first cut. Arrow points to coronary sinus opening.

2 Second cut is started in left atrium and extended down into left ventricle.

3 Interior of heart as seen after second cut. Compare wall thickness with illustration 1.

Figure 46.4   Sheep heart dissection.

47

Cardiovascular Sounds

Cardiovascular sounds are created by myocardial contraction and valvular movement. Three distinct sounds are heard.

The **first sound** (lub) occurs simultaneously with the contraction of the ventricular myocardium and the closure of the mitral and tricuspid valves.

The **second sound** (dup) follows the first one after a brief pause. It differs from the first sound in that it has a snapping quality of higher pitch and shorter duration. This sound is caused by the virtually simultaneous closure of the aortic and pulmonic valves at the end of ventricular systole.

The **third sound** is low pitched and is loudest near the apex of the heart. It is difficult to detect, but can be heard most easily when the subject is recumbent. It appears to be caused by the vibration of the ventricular walls and the AV valves during systole.

Abnormal heart sounds are collectively referred to as **murmurs.** They are usually due to damaged valves. Mitral valve damage due to rheumatic fever is a most frequent cause of heart murmur; the aortic valve is next in susceptibility to damage. Damaged valves may result in *stenosis* or *regurgitation* of blood flow.

Examination Procedures

Listening to sounds generated within the body is called *auscultation.* For maximum results in monitoring valvular sounds it is necessary to be familiar with four specific spots in the thorax. Illustration 1, figure 47.1, shows their locations. Note that these auscultatory areas do not coincide with the anatomical locations of the various valves; this is because valvular sounds are projected to different spots on the rib cage.

Three methods can be used for monitoring heart sounds: (1) the conventional stethoscope, (2) electronic recording, and (3) the use of an audio monitor. Only the procedures for the first two methods are presented here. The audio monitor, if available, will be used only as a demonstration. *Since noise is distracting in these experiments, it is imperative that all students keep voices low.*

Audio Monitor Demonstration (By Instructor)

Using a student as a subject in a recumbent position, the instructor will use an audio monitor to demonstrate heart sounds to the class. If a sensitive unit, such as the Grass AM7 is used, it will not be necessary to bare the subject's chest.

When attempting to demonstrate second sound splitting, hand signals should be used instead of verbalization. The subject can be instructed to inspire when the instructor's hand is raised and expire when the hand is lowered.

Stethoscope Auscultation

Work with your laboratory partner to auscultate each other's heart with a stethoscope. Use the following procedure:

Materials:
   stethoscope
   alcohol and absorbent cotton

1. Clean and disinfect the earpieces of a stethoscope with cotton saturated in alcohol. Fit them to your ears, directing them inward and upward.
2. Try first to maximize the **first sound,** by locating the stethoscope on the tricuspid and mitral auscultatory areas. See illustration 1, figure 47.1.

   Note that the mitral area coincides with the apex of the heart (fifth rib), and the tricuspid area is about 2 to 3 inches medial to this spot.
3. Now move the stethoscope to the aortic and pulmonic areas to hear the **second sound** more clearly.

   Observe that the aortic area is on the right border of the sternum in the space between the second and third ribs, and the pulmonic area is 2 to 3 inches to the left of it.
4. While listening to the second sound, attempt to detect the splitting of this sound while the subject is inhaling. Use hand signals instead of verbalization to communicate with the subject. Have the subject inspire when you raise your hand and expire when you lower your hand.

This splitting characteristic of the second sound during inspiration is normal. It is due to the fact that during inspiration more venous blood is forced into the right side of the heart, causing delayed closure of the pulmonic valve during systole. Result: aortic and pulmonic closure sounds are heard separately.

5. After listening to the sounds with the subject sitting erect, listen to them with the subject lying down, face upward.
6. Can you detect the **third sound?**
7. Is there any evidence of **murmurs?**
8. Compare the sounds before and after exercise, recording your observations on the Laboratory Report. For exercise, leave the room to run up and down stairs or jog for a short distance.
9. Before ending the examination, check all the auscultatory areas in figure 47.1 to make sure that you have identified all valvular sounds.

Electronic Recording (*Unigraph Setup*)

Produce a phonocardiogram of heart sounds of a subject lying down, face upward.

Materials:
Unigraph
Trans/Med 6605 adapter
Statham P23AA pressure transducer

1. Attach a Trans/Med 6605 adapter to the input end of the Unigraph.

2. Attach a Statham pressure transducer and microphone jack to the 6605 adapter. Refer to illustration 3, figure 47.1.
3. Turn on the Unigraph and set the following controls: chart control (c.c.) at STBY, heat control at two o'clock position, speed control at slow position, Gain at 1 MV/CM, and Mode on TRANS.
4. Place the microphone against one of the auscultatory areas and start the chart moving by placing the c.c. lever at Chart On. Adjust the temperature control knob to get a good recording. Center the tracing with centering control knob and adjust sensitivity to produce at least 5 mm stylus displacement.
5. Move the speed control lever to the high speed position.
6. Make recordings with microphone over each auscultatory area. Be sure to mark the chart to identify the areas being monitored.
7. Attempt to record splitting of second sound, using the same techniques that were used with a stethoscope.
8. Attach phonocardiograms to the Laboratory Report.
9. Deactivate all controls on the Unigraph at the end of the experiment.

Laboratory Report

Complete the first portion of Laboratory Report 47, 48.

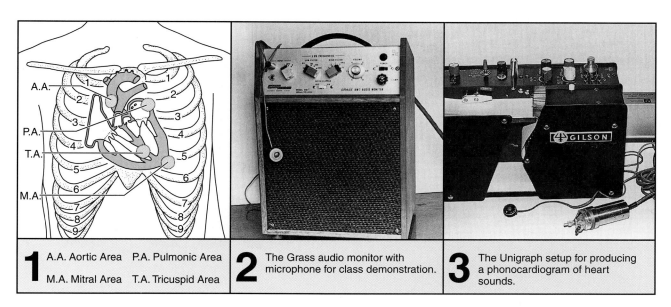

1 A.A. Aortic Area   P.A. Pulmonic Area
   M.A. Mitral Area   T.A. Tricuspid Area

2 The Grass audio monitor with microphone for class demonstration.

3 The Unigraph setup for producing a phonocardiogram of heart sounds.

Figure 47.1   Auscultatory areas and two electronic setups.

48

Electrocardiogram Monitoring:
Using Chart Recorders

In this laboratory period we will learn the procedures that are involved in making an electrocardiogram (ECG), utilizing a chart recorder such as the Gilson Unigraph or the Narco Physiograph®. Separate instructions will be provided for both types of equipment. If the Cardiocomp™ (Exercise 49) is to be used instead of one of these chart recorders, it will be necessary for you to read over the applications provided here in this exercise since this background information is lacking in the Cardiocomp exercise.

The ECG Wave Form

Myocardial contractions of the heart originate with depolarization of the sinoatrial (SA) node of the right atrium. As the myocardium of the right atrium is depolarized and contracts in response to depolarization of the SA node, the atrioventricular (AV) node is activated, causing it to send a depolarization wave via the atrioventricular bundle and conduction myofibers to the ventricular myocardium. The electrical potential changes that result from this depolarization and repolarization can be monitored to produce a record called an **electrocardiogram,** or **ECG.**

To produce such a record it is necessary to attach a minimum of two and as many as ten electrodes to different portions of the body to record the differences in electrical potential that occur. If one electrode is placed slightly above the heart and to the right, and another is placed slightly below the heart to the left, one can record the wave form at its maximum potential.

Figure 48.1 reveals a typical ECG as recorded on a Unigraph, and figure 48.2 depicts the characteristics of an individual normal ECG wave form. The initial depolarization of the SA node, which causes atrial contraction, manifests itself as the **P wave.** This depolarization is immediately followed by repolarization of the atria. Atrial repolarization, not usually seen with surface electrodes, can be demonstrated with implanted electrodes; such a wave is designated as the *TA wave.*

Depolarization of the ventricles via the atrioventricular bundle and conduction myofibers results in the production of the **QRS complex.** This depolarization causes ventricular systole.

As soon as depolarization of the ventricular muscle is completed, repolarization takes place, producing the **T wave.** Some ECG wave forms show an additional wave occurring after the T wave. It is designated as the **U wave.** This small wave is attributed to the gradual repolarization of the papillary muscles.

The final result of this depolarization wave is that the myocardium of the ventricles contracts as a

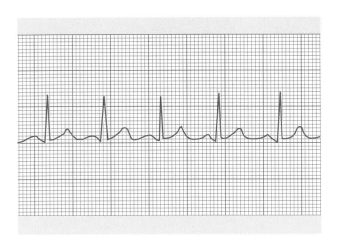

Figure 48.1   An electrocardiogram (Unigraph tracing).

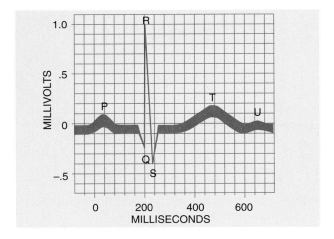

Figure 48.2   The ECG wave form.

unit, forcing blood out through the pulmonary trunk and aorta. Repolarization of the affected tissues quickly follows in preparation for the next contraction cycle.

The value of the ECG to the cardiologist is immeasurable. Almost all serious abnormalities of the heart muscle can be detected by analyzing the contours of the different waves. Interpretation of the ECG wave forms to determine the nature of heart damage is a highly developed technical skill involving vector analysis and goes considerably beyond the scope of this course; however, figure 48.4 illustrates three examples of abnormal ECGs in which heart damage has occurred.

Electrodes and Leads

As indicated earlier, as few as two and as many as ten electrodes can be used to produce an ECG. An understanding of Einthoven's triangle (figure 48.3) will clarify the significance of the different electrode positions and the meaning of leads I, II, and III. Einthoven was the first person to produce an electrocardiogram around the turn of the twentieth century.

The number of **electrodes** used for the Unigraph is three: the two wrists and the left ankle as shown in figure 48.7. For the Physiograph® setup, five electrodes will be used (figure 48.6): the four appendages and one on the chest over the heart. The right leg acts as a ground in the five-electrode setup.

Three leads, designated as leads I, II, and III, are derived from these electrode hookups. A **lead,** as defined by Einthoven, is *the electrical potential in the fluids around the heart between two electrodes during depolarization.* On the Physiograph® and Cardiocomp setups you will have an opportunity to make lead selections.

Note in figure 48.3 that the apexes of the upper part of the triangle represent the points at which the two arms connect electrically with the fluids around the heart. The lower apex is the point at which the left leg (+) connects electrically with the pericardial fluids. **Lead I** records the electrical potential between the right (−) and left (+) arm electrodes. **Lead II** records the potential between the right arm (−) and the left leg (+). **Lead III** records the potential between the left arm (−) and the left leg (+). These potentials are recorded graphically on the ECG in millivolts.

When we put leads I, II, and III together on one diagram, as in figure 48.3, we see that the three leads form a triangle around the heart, known as **Einthoven's triangle.** The significance of this trian-

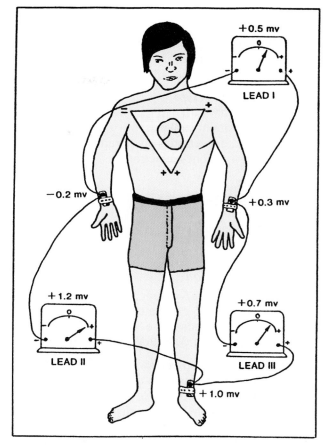

Figure 48.3 In the conventional three-lead hookup, the two arms and left leg form apexes of a triangle (Einthoven's) surrounding the heart. Note that the sum of the voltages in leads I and III equals the voltage in lead II.

gle is that if the potentials of two of the leads are recorded, the third one can be determined mathematically. Or, stated another way: the sum of the voltages in leads I and III equals the voltage of lead II. This principle is known as **Einthoven's law:**

$$\text{Lead II} = \text{Lead I} + \text{Lead III}$$

Procedure

Students will work in teams of three or four, with one individual being the subject and the others making the electrode hookups and operating the controls. It will be necessary to calibrate the recorder prior to recording the ECG; separate calibration instructions must be followed for the type of recorder being used.

Failure in this experiment usually results from (1) poor skin contact, (2) damaged electrode cables, or (3) incorrect hookup. Proceed as follows:

Materials:

for Unigraph setup:
    Unigraph
    three-lead patient cable
    ECG plate electrodes (3) and straps

for Physiograph® setup:
    Physiograph® with cardiac coupler
    five-lead patient cable
    4 ECG plate electrodes and straps
    1 suction-type chest electrode

for both setups:
    Scotchbrite pad
    70% alcohol
    electrode paste

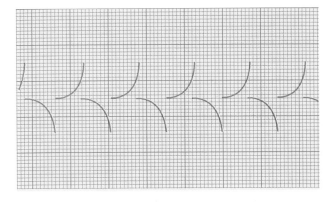

Figure 48.5 Calibration tracing for Unigraph.

Unigraph Calibration

While one member of your team attaches the electrodes to the subject, calibrate the Unigraph so that one millivolt produces one centimeter of stylus deflection (1 mv/1 cm). Follow these steps:

1. With the main power switch still in the OFF position, set the controls as follows: c.c. switch at STBY, speed control lever at the slow position, Gain at 1 MV/CM, sensitivity knob completely counterclockwise, Mode at CC-Cal, and the Hi Filter and Mean switches on NORM.
2. Turn on the power switch and rotate the stylus heat control to the two o'clock position.
3. Place the c.c. switch on Chart On, observe the trace and center the stylus, using the centering control.
4. Increase the sensitivity by rotating the sensitivity control clockwise one-half turn.

5. Depress the 1 MV button, hold it down for about 2 seconds, and then release it. The upward deflection should measure 1 cm. The downward deflection should equal the upward deflection. Refer to figure 48.5. If the travel is not 1 cm, adjust the sensitivity control and depress the 1 MV button until exactly 1 cm is achieved.
6. Place the c.c. switch on STBY and label the chart with the date and subject's name.
7. Place the Mode control on ECG and move the speed selector level to the high speed (25 mm per sec) position. The Unigraph is now ready to accept the patient cable.

Physiograph® Calibration

While one member of your team attaches the electrodes to the subject, calibrate the channel sensitivity

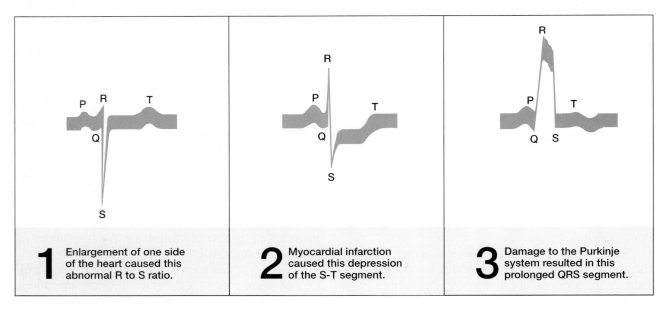

1 Enlargement of one side of the heart caused this abnormal R to S ratio.

2 Myocardial infarction caused this depression of the S-T segment.

3 Damage to the Purkinje system resulted in this prolonged QRS segment.

Figure 48.4 Abnormal electrocardiograms.

of the Physiograph® so that one millivolt produces two centimeters of pen deflection (1 mv/ 2 cm). Follow these steps:

1. Place the Time Constant switch of the cardiac coupler in the 3.2 position.
2. Place the Gain switch on the coupler to the X2 position.
3. Place the Lead Selector Control in the calibrate (Cal) position.
4. Set the outer knob of the channel Amplifier Sensitivity Control on the 10 MV/CM position, and rotate the inner sensitivity knob completely clockwise until it clicks.
5. Start the chart moving at 0.25 cm/sec and lower the pens onto the paper. Adjust the pen to the center of its arc with the Position Control.
6. Place the channel amplifier record button into the ON position.
7. Place the CAL toggle switch on the cardiac coupler into the 1 MV position. Observe that the channel recording pen will deflect *upward* 2 centimeters.
8. Hold the CAL switch in the 1 MV position until the recording pen returns to the center line, then release the switch. Observe that upon release, the channel recording pen will now deflect *downward* 2 cm.
9. If the upward and downward deflections are not exactly 2 centimeters, make adjustments with the amplifier sensitivity controls.
10. Stop recording. The channel is now calibrated.

Preparation of the Subject

While the recorder is being calibrated, the electrodes should be attached to the subject by other members of the team. Although it is not absolutely essential, it is desirable to have the individual lying on a comfortable cot.

Note in figure 48.7 that the two wrists and left ankle will be used. As shown in figure 48.6, both wrists, both ankles, and the heart region will have electrodes attached if the Physiograph® is to be used. Proceed as follows to attach the electrodes to the subject:

1. To achieve good electrode contact on the wrists and ankles, rub the contact areas with a Scotchbrite pad, disinfect the skin with 70% alcohol, and add a little electrode paste to the contact surface of each electrode.
2. Attach each of the plate electrodes with a rubber strap, as shown in figure 48.8.
3. If you are using the five-electrode setup, affix the suction-type electrode over the heart region, using an ample amount of electrode paste on the contact surface.
4. Attach the leads of the patient cable to the electrodes as follows:
   Unigraph Setup: Attach the black unnumbered lead to the left ankle, one of the numbered leads to the right wrist, and the other numbered lead to the left wrist.
   Physiograph® Setup: Attach each lead according to its label designation: LL—left leg,

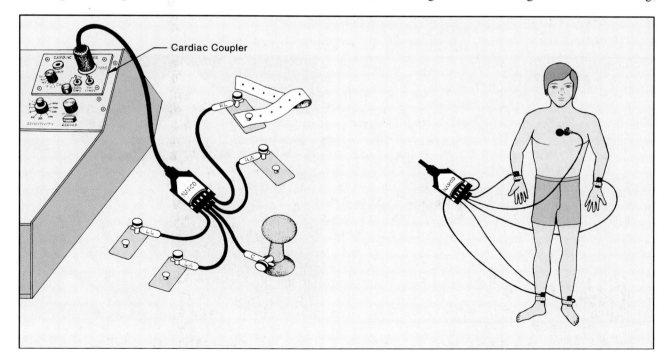

Figure 48.6 Physiograph® setup for ECG monitoring.

RL—right leg, LA—left arm, RA—right arm, and C—cardiac.

5. Make certain that the leads are securely connected to the electrode binding posts by firmly tightening the knobs on the binding posts. If the patient cable has alligator clamps instead of straight jacks, take care to see that the clamps can't slip off easily.

6. If the subject is sitting in a chair instead of lying down, *make sure that the chair is not metallic.*

7. Connect the free end of the patient cable to the appropriate socket on the Unigraph or Physiograph® cardiac coupler.

8. The subject is now ready for monitoring.

Monitoring

Tell the subject to relax and be still. Turn on the recorder and observe the trace. Follow the steps outlined below for the type of recorder you are using.

Unigraph   Adjust the stylus heat control so that optimum trace is achieved.

1. Compare the ECG with figures 48.1 and 48.2 to see if the QRS spike is up or down (positive or negative). If it is negative (downward), reverse the leads to the two wrists.

2. After recording for about 30 seconds, stop the paper and study the ECG. Does the wave form appear normal? Determine the pulse rate by counting the spikes between two margin lines and multiplying by 20.

3. Disconnect the leads to the electrodes and have the subject exercise for 2–5 minutes by leaving the room to run up and down stairs or jog outside the building.

4. After reconnecting the electrodes, record for another 30 seconds and compare this ECG with the previous one. Save the record for attachment to the Laboratory Report.

Physiograph®   Place the lead selector control at the LEAD I position. With the paper moving, lower the pens onto the recording paper. Make certain that the timer is activated.

1. Place the RECORD button in the ON position and record one sheet of ECG activity.

2. If the pen goes off scale, simply press the TRACE RESET switch on the cardiac coupler and the pen will return to the center line.

3. Place the RECORD button in the OFF position, turn the lead selector control to the LEAD II position.

4. Place the RECORD button back to the ON position and record another sheet of ECG activity.

5. Repeat step 4 above for the LEAD III position on the lead selector control.

6. Stop the recording and compare your results with the sample tracing in figure 48.9.

Laboratory Report

Complete the last portion of Laboratory Report 47, 48.

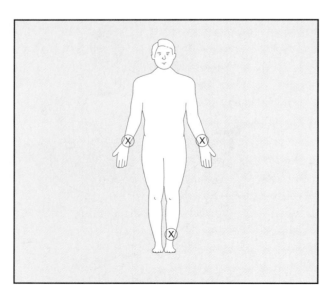

Figure 48.7 Three-lead hookup used for Unigraph setup.

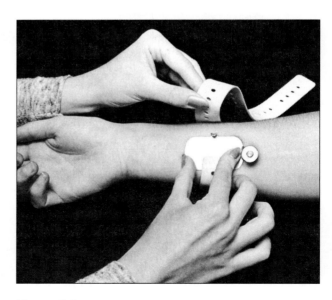

Figure 48.8 After applying electrode paste to electrode, strap it in place.

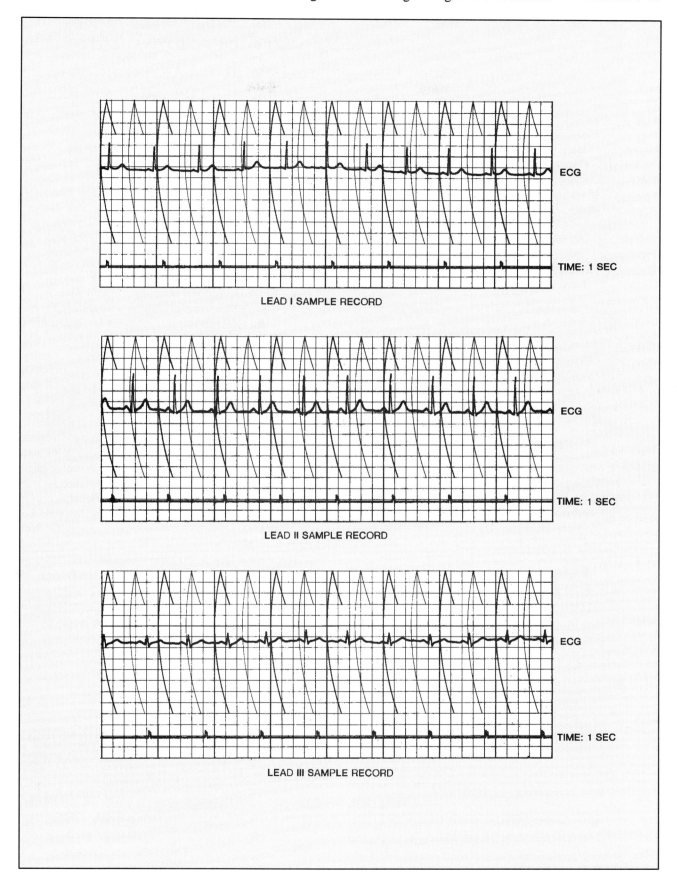

LEAD I SAMPLE RECORD

LEAD II SAMPLE RECORD

LEAD III SAMPLE RECORD

Figure 48.9 Physiograph® electrocardiogram sample recordings at different lead settings.

49 Electrocardiogram Monitoring: Using the Intelitool Cardiocomp™

In this laboratory period we will use the Intelitool Cardiocomp with a computer to produce electrocardiograms of members of the class. Figure 49.1 illustrates the setup. Before proceeding with this experiment, however, it will be necessary for you to read pages 278 and 279 in the last exercise so that you understand the characteristics of the ECG wave form; Einthoven's law; and leads I, II, and III.

Augmented Unipolar Leads

It was pointed out in the last exercise that there are three bipolar leads (I, II, and III), which are based on Einthoven's triangle. Figure 49.2 illustrates the relationship of these leads to the triangle. Arrows in the diagram represent the potential gradient direction for a positive deflection on the screen.

In addition to these three bipolar leads are three augmented unipolar leads. An **augmented unipolar lead** is *one in which the electrical potential is between one of the positively charged limbs and the average of the other two limbs that are negative.*

Figure 49.3 illustrates how these three leads are superimposed on the triangle with bipolar Leads I, II, and III. Note that the arrowhead of each lead points to the positive (+) electrode of an appendage and that the line is perpendicular to one side of Einthoven's triangle, bisecting that side. The three augmented leads are as follows:

aVR: Note in figure 49.3 that this lead bisects the side of the triangle that goes from the left arm (LA) to the left leg (LL). It is directed toward the electrode of the right arm (RA).

aVL: This unipolar augmented lead bisects the side of the triangle that goes from the right arm (RA) to the left leg (LL). It is directed toward the positive electrode of the left arm (LA).

aVF: This lead is formed by a line perpendicular to the side of the triangle that extends from RA to LA and is directed downward to LL.

Figure 49.4 is an attempt to clarify the directional nature of the six leads. Note that by using six leads, the direction of the electrical potential through the heart can be monitored, radially, in 30 degree segments. By scanning the six leads, a cardiologist can determine the direction of the potential *by simply noting in which lead the R wave has the strongest deflection.* If the strongest deflection

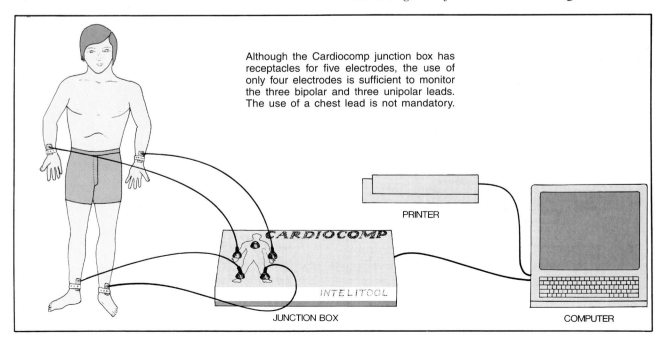

Although the Cardiocomp junction box has receptacles for five electrodes, the use of only four electrodes is sufficient to monitor the three bipolar and three unipolar leads. The use of a chest lead is not mandatory.

PRINTER

CARDIOCOMP
INTELITOOL
JUNCTION BOX

COMPUTER

Figure 49.1 Cardiocomp setup for ECG monitoring.

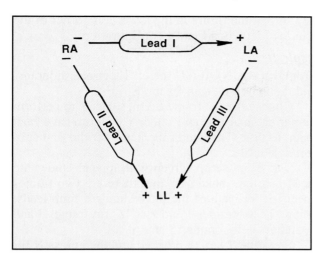

Figure 49.2 Bipolar lead orientation according to Einthoven's triangle.

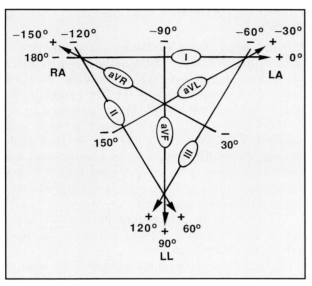

Figure 49.3 Diagram illustrating the relationship of bipolar to unipolar augmented leads.

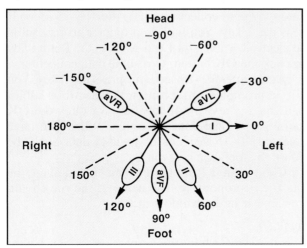

Figure 49.4 Directional orientation of the six most commonly used leads in ECG monitoring.

happens to be in lead II, as is usually the case, one knows that the potential is in the direction from the right shoulder to the left foot, or about 60°. Another reason for using six leads instead of only three bipolar leads is that there is less chance of missing some unusual cardiac event.

Chest Electrodes

Note in figure 49.1 that only four electrodes are used in our experiment and that the Cardiocomp has one receptacle for a chest electrode. In clinical practice most cardiologists rely on six standardized chest electrodes in addition to the ones studied here. While chest electrodes are invaluable to the cardiologist, they reveal nothing about the wave form or arrhythmias that cannot be determined with the bipolar and unipolar augmented leads; thus, for simplicity in our situation, we will not use a chest electrode.

Objectives of This Experiment

In this experiment there are four things that we will attempt to accomplish. They are as follows:

- Acquire data from each of the three standard bipolar leads (I, II, and III) and identify the various wave forms (P, Q, R, S, and T).
- Measure the PR, QRS, and QT intervals and determine the ventricular rate.
- Determine the approximate angle of the R-wave axis by visual inspection of the signal amplitude in each of the three bipolar leads.
- Utilize Einthoven's triangle in order to more accurately determine the R-wave axis.

Procedure

Working together in groups of two or more students, proceed as follows to do a series of ECG measurements on all members of the class, using the Cardiocomp. While one student is being monitored, another student will be wired up for testing.

Cardiocomp can be run on three computer platforms: (1) IBM or compatible PC (MS-DOS 3.3 or higher), (2) Macintosh, and (3) Apple II (ProDOS). For whichever platform you are using there will be separate instructions. It is assumed here that you have a basic knowledge of the platform you are using. Note also that in the materials list two Cardiocomp manuals are available for reference. Proceed as follows:

Materials:
  computer and printer
  Cardiocomp junction box
  Cardiocomp program diskette
  blank initialized diskette
  4 electrode wires and 4 flat-plate electrodes
  Scotchbrite pad, alcohol swabs, electrode gel

protractor (optional)
Cardiocomp User Manual
Cardiocomp Lab Manual

1. Get the Subject Ready.

Most of the problems that you might encounter can be avoided with a good electrode application procedure. With that in mind, follow these steps:

- Lightly rub the contact areas with a Scotchbrite pad, disinfect the skin with an alcohol swab, and add an ample amount of electrode gel to the flat contact surface of each electrode.
- Attach each of the plate electrodes to the limbs with the rubber straps. Be sure that they are snugly attached but not too tight.
     Note in figure 49.1 that both wrists and both ankles have electrodes attached to them. *The right ankle is the ground.*
- Snap the wires to the electrodes. Be sure to match the labels to the limbs.
- Plug the electrode wires into the proper receptacles of the Cardiocomp junction box.
- Have the subject lie down on a table, face up. Movement must be minimal. **It is very important that you arrange the wires parallel to the body, separate from each other and away from any power cables to avoid "noise."**

2. Set the Acquisition Parameters.

The Cardiocomp software must be told which leads you wish to monitor and for how long. For this procedure we will acquire two frames of data for each of the three bipolar leads (six frames total). Specify these parameters as follows:

MS-DOS:

Select "Experiment Menu" from the main CARDIOCOMP MENU and then select "Quick Exam Mode."

Press "6" to specify six frames for the exam, and then press Y to indicate that you wish to save each exam to disk. Indicate whether or not you wish to print a copy of each exam and then accept your choices by pressing Y.

The "Quick Exam Setup" screen will appear. Under the Lead column, make sure that there are two frames specified for each of the three standard limb leads: i.e., a "1" for frames 1 and 2, a "2" for frames 3 and 4, and a "3" for frames 5 and 6. You are now ready to proceed to step 3. Press the S key to proceed.

Macintosh:

If a new General ECG window is not already visible choose **New** from the **File** menu, and then click the **General ECG** button. Set the acquisition parameters such that **AutoStop** is enabled (checked), duration is 60 seconds, and acquisition rate is 240 Hz. Then press the 4 key and the return

key so that the figure in the lower left corner of the window depicts a Lead I configuration.

Apple II:

From the MAIN MENU select "Electrocardiogram" and then choose "Run Exam 'Strip'."

Choose the six-frame exam, select Y to perform numerical analysis, and select Y again to save each exam to disk. Then enter the name for the first subject's data file.

The exam setup screen will appear. Under the Lead column, make sure that there are two frames specified for each of the three standard limb leads: i.e., a "1" for frames 1 and 2, a "2" for frames 3 and 4, and a "3" for frames 5 and 6.

Press the S key to proceed and then press N for Normal Speed acquisition.

3. Collect ECG Data.

MS-DOS:

With the subject lying motionless, press any key to begin collecting data. The computer will collect two frames of data for each of the three standard leads. (Cardiocomp I users will be prompted to reposition the electrode cables for each lead.)

After collecting the data, you will be prompted to save the file. You can either accept the default name or enter a different name, such as the subject's initials. (You may wish to save each subject's data to a separate diskette if analysis is to be performed later at a different computer.)

After saving the data, the computer will, if specified in step 2, print a copy of the exam data and perform an interval analysis.

Repeat these procedures in step 3 to acquire ECG data for each subject.

Macintosh:

With the subject lying motionless, click the **Start** button to begin collecting data. After 10–15 seconds press the 2 key. You will be prompted to reposition the electrode cables for Lead 2. Press the Return key (or click the **OK** button) to resume data collection.

Again after 10–15 seconds, press the 3 key. You will be prompted to reposition the electrode cables for Lead 3. Press the Return key (or click the **OK** button) to resume data collection. When you have collected 10–15 seconds of Lead 3 data click the mouse to stop acquiring data.

Choose **Save** from the **File** menu and name the data file appropriately. The name of the file should now appear in the window title.

Apple II:

With the subject lying motionless, press any key to begin collecting data. The computer will collect two frames of data for each of the three standard leads.

(Cardiocomp I users will be prompted to reposition the electrode cables for each lead.)

After collecting the data, a printout and, if specified in step 2, an interval analysis will automatically be performed.

You will then be prompted for the file name under which to save the next subject's data. (You may wish to save each subject's data to a separate diskette if analysis is to be performed later at a different computer.)

Repeat these procedures in step 3 to acquire ECG data for each subject.

4. Identify ECG Wave Forms.

MS-DOS:

If your data file is not already loaded into memory, do so by selecting "Data File/Disk Commands" from the main CARDIOCOMP MENU.

Then choose "OPEN" File and enter the full path and name of your data file. The name of your file should now appear at the bottom of the screen.

Next, select "Review/Analysis Menu" and then choose "Review/Time Analysis." Press the S key to stop the trace at frame 3 or 4, which contains Lead 2 data.

By referring to figure 49.5, identify the **P wave, QRS complex,** and **T wave.** The U wave may not be visible. It is due to gradual repolarization of the papillary muscles. Advance through the data frames one at a time and note any differences in the appearance of the various waves for the different leads.

Macintosh:

If your data file is not already loaded into memory, do so by selecting **Open** from the **File** menu, and then select the name of the file you wish to open. Otherwise select **Time/Voltage** from the **Analysis** menu to convert the data acquisition window into an analysis window.

By referring to figure 49.5, identify the **P wave, QRS complex,** and **T wave.** The U wave may not be visible. It is due to gradual repolarization of the papillary muscles. Scroll through the data and note any differences in the appearances of the various waves for the different leads.

Apple II:

If your data file is not already loaded into memory, do so by selecting "Load Old Data File" from the MAIN MENU and then enter the name of the file you wish to load. (On single disk drive systems you must insert the data disk into drive 1 before you press the return key.) The name of your file should now appear at the bottom of the screen.

Next select "Review/Analysis Menu" and then choose "Review/Time Analysis." Press the S key to stop the trace.

By referring to figure 49.5, identify the **P wave, QRS complex,** and **T wave.** The U wave may not be visible. It is due to gradual repolarization of the papillary muscles.

Advance through the data frames one at a time and note any differences in the appearance of the various waves for the different leads.

5. Measure the ECG Intervals

MS-DOS:

Using the S, Z, and arrow keys to position the data markers, measure the PR, QRS, and QT intervals. Compare your values with the normative values shown in figure 49.6.

Macintosh:

Position the data markers by using the arrow keys or by clicking and dragging them with the mouse. Measure the PR, QRS, and QT intervals. Compare your values with the normative values shown in figure 49.6.

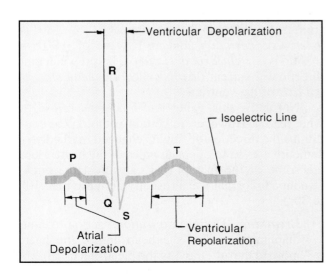

Figure 49.5 The ECG wave form.

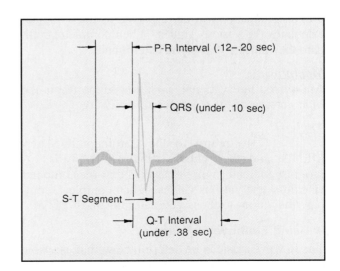

Figure 49.6 ECG intervals.

Apple II:

Using the D key and the cursor keys to position the data markers, measure the PR, QRS, and QT intervals. Compare your values with the normative values shown in figure 49.6.

6. Measure the Ventricular Rate.

Ventricular rate (heart rate) can be determined by measuring the elapsed time for one or more consecutive cardiac cycles. Divide the number of cycles by the elapsed time in seconds and then multiply by 60 to get a value in beats per minute.

Moving the data markers as done in step 5, position one of them on the peak of an R wave and position the other on the peak of the second or third R wave from the first. Record this time interval and then use the procedure described in the preceding paragraph to calculate the ventricular rate.

Ventricular rate _____

7. Approximate the R-Wave Axis.

By quickly scanning the three leads, you should see that in one of the leads the R wave has the greatest magnitude (deflection). Look at figure 49.7 and notice the angle of that particular lead.

The angle of the lead in which the R-wave deflection is the greatest is close to the angle of the R wave: i.e., each lead (electrode orientation) can only detect current that flows parallel to it. If the magnitude of the R-wave deflection is about the same in two leads, then the R-wave axis is between the angles described by those two leads. It is in these situations where using the augmented unipolar leads comes into play.

8. Perform Interval Analysis.

MS-DOS:
Press ESC to return to the main Review/Analysis Menu, and then choose "Interval Analysis." A progress indicator will be displayed as the computer calculates the various values. When complete, compare the values to those you determined.

Macintosh:
Automated interval analysis is not supported in the Mac version of the Cardiocomp software.

Apple II:
Press ESC to return to the Cardiocomp MAIN MENU, and then choose "Interval Analysis." A progress indicator will be displayed as the computer calculates the various values. When complete, compare the values to those you determined.

9. Using Einthoven's Triangle.

The R-wave axis can be determined with reasonable accuracy using Einthoven's triangle in figure 49.8. The procedure is as follows:

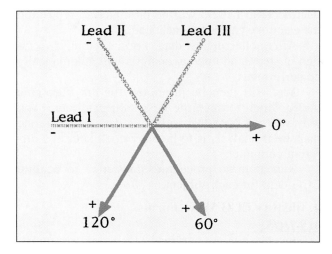

Figure 49.7 Axes of the three standard (bipolar) leads. Compare this illustration with figure 49.4.

First measure the voltage of the R wave for the two leads that have the largest R-wave deflection (i.e., the highest voltage).

Plot these two points on the appropriate side of the triangle representing the corresponding lead. (The numbers on each side of the triangle represent voltage in millivolts.)

Next, from each of these two points draw a line that is perpendicular to that side of the triangle that extends across the circle. Use a protractor for accuracy.

Finally, draw a line from the center of the circle through the intersection of the two lines that you just drew.

The point at which this last line intersects the circle indicates the angle of the R-wave axis.

Compare your result with the value calculated by the computer previously.

10. Causes of Deviations.

The typical range for a "normal" R-wave axis is from 0° to +90°. An R-wave axis of 0° to −180° is a *left axis deviation,* and an R-wave of +90° to +180° is a *right axis deviation.* The preponderant direction of current flow is called the *mean electrical axis* of the ventricle.

Normally, this value is 59 degrees (close to what lead?). However, in certain pathological conditions the direction of flow is changed markedly—sometimes even to opposite poles of the heart. Here are some of the factors that can affect the angle in nonpathological ways (usually no more than 10° to 30°):

• *Breathing.* During *expiration,* action of the diaphragm against the heart forces the heart more to the left, moving the mean electrical axis to the left. During *inspiration* the axis will move to the right.

- **Change in body position.** When a person lies down, pressure of the abdominal organs against the diaphragm forces the heart more to the left, causing the axis to shift to the left. The reaction here is much more pronounced in short, fat, and stocky individuals. Rising from a recumbent position can also cause a shift to the right.

Pathological conditions that manifest themselves in axis deviation are as follows:

- **Hypertrophy of left ventricle.** If the left ventricle is enlarged, there is usually a shift of the mean axis to the left (of as much as 20° or more). This type of hypertrophy is characteristic in *hypertension.*
- **Hypertrophy of right ventricle.** If the right ventricle is enlarged the axis deviation will be to the right. This is usually due to pulmonary valve stenosis.

- **Congenital heart diseases.** Tetralogy of Fallot and interventricular septal defects can cause right axis deviation.
- **Bundle branch block.** If one of the major bundle branches in the heart muscle is blocked, depolarization of the two ventricles does not occur simultaneously. This can cause axis deviation to the right or left depending on which bundle branch is involved.
- **Heart muscle destruction.** When heart muscle is damaged due to blocked coronary vessels, fibrous tissue replaces the functioning cardiac tissue. This replacement of viable cardiac tissue will greatly affect the axis.

Laboratory Report

Since there is no Laboratory Report for this exercise, your instructor will indicate what form of report is required.

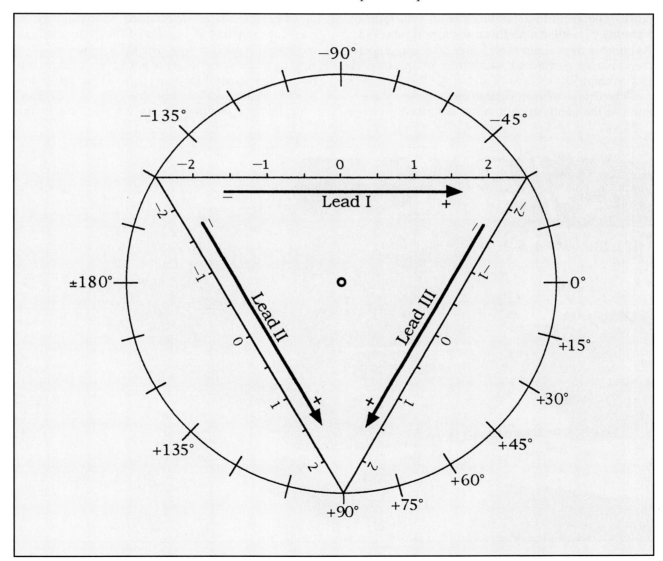

Figure 49.8   Template for utilizing Einthoven's Triangle.

50

Pulse Monitoring

When the ventricles of the heart undergo systole, a surge of blood flows into the arterial tree, manifesting itself as the **pulse** in the extremities. Since the rate and strength of the pulse indicate cardiovascular function, physicians always monitor it in routine medical examinations. This physiological parameter is usually determined simply by placing the fingers over one of the patient's arteries in the wrist or neck region.

In this exercise we will utilize a photoelectric plethysmograph on the index finger to record the pulse. The recording produced with this type of transducer is called a **plethysmogram** (figure 50.2). As indicated in figures 50.1 and 50.3 this experiment can be performed with either a Unigraph or Physiograph®.

The details of construction and manner of operation of the plethysmograph are described on page 73, Exercise 16. It operates on the principle of a light beam that is projected into the soft tissues of the finger. When the beam is altered by a surge of blood passing through the finger, a photoresistor is activated by the scattered light to produce a signal that can be recorded on an oscilloscope or chart paper.

In addition to determining the pulse rate, this type of transducer has merit in detecting indirect and relative blood pressure differences, but not precise blood pressures. The reason that precise blood pressures cannot be determined with this type of transducer is that it is very difficult to calibrate recorders in such a way that the volume of blood surging through the finger at any given moment represents an exact blood pressure. This limitation, however, does not prevent this type of transducer from being a useful tool. As we shall see in this

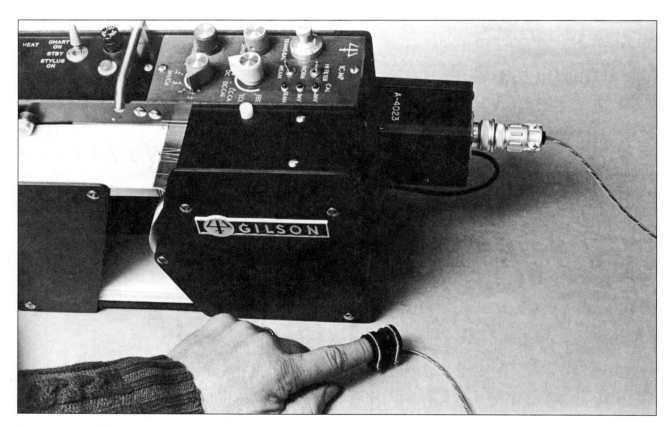

Figure 50.1   Pulse monitoring with Unigraph.

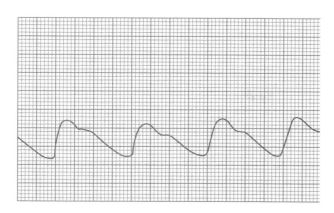

Figure 50.2   A plethysmogram (Unigraph tracing).

experiment, much can be learned about the mechanics of blood pressure by using it.

In this experiment we hope to accomplish the following: (1) determine the exact pulse rate, (2) explore some of the factors that affect the pulse rate, (3) identify the dicrotic notch and its significance, and (4) determine the range of pulse rates among class members. Proceed as follows:

Materials:
for Unigraph setup:
    Unigraph
    A4023 adapter (Gilson)
    Photoresistor pulse pickup (Gilson T4020)

for Physiograph® setup:
    Physiograph® with transducer coupler
    pulse transducer (Narco 705–0050)

Preliminaries

Prepare the subject and make the preliminary adjustments on the recorder according to the following suggestions.

Subject Preparation

1. Position the subject on a laboratory tabletop, face upward. Keep both arms parallel to the body.
2. Attach the pulse transducer to the index finger of one hand. Be sure that the light source faces the pad of the fingertip. This hand should be near the edge of the table so that it can be lowered toward the floor or extended upward toward the ceiling.
3. Make certain that the subject is comfortable and relaxed. The laboratory must be free of distracting stimuli.

Unigraph Setup

1. Attach the A4023 adapter to the receiving end of the Unigraph. Make certain that the locknut is tightened securely and that the phone jack is also plugged in.

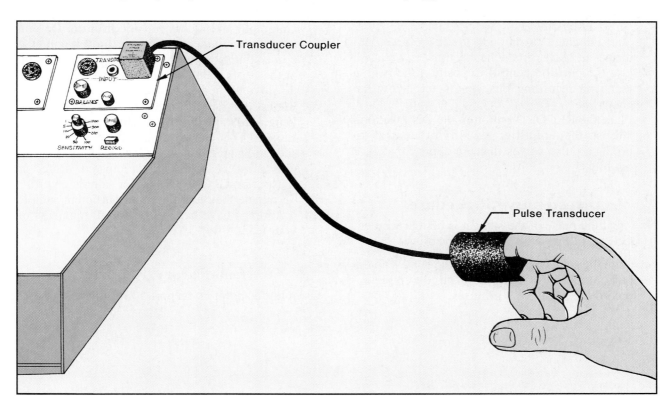

Figure 50.3   Pulse monitoring setup with Physiograph®.

2. Secure the jack of the pulse pickup to the adapter. Plug in the power cord.

3. Set the controls as follows: speed control lever at slow position, stylus heat control at two o'clock position, Gain control at 2 MV/CM, sensitivity control completely counterclockwise, and Mode selector on DC.

4. Turn on the power switch and set the c.c. switch at Stylus On. Wait about 15 seconds for the stylus to warm up and then place the c.c. switch at Chart On.

5. Observe the trace and adjust the centering control to position the trace near the center of the paper.

6. Increase the sensitivity by turning the sensitivity control clockwise until you get approximately 2 cm pen deflection. If this amount of deflection cannot be achieved, turn the sensitivity control down again and increase the gain to 1 MV/CM.

   Now increase the sensitivity with the sensitivity control again to get the desired deflection.

7. Proceed to the instructions under "Investigative Procedure."

Physiograph® Setup

1. Plug the end of the transducer cable into the proper socket on the transducer coupler.

2. Turn on the power switch and set the paper speed at 0.5 cm/sec.

3. Start the chart moving and lower the pens onto the paper. Make certain that the timer is activated.

4. Use the position control to position the pen so that it is recording one centimeter below the centerline.

5. Place the RECORD button in the ON position.

6. Adjust the amplifier sensitivity control to produce pulse waves that are approximately 2 cm high.

Investigative Procedure

Now that the pulse of the subject is being recorded, proceed as follows with your observations:

1. With the subject lying still and the hand with the pulse transducer lying flat on the tabletop, record the pulse for **2 minutes.**

2. Stop the recording and calculate the pulse rate. (On the Unigraph chart, the space between two margin marks represents 30 seconds; thus, it is necessary to count the spikes and fractions thereof between two marks and multiply by two.)

3. Refer to illustration A, figure SR-2, appendix C, for a Physiograph® sample record.

4. Resume recording, and have the subject slowly raise his or her arm to a fully vertical position above the body, hold that position for **15 seconds,** and then slowly return the arm to its original position on the table. Allow the pulse to stabilize. Refer to illustration B, figure SR-2, for sample Physiograph® record.

5. Now, have the subject slowly lower his or her arm over the edge of the table until it hangs down vertically. Hold this position for **15 seconds,** and then slowly return the arm to its former resting position. Allow the pulse to stabilize. Refer to illustration C, figure SR-2 for sample record.

6. Have the subject take a deep breath and hold it for **15 seconds,** before exhaling and resuming normal breathing. Refer to illustration A, figure SR-3, for sample record.

7. Locate the brachial artery below the biceps muscle. Occlude this vessel for **15–20 seconds,** and then release the pressure. Allow the pulse to stabilize. Refer to illustration B, figure SR-3, for sample record.

8. Increase the speed to 2.5 cm/sec (fast speed on Unigraph) and record for **one minute.** Do you see a pronounced **dicrotic notch** on the descending slope of the curve? This interruption of the curve is due to sudden closure of the aortic valve of the heart during diastole.

9. Return the speed to 0.25 cm/sec (slow on Unigraph), and mildly frighten or startle the subject. This can be done with an unexpected sudden handclap or other loud noise. The stimulus must be totally unanticipated. Refer to illustration C, figure SR-3, for sample record.

10. Terminate the recording after sufficient sample recordings have been made for all members of your team.

Laboratory Report

Complete Laboratory Report 50 for this exercise.

Blood Pressure Monitoring

51

The contractions of the ventricles of the heart exert a propelling force on the blood that manifests itself as blood pressure in vessels throughout the body. Since the heart acts as a pump that forces out spurts of blood at approximately seventy times a minute while at rest, the pressure during a short interval of time will fluctuate up and down.

The force on the walls of the blood vessels is greatest during contraction. This pressure during maximum contraction is called the **systolic pressure.** When the heart relaxes and fills up with blood in preparation for another contraction, the pressure falls to its lowest value. The pressure during this phase is called the **diastolic pressure.**

There are several different methods that one might use for determining blood pressure. Probably the most sensitive and precise way is to insert a hollow needle into a vessel, which allows the force of the blood pressure to act on a pressure transducer. The signal from such a device can be fed into an amplifier and recorder for monitoring. The impracticality of this method for us in this laboratory precludes its use here.

Two methods will be outlined here for student participation: (1) the conventional stethoscope–pressure cuff method used routinely by medical personnel, and (2) an electronic recording method utilizing a Physiograph® or Unigraph to record pressure changes via a pressure cuff. A third method, using an audio monitor, may be used for instructor demonstration. Your instructor will indicate which procedures will be followed.

A Demonstration

When one takes the blood pressure of a patient with a sphygmomanometer and stethoscope, he or she listens for the sound of blood rushing through a partially occluded brachial artery of the arm. The sounds, which are called **Korotkoff** (Korotkow) **sounds,** can readily be demonstrated with an audio monitor, utilizing a setup similar to the one shown in figure 51.1.

Materials:
    microphone
    audio monitor
    pressure cuff (sphygmomanometer)

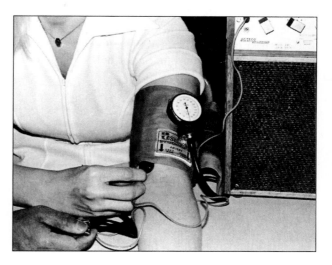

Figure 51.1 Auscultation of Korotkoff sounds with an audio monitor is performed by slipping a microphone under the cuff over the brachial artery.

To demonstrate these sounds, the instructor will wrap a pressure cuff around the upper arm of a subject, insert a microphone under the cuff over the brachial artery, and demonstrate the Korotkoff sounds that one listens for when systolic and diastolic pressures are determined.

To initiate the process, pressure will be built up in the cuff by pumping enough air into it to a level where all blood flow is shut off in the brachial artery.

While watching the pressure gauge, air will be released from the cuff *slowly* until the first sound of blood rushing through the occluded artery is heard. The pressure at which the sound is *first heard* is called the **systolic pressure.** As the sound slowly diminishes, it eventually disappears. The pressure at which the sound just disappears is the **diastolic pressure.**

By correlating the sounds heard through the audio monitor with the pressure gauge, a large number of students can, in a short period of time, quickly grasp the concept.

Stethoscope Method

Students, working in pairs, will take each other's blood pressure. To determine the broad range of normal blood pressures, individual systolic pressures will be recorded on the chalkboard. From this tabulation a median systolic pressure for the class will be determined. The effects of exercise on blood pressure will also be studied.

Materials:

stethoscope
sphygmomanometer

1. Wrap the cuff around the right upper arm. The method of attachment will depend on the type of equipment. Your instructor will show you the best way to use the specific type that is available in your laboratory. See figure 51.2.

    Make certain that the subject's forearm rests comfortably on the table.
2. Close the metering valve on the neck of the rubber bulb. *Don't twist it so tight that you won't be able to open it!*
3. Pump air into the sleeve by squeezing the bulb in your right hand. Watch the pressure gauge. Allow the pressure to rise to about 180 mm Hg and continue to hold the metering valve closed.
4. Position the bell of the stethoscope just below the cuff at a point that is *midway between the epicondyles of the humerus.* This is the lowest extremity of the brachial artery. About 2.5 cm distal to this point, the artery bifurcates to become the radial and ulnar arteries. See figure 51.3
5. Slowly release the valve so that the pressure goes down gradually. Listen carefully as you watch the pressure fall. When you just begin to hear the Korotkoff sounds, note the pressure on the gauge. **This is the systolic pressure.**

6. Continue listening as the pressure falls. Just when you are unable to hear the Korotkoff sounds anymore, note the pressure on the gauge. **This is the diastolic pressure.**
7. Record the systolic pressure on the chalkboard and the systolic/diastolic pressures on the Laboratory Report.
8. Repeat taking the pressure two or three times to see if you get consistent results.
9. Now, have the individual do some exercise, such as running up and down stairs a few times. Measure the blood pressure again and record your results on the Laboratory Report.

Using a Chart Recorder

A chart recording of blood pressure can be made with either a Unigraph or a Physiograph®. The Physiograph® setup is illustrated in figure 51.4. If a Unigraph is used, one needs a Statham pressure transducer similar to the one shown in figure 59.3. Procedures are provided here for using either type of recorder.

In this portion of the exercise we will (1) record systolic and diastolic pressures, (2) determine mean arterial and mean pulse pressures, (3) observe the effects of postural changes on arterial pressure, and (4) study the effects of physical exertion on arterial pressure.

Materials:

for Unigraph setup:
Unigraph
adult pressure cuff
Statham T-P231D pressure transducer
transducer stand and clamp for transducer

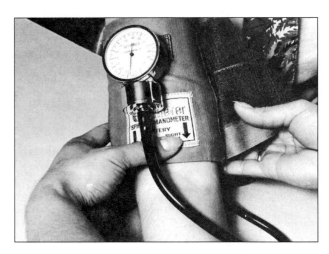

Figure 51.2 Sphygmomanometer cuff is wrapped around the upper arm, keeping lower margin of cuff above line through the epicondyles of the humerus.

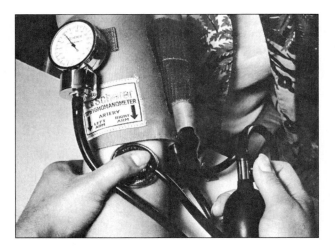

Figure 51.3 The inflated cuff shuts off the flow of blood in the brachial artery. Korotkoff sounds are listened for as pressure is gradually lowered.

for Physiograph® setup:
>Physiograph® with ESG coupler
>adult pressure cuff (Narco PN 712-0016)

Preliminary Preparations

As has been true for all previous experiments, separate instructions will be provided here for the Physiograph® and Unigraph. One difference between the two setups is that the Gilson setup utilizes a separate pressure transducer. The Physiograph® setup, on the other hand, doesn't need one because the ESG coupler has a pressure transducer built into it.

Unigraph Setup Refer to figure 59.3 to see how the Statham pressure transducer is mounted to a transducer stand with a transducer clamp. The transducer is seen in the background with a hose attached to it. If no support is available for this unit, it can be placed on its side on the table. Make sure, however, that it is in a spot where it can't roll off onto the floor.

1. Insert the jack of the pressure transducer cord into the receptacle of the Unigraph. No adapter is necessary.
2. Place the pressure cuff on the upper right arm, securing it tightly in place.
3. Attach the open end of the tube from the cuff to the fitting on the pressure transducer.

4. Test the cuff by pumping some air into it and noting if the pressure holds when the metering valve is closed on the bulb. Release the pressure for the comfort of the subject while other adjustments are being made.
5. Plug in the Unigraph and turn on the power. Set the speed control lever at the slow position and the stylus heat control knob in the two o'clock position.
6. Set the Gain at 1 MV/CM, and turn the sensitivity control completely counterclockwise to its lowest value. Set the Mode on TRANS.
7. **Calibrate the Statham transducer** by depressing the TRANS button and simultaneously adjusting the sensitivity knob to get 2 cm deflection. The calibrated line represents 100 mm Hg.
8. Proceed to the "Investigative Procedure" portion of this experiment.

Physiograph® Setup Refer to figure 51.4 to see how the various components of this setup are hooked up. Note that the pressure bulb tube is affixed to the "Bulb" fitting of the ESG coupler, the pressure cuff to the "Cuff" fitting on the coupler, and the microphone cable is inserted into the "Mic" receptacle of the ESG coupler. Before any tests can be run it will be necessary to balance and calibrate the Physiograph®.

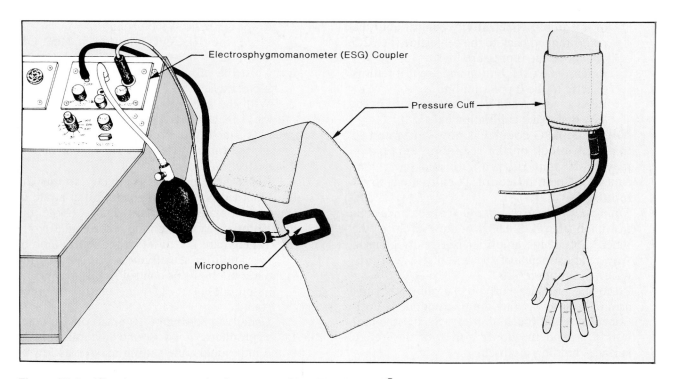

Figure 51.4 Blood pressure monitoring setup with a Physiograph®.

295

1. **Balance the ESG coupler** as follows:
   a. Start the chart moving at 0.25 cm/sec and lower the pens onto the paper.
   b. Keep the amplifier RECORD button in the OFF position.
   c. Orient the recording pen to the centerline by using the POSITION control knob.
   d. Start recording by pressing the RECORD button to the ON position (down).
   e. Using the balance control knob on the ESG coupler, return the recording pen to the centerline established in step *c* above.
   f. Check the balance by placing the amplifier RECORD button in the OFF position. If the system is balanced, the pen will remain on the centerline.
   g. If the pen does not remain on the centerline, repeat steps *e* and *f* until balancing is achieved.

2. **Calibrate the coupler** to produce 2.5 cm of pen deflection, which is equivalent to 100 mm Hg. Figure 51.5 is a sample recording of the procedure. Proceed as follows:
   a. Start recording by pressing the RECORD button to the ON position.
   b. With the POSITION control knob, set the recording pen on a baseline that is exactly 2.5 cm below the centerline; i.e., 2.5 cm above the 0 pressure line.
   c. While depressing the "100 mm Hg CAL" button on the coupler, rotate the inner knob of the amplifier sensitivity control until the pen is moved back to the centerline (i.e., 2.5 cm above the 0 pressure line).
   d. Release the CAL button and see if it returns exactly to the 0 pressure line.
   e. Repeat steps *c* and *d* several times to make sure that exact calibration exists.

   Note: The ESG coupler is now calibrated so that each block on the recording paper represents 20 mm Hg pressure and each centimeter of pen deflection is equivalent to 40 mm Hg.

3. *From this point on do not touch* the balance control knob, the POSITION control knob, or the inner knob of the amplifier sensitivity control. Any readjustments of these controls will produce inaccurate results.

4. Place the pressure cuff on the subject's right arm in such a way that the microphone is positioned over the brachial artery. Secure the cuff tightly around the upper arm with the Velcro surfaces holding it closed.

5. **Calibrate the cuff microphone sensitivity** as follows:

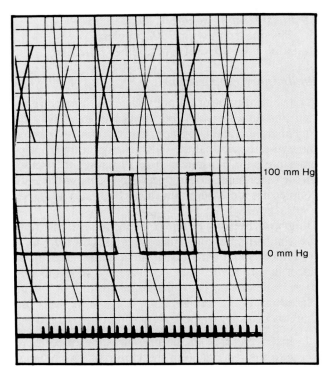

Figure 51.5 **Pressure Calibration:** Zero pressure line "0" is calibrated so that it is 2.5 cm below the centerline. Each block on the recording paper represents 20 mm Hg; thus, the centerline represents 100 mm Hg.

   a. Start the chart moving at 0.5 cm/sec and lower the pens to the paper.
   b. Place the RECORD button in the ON position.
   c. Close the valve on the hand bulb by rotating it clockwise.
   d. With the subject's forearm resting on the table in a relaxed position, pump up the cuff until the recording pen indicates approximately 160 mm Hg (4 cm of pen deflection).
   e. Release the bulb valve slightly so that the recorded pressure begins to fall at a rate of approximately 10 mm Hg/sec. Note the Korotkoff sounds on the recording.
   f. Taking care not to touch the inner knob of the amplifier sensitivity control, adjust the outer knob of this control so that you get an amplitude of 1–1.5 cm on the recorded sound.

   Usually, a setting of 10 or 20 on the outer knob produces good sound recordings. On some individuals, the setting may be as low as 5 or 2; on others it may be as high as 100.

6. You are now ready to perform your investigation.

Investigative Procedure

Four recordings will be made of the subject's blood pressure under the following conditions: (1) while seated in a chair, (2) after standing for 4 or 5 minutes, (3) while lying down, and (4) after exercising for 5 minutes. Use the following procedure:

With Subject in Chair

1. Start the paper advance at 0.5 cm/sec, and lower the pens onto the chart. Activate the timer.
2. Press the amplifier RECORD button to the ON position.
3. Tighten the pressure bulb valve (turn fully clockwise) and inflate the cuff until you are recording a pressure of approximately 40 mm Hg above the expected systolic pressure. Usually, 160–180 mm Hg is sufficient.
4. Release the hand bulb valve slightly so that the recorded pressure begins to fall at a rate of approximately 10 mm Hg/sec.
5. Watch for the appearance of the **first** and **last Korotkoff sounds** on the recording. These pressures can be considered to be the "normal" systolic and diastolic pressures of your subject.
6. Terminate the recording and refer to appendix C, illustration A, figure SR-4, for Physiograph® sample record.

With Subject Standing Up

Allow the subject to stand up for 4 or 5 minutes and repeat steps 1 through 6 above while the individual is still standing. Refer to illustration B, figure SR-4, for sample record.

With Subject Reclining

Allow the subject to lie on a laboratory table or cot for 4 or 5 minutes and then follow steps 1 through 6 again to record the blood pressure in this position. Refer to illustration C, figure SR-4 for sample record.

After Exercising

Remove the cuff from the subject and allow him or her to perform 5 minutes of strenuous exercise, such as running, jumping, or climbing stairs.

After exercising, seat the subject in a chair and *immediately* attach the cuff. Record the systolic and diastolic pressures, using the same procedures as above. Remember, it will be necessary to inflate the cuff to a higher pressure (240 mm Hg) to occlude the brachial artery. Refer to illustration A, figure SR-5, for Physiograph® sample record.

Make a series of recordings one minute apart until the pressure has returned to what it was in the original (seated) test. Refer to illustrations B and C, figure SR-5 for sample records.

Laboratory Report

Complete the first portion of Laboratory Report 51, 52.

Answers to Histology Self-Quiz No. 3

1. Basophil
2. Erythrocyte
3. Lymphocyte
4. Neutrophil
5. Lymphocyte
6. Monocyte
7. Immature
8. Esophagus
9. Serosa
10. Smooth muscle
11. Epithelium
12. Mucus
13. Secretory duct
14. Tooth bud
15. Dental papilla
16. Oral epithelium
17. Developing tooth
18. Oral epithelium
19. Enamel
20. Dentin
21. Dental pulp
22. Tongue
23. Taste buds
24. Trench

25. Stomach wall (fundus)
26. Parietal cells
27. Produce precursor of HCl
28. Chief cells
29. Produce precursor of pepsin
30. Odontoblasts
31. Predentin
32. Predentin
33. Tomes' enamel process
34. Ameloblasts
35. Enamel of tooth
36. Vallate papilla
37. Posterior dorsum of tongue
38. Jejunum
39. Muscularis
40. Mucosa
41. Intestinal glands
42. Secrete digestive enzymes
43. Villus
44. Lacteal
45. Absorb fats
46. Liver
47. Hepatic sinusoids

48. Kupffer cells
49. Phagocytosis
50. Stomach (fundus)
51. Gastric pit
52. Plain columnar epithelium
53. Pancreas
54. Pancreatic digestive enzymes
55. Duodenum
56. Duodenal glands
57. Secrete mucus
58. Colon
59. Lamina propria
60. Intestinal gland
61. Palatine tonsil
62. Lymphoid nodule
63. Tonsillar crypt
64. Gallbladder
65. Smooth muscle
66. Plain columnar
67. Lung
68. Alveolus
69. Blood vessel
70. Bronchiole

Answers to Histology Self-Quiz No. 4

1. Cervical crypts
2. Mucus
3. Cyst
4. Crypt was sealed off
5. Vaginal epithelium
6. Prostate gland
7. Prostatic concretions
8. Prostatic secretion
9. Thymus
10. Cortex
11. T-lymphocytes
12. Chief cells
13. Parathyroid hormone
14. Oxyphil cells
15. Unknown
16. Seminal vesicle
17. Two
18. Energy for sperm and coagulation
19. Kidney
20. Collecting tubule
21. Papilla
22. Calyx

23. Vagina
24. Stratified squamous
25. Ureter
26. Transitional
27. Lamina propria
28. Uterus
29. Uterine glands
30. Vas deferens
31. Spermatic artery
32. Glomerulus
33. Glomerular capsule
34. Distal convoluted tubule
35. Cortex
36. Spermatozoa
37. Spermatid
38. Haploid
39. Spermatogonial cell
40. Diploid
41. Seminiferous tubule
42. Interstitial cells
43. Produce testosterone
44. Thymus gland
45. No

46. Thymic corpuscle (Hassall's body)
47. Adrenal
48. Capsule
49. Zona glomerulosa
50. Zona fasciculata
51. Zona reticularis
52. Sinusoids
53. Aldosterone
54. Cortisol
55. Pineal gland
56. Pineal sand
57. Thyroid
58. Follicles
59. Thyroid
60. Adenohypophysis
61. Basophil
62. Acidophil
63. Chromophobe
64. Graafian follicle
65. Oocyte (ovum)
66. Estrone (estrogen)

Peripheral Circulation Control (Frog)

52

The arterial and venous divisions of the circulatory system are united by an intricate network of capillaries in the tissues. **Capillaries** are short thin-walled vessels of approximately 9 micrometers in diameter. They are so numerous that all body cells are no more than two or three cells away from one of these vessels.

The combined diameter of all the capillaries causes the blood to slow down as it passes through them. This slow flow of blood is of great importance because it allows sufficient time for the exchange of materials between the blood and tissue cells. The pathway of exchange is as follows:

$$\text{capillary blood} \rightleftharpoons \text{tissue fluid} \rightleftharpoons \text{cells}$$

Whether or not blood enters a capillary is determined by the size of its preceding arteriole and the action of its precapillary sphincter muscle. The arteriole has the capacity of vasodilation and vasoconstriction to affect the amount of blood flow through it. The **precapillary sphincter** that guards the entrance to the capillary can also restrict or permit blood flow. Both the arterioles and sphincters are under neural control of the vasomotor center of the medulla oblongata through the autonomic nervous system. Although the vasomotor center is influenced by many factors, carbon dioxide concentration of the blood is of major importance.

Locally, carbon dioxide, histamine, and epinephrine may influence the volume of blood flow. High CO_2 concentration in certain tissues will result in increased blood flow in these tissues and a decrease of blood to other tissues that lack the high CO_2 concentration. Increased epinephrine levels in the blood cause vasodilation in some areas (muscles, heart, lungs) and vasoconstriction in other areas. Histamine, a hormone that is present in large quantities in allergic reactions, causes extensive vasodilation in peripheral circulation. Such reactions may cause a precipitous drop in blood pressure, and death.

In this exercise we will study some of these factors that influence capillary circulation. A frog will

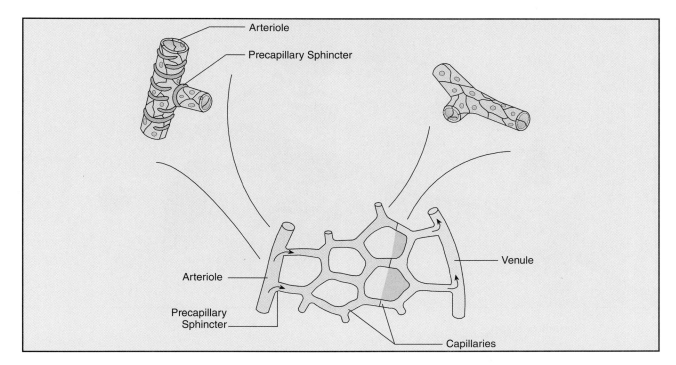

Figure 52.1 Capillary circulation.

be strapped to a board so that the capillaries in the webbing between its toes can be studied under a microscope.

Materials:

small frog, frog board, 4″ × 12″ cloth straps
string, pins, rubber bands
dropping bottles of epinephrine (1:1000)
dropping bottles of histamine (1:10,000)

1. Obtain a frog from the stock table and strap it to a frog board as illustrated in figure 52.2. Note that the webbing of one foot is positioned over the hole in the board.
2. Pin the webbing of the foot in such a way that it spreads over the viewing hole. Keep the webbing moistened with water.
3. If necessary, secure the board to the microscope stage with a large rubber band.
4. Position the foot over the light source and focus on it with the 10× objective.
5. Observe the blood flowing through the vessels. Locate a pulsating **arteriole,** with its rapid blood flow, and a **venule** with its slower steady flow of blood.
6. Identify a **capillary,** with its smaller diameter. Note that blood cells move through it slowly and in single file.
7. Study a capillary closely to see if you can locate the juncture of the capillary and arteriole, which is the site of the **precapillary sphincter.** You probably won't be able to see the sphincter at this low magnification, but you can observe the irregular flow of blood cells into the capillary, which results partly from the sphincter muscle's action.
8. Now remove most of the water from the frog's foot by blotting it with paper toweling and add several drops of 1:10,000 **histamine** solution. Observe the change in blood flow and record your observations on the Laboratory Report.
9. Wash the foot with water, blot it dry, and then add several drops of 1:1000 epinephrine solution. Observe any change in rate of blood flow and record your results on the Laboratory Report.
10. Wash the foot again, remove the frog from the board, and return it to the stock table. Clean the microscope stage, if necessary.

Laboratory Report

Complete the last portion of combined Laboratory Report 51, 52.

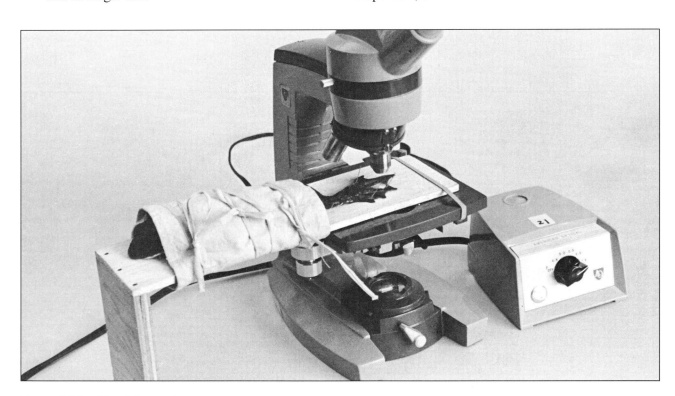

Figure 52.2 Blood flow setup.

The Arteries and Veins

53

The principal arteries and veins of the cardiovascular system will be studied in this exercise. A cat, in which the arteries and veins have been injected with colored latex, will be used for dissection. Before doing the cat dissection, however, it is desirable that all human illustrations be labeled.

The Circulatory Plan

Figure 53.1 is an incomplete flow diagram of blood through the heart and major regions of the body. If you can complete this diagram by filling in the proper blood vessels and providing correct labels, it can be assumed that you understand the overall plan of the circulatory system as described below. The area where students have most difficulty with this diagram is the flow of blood to the intestines and liver.

Starting at the heart, the blood enters the right atrium (left side of illustration) from the **superior** and **inferior venae cavae,** which have collected blood from all parts of the body. This blood is dark because it is low in oxygen and high in carbon dioxide content.

From the right atrium the blood passes to the right ventricle. When the heart contracts, blood leaves the right ventricle through the **pulmonary trunk** and pulmonary arteries (one vessel on diagram) to the lungs, where it picks up oxygen and gives off carbon dioxide. The blood leaves the lungs by way of the **pulmonary veins** (one vessel on diagram) and returns to the left atrium of the heart. Blood in the pulmonary veins is brightly colored due to its high oxygen content. The pulmonary arteries, veins, and lung capillaries constitute the *pulmonary system.*

From the left atrium the blood passes to the left ventricle. When the heart contracts, blood leaves the left ventricle through the **aortic arch.** This blood, rich in oxygen, passes to all parts of the body. The aortic arch has branches (one vessel on diagram) that go to the head and arms.

Passing downward on the right side of the illustration, the aortic arch becomes the **descending aorta,** which has branches going to the liver, digestive organs, kidneys, pelvis, and legs. The branch that enters the liver is called the **hepatic artery.** The intestines are supplied by the **superior mesenteric**

artery, and the kidneys receive blood through the **renal arteries.** The pelvis and legs are supplied by several arteries, but only one is shown for simplicity.

The blood leaving most of the organs of the trunk (including the pelvis and legs) empties directly into a large collecting vein, the inferior vena cava. The **renal veins** (one vessel on diagram) pass from the kidneys, and the **hepatic vein** leads from the liver to empty directly into the inferior vena cava.

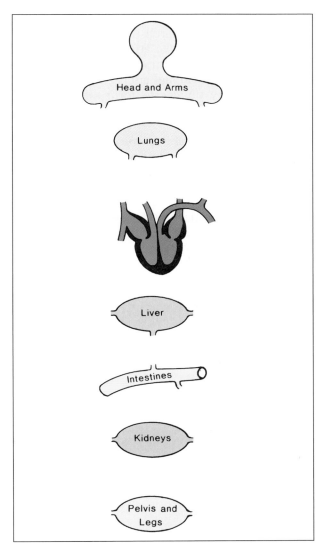

Figure 53.1 The circulatory plan.

Note that blood from the intestines does not go directly into the vena cava; instead, all blood from this region passes to the liver by way of the **portal vein.** This route of the blood from the intestines to the liver is called the *portal circulation.*

Assignment:
After drawing in all the necessary blood vessels, color the oxygenated vessels **red** and the deoxygenated ones **blue.** Provide arrows to show direction of blood flow, and label the vessels.

The Arterial Division

The principal arteries of the body are shown in figure 53.2. Starting at the heart, blood leaves it through the **ascending aorta,** which curves to the left forming the **aortic arch.** From the upper surface of the aortic arch emerge the left subclavian, left common carotid, and brachiocephalic arteries.

The **left subclavian** artery is that vessel that passes from the aortic arch into the left shoulder behind the clavicle. In the armpit *(axilla)* the subclavian becomes the **axillary** artery. The axillary in turn, becomes the **brachial** artery in the upper arm. This latter artery divides to form the **radial** and **ulnar** arteries that follow the radial and ulnar bones of the forearm.

The **brachiocephalic** *(innominate)* artery is the short vessel coming off the aortic arch on the right side of the body. It gives rise to two vessels: the right common carotid and the right subclavian. The branch that extends upward to the head is the **right common carotid** artery. It furnishes the right side of the head with blood. The **right subclavian** artery

is an outward extension of the brachiocephalic from the base of the right common carotid. It passes behind the right clavicle into the shoulder.

The third branch of the aortic arch is the **left common carotid.** It supplies the left side of the head with oxygenated blood.

As it passes down through the thorax, the aortic arch becomes the **descending** *(thoracic)* **aorta.** Below the diaphragm it is called the **abdominal aorta.** Although the aorta has many branches leading to various organs, only the larger ones are shown in figure 53.2.

The first branch of the abdominal aorta emerging just below the diaphragm is the **celiac trunk,** a very short artery, about 1.25 cm long. It has three branches: the *left gastric,* the *hepatic,* and the *splenic,* which supply blood to the stomach, liver, and spleen. None of these branches are shown in figure 53.2.

Just below the celiac trunk is the **superior mesenteric** artery, which supplies most of the small intestine and part of the large intestine. Inferior to the superior mesenteric are a pair of **renal** arteries that supply the kidneys, and emerging from the anterior surface of the aorta inferior to the renal arteries are two **gonadal** *(spermatic or ovarian)* arteries, which supply blood to the testes or ovaries.

Inferior to the gonadal arteries is seen a single artery, the **inferior mesenteric,** which supplies part of the large intestine and rectum with blood.

In the lumbar region the abdominal aorta divides into the right and left **common iliac** arteries. Each common iliac passes downward a short distance and then divides into a smaller inner branch,

---

Legend for Figure 53.2

_____ Abdominal Aorta	_____ Deep Femoral	_____ Posterior Tibial
_____ Anterior Tibial	_____ Descending (Thoracic) Aorta	_____ Radial
_____ Aortic Arch	_____ External Iliac	_____ Renal
_____ Ascending Aorta	_____ Femoral	_____ Right Common Carotid
_____ Axillary	_____ Inferior Mesenteric	_____ Right Gonadal
_____ Brachial	_____ Internal Iliac	_____ Right Subclavian
_____ Brachiocephalic	_____ Left Common Carotid	_____ Superior Mesenteric
_____ Celiac Trunk	_____ Left Subclavian	_____ Ulnar
_____ Common Iliac	_____ Popliteal	

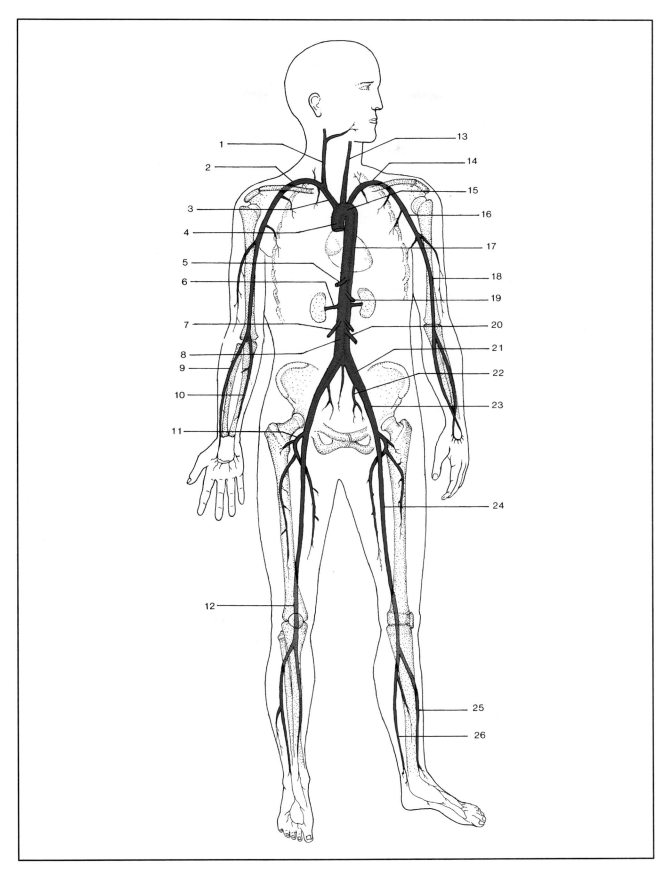

Figure 53.2 Major arteries of the body.

the **internal iliac** (*hypogastric*) artery and a larger branch, the **external iliac** artery, which continues on down into the leg.

The external iliac becomes the **femoral** artery in the upper three-fourths of the thigh. Near the origin of the femoral artery, the **deep femoral** branches off and passes backward and downward along the medial surface of the femur. In the knee region the femoral becomes the **popliteal** artery. Just below the knee the popliteal divides to form the **posterior tibial** and **anterior tibial** arteries.

Assignment:
Label the arteries in figure 53.2 by recording the numbers in the legend box on page 302.

The Venous Division

Figure 53.3 illustrates most of the larger veins of the body. Starting at the heart in figure 53.3, we see the **superior vena cava** emptying into the upper part of the right atrium and the **inferior vena cava** leading into the lower part of the right atrium.

In the neck region are seen four veins: two medial **internal jugular** veins and two lateral **external jugulars.** The internal jugulars empty into the brachiocephalic veins, and the smaller externals empty into the **subclavian** veins, which pass behind the clavicles. The **brachiocephalic** (*innominate*) veins are the short veins that empty directly into the superior vena cava.

Three veins, the basilic, brachial, and cephalic, collect blood in the upper arm. The **basilic** is on the medial side of the arm, and the **brachial** vein lies along the posterior surface of the humerus. The basilic and brachial veins unite to form the **axillary** vein in the armpit region. The **cephalic** vein courses along the lateral aspect of the arm and enters the axillary vein at its proximal end near the subclavian vein.

Between the basilic and cephalic veins in the elbow region is seen the **median cubital** vein. An **accessory cephalic** vein lies on the lateral portion of the forearm and empties into the cephalic in the elbow region.

Blood in the legs is returned to the heart by superficial and deep sets of veins. The superficial veins are just beneath the skin. The deep veins accompany the arteries. Both sets are provided with valves that are more numerous in the deep ones.

The presence of valves in veins can be vividly demonstrated on the back of your own hand. If you apply pressure on a prominent vein on the back of your left hand near the knuckles with the middle finger of your right hand and then strip the blood

in the vein away from the point of pressure with your index finger, you will note that the blood does not flow back toward the point of pressure. This indicates the presence of valves that prevent backward flow.

The **posterior tibial** vein (label 24) is one of the deep veins that lies behind the tibia. It collects blood from the calf and foot. In the knee region the posterior tibial becomes the **popliteal** vein, and above the knee this vessel becomes the **femoral** vein, which in turn, empties into the **external iliac** vein. The external and **internal** iliac (label 10) empty into the **common iliac** vein.

A large superficial vein of the leg, the **great saphenous,** originates from the **dorsal venous arch** on the superior surface of the foot; it enters the femoral vein at the top of the thigh.

Additional veins shown in figure 53.3 that empty into the inferior vena cava are the right gonadal, renal, and hepatic veins. The **hepatic** vein is the short one just under the heart that carries blood from the liver to the inferior vena cava. The two **renals** that drain the kidneys are inferior to the hepatic. Note that the left renal vein has a downward-extending branch, the **left gonadal** (*spermatic* or *ovarian*), which drains blood from the left testis. Note, also, that the **right gonadal** empties directly into the inferior vena cava.

Assignment:
Label the veins in figure 53.3.

Portal Circulation

Figure 53.4 reveals the venous system that constitutes the human *portal circulation.* All the veins shown here that lead from the stomach, spleen, pancreas, and intestine drain into the **portal vein** (label 2), which, in turn, empties into the liver. As a result of this drainage system, all blood from the digestive organs passes through the liver before entering the general circulation. This arrangement enables the liver, with its multiplicity of metabolic functions, to balance out blood composition before allowing the blood to be transported throughout the body.

The large vessel that collects blood from the ascending colon (label 4) and the ileum (label 5) is the **superior mesenteric** vein. It empties directly into the portal vein. Blood from the descending colon (label 7) and rectum is collected by the **inferior mesenteric,** which empties into the **splenic** (*lienal*) vein. The latter blood vessel parallels the length of the pancreas, receiving blood from the pancreas via several short **pancreatic** veins. Near

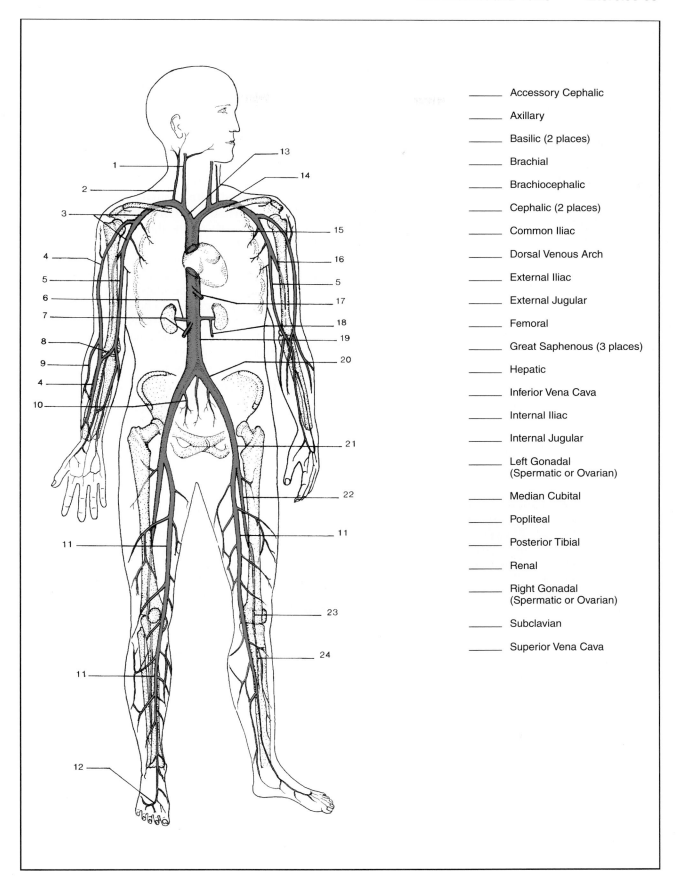

Figure 53.3 Major veins of the body.

Accessory Cephalic

Axillary

Basilic (2 places)

Brachial

Brachiocephalic

Cephalic (2 places)

Common Iliac

Dorsal Venous Arch

External Iliac

External Jugular

Femoral

Great Saphenous (3 places)

Hepatic

Inferior Vena Cava

Internal Iliac

Internal Jugular

Left Gonadal
(Spermatic or Ovarian)

Median Cubital

Popliteal

Posterior Tibial

Renal

Right Gonadal
(Spermatic or Ovarian)

Subclavian

Superior Vena Cava

the spleen, the splenic vein also receives blood from the stomach through the **short gastric** vein.

From the stomach a venous loop empties blood from the medial surface of the stomach into the portal vein. The upper portion of the loop is the **coronary** vein; the lower portion is the **pyloric** vein.

Assignment:
Label figure 53.4.

Cat Dissection

To facilitate differentiation of arteries and veins in the cat, the animal has been injected with red and blue latex: red for arteries and blue for veins. When tracing blood vessels it is necessary that each vessel be freed from adjacent tissue so that it is clearly visible. A sharp dissecting needle is indispensable for this purpose. It is essential, however, that considerable care be taken to avoid accidental severance of the blood vessels. Once they are cut they become difficult to follow.

In this study of the arteries and veins it will be necessary to refer to illustrations in other portions of this manual that pertain to other organ systems. By the time you have completed this circulatory study you should be familiar with the respiratory, diges-

tive, excretory, and reproductive organs of the cat. Great care should be taken, however, to avoid damaging the organs of the various systems at this time since they will be required for later study.

Exposing the Organs

Open up the ventral surface of the animal by starting a longitudinal incision in the thoracic wall, which is about one centimeter to the right or left of midline. This cut should pass through the rib cartilages and is easily performed with a sharp scalpel or scissors. Extend the incision, anteriorly to the apex of the thorax and, posteriorly, to the pubic region. In addition, make two lateral cuts posterior to the diaphragm.

Spread apart the walls of the thorax to expose its viscera. Greater exposure can be achieved by cutting away part of the chest wall on each side, using heavy-duty scissors to break through the ribs. In addition, trim away portions of the abdominal wall on each side of the original longitudinal incision to expose the abdominal organs.

Organ Identification

By comparing your dissection with figure 56.4 identify the **lungs, thymus gland, thyroid gland,**

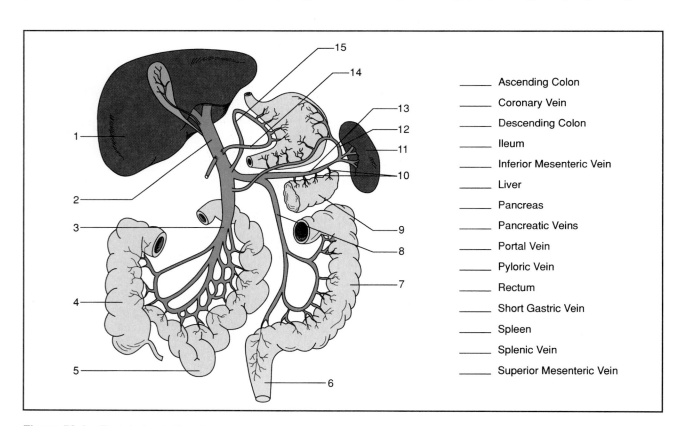

_____ Ascending Colon
_____ Coronary Vein
_____ Descending Colon
_____ Ileum
_____ Inferior Mesenteric Vein
_____ Liver
_____ Pancreas
_____ Pancreatic Veins
_____ Portal Vein
_____ Pyloric Vein
_____ Rectum
_____ Short Gastric Vein
_____ Spleen
_____ Splenic Vein
_____ Superior Mesenteric Vein

Figure 53.4 Portal circulation, human.

diaphragm, trachea, and **larynx.** In addition, squeeze the surface of the **heart** between your thumb and forefinger, noting the thickness and texture of the **pericardium** that envelops it. Now refer to figure 63.9 to identify the **liver, stomach, spleen, pancreas,** and the various portions of the **small** and **large intestines.**

Veins of the Thorax

Since the veins of the thoracic region lie over most of the arteries, it is best to study them first. By referring to figures 53.6 and 53.8 identify the veins in the following sequence.

Veins Entering the Heart

The **superior vena cava, inferior vena cava,** and three groups of **pulmonary veins** empty blood into the heart. Only the superior vena cava is shown in figure 53.8. Probe around the heart to find all these vessels. Each group of pulmonary veins, which empty into the left atrium, is composed of two or three veins.

Veins Emptying into Superior Vena Cava

Locate the **azygous** vein. It is a large vein that empties into the dorsal surface of the superior vena cava (s.v.c.) close to where the s.v.c. enters the right atrium of the heart. Since it is obscured by the heart and lungs, and not shown in figure 53.8, it will be necessary to force these organs to the left to locate it. Note that this vein lies along the right side of the vertebral column, collecting blood from the **intercostal, esophageal,** and **bronchial** veins. Trace the azygous vein to these branches. All of these veins are present in humans.

Identify the **internal mammary** *(sternal)* vein, which enters the s.v.c. opposite the third rib.

Examine the dorsal surface of the s.v.c. to identify the spot where the **right vertebral** and **right costocervical** veins enter it. These two veins unite to form a short vessel before entering the s.v.c. A stub of this vessel, labeled VE, is shown in figure 53.8. Note that the **left costocervical** vein is shown entering the left brachiocephalic in figure 53.8. A short stub of the left vertebral is also shown joining the left costocervical.

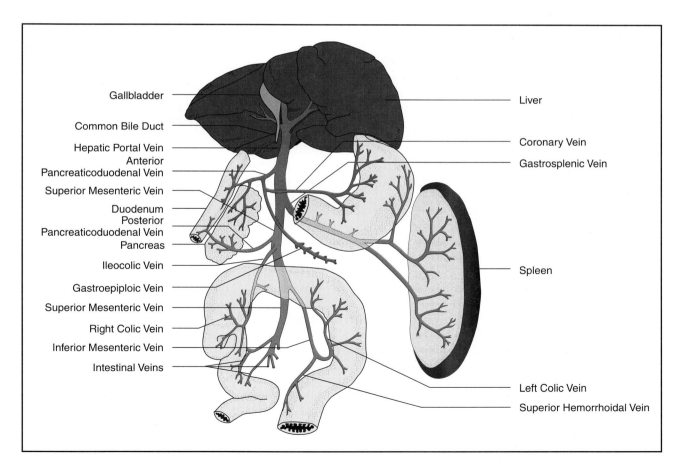

Figure 53.5 Portal circulation, cat.

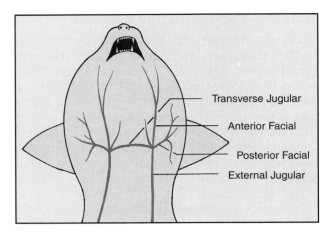

Figure 53.6 Branches of the external jugular vein.

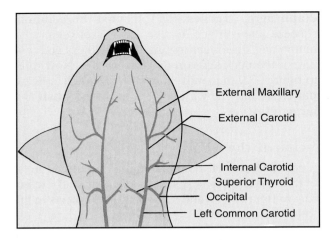

Figure 53.7 Branches of the carotid artery.

Identify the right and left brachiocephalic veins (BC), which empty into the s.v.c. These veins receive blood from the head and arms.

Drainage into the Brachiocephalic Veins

Note that the right brachiocephalic vein receives blood from two vessels, and the left brachiocephalic accepts blood from four veins. The **external jugular** and **subclavian** veins empty into both brachiocephalics. The left brachiocephalic also receives blood from the **left vertebral** and **left costocervical.** Identify all these vessels on your specimen.

Drainage into the External Jugulars

The principal veins that empty into the external jugulars are the **transverse scapular** and **internal jugulars.** Locate them first in figure 53.8 and then on your specimen. Note that the internal jugulars parallel the common carotid arteries.

Refer to figure 53.6 to locate the **transverse jugular** vein, which connects the two external jugular veins in the chin region; then, identify it on your specimen. Locate also the **posterior** and **anterior facial** veins in figure 53.6.

The **thoracic duct,** a part of the lymphatic system, also empties into the external jugular vein. Its point of entrance is very close to where the left external jugular enters the left subclavian. This vessel is difficult to find. It will be necessary to push the left lung to the right and look at the back of the left thoracic cavity near the vertebral column to find it. It has a beaded structure due to the presence of valves.

Drainage into the Subclavian Vein

The principal veins that empty into the subclavian are the **subscapular** and **axillary.** Locate them on your specimen. Note that the axillary receives blood

from the **thoracodorsal** and **brachial** veins. Locate them, also.

Arteries of the Thorax

Once you have identified the veins of the thorax, examine the arteries by gingerly manipulating the veins to one side. Do not cut away the veins, however. By referring to figures 53.9 and 53.10 identify the arteries in the following sequence.

Arteries Emerging from the Heart

Locate the **aortic arch** and **pulmonary trunk** by referring to figure 53.9. Trace the pulmonary trunk to where it divides into the **right** and **left pulmonary arteries.**

Dissect away the connective tissue at the base of the aorta to locate the **coronary arteries.** Note their pathways on the surface of the heart.

Branches of the Aortic Arch

Observe that in the cat there are only two branches emerging from the upper surface of the aortic arch: the **brachiocephalic** *(innominate)* and the **left subclavian.** How does this compare with humans? Note, also, that the **right common carotid, right subclavian,** and **left common carotid** arteries branch off the brachiocephalic.

Branches of the Subclavian Arteries

Locate the vertebral, costocervical, internal mammary, and thyrocervical arteries, which branch off the subclavian arteries. The **vertebral** carries blood to the brain, passing through the transverse foramina of the cervical vertebrae.

The **costocervical** supplies blood to deep muscles of the back and neck. The **internal mammary** arteries

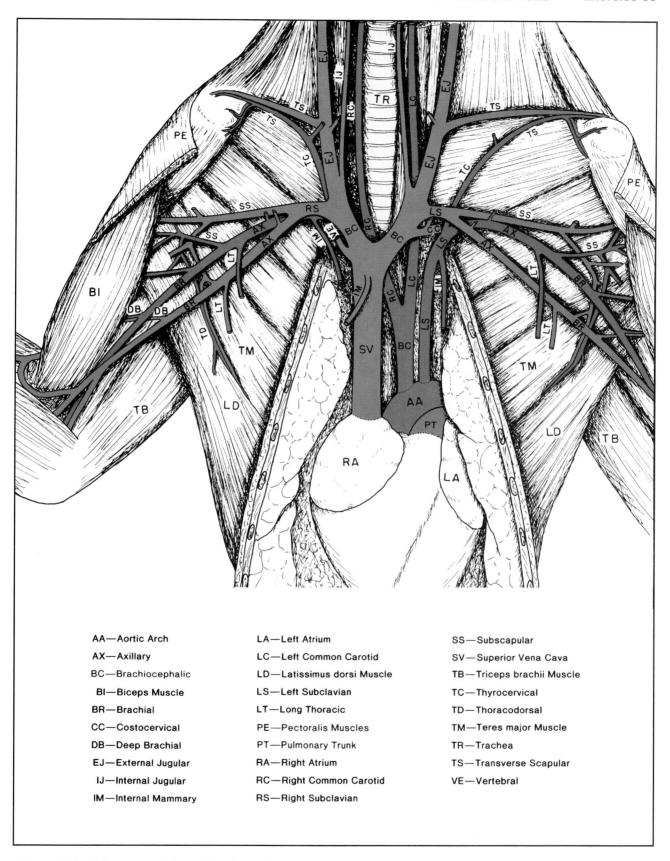

Figure 53.8 Veins and arteries of the thorax.

AA—Aortic Arch
AX—Axillary
BC—Brachiocephalic
BI—Biceps Muscle
BR—Brachial
CC—Costocervical
DB—Deep Brachial
EJ—External Jugular
IJ—Internal Jugular
IM—Internal Mammary

LA—Left Atrium
LC—Left Common Carotid
LD—Latissimus dorsi Muscle
LS—Left Subclavian
LT—Long Thoracic
PE—Pectoralis Muscles
PT—Pulmonary Trunk
RA—Right Atrium
RC—Right Common Carotid
RS—Right Subclavian

SS—Subscapular
SV—Superior Vena Cava
TB—Triceps brachii Muscle
TC—Thyrocervical
TD—Thoracodorsal
TM—Teres major Muscle
TR—Trachea
TS—Transverse Scapular
VE—Vertebral

supply the ventral body wall, and the **thyrocervical** arteries supply blood to the neck and shoulder.

Note that the thyrocervical becomes the **transverse scapular** and that the subclavian becomes the **axillary.** The axillary, in turn, eventually becomes the **brachial** artery.

Branches of the Axillary Artery

Identify the **ventral thoracic, long thoracic,** and **subscapular** arteries, which branch off the axillary artery. These branches supply blood to the latissimus dorsi and pectoral muscles. Note also that the subscapular becomes the **thoracodorsal,** and the axillary becomes the **brachial.** The latter becomes the **radial** artery at the elbow.

Branches of Each Common Carotid

Each common carotid has an inferior and a superior thyroid artery. The **inferior thyroid** is a small vessel that originates near the base of the common carotid. Probe around this area to locate it. The **superior thyroid** is shown in figure 53.7. Locate this one also.

If your specimen is properly injected you should be able to locate the **occipital, external carotid,** and **internal carotid,** which branch off the common carotid.

Branches of the Thoracic Aorta

Pull the viscera of the thorax to the right to expose the aorta in the thoracic cavity. Since the aorta lies dorsal to the parietal pleura, peel away this membrane to expose it.

Locate the ten pairs of **intercostal** arteries, which emerge from the thoracic portion of the aorta to supply the intercostal muscles. Also, locate the **bronchial** arteries, which emerge from the aorta opposite the fourth intercostal space or from the fourth intercostal artery. They accompany the bronchi to the lungs. Next, identify the **esophageal** arteries, which supply blood to the esophagus.

Branches of the Abdominal Aorta

Remove the peritoneum from the dorsal abdominal wall just below the diaphragm to expose the **celiac trunk.** Note in figure 53.10 that this is the first

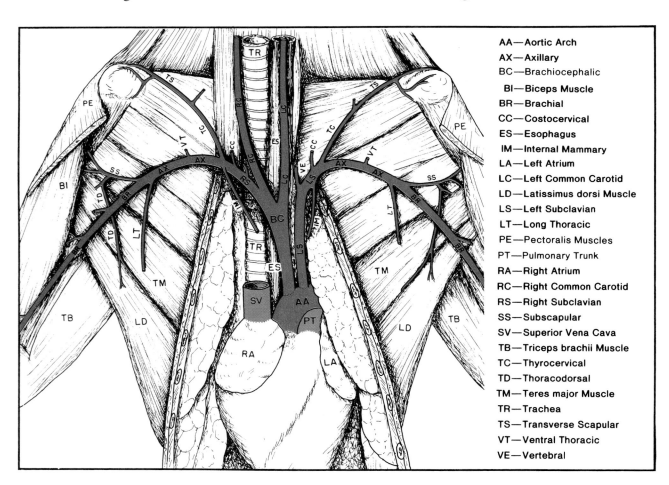

AA—Aortic Arch
AX—Axillary
BC—Brachiocephalic
BI—Biceps Muscle
BR—Brachial
CC—Costocervical
ES—Esophagus
IM—Internal Mammary
LA—Left Atrium
LC—Left Common Carotid
LD—Latissimus dorsi Muscle
LS—Left Subclavian
LT—Long Thoracic
PE—Pectoralis Muscles
PT—Pulmonary Trunk
RA—Right Atrium
RC—Right Common Carotid
RS—Right Subclavian
SS—Subscapular
SV—Superior Vena Cava
TB—Triceps brachii Muscle
TC—Thyrocervical
TD—Thoracodorsal
TM—Teres major Muscle
TR—Trachea
TS—Transverse Scapular
VT—Ventral Thoracic
VE—Vertebral

Figure 53.9 Arteries of the thorax.

branch of the abdominal portion of the aorta. Follow it out to its three branches: **hepatic, left gastric,** and **splenic** arteries.

Identify the **superior mesenteric, renal, internal spermatic** or **ovarian, inferior mesenteric,** and **iliolumbar** arteries, in that order, noting the organs that they supply. Note that a pair of **adrenolumbar** arteries emerge caudad to the superior mesenteric. These arteries supply the adrenal glands, diaphragm, and muscles of the body wall.

There are also seven pairs of small **lumbar** arteries that assist the iliolumbar arteries in providing the abdominal body with blood. Can you locate them?

Observe that in the cat there are no common iliac arteries; instead, two large branches, the **external iliac** arteries, pass into the hind legs. Each external iliac eventually becomes the **femoral** in the thigh region.

Caudad to the external iliacs are a pair of **internal iliac** arteries branching off the aorta. The aorta terminates as the **caudal** artery.

Veins of the Abdomen

Now that you have identified all the arteries and the veins of the upper part of the cat, let's move on to the veins that empty into the inferior vena cava.

Drainage into the Inferior Vena Cava

Figure 53.11 illustrates the ramifications of the venous system, which drains into the **inferior vena cava** (i.v.c.). Note that the i.v.c. passes through the diaphragm and lies to the right of the abdominal aorta. The tributaries of this vein usually parallel similarly named arteries.

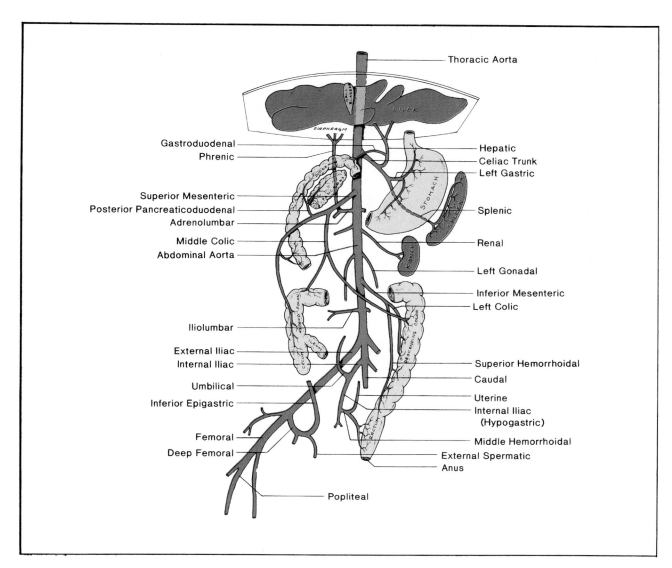

Figure 53.10 Arteries of the abdomen and leg.

Scrape away some of the liver tissue on its anterior surface to locate the **hepatic** veins that empty into the i.v.c.

Study the **adrenolumbar** and **renal** veins. Note that the right adrenolumbar empties into the i.v.c., while the left one empties into the left renal vein. Observe, also, that the right and left **spermatic (ovarian)** veins differ in their drainage, with only the right one emptying directly into the i.v.c.

Locate the **iliolumbar** veins, which convey blood from the body wall in the lumbar region to the i.v.c.

Identify the right and left **common iliac** veins, which convey blood from the legs to the inferior vena cava.

Tributaries of the Common Iliac Veins

Note in figure 53.11 that the **caudal** vein empties into the left common iliac instead of being a direct extension of the i.v.c. This is not always the case, however.

Trace the common iliac vein in one leg down to where it branches to form the **external iliac** and **internal iliac.** Note that, as in humans, the external iliac becomes the **femoral** in the thigh region. Follow the femoral farther into the leg to locate the **greater saphenous** and **popliteal** veins.

The Hepatic Portal System

The veins of this drainage system are shown in figures 53.5 and 53.11. Using either illustration, proceed as follows:

Observe that, as in humans, blood from the stomach, spleen, pancreas, small intestine (duodenum), and large intestine (colon), empties into the **hepatic portal,** or simply, **portal,** vein.

Locate the following tributaries that feed into this large vessel: **gastrosplenic, superior mesenteric, inferior mesenteric,** and **pancreaticoduodenal** veins. Pay particular attention to the organs that supply them with venous blood.

Laboratory Report

Complete Laboratory Report 53 for this exercise.

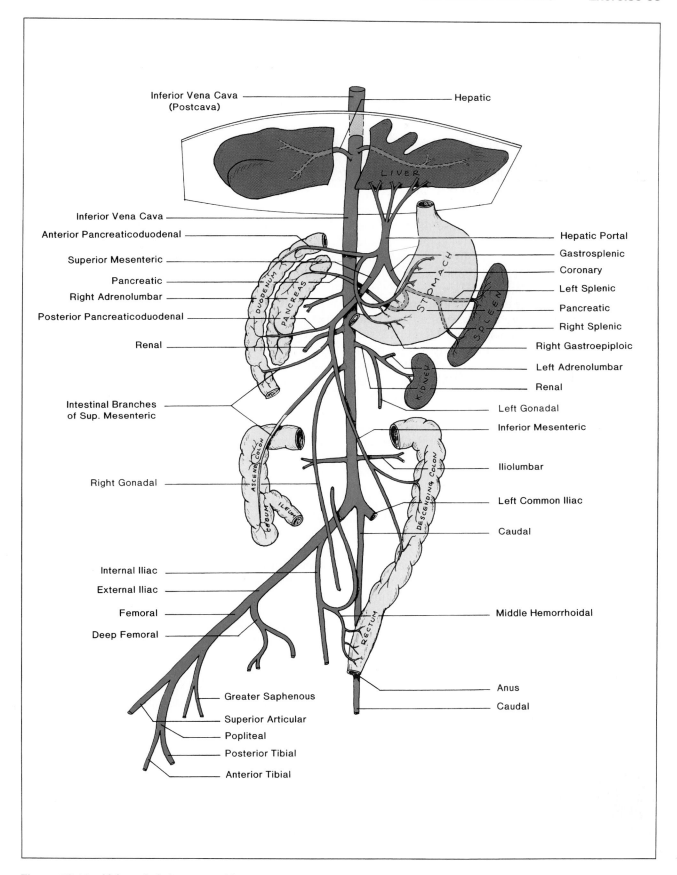

Figure 53.11 Veins of abdomen and leg.

54

Fetal Circulation

During human embryological development the circulatory system is necessarily somewhat different from that after birth. The fact that the lungs are nonfunctional before birth dictates that less blood should go to the lungs. Also, because the food and oxygen supply for the fetus must come from the mother by way of the placenta, blood vessels to and from the placenta through the umbilical cord must be present and functional.

Figure 54.1 illustrates, diagrammatically, the pathway of blood through the heart of a fetus. The purpose of this exercise is threefold: (1) to see how the blood flow pattern in a fetus differs from an adult, (2) to see what changes occur in the flow pattern after birth, and (3) to understand what forces control the changes that occur after birth.

The blood supply of a fetus passes from the placenta in the uterus through the **umbilical cord** (label 16) via the **umbilical vein.** Since the blood in this vein is rich in nutrients and oxygen it is colored red in figure 54.1. After circulating throughout the body of the fetus, the blood, ladened with carbon dioxide and metabolic wastes, is returned to the placenta via a pair of **umbilical arteries,** which are shown wrapped around the umbilical vein in the umbilical cord.

Where the umbilical vein approaches the liver it divides: one branch enters the liver, and the other branch becomes the **ductus venosus,** which continues upward to join a branch from the liver and enter the **inferior vena cava.** Note that the entrance to the ductus venosus has a constriction in it caused by the presence of a **sphincter muscle** in the wall of the vessel. This sphincter does not restrict the flow of blood during growth and development; its force comes into play only immediately after parturition (birthing process).

Observe that oxygen-rich blood of the ductus venosus mixes with venous blood in the inferior vena cava before it enters the right atrium of the heart. In the right atrium the blood can move in two directions. Some of this blood will pass to the right ventricle as it would in an adult, and the remainder passes through a valve, the **foramen ovale,** into the left atrium. The reason that blood is shunted to the left side of the heart through this valve is that less

blood is needed in the lungs since they are nonfunctional in the fetus.

When ventricular systole occurs, blood in the right ventricle exits the heart through the **pulmonary trunk** and blood in the left ventricle exits through the **aortic arch.** Observe that blood flowing from the right ventricle through the pulmonary trunk flows in three directions: some blood to the right lung through the **right pulmonary artery,** some to the left lung through the **left pulmonary artery,** and the remainder to the aortic arch through a very short vessel, the **ductus arteriosus.** Thus, we see that the blood in the left atrium is a mixture of unoxygenated blood from the lungs (via four pulmonary veins) and partially oxygenated blood from the right atrium. It is this mixture that passes from the left atrium to the left ventricle and out to all parts of the body through the aortic arch.

Blood flowing through the aortic arch passes down the **abdominal aorta** to the pelvic region, where it divides into two **common iliac arteries.** Each common iliac artery, in turn, branches into **external iliac** and **hypogastric arteries.** The latter arteries pass by the **urinary bladder** to the umbilical cord, where they become the umbilical arteries that return blood to the placenta.

Once the umbilical cord is cut, separating the fetus from the placenta at birth, changes occur in the heart, veins, and arteries to provide a new route for the blood. One of the first things that happens is the closure of the umbilical vein by the sphincter muscle at the ductus venosus to minimize blood loss. Coagulation of blood also occurs in this vessel. In addition, the attending physician prevents further blood loss by ligature. Within five days after birth the umbilical vein is converted to the *ligamentum teres hepatis* (round ligament), which extends from the umbilicus to the liver in the adult.

As soon as the newly born infant begins to breathe, more blood is immediately drawn to the lungs via the pulmonary arteries. The abandonment of the path through the ductus arteriosus causes this vessel to collapse and begin its gradual transformation to connective tissue of the *ligamentum arteriosum.*

With the increase in blood volume and blood pressure in the left atrium, due to a greater blood flow from the lungs, the flap on the foramen ovale closes this opening. Eventually, connective tissue permanently seals off this valve.

The ductus venous becomes a fibrous band, the *ligamentum venosum* of the liver. Those portions of the hypogastric arteries that lie along the bladder form into fibrous cords *(lateral umbilical ligaments)* and *vesical arteries.* The lateral umbilical ligaments extend anteriorly and upward from the bladder to the inner abdominal wall. The superior, middle, and inferior vesical arteries supply blood to the urinary bladder.

Laboratory Report

Label figure 54.1 and answer the questions on Laboratory Report 54, 55 that pertain to this exercise.

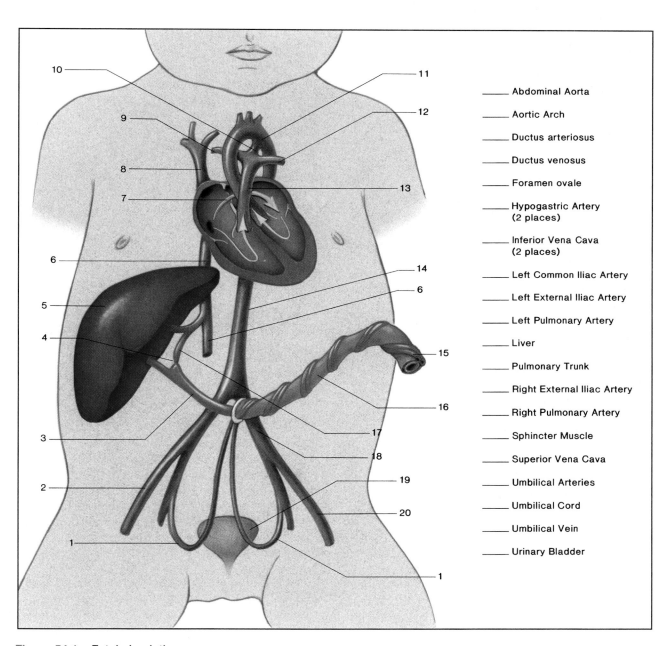

_____ Abdominal Aorta

_____ Aortic Arch

_____ Ductus arteriosus

_____ Ductus venosus

_____ Foramen ovale

_____ Hypogastric Artery (2 places)

_____ Inferior Vena Cava (2 places)

_____ Left Common Iliac Artery

_____ Left External Iliac Artery

_____ Left Pulmonary Artery

_____ Liver

_____ Pulmonary Trunk

_____ Right External Iliac Artery

_____ Right Pulmonary Artery

_____ Sphincter Muscle

_____ Superior Vena Cava

_____ Umbilical Arteries

_____ Umbilical Cord

_____ Umbilical Vein

_____ Urinary Bladder

Figure 54.1   Fetal circulation.

55

The Lymphatic System and the Immune Response

The lymphatic system consists of a drainage system of lymphatic vessels that returns tissue fluid from the interstitial cell spaces to the blood. On its journey back to the blood, the lymph in these vessels is cleansed of bacteria and other foreign material by macrophages in the lymph nodes. In addition to macrophages, the lymph nodes are packed with lymphocytes that play a leading role in cellular and humoral immunity. In this exercise we will explore the genesis and physiology of lymphocytes as well as the overall anatomy of the lymphatic system.

Lymph Pathway

The smallest vessels of the lymphatic system are called *lymph capillaries.* These tiny vessels have closed ends, are microscopic, and are situated among the cells of the various tissues of the body. The lymph capillaries unite to form the **lymphatic vessels** *(lymphatics),* which are the visible vessels shown in the arms and legs in figure 55.1. The irregular appearance of the walls of these vessels is due to the fact that they contain many valves to restrict backward flow of lymph.

As lymph moves toward the center of the body through the lymphatics, it eventually passes through one or more of the small oval **lymph nodes.** Note that these nodes are clustered in the neck, groin, axillae, and abdominal cavity; they are also found in smaller clusters in other parts of the body. They may be as small as a pinhead or as large as an almond.

Two collecting vessels, the thoracic and right lymphatic ducts, collect the lymph from different regions of the body. The largest one is the **thoracic duct,** which collects all the lymph from the legs, abdomen, left half of the thorax, left side of the head, and left arm. Its lower extremity consists of a saclike enlargement, the **cisterna chyli.** Lymph from the intestines passes through lymphatics in the mesentery to the cisterna chyli. The lymph from this region contains a great deal of fat and is usually referred to as *chyle.* The thoracic duct empties into the left subclavian vein near the left internal jugular vein. The **right lymphatic duct** is a short vessel that drains lymph from the right arm and right side

of the head into the right subclavian vein near the right internal jugular.

Lymph Node Structure

The enlarged node shown in figure 55.1 reveals the structure of a typical lymph node. Note that it has a slight depression at its upper end that is called the **hilum.** It is through this depression that blood vessels and an **efferent lymphatic vessel** emerge. Although each node has only one hilum and one efferent lymphatic vessel, there are usually two or more **afferent lymphatic vessels** that carry lymph into the node (two are shown in figure 55.1).

Surrounding the entire node is a **capsule** of fibrous connective tissue, and just inside the capsule is a **cortex** that consists of sinuses reinforced with reticular tissue, and many **germinal centers** *(nodules).* Approximately thirteen ovoid germinal centers are shown in figure 55.1. These centers are the basic structural units of a lymph node. They arise from small nests of lymphocytes or lymphoblasts and reach about one millimeter in diameter. The central portion of a lymph node is called the **medulla.**

Assignment:
Label figure 55.1.

Role of Lymphocytes in Immunity

The immune response to foreign invaders involves the integrated action of an army of different cell types, including monocytes, macrophages, eosinophils, basophils, and lymphocytes. Although each of these cells plays a different role, they interact with each other, even to the extent of regulating each other's activities. The "commander" and predominant "foot soldier" of this defensive force is the lymphocyte.

Although all lymphocytes are morphologically similar, there are considerable differences between individual cells. In addition to being present in blood, lymph, and lymph nodes, they are also seen in the bone marrow, thymus gland, spleen, and lymphoid masses associated with the digestive, respiratory, and urinary passages.

Lymphocytes are subdivided into two subclasses: *B-cells* and *T-cells*. The *B-cells* (*B* for bursa) govern what is called *humoral,* or *antibody-mediated,* immunity. The *T-cells* (*T* for thymus) are responsible for *cell-mediated immunity.*

Each *B*-cell is able to recognize only a single foreign antigen, which may be on a bacterium, virus, or some other invader. In other words, all the antibodies on a *B*-cell are of a single type. Once a *B*-cell receptor recognizes the foreign antigen, it does two things: it begins to secrete antibodies into the blood, and it begins to multiply very rapidly, increasing the number of identical *B*-cells. Once these *circulating antibodies* bind to antigens or antigen-bearing targets, they mark them for destruction by other components of the immune system. Since there is a preponderance of this type of lymphocytes in the spleen, it is believed by some that one of the functions of this organ is to generate new *B*-cells.

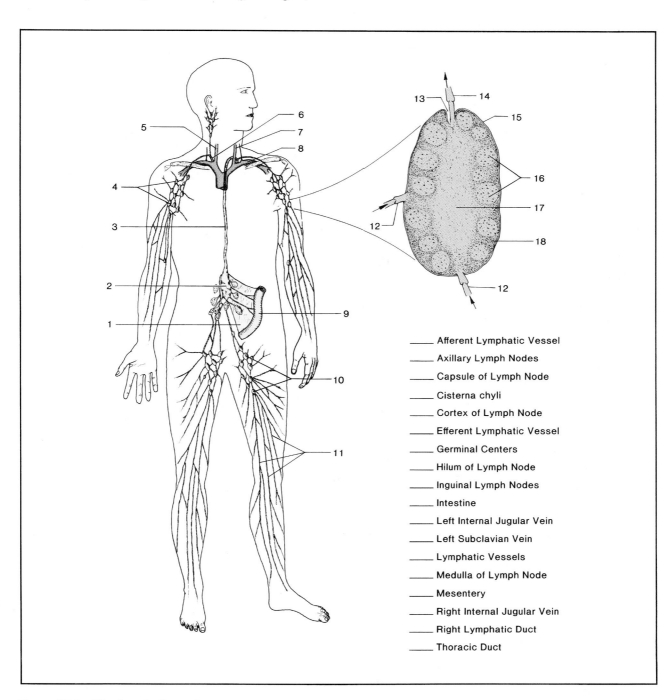

_____ Afferent Lymphatic Vessel

_____ Axillary Lymph Nodes

_____ Capsule of Lymph Node

_____ Cisterna chyli

_____ Cortex of Lymph Node

_____ Efferent Lymphatic Vessel

_____ Germinal Centers

_____ Hilum of Lymph Node

_____ Inguinal Lymph Nodes

_____ Intestine

_____ Left Internal Jugular Vein

_____ Left Subclavian Vein

_____ Lymphatic Vessels

_____ Medulla of Lymph Node

_____ Mesentery

_____ Right Internal Jugular Vein

_____ Right Lymphatic Duct

_____ Thoracic Duct

Figure 55.1 The lymphatic system.

There are three principal types of *T*-cells: *helper, suppressor,* and *cytotoxic* or *"killer" cells.* A *T*-cell that recognizes and binds to an antigen on the surface of another cell becomes *activated.* An activated *T*-cell can multiply, and if it is a cytotoxic cell, it can kill the bound cell. For example, cancer cells that display antigens not found on healthy cells can potentially activate *T*-cells, resulting in their destruction.

The way that cells of the immune system regulate each other is by secreting potent hormones called **cytokines.** If they are produced by lymphocytes they are called *lymphokines.* Hormones produced by monocytes and macrophages are called *monokines.* These hormones differ from hormones of the endocrine gland system in that they act *locally* and do not circulate in the blood.

Lymphocytes, as well as *all* other blood cells, develop from stem cells in the bone marrow before birth and shortly after birth. Stem cells are released into the blood and follow one of two pathways for "processing," as illustrated in figure 55.2.

Cells that settle in the thymus for modification become *T*-cells. Those that become *B*-cells are processed in the spleen or some other unknown area. It may well be that they form spontaneously, at random, as suggested by some workers. The reason they were originally named *B*-cells is that their place of processing in birds was found to be in the *bursa of Fabricius,* a lymphoidal structure on the hindgut. At the present time an equivalent structure in humans may be the spleen.

Once the *T*-cells have matured in the thymus, they leave the thymus via the blood and colonize in specific loci of the lymph nodes and spleen. *T*-cells have a long life, possibly years or a lifetime. Once activated by antigens, they are transformed to **lymphoblasts** to perform their specific functions.

Like the *T*-cells, the *B*-cells also colonize in specific regions of the lymph nodes and spleen. When *B*-cells come in contact with the appropriate antigens, they are transformed into **plasma cells** that produce the antibodies specific for the foreign protein.

Laboratory Assignment

Materials:
> prepared slide of a lymph node (H13.11)
> microscope

1. Examine a prepared slide of a lymph node and compare it with the photomicrographs in figure HA-29 of the Histology Atlas. Locate a **germinal center.**
2. Examine the **lymphocytes** under high-dry or oil immersion to see if you can identify cells that are dividing.
3. Make drawings, if required.
4. Answer all questions on Laboratory Report 54, 55 that pertain to this exercise.

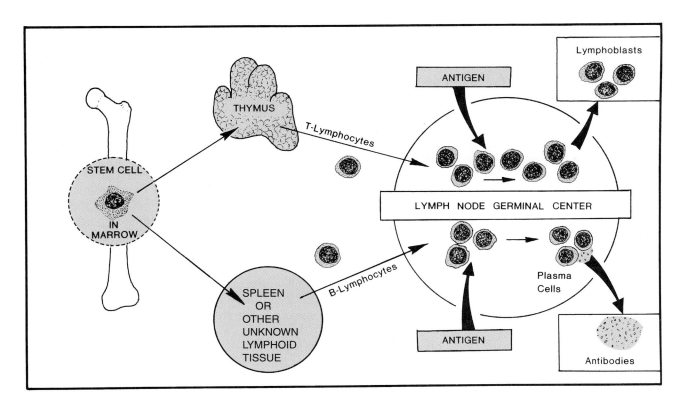

Figure 55.2   The genesis of lymphoblasts and plasma cells.

56

The Respiratory Organs

Both gross and microscopic anatomy of the respiratory system will be studied in this exercise. Although the cat will be the primary dissection specimen here, sheep and frog materials are also available for certain observations. Before any laboratory dissections are performed, however, it is desirable that figures 56.1, 56.2, and 56.3 be labeled.

Materials:
>models of median section of head and larynx
>frog, pithed
>sheep pluck
>dissecting instruments and trays
>Ringer's solution
>bone or autopsy saw

The Upper Respiratory Tract

In addition to breathing, the upper respiratory tract also functions in eating and speech; thus, a study of the respiratory organs in this region must include the organs concerned with those other activities. Figure 56.1 is of this area.

The two principal cavities of the head are the nasal and oral cavities. The upper **nasal cavity,** which serves as the passageway for air, is separated from the lower **oral cavity** by the *palate.* The anterior portion of the palate is reinforced with bone and is called the **hard palate.** The posterior part, which terminates in a fingerlike projection, the **uvula,** is the **soft palate.**

The oral cavity consists of two parts: the **oral cavity proper,** which contains the tongue, and the **oral vestibule,** which is between the lips and teeth.

During breathing, air enters the nasal cavity through the nostrils, or **nares** (*naris,* singular). The lateral walls of this cavity have three pairs of fleshy lobes, the superior, middle, and inferior **nasal conchae.** These lobes serve to warm the air as it enters the body.

Above and behind the soft palate is the **pharynx.** That part of the pharynx posterior to the tongue, where swallowing is initiated, is called the **oropharynx.** Above it is the **nasopharynx.** Two openings to the **auditory** *(Eustachian)* **tubes** are seen on the walls of the nasopharynx.

After air passes through the nasal cavity, nasopharynx, and oropharynx, it passes through the larynx and trachea to the lungs. The cartilages of the larynx (labels 8, 10, and 12) vary considerably in size and histology. The **epiglottis,** which is the uppermost cartilage, is composed of elastic cartilage. Its function is to prevent food from entering the respiratory passages during swallowing. Inferior to the epiglottis is the **thyroid cartilage,** which forms the side walls of the larynx and protrudes externally as the "Adam's apple" in the throat region. Just below the thyroid cartilage is the **cricoid cartilage,** a signet ring-shaped structure. The thyroid and cricoid cartilages are composed of hyaline cartilage.

Note that on the inner wall of the larynx is depicted a **vocal fold,** which is, essentially, the true vocal cord of the larynx. The paired vocal cords are actuated by muscles through the **arytenoid cartilages** (label 5, figure 56.3) to produce sounds. One of these vocal folds exists on each side of the larynx. The vocal folds should not be confused with the **ventricular folds** (label 9, figure 56.1), which lie superior to the vocal folds. The ventricular folds are also known as "false vocal cords." Posterior to the larynx and trachea lies the **esophagus,** the tube that carries food from the pharynx to the stomach.

In several areas of the oral cavity and nasopharynx are seen islands of lymphoidal tissue called *tonsils.* The **pharyngeal tonsils,** or *adenoids,* are situated on the roof of the nasopharynx; the **lingual tonsils** are on the posterior inferior portion of the tongue; and the **palatine tonsils** are on each side of the tongue on the lateral walls of the nasopharynx. These masses of tissue are as important as the lymph nodes in protection against infection. The palatine tonsils are the ones most frequently removed by surgical methods.

Assignment:
Label figure 56.1.
Study models of the head and larynx. Be able to identify all structures.

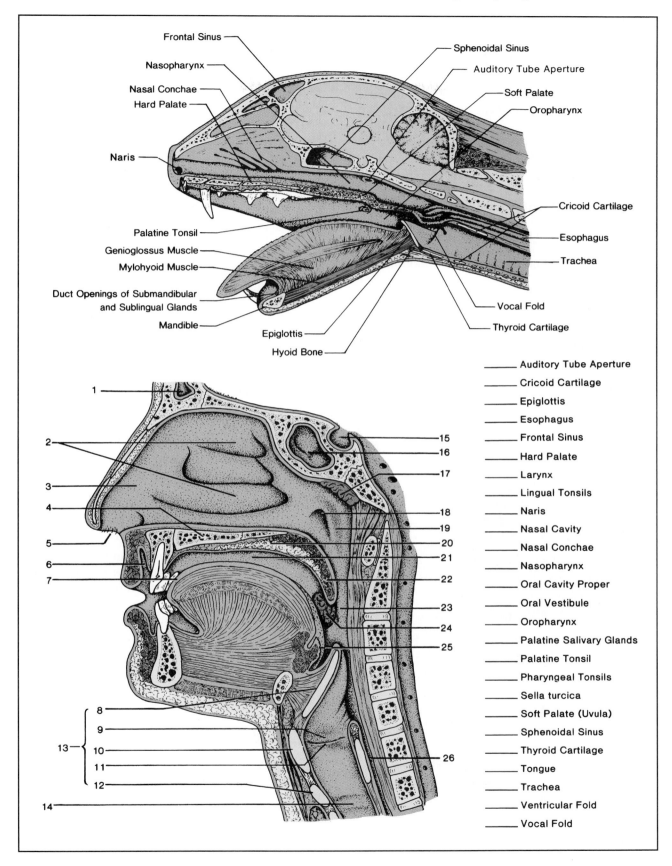

Figure 56.1   Upper respiratory passages, cat and human.

_____ Auditory Tube Aperture
_____ Cricoid Cartilage
_____ Epiglottis
_____ Esophagus
_____ Frontal Sinus
_____ Hard Palate
_____ Larynx
_____ Lingual Tonsils
_____ Naris
_____ Nasal Cavity
_____ Nasal Conchae
_____ Nasopharynx
_____ Oral Cavity Proper
_____ Oral Vestibule
_____ Oropharynx
_____ Palatine Salivary Glands
_____ Palatine Tonsil
_____ Pharyngeal Tonsils
_____ Sella turcica
_____ Soft Palate (Uvula)
_____ Sphenoidal Sinus
_____ Thyroid Cartilage
_____ Tongue
_____ Trachea
_____ Ventricular Fold
_____ Vocal Fold

The Lungs, Trachea, and Larynx

Figure 56.2 illustrates the respiratory passages from the larynx to the lungs. Note that the larynx consists of three cartilages: an upper **epiglottis**, a large middle **thyroid cartilage**, and a smaller **cricoid cartilage.** Figure 56.3 reveals these cartilages in greater detail.

Below the larynx extends the **trachea** to a point in the center of the thorax where it divides to form two short **bronchi.** Note that both the trachea and bronchi are reinforced with rings of cartilage of the hyaline type. Each bronchus divides further into many smaller tubes called **bronchioles.** At the terminus of each bronchiole is a cluster of tiny sacs, the **alveoli,** where gas exchange with the blood takes place. Each lung is made up of thousands of these sacs.

Free movement of the lungs in the thoracic cavity is facilitated by the pleural membranes. Covering each lung is a **pulmonary pleura,** and attached to the thoracic wall is a **parietal pleura.** Between these two pleurae is a potential cavity, the **pleural** *(intrapleural)* **cavity.** Normally, the lungs are firmly pressed against the body wall with little or no space between the two pleurae. The bottom of the lungs rests against the muscular **diaphragm,** the

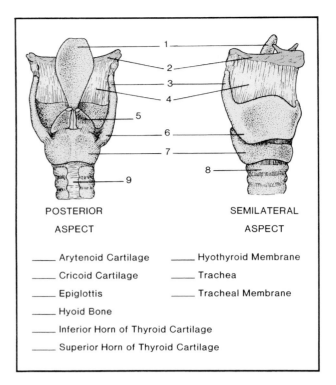

_____ Arytenoid Cartilage   _____ Hyothyroid Membrane

_____ Cricoid Cartilage   _____ Trachea

_____ Epiglottis   _____ Tracheal Membrane

_____ Hyoid Bone

_____ Inferior Horn of Thyroid Cartilage

_____ Superior Horn of Thyroid Cartilage

Figure 56.3   Human larynx.

principal muscle of respiration. The parietal pleura is attached to the diaphragm also.

Assignment:
Label figures 56.2 and 56.3.

Cat Dissection

Examination of the upper and lower respiratory passages of the cat will require some dissection. The amount of dissection to be performed will depend on what systems have been previously studied on your specimen. Figures 56.1, 56.4, and 56.5 will be used for reference.

Materials:
    autopsy saw
    embalmed cat
    dissecting tray and dissecting instruments

Upper Respiratory Tract

To identify all the cavities and structures of the cat's head that are shown in figure 56.1, it is necessary to use a mechanical bone saw or an electric autopsy saw to cut the head down the median line. The instructor may designate certain students to make sagittal sections that can be studied by all members of the class. Most specimens will be left intact to facilitate the study of other systems.

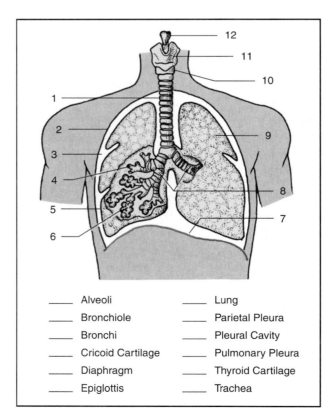

_____ Alveoli   _____ Lung

_____ Bronchiole   _____ Parietal Pleura

_____ Bronchi   _____ Pleural Cavity

_____ Cricoid Cartilage _____ Pulmonary Pleura

_____ Diaphragm   _____ Thyroid Cartilage

_____ Epiglottis   _____ Trachea

Figure 56.2   The respiratory tract.

If an autopsy saw is used, it is essential that the head of the specimen be held by one student while another student does the cutting. The instructor will demonstrate the procedure. Although an electric autopsy saw is relatively safe because of its reciprocating action, there are certain precautions that must be observed.

The principal rules of safety are: (1) cut only away from the assistant's hands; (2) while cutting, brace your hands against the table for steadiness: don't try to "free-hand" it; and (3) the cutter's hands should never be allowed to tire: frequent rests are essential.

Use the large cutting edge for straight cuts and the small cutting edge for sharp curves. Cut only deep enough to get through the bone; use the scalpel for soft tissues. Once the head has been cut through, it should be washed free of all loose debris. After completing the cutting, identify the structures using figure 56.1 for reference.

1. First, identify the **nares** and **nasal cavity.** As in humans, the nasal cavity lies superior to the palate.

2. Locate the **nasal conchae,** which are shaped somewhat differently in cats than humans.
3. Identify the region designated as the **nasopharynx.** Note that a small **auditory tube aperture** is located in this region. Insert a probe into it.
4. Press against the palate with a blunt probe to note where the **hard palate** ends and the **soft palate** begins.
5. Locate the **oropharynx,** which is located at the back of the mouth near the base of the tongue.
6. Observe that the cat has a very small **palatine tonsil** on each side of the oropharynx. Does it appear to lie in a recess, as is true of the same tonsil in humans?

Lower Respiratory Passages

After identifying the aforementioned structures on the sagittal section, open the thoracic cavity if it has not already been done. If the circulatory system has been studied, it will already be open. The exposed organs should appear as in figure 56.4.

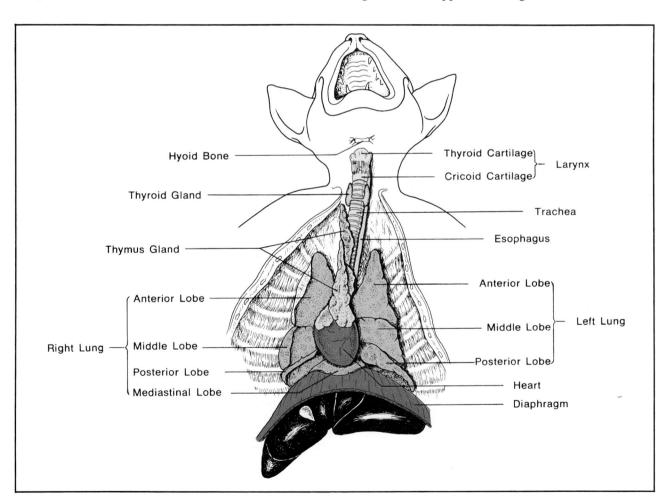

Figure 56.4 Thoracic organs of the cat.

To open the chest cavity, make a longitudinal incision about one centimeter to the right or left of the midline. Such a cut will be primarily through muscle and rib cartilage rather than bone. The incision may be cut partially through with a scalpel and completed with scissors. Avoid damaging the internal organs. Extend the cut up the throat to the mandible.

Continue the incision down the abdominal wall past the liver. Make two cuts laterally from the midline in the region of the liver. Spread apart the thoracic walls and sever the diaphragm from the wall with scissors. To enable the thoracic walls to remain open, make a shallow longitudinal cut with a scalpel along the inside surface of each side that is sufficiently deep to weaken the ribs. When the walls are folded back, the thoracic wall on each side should break at the cut.

Proceed to identify the organs of the lower respiratory passages according to the following procedure. Use figures 56.4 and 56.5 for reference.

1. Examine the larynx to identify the large flap-like **epiglottis,** the **thyroid cartilage,** and the **cricoid cartilage.** Consult figure 56.5.
2. Locate the **arytenoid cartilages** and the paired **vocal cords** on the larynx.
3. Trace the **trachea** down into the **lungs.** In humans the right lung has three lobes and the left lung has only two. How does this compare with the cat?
4. Cut out a short section of the trachea and examine a cross section through the **cartilaginous rings.** Are they continuous all the way around the trachea?
5. Can you see where the trachea branches to form the **bronchi?** Make a frontal section through one lung and look for further branching of the bronchus.

Sheep Pluck Dissection

A "sheep pluck" consists of the trachea, lungs, larynx, and heart of a sheep as removed during routine slaughter. Since it is fresh material rather than formalin-preserved, it is more lifelike than the organs of an embalmed cat. If plucks are available, proceed as follows to dissect one.

Materials:
    sheep plucks, less the liver
    dissecting tray and instruments
    drinking straws or sterile serological pipettes

1. Lay out a fresh sheep pluck on a tray. Identify the major organs, such as the **lungs, larynx, trachea, diaphragm** remnants, and **heart.**
2. Examine the larynx more closely. Can you identify the **epiglottis, thyroid,** and **cricoid cartilages?** Look into the larynx. Can you see the **vocal folds** (cords)?
3. Cut a cross section through the upper portion of the trachea and examine a sectioned cartilaginous ring. Is it similar to the cat in structure?
4. Force your index finger down into the trachea, noting the smooth and slimy nature of the inner lining. What kind of tissue lines the trachea that makes it so smooth and slippery?
5. Note that each lung is divided into lobes. How many lobes make up the right lung? The left lung? Compare with the human lung (figure 56.2).

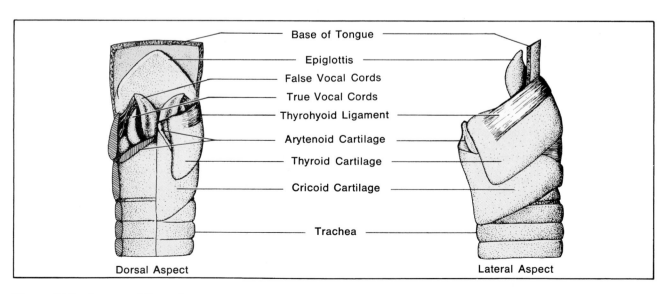

Base of Tongue

Epiglottis

False Vocal Cords

True Vocal Cords

Thyrohyoid Ligament

Arytenoid Cartilage

Thyroid Cartilage

Cricoid Cartilage

Trachea

Dorsal Aspect

Lateral Aspect

Figure 56.5   Larynx of the cat.

6. Rub your fingers over the surface of the lung. What membrane on the surface of the lung makes it so smooth?

7. Free the connective tissue around the **pulmonary trunk** and expose its branches leading into the lung. Also, locate the **pulmonary veins** that empty into the left atrium. Can you find the **ligamentum arteriosum,** which is between the pulmonary trunk and aorta?

8. With a sharp scalpel, free the trachea from the lung tissue and trace it down to where it divides into the **bronchi.**

9. With a pair of scissors, cut off the trachea at the level of the top of the heart.

10. Now, cut the trachea down its posterior surface on the median line with a pair of scissors. The posterior surface of the trachea is opposite the heart.

    Extend this cut down to where it branches into the two primary bronchi. (Observe that the upper right lobe has a separate bronchus leading into it. This bronchus branches off some distance above where the primary bronchi divide from the trachea.)

11. Continue opening up the respiratory tree until you get down deep into the center of the lung. Note the extensive branching.

12. Insert a plastic drinking straw or a serological pipette into one of the bronchioles and blow into the lung.

    If a pipette is used, put the mouthpiece of the pipette into the bronchiole and blow on the small end.

    Note the great expansive capability of the lung.

13. Cut off a lobe of the lung and examine the cut surface. Note the sponginess of the tissue.

14. Record all your observations on the Laboratory Report.

Important: Wash your hands with soap and water at the end of the period.

Frog Lung Observation

To study the nature of living lung tissue in an animal, we will do a microscopic study of the lung of a frog shortly after pithing and dissection. A frog lung is less complex than the human lung, but it is made of essentially the same kinds of tissue. Proceed as follows:

Materials:
　frog, recently double pithed
　dissecting instruments and dissecting trays
　medicine droppers (optional)
　dissecting microscope
　Ringer's solution in wash bottle

1. Pin the frog, ventral side up, in a wax-bottomed dissecting pan.

2. With your scissors, make an incision through the skin along the midline of the abdomen.

3. Make transverse cuts in both directions at each end of the incision and lay back the flaps of skin.

4. Carefully, make an incision through the right or left side of the abdomen over the lung, parallel to the midline. Take care not to cut too deeply. The inflated lung should now be visible.

5. With a probe, gently lift the lung out through the incision. *Do not perforate the lung!* From this point on keep the lung moist with Ringer's solution.

6. If the lungs of your frog are not inflated, insert a medicine dropper into the slitlike glottis on the floor of the oral cavity and pump air into them by squeezing the bulb. Deflated lungs may be the result of excessive squeezing during pithing.

7. Observe the shape and general appearance of the frog lung. Note that it is basically saclike.

8. Place the frog under a dissecting microscope and examine the lung surface.

9. Locate the network of ridges on the inner walls. The thin-walled regions between the ridges represent the **alveoli.**

10. Look carefully to see if you can detect blood cells moving slowly across the alveolar surface. Careful examination of the lungs may reveal the presence of parasitic worms. They are quite common in frogs.

11. When you have completed this study, dispose of the frog, as directed, and clean the pan and instruments. Wash your hands with soap and warm water.

Histological Study

Prepared slides of the nasal cavity, trachea, and lung tissue will be available for study. If Turtox slides are used, the tissues will probably be from monkey or human organs. Tissues from other mammals, such as the mouse or rabbit, may also be available.

　Use figures HA-30 and HA-31 in the Histology Atlas for reference, and follow these suggestions in doing your study:

Materials:
　prepared slides of nasal septum (H6.1), trachea (H6.41), and lung tissue (H6.52)

Nasal Septum　Scan this slide first with low power to locate the **nasal septum** and **nasal concha.** See illustration A, figure HA-30 for reference. Identify the bony tissue and nasal epithelium.

Study the ciliated epithelium and the tissues beneath it with high-dry magnification. Look for the **olfactory** (Bowman's) **glands** (illustration B), which are located in the lamina propria. The secretion produced by these glands helps to keep the epithelial surface moist and facilitates the solution of substances that stimulate the olfactory receptors.

Observe that a large number of unmyelinated **olfactory fibers** lie between the glands and the ciliated epithelium. These fibers carry nerve impulses from the olfactory receptors that are dispersed among the cells of the epithelium.

Trachea Identify the hyaline cartilage, ciliated epithelium, lamina propria, and muscularis mucosae.

Look for goblet cells on the epithelial layer. Refer to illustrations C and D, figure HA-30.

Lung Tissue Examine with the high-dry objective. Look for the thin-walled **alveoli,** a **bronchiole,** and blood vessels. Consult illustrations A and B of figure HA-31 for normal lung tissue.

After comparing "smokers lungs" (illustration C) with normal lung tissue, can there be any justification for smoking?

Laboratory Report

Complete Laboratory Report 56 for this exercise.

Hyperventilation and Rebreathing

57

Breathing is a vital activity in which the respiratory muscles are stimulated to increase and decrease the volume of the thoracic cavity to ventilate the alveoli sufficiently to satisfy the exact oxygen requirements of all cells in the body. The center of control for this activity is the respiratory center, which is located in the medulla oblongata. The amounts of CO_2, H^+, and O_2 in the blood and cerebrospinal fluid are the chemical stimuli that act directly, or indirectly, on the respiratory center to regulate the muscles of respiration.

Of the three, the most important factor influencing the respiratory rate is the hydrogen ion (H^+) concentration. The least important factor is the amount of oxygen in the blood. Carbon dioxide facilitates the hydrogen ion concentration by forming carbonic acid in the blood and cerebrospinal fluid.

The pH of blood and tissue fluid during normal breathing is around 7.4. If forced deep breathing (hyperventilation) takes place, the pH of these fluids may be elevated to 7.5 or 7.6 as carbon dioxide is blown off. The reduced hydrogen ion concentration depresses the respiratory center, lessening the desire for increased alveolar ventilation.

Hyperventilation in individuals caused by unconscious deep breathing or sighing, can cause a drop in blood pressure, extreme discomfort, dizziness, and even unconsciousness. All the symptoms are due to the washing out of CO_2 from the blood, causing *alkalosis*. The carbon dioxide depletion can be quickly restored to the blood by **rebreathing** into a paper bag for several minutes. It is these two phenomena that will be studied here.

The symptoms of hyperventilation, and the means of compensating for it, are studied here by two methods: by recording on a Physiograph® or Duograph or by not recording. The nonrecording procedures are presented first. Separate procedures are provided for the Physiograph® and Duograph. If recording equipment is to be used, it is recommended that students do the nonrecording portion as well. Both experiments have unique characteristics.

Nonrecording Experiment

Proceed as follows to observe the characteristics of hyperventilation, and to compensate for it with proper rebreathing techniques.

Materials:
 paper bag

1. Breathe very deeply at a rate of about 15 inspirations per minute for 1 or 2 minutes. Do not hurry the rate; concentrate on breathing deeply. Observe that the following symptoms occur:
   - It will become increasingly difficult to breathe deeply.
   - A feeling of dizziness will develop.

   Two events probably account for this: first, the blood pressure drops due to dilation of the splanchnic vessels, which lessens blood to the brain; and second, the reduced CO_2 content of the blood causes vasoconstriction of the cerebral blood vessels, further reducing the blood supply.
2. Now, place a paper bag over your nose and mouth, holding it tightly to your face and breathing deeply into it for about **3 minutes.** Note how much easier it is to breathe into the bag than into free air. Why is this?
3. Allow your breathing rate to return to normal and breathe, normally, for 3 or 4 minutes. At the end of a normal inspiration, *without deep inhaling,* pinch your nose shut with the fingers of one hand, and hold your breath as long as you can. Do this three times, timing each duration carefully. Record these times and calculate the average on the Laboratory Report.
4. Now, breathe deeply, as in step 1, for **2 minutes,** and then hold your breath as long as you can. Record your results on the Laboratory Report.
5. Exercise by running in place for **2 minutes,** and then hold your breath as long as you can. Record your results on the Laboratory Report.

Recording Experiment

To produce a record of ventilation during hyperventilation and rebreathing, it will be necessary to use a bellows transducer as shown in figures 57.1 and 57.2. Separate instructions will be provided for the Physiograph® and Duograph. The Duograph is preferable to the Unigraph here for Gilson equipment users since the experiment can be expanded to two-channel recording experiments in the next two exercises. It is conceivable that, where time permits, more than one of these experiments may be performed in a single laboratory period.

Materials:

for Physiograph® setup:
> Physiograph® with transducer coupler
> bellows pneumograph transducer
> > (PN 705-0190)
> cable for bellows pneumograph transducer

for Duograph setup:
> Duograph
> Statham T–P231D pressure transducer
> Gilson T–4030 chest bellows
> stand and clamp for transducer

for both setups:
> nose clamp
> paper bag

Physiograph® Setup

Set up one channel, using the transducer coupler and bellows, as shown in figure 57.1.

1. Connect the bellows pneumograph transducer to the transducer coupler with a transducer cable. Be sure to match up the yellow dots on the connectors.
2. Balance the recording channel according to the instructions in appendix B.
3. Attach the bellows pneumograph to the subject as follows:
   a. With the channel amplifier RECORD button in the OFF position, and the valve at the end of the rubber bellows open (turned counterclockwise), place the transducer on the subject's chest and fasten it with the attached leather strap.
   b. Locate the bellows as high as possible where it will get the greatest chest movement during breathing. Note that the transducer cable should be positioned over the shoulder to help support the weight of the transducer.
4. Start the paper advance at 0.25 cm/sec and lower the pens to the recording paper. Activate the timer.
5. Position the recording pen with the POSITION control to a point that is 2 cm below the center line.

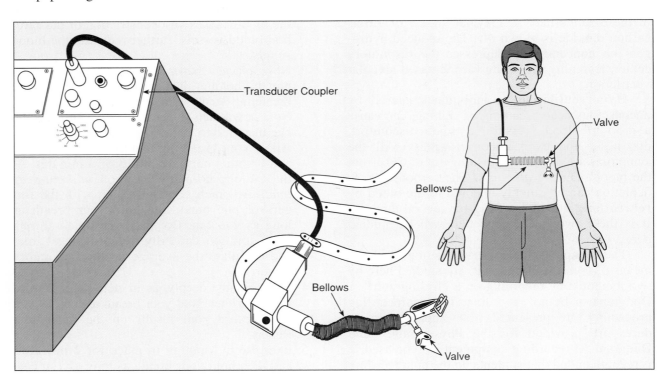

Figure 57.1 Physiograph® setup for monitoring respiration.

6. Close the valve on the bellows, while pinching the bellows slightly. The valve is closed by rotating it completely clockwise.
7. Place the channel amplifier RECORD button in the ON (down) position.
8. Starting at the 1000 position, rotate the outer channel amplifier sensitivity control knob to higher sensitivities (lower numbers) until normal shallow ventilation by the subject gives pen deflections of approximately 2 centimeters. Proceed to perform the following investigations.

Normal Respiration With the subject seated and relaxed, start the paper moving at 0.25 cm/sec, lower the pens, and activate the timer.

1. Place the amplifier RECORD button in the ON position.
2. Record one page of normal respiratory activity.
3. Increase the paper speed to 5 cm/sec and record several pages of normal respiratory activity.
4. Compare your tracings with sample records A and B in figure SR-6 of appendix C.

Rebreathing Experiment Place a nose clamp on the subject and proceed as follows to note the effects of rebreathing into a paper bag.

1. Set the paper speed at 0.25 cm/sec and continue recording.
2. Record about a half page of normal rebreathing, and then place a paper bag tightly over the subject's mouth.
3. Allow the subject to breathe into and out of the bag until visible changes occur in the inspiratory rate and depth.
4. Remove the bag and allow the respiratory rate to return to normal.
5. Compare your tracings with sample records C and D of figure SR-6, appendix C.

Hyperventilation While recording at 0.25 cm/sec, have the subject hyperventilate by inhaling and exhaling as deeply as possible for 1½ to 2 minutes.

1. **When the subject begins to feel severe discomfort,** tell the subject to stop the deep breathing and allow the respiratory movements to normalize.
2. During normalization, look for changes that occur in the rate and amplitude of respiratory movements as compared with prehyperventilation recordings.
3. Terminate the recordings.
4. Compare your tracings with the Physiograph® sample records in figure SR-7, appendix C.

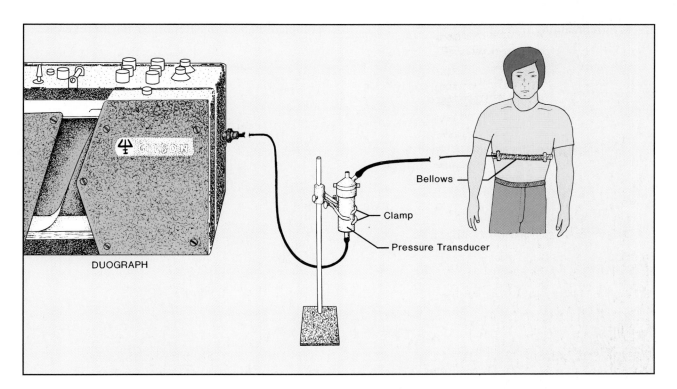

Figure 57.2 Duograph setup for monitoring respiration.

Duograph Setup

Utilize one channel of the Duograph to hook up a subject according to the following instructions. Figure 57.2 illustrates the setup.

1. Mount the Statham pressure transducer on the transducer stand with a clamp and plug the jack from the transducer into the receptacle of channel 1 on the Duograph.
2. Attach the chest bellows to the subject at the highest point where the greatest respiratory movements occur.
3. Attach the tubing from the bellows to the Statham transducer.
4. Activate the Duograph at slow speed, noting the trace.
5. Adjust the excursion for normal breathing at about 1 centimeter. Proceed as follows to record normal, rebreathing, and hyperventilation activities.

Normal Respiration   With the subject seated and relaxed, make the following recordings of normal respiratory activity.

1. For about 1½ to 2 minutes record normal breathing at the slow speed (2.5 mm/sec).
2. Increase the speed to fast (25 mm/sec) and record for another 1½ to 2 minutes.

Rebreathing Experiment   Place a nose clamp on the subject and proceed as follows to note the effects of rebreathing into a paper bag.

1. Return the paper speed to slow and continue recording.
2. For about 30 seconds, record normal ventilation and then place a paper bag tightly over the subject's mouth.
3. Allow the subject to breathe into and out of the bag until visible changes occur in the respiratory rate and depth.
4. Remove the bag and allow the respiratory rate to return to normal.

Hyperventilation   Continue recording at slow speed and have the subject hyperventilate by inhaling and exhaling, as deeply as possible, at a rate of one respiratory movement per second.

1. **When the subject begins to feel severe discomfort,** tell the subject to stop the deep breathing and continue the recording as the respiratory movements normalize.
2. During normalization, look for changes that occur in the rate and amplitude of respiratory movements as compared with prehyperventilation recordings.
3. Terminate the recording.

Laboratory Report

Complete Laboratory Report 57 for this exercise.

The Diving Reflex

58

Many aquatic air-breathing animals have evolutionary adaptations that enable them to remain submerged under water for quite long periods of time. The "diving reflex" (or dive reflex) is one of these adaptations. In these animals there are prompt cardiovascular and respiratory adjustments that are triggered by partial entry of water into the air passages. These adjustments include a change in the heart rate, the shunting of blood from less essential tissues, and a cessation of breathing. In humans, these changes are less pronounced, but some adjustments are made. It is this series of changes that have accounted for the survival of some individuals, usually young children, who have fallen through the ice and remained under water for unusually long periods of time.

In an attempt to observe the existence of this reflex in humans, we will monitor cardiovascular changes with a pulse transducer while a subject's face is immersed in cold water. Needless to say, the subject for this experiment should be one who has no aversion to facial submersion.

The equipment setup for this experiment requires two channels on a recorder. To record the respiratory rate on one channel, a chest bellows will be used as in the last experiment (hyperventilation). To detect changes in the heart rate, a pulse transducer will be used on the second channel. Figures 58.1 and 58.2 illustrate the setups for a Physiograph® and a Duograph. Separate instructions are provided for setting up each system.

The immersion maneuver may take place at your laboratory station if large pans are available, or it may be performed in a sink at a perimeter counter. In either case the recording equipment will have to be set up wherever the immersion is to take place.

In this experiment the subject's face will be submerged under water at two different temperatures: 70° F and 60° F (cooled with ice). Before submersion, however, recordings will be made to establish pulse rates (1) during normal breathing and (2) while holding one's breath.

After norms are established, the subject will submerge his or her face in water, first at 70° F and then at 60° F to note changes in the heart rate.

Hyperventilation will also be performed. The purpose of these tests is to determine if any cardiovascular changes are detectable during submersion, and, to what extent, if any, does temperature affect heart rate.

Equipment Setup

If this experiment is being performed along with Exercise 57, the only additional equipment needed is a pulse transducer.

Physiograph® Setup

By referring to figure 58.1 follow these instructions to hook up the equipment to a subject.

Materials:
> Physiograph® with two transducer couplers
> bellows pneumograph transducer (Narco PN 705–0190)
> cable for bellows pneumograph transducer
> pulse transducer (Narco 705–0050)
> large pan of water (or sink basin)
> ice, towel, thermometer

1. Attach the finger pulse transducer and chest bellows to the subject. As in the previous experiment, locate the bellows high enough to get maximum excursion during breathing.
2. Make all connections in such a way that immersion movement does not tug on the cables.
3. While the subject is being prepared, fill the pan or sink with tap water, adjusting the temperature to 70° F.
4. Turn on the power switch and set the paper speed at 0.25 cm/sec.
5. Start the chart paper moving and lower the pens to the paper. Make certain that the timer is activated.
6. Use the POSITION controls on both channels to position the pens so that they are recording about one centimeter below the center line.
7. Place the RECORD button in the ON position.
8. Adjust the amplifier sensitivity controls so that the pulse wave amplitude is 2 cm and the tidal ventilation amplitude is 1 cm. You are now

ready to perform the tests. Proceed to instructions under "Test Procedure."

Duograph Setup

By referring to figure 58.2 follow these instructions to hook up the equipment to a subject.

Materials:
    Duograph
    Statham T-P231D pressure transducer
    stand and clamp for Statham transducer
    Gilson T-4030 chest bellows
    Gilson pulse pickup (T4020)
    Gilson A-4023 adapter
    large pan of water (or sink basin)
    ice, large bath towel, thermometer

1. Attach the tubing from the chest bellows to the Statham transducer, and plug the jack from the transducer into the receptacle for channel 1.
2. Attach the A-4023 adapter to the receptacle for channel 2 of the Duograph. Make certain that the lock nut is tightened securely, and that the phone jack of the pulse pickup is plugged into it.

3. Attach the finger pulse pickup and chest bellows to the subject. Keep the chain of the bellows as high into the armpits as possible.
4. Make all connections in such a way that immersion movement does not tug on the cables.
5. While the subject is being prepared, fill the pan or sink with tap water, adjusting the temperature to 70° F.
6. Set the speed control lever in the slow position, and the stylus heat control knobs of both channels in the two o'clock position.
7. Set the Gain for both channels at 2 MV/CM, and turn the sensitivity control knobs of both channels completely counterclockwise (lowest value).
8. Set the Mode for channel 1 on TRANS and for channel 2 on DC.
9. Activate both channels on the Duograph to observe the traces.
10. Adjust the excursion for tidal breathing on channel 1 to 1 cm, using the sensitivity and Gain controls.
11. Adjust the pulse amplitude on channel 2 to about 2 cm.
12. You are now ready to perform the tests. Proceed to instructions under "Test Procedure."

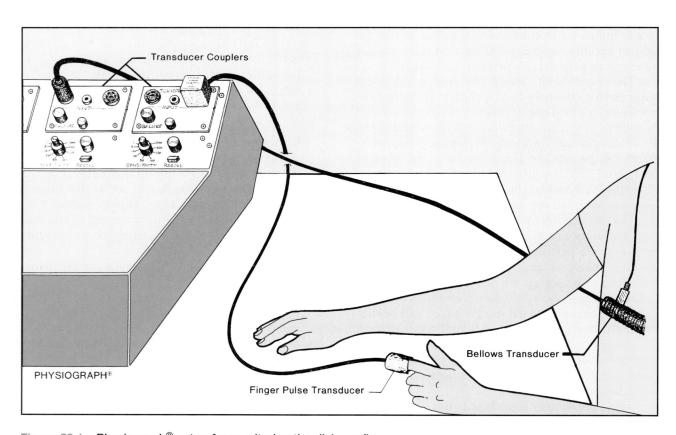

Figure 58.1 Physiograph® setup for monitoring the diving reflex.

Test Procedure

Now that the equipment is properly adjusted and activated, proceed as follows:

1. Establish a reference by recording for at least **one minute** before facial submersion. During this time, the subject will breathe regularly (tidal inspirations) without speaking or hyperventilating. **Do not readjust the Gain or sensitivity controls from this point on in the experiment.**

2. To further establish reference values, direct the subject to hold his or her breath for **one minute** after a tidal inspiration, without prior deep breathing, or hyperventilating.

3. Allow the subject to recover for **2 to 3 minutes** by normal breathing.

4. After recovery, have the subject completely submerge his or her face into the 70° F water. Mark the chart at this point. On the Duograph the event button should be depressed at the exact moment of immersion.
   Immersion should last 15 to 30 seconds. Use a large towel to protect the subject's clothing from getting wet.

5. Repeat the immersion procedure to confirm the results.

6. Chill the water to 60° F with ice cubes and remove the ice cubes once the correct temperature is reached.

7. Have the subject completely submerge his or her face again in the colder water for **15 to 20 seconds.**

8. After approximately 2 to 3 minutes recovery, have the subject repeat step 7.

9. Have the subject hyperventilate vigorously for **30 seconds** and then hold his or her breath for **as long as possible.** This is done without submersion. Mark the chart for this activity.

10. Once regular breathing has resumed, have the subject hyperventilate again for **30 seconds,** and then submerge his or her face for **as long as possible.** Be sure to mark the chart.

11. Terminate the recording, disconnect the subject, put away the apparatus, clean the work area, and return all materials to their proper places.

12. Analyze the tracing as a team and label it with additional information, as needed.

Laboratory Report

Complete the first portion of combined Laboratory Report 58, 59.

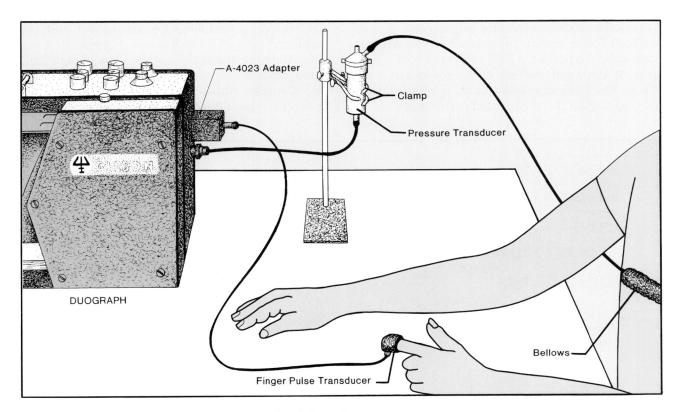

Figure 58.2 Duograph setup for monitoring the diving reflex.

59

The Valsalva Maneuver

Antonio Valsalva, an Italian anatomist, discovered in 1723 that the middle ear could be filled with air if one expires forcibly against closed mouth and nostrils. This action became known as the *Valsalva maneuver*. The term also signifies attempting to expire air from the lungs against a closed glottis. In this maneuver, the abdominal muscles and internal intercostal muscles greatly increase intraabdominal and intrathoracic pressures. This momentary increase in intrathoracic pressure affects both the pulse and blood pressure.

Figure 59.1 illustrates what happens to the blood pressure and pulse in this maneuver. Note that during the maneuver, the blood pressure goes up at the start of the straining, and then it falls a short time later. After the maneuver, the blood pressure increases considerably, and the heart rate slows somewhat.

The initial rise in blood pressure is caused by the application of intrathoracic pressure on the thoracic aorta. The subsequent drop in blood pressure is caused by decreased cardiac output resulting from reduced venous return. The reduced venous return is caused by the restricted flow of blood into the right side of the heart through the compressed venae cavae. The final surge in increased cardiac output is a compensation for the reduced prior output.

The Valsalva maneuver is more than just a stunt to observe in the laboratory. Its real merit is in the diagnosis of two pathological conditions: autonomic insufficiency and hyperaldosteronism.

For example, patients who suffer from *autonomic insufficiency* will exhibit no pulse changes during this maneuver. The cause of this condition is not well understood.

Patients who have *hyperaldosteronism* (excess production of aldosterone by the adrenal cortex) fail to show pulse rate changes and blood pressure rise after this maneuver. This test is often used, diagnostically, on patients suspected of having adrenal cortex tumors that are producing excessive amounts of aldosterone.

Students will work in teams of four or five to monitor blood pressure and pulse of one of the team members during this maneuver. Note that the equipment setups for this experiment are similar to those used in the last experiment, except that the bellows pneumograph transducer will be replaced by a pulse transducer. Figures 59.2 and 59.3 illustrate the setups for the Physiograph® and Duograph.

Caution: Although the Valsalva maneuver is a perfectly safe, normal physiological phenomenon for

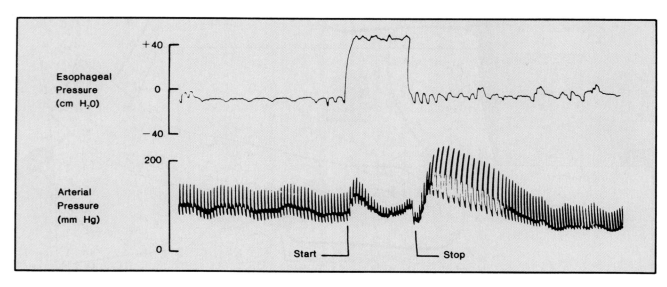

Figure 59.1   Blood pressure and pulse response in the Valsalva maneuver.

healthy individuals, any team member who has a history of heart disease should not be used as a subject.

Equipment Setup

In the last experiment a chest bellows and pulse transducer were used. In this experiment the chest bellows will be replaced with an arm cuff. Make the following hookups and adjustments.

Physiograph® Setup

By referring to figure 59.2, follow these instructions to hook up the equipment to a subject. Note that an ESG coupler replaces one of the transducer couplers that was used in the last experiment.

Materials:

Physiograph® with transducer and ESG couplers
adult pressure cuff (Narco PN 712-0016)
pulse transducer (Narco 705-0050)

1. Hook up the pressure cuff to the ESG coupler according to figure 59.2.
2. Turn on the power switch, and set the paper speed at 0.25 cm/sec.
3. Balance and calibrate the ESG coupler according to the instructions on page 296.

4. From this point on do not touch the balance and position control knobs.
5. Place the cuff on the upper right arm, making certain that the microphone is positioned over the brachial artery.
6. Calibrate the cuff microphone according to the instructions on page 296.
7. Insert the subject's left index finger into the finger pulse pickup, and plug the jack of the pulse transducer into the transducer channel.
8. Start the chart paper moving, again, lower the pens to the paper, and make certain the timer is activated.
9. Adjust the amplifier sensitivity controls on the transducer channel so that the pulse wave amplitude is 1.5 to 2 cm.
10. You are now ready to start monitoring. Proceed to "Test Procedure."

Duograph Setup

Set up the equipment according to the following instructions, using figure 59.3 for reference. Note that the same Statham pressure transducer that was used in the last experiment with a chest bellows will be used here with an arm cuff.

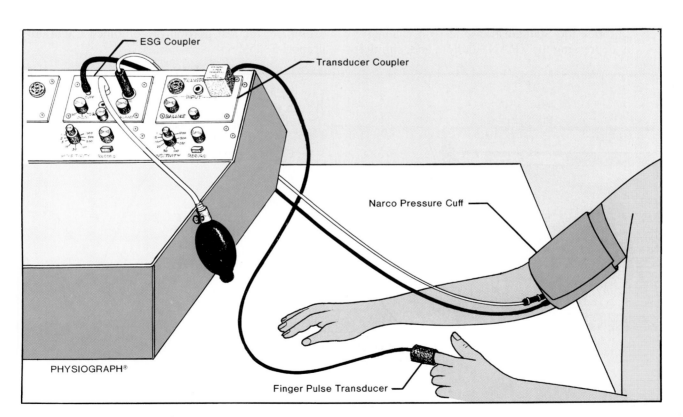

Figure 59.2 Physiograph® setup for monitoring effects of the Valsalva maneuver.

Materials:
 Duograph
 Statham T-P231D pressure transducer
 adult arm cuff
 stand and clamp for T-P231D transducer
 Gilson T-4020 finger pulse pickup
 Gilson A-4023 adapter for pulse pickup

1. Place the cuff on the upper right arm, securing it tightly in place. Attach the tubing from the cuff to the Statham transducer.
2. Test the cuff by pumping some air into it and noting if the pressure holds when the metering valve is closed on the bulb. Release the pressure for the comfort of the subject, while other adjustments are made.
3. Insert the subject's left index finger into the finger pulse pickup.
4. Plug in the Duograph and turn on the power switch.
5. Set the speed control lever in the slow position and the stylus heat control knobs of both channels in the two o'clock position.
6. Set the Gain for channel 1 at 2 MV/CM, and for channel 2 at 1 MV/CM.
7. Turn the sensitivity control knobs of both channels completely counterclockwise (lowest values).
8. Set the Mode for channel 1 on DC, and for channel 2 on TRANS.
9. Calibrate the Statham transducer on channel 2 by depressing the TRANS button, and, simulta-

neously, adjusting the sensitivity knobs to get 2 cm deflection. The calibration line represents 100 mm Hg pressure.
10. Adjust the sensitivity on channel 1 so that you get 1.5 to 2 cm deflection on the pulse.
11. You are now ready to start monitoring. Proceed to "Test Procedure."

Test Procedure

Now that the equipment is properly adjusted and activated, proceed as follows:

1. Pump air into the cuff and establish a baseline for one minute.
2. Have the subject take a deep breath, and perform a Valsalva maneuver, exerting as much internal pressure as possible.
3. Mark the chart at the instant the maneuver is begun. On the Duograph a mark can be made by pressing the event marker button and holding it down until the maneuver ends.
4. Repeat the maneuver several times to provide enough chart material for each team member.
5. Repeat the experiment, using other team members as subjects.

Laboratory Report

Complete the last portion of combined Laboratory Report 58, 59.

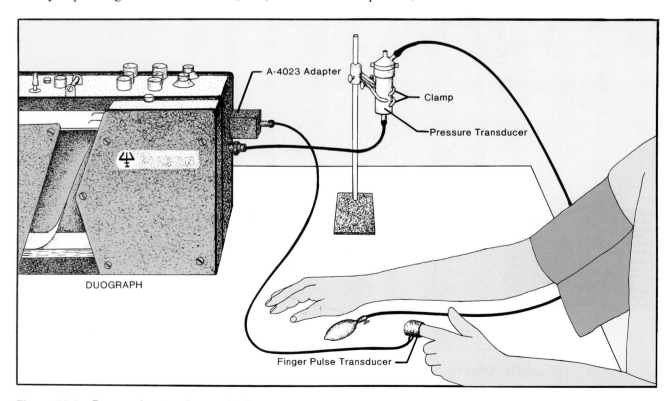

Figure 59.3 Duograph setup for monitoring the effects of the Valsalva maneuver.

Spirometry:
Lung Capacities

60

The volume of air that moves in and out of the lungs during breathing is measured with an apparatus called a **spirometer.** Two types are available: the hand-held (Propper) type shown in figure 60.2, and the tank-type recording spirometer (figure 61.1). When a recording spirometer is used, a **spirogram** similar to figure 60.1 is made. The Propper spirometer is designed, primarily, for screening measurements of vital capacity.

In this exercise we will use the Propper spirometer to determine individual respiratory volumes. Although this convenient device cannot measure inhalation volumes, it is possible to determine most of the essential lung capacities. While most members of the class are working with Propper spirometers, some students can be working with a recording spirometer as directed in Exercise 61.

Proceed as follows to measure or calculate the various respiratory volumes using a Propper spirometer.

Materials:
Propper spirometer
disposable mouthpieces (Ward's Nat'l Science Est., Catalog No. 14W5070)
70% alcohol

Tidal Volume (TV)

The amount of air that moves in and out of the lungs during a normal respiratory cycle is called the *tidal volume.* Although sex, age, and weight determine this and other capacities, the average normal tidal volume is around 500 ml. Proceed as follows to determine your tidal volume.

1. Swab the stem of the spirometer with 70% alcohol and place a disposable mouthpiece over the stem.
2. Rotate the dial of the spirometer to zero as illustrated in figure 60.2.
3. After three normal breaths, expire three times into the spirometer while inhaling through your nose. Do not exhale forcibly. **Always hold the spirometer with the dial upward.**
4. Divide the total volume of the three breaths by 3. This is your tidal volume. Record this value on the Laboratory Report.

Minute Respiratory Volume (MRV)

Your *minute respiratory volume* is the amount of tidal air that passes in and out of your lungs in one minute. To determine this value, count your respirations for

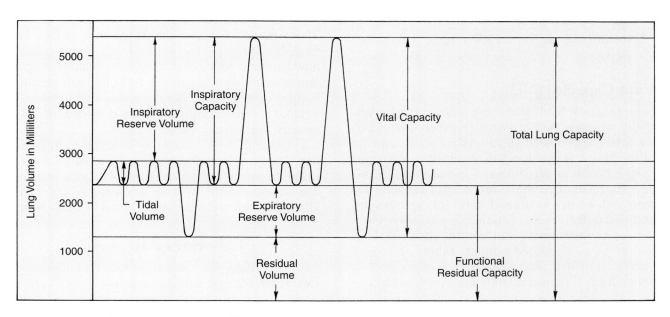

Figure 60.1   Spirogram of lung capacities.

337

Figure 60.2 Dial face of spirometer is rotated to zero prior to measuring exhalations.

one minute and multiply this number by your tidal volume. Record this volume on the Laboratory Report.

Expiratory Reserve Volume (ERV)

The amount of air that one can expire beyond the tidal volume is called the *expiratory reserve volume*. It is usually around 1100 ml. To determine this volume, do as follows:

1. Set the spirometer dial on 1000.
2. After making three normal expirations, expel all the air you can from your lungs through the spirometer.
3. Subtract 1000 from the reading on the dial to determine the exact volume.
4. Repeat steps 1–3 two more times to get an average ERV.
5. Record the average ERV on the Laboratory Report.

Vital Capacity (VC)

Note in figure 60.1, that if we add the tidal, expiratory reserve, and respiratory reserve volumes we arrive at the total functional or *vital capacity* of the lungs. This value is determined by taking as deep a breath as possible and exhaling all the air possible. Although the average vital capacity for men and women is around 4500 ml, age, height, and sex do affect this volume appreciably. Even the established norms can vary as much as 20% and still be considered normal. Tables of normal values for men and women are provided in appendix A. Proceed as follows:

1. Set the spirometer dial on zero.
2. After taking two or three deep breaths and exhaling completely after each inspiration, take one final deep breath and exhale all the air through the spirometer. A slow, even, forced exhalation is optimum.
3. Repeat two or more times to see if you get approximately the same readings. Your VC should be within 100 ml each time.
4. Record the average on the Laboratory Report.
5. Consult tables III and IV in appendix A for predicted (normal) values.

Inspiratory Capacity (IC)

If you take a deep breath to your maximum capacity after emptying your lungs of tidal air, you will have reached your maximum inspiratory capacity. This volume is usually around 3000 ml. Note in the figure 60.1 spirogram that this volume is the sum of the tidal and inspiratory reserve volumes.

Since this type of spirometer cannot record inhalations, it will be necessary to calculate this volume, using the following formula:

$$IC = VC - ERV$$

Inspiratory Reserve Volume (IRV)

This is the amount of air that can be drawn into the lungs in a maximal inspiration after filling the lungs with tidal air. Since the IRV is the inspiratory capacity less the tidal volume, make this subtraction and record your results on the Laboratory Report.

Residual Volume (RV)

The volume of air in the lungs that cannot be forcibly expelled is the *residual volume*. No matter how hard one attempts to empty one's lungs, a certain amount, usually around 1200 ml, will remain trapped in the alveoli. The magnitude of the residual volume is often significant in the diagnosis of pulmonary impairment disorders.

Although the RV cannot be determined by ordinary spirometric methods, it can be done by washing all the nitrogen from the lungs with pure oxygen and measuring the volume of nitrogen expelled. Obviously, this measurement will not be made here.

Laboratory Report

After recording your lung capacities on Laboratory Report 60, 61, evaluate your results.

Spirometry:
The FEV$_T$ Test

Impairment of pulmonary function in the form of asthma, emphysema, and cardiac insufficiency (left-sided heart failure) can be detected with a spirometer. The symptom common to all these conditions is shortness of breath, or *dyspnea*. The lung capacity that is pertinent in these conditions is the vital capacity.

In several types of severe pulmonary impairment, the vital capacities of patients may exhibit nearly normal values. However, if the rate of expiration is recorded on a kymograph and timed, the extent of pulmonary damage becomes quite apparent. A timed expiratory test is called the *Timed Vital Capacity* or *Forced Expiratory Volume* (*FEV$_T$*). Figure 61.1 illustrates a spirometer (Collins Recording Vitalometer) that will be used in this experiment. The sample record produced on such a kymograph is shown in figure 61.2. A record of this type is called an *expirogram*.

To perform a timed vital capacity test, one makes a maximum inspiration and then expels all the air from the lungs into the spirometer *as fast as possible*. The moving kymograph drum has chart paper on it that is calibrated vertically in liters and horizontally in seconds. A pen poised on the drum produces a record on the chart paper that reveals how much air is expelled in a given period of time.

Approximately 95% of a normal person's vital capacity can be expelled within the first three seconds. From a diagnostic standpoint, however, the *percentage of vital capacity that is expelled during the first second* is of paramount importance. An individual with no pulmonary impairment should be able to expel 75% of the lung's total capacity during the first second. Individuals with emphysema and asthma, however, will expel a much lower percentage due to "air entrapment." Given sufficient time to exhale, they might be able to do much better.

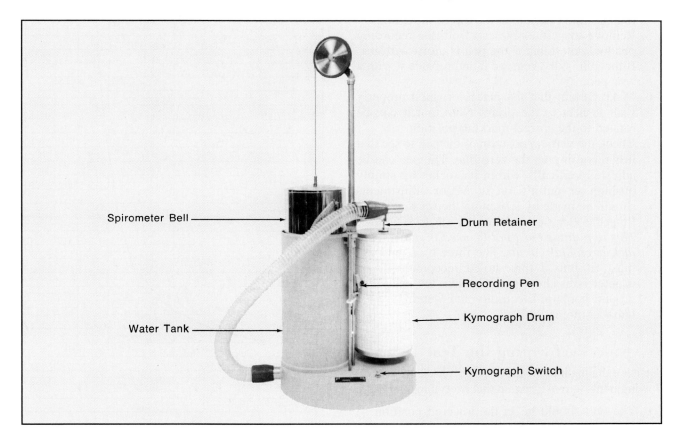

Figure 61.1   A Collins recording spirometer.

In this experiment, each member of the class will have an opportunity to determine his or her FEV$_T$. Disposable mouthpieces will be available for each person doing the test. Follow this procedure:

Materials:
spirometer (Collins Recording Vitalometer)
disposable mouthpieces (Collins #P612)
kymograph paper (Collins #P629)
noseclip
Scotch tape, dividers, and ruler

Preparation of Equipment

Prior to performing this test, prepare the kymograph in the following manner.

1. Remove the spirometer bell and fill the water tank to within 1½″ of the top. *Before filling, be sure to close the drain petcock.*
2. Replace the bell, making certain that the bead chain rests in the pulley groove.
3. Remove the kymograph drum by lifting the drum retainer first and wrap a piece of chart paper around the drum. Attach paper to the drum with Scotch tape. Make certain that the right edge overlaps the left edge.
4. Replace the kymograph drum, checking to see that the bottom spindle is in the hole in the bottom of the drum. Lower the kymograph drum retainer into the hole in the top of the drum.
5. Remove the protective cap from the recording pen and determine if the pen is moist with ink. If the pen is dry, remove it and replace it with a new pen.
6. Make certain that the pen is oriented properly with respect to the drum. Note that it can be rotated to the correct marking position.
7. Check the vertical position of the pen to see that it is recording on the zero line. The pen can be adjusted vertically within its socket by simply pushing or pulling on it. Major adjustments should be made by loosening the set screw.
8. Plug the electric cord into an electrical outlet. *This apparatus must not be used in outlets that lack a ground circuit.* The three-pronged plug must fit into a three-holed receptacle. If an adapter is used, be sure to make the necessary ground hookup. Electricity and water can be a lethal combination!

Performing the Test

Before exhaling into the mouthpiece, check out the following items to make certain that everything is ready:

• The bell should be in the lowered position so that the recording pen is exactly on 0.

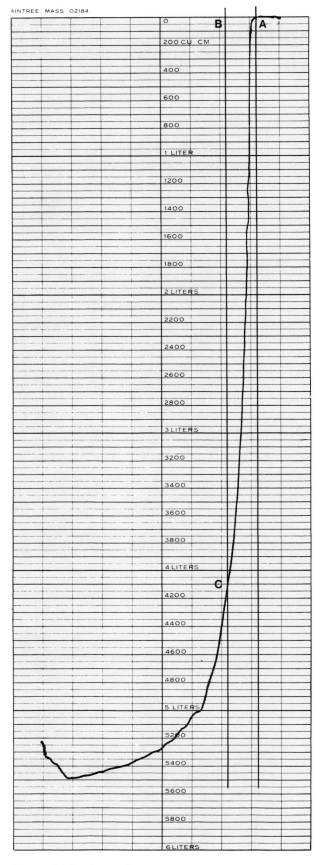

Figure 61.2   An expirogram.

340

- The drum should have a clean sheet of paper on the kymograph drum.
- A sterile mouthpiece should be in place on the end of the breathing tube.

Proceed as follows:

1. Apply a noseclip to the subject to prevent air leakage through the nose. Allow the subject to hold the tube.
2. Before turning on the instrument explain to the subject that:
   - he or she will be taking as deep a breath as possible,
   - he or she will expel as much air as possible into the mouthpiece, *as rapidly and completely as possible,* and
   - inhalation should take place *before* the mouthpiece is placed in his or her mouth.
3. Turn on the kymograph and tell the subject to perform the test as described above.
4. As the subject blows into the tube, encourage him or her to *push, push, push,* to get all the air out.
5. Turn off the kymograph.
6. After recording as many tracings as desired, remove the paper from the drum.

Analysis of Tracing

To determine the FEV$_T$ from this expirogram, proceed as follows:

1. Draw a vertical line through the starting point of exhalation (see point A, figure 61.2).
2. Set a pair of dividers to the distance between two vertical lines. This distance represents one second.
3. Place one point of the dividers on point A and establish point B to the left of point A. Draw a parallel vertical line through point B. This line intersects the expiratory curve at point C.

   The one-second volume (FEV$_1$), or one-second timed vital capacity, is represented by the distance B-C. This volume is read directly from the volume markings. In figure 61.2, it is 4100 ml. The total vital capacity is also read directly from the volume markings. It is 5500 ml.
4. Next, correct the recorded total vital capacity for body temperature, ambient pressure, and water saturation (BTPS). To do this, consult table 61.1 for the conversion factor and multiply the recorded vital capacity by this factor.

   Example: If the temperature in the spirometer is 25° C, the conversion factor is 1.075. Thus:

$$5500 \text{ ml} \times 1.075 = 5913 \text{ ml}$$

Table 61.1 Conversion factors for temperature differentials.

°C	°F	CONVERSION FACTOR
20	68.0	1.102
21	69.8	1.096
22	71.6	1.091
23	73.4	1.085
24	75.2	1.080
25	77.0	1.075
26	78.8	1.068
27	80.6	1.063
28	82.4	1.057
29	84.2	1.051
30	86.0	1.045
31	87.8	1.039
32	89.6	1.032
33	91.4	1.026
34	93.2	1.020
35	95.0	1.014
36	96.8	1.007
37	98.6	1.000

5. Also, convert the 1-second timed vital capacity (FEV$_1$) in the same manner.

$$4100 \text{ ml} \times 1.075 = 4408 \text{ ml}$$

6. Divide the FEV$_1$ by the total vital capacity.

$$\frac{4408}{5913} = 75\%$$

7. Determine how the subject's vital capacity compares with the predicted (normal) values in tables II and III of appendix A.

   Example: Individual in the above test was a 26-year-old male, 6'3" (184 cm) tall. From table III his predicted vital capacity is 4545 ml.

$$\frac{\text{Actual VC}}{\text{Predicted VC}} \times 100\% = \frac{5913}{4545} \times 100 = 131\%$$

Laboratory Report

Complete the last portion of Laboratory Report 60, 61.

62 Spirometry:
Using the Intelitool Spirocomp™

In this laboratory period we will use the Intelitool Spirocomp™ to perform all the spirometry experiments that are performed in Exercises 60 and 61. Figure 62.1 is of a spirogram that illustrates all of the lung respiratory volumes. Figure 62.2 illustrates the experimental setup that we will use. Note that a wet spirometer is used, which has an interface box that enables a computer to communicate with movements of the spirometer bell when an individual breathes into it. The following respiratory volumes will be measured or computed:

- Tidal Volume (TV)
- Vital Capacity (VC)
- Expiratory Reserve Volume (ERV)
- Inspiratory Reserve Volume (IRV)
- Residual Volume (RV)
- Total Lung Capacity (TLC)
- Forced Expiratory Volume (FEV_T)

The principal advantage of this computerized setup over the methods used in Exercises 60 and 61 is that all of the calculations are made *automatically* from data that are fed into the program. In addition, the program allows one to compute and average the values of up to 30 subjects. Respiratory efficiency comparisons of smokers with nonsmokers, athletes with nonathletes, and males with females can also be made easily.

Spirocomp can be run on four computer platforms: (1) IBM or compatible PC (MS-DOS 3.3 or higher), (2) Macintosh, (3) Windows 95, and (4) Apple II (ProDOS). For whichever platform you are using there will be separate instructions. It is assumed here that you have a basic knowledge of the platform you are using. Note also that in the materials list two Spirocomp manuals are available for reference.

In performing these experiments you will be working with a laboratory partner. While one acts as the subject, breathing into the spirometer, the other person can run the computer and prompt the subject as to the procedures.

Materials:
computer with monitor and printer
Phipps & Bird wet spirometer
Spirocomp scale arm and interface box
computer game port to transducer cable
Spirocomp program diskette
Spirocomp User Manual

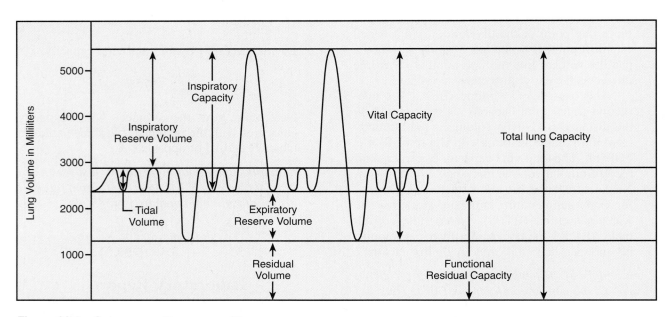

Figure 62.1   Spirogram of lung capacities.

Spirocomp Lab Manual
blank diskette for saving data
valve assembly (Spirocomp)
4 reusable mouthpieces
noseclip
stopwatch or watch with second hand
Lysol disinfectant (or Clorox bleach)
1 beaker (1 liter size) of 70% alcohol
container of mild soap solution large enough to
  immerse one-way valve assembly

Equipment Hookup

The Spirocomp scale arm and interface box will already be attached to the spirometer post (see figure 62.2), so all you have to do is hook up the components, fill the spirometer with water, and calibrate the Spirometer prior to starting any tests. Proceed as follows:

1. Fill the spirometer to a little over the fill line with fresh tap water, and add Lysol or Clorox to the water for disinfection.
2. Plug in the computer and printer power cords to *grounded* 110-volt electrical outlets.
3. *With the computer turned off,* connect the transducer to computer cable between the interface box and the game port of the computer.
4. Insert the program diskette and turn on the computer.

Preliminary Preparations

Before you start the experiments make the following preparations:

1. Place the four reusable mouthpieces, valve assembly, and noseclip in a beaker of 70% alcohol. Between subjects, always swish the valve assembly through the mild soap solution, a water rinse, and immerse in 70% alcohol before reusing.
2. Seat the subject in a straight high-backed chair and allow him or her to rest for five minutes before starting the test. Unless the subject is relaxed, calm, and motionless, normal respiratory volumes are unattainable. (Even slight shifting of the body during data acquisition can alter results.)

   In addition, the subject should not face his or her lab partner or the computer screen. Watching the results on the monitor can cause the subject to, consciously or unconsciously, alter the breathing pattern. (This is only a problem when measuring the TV and ERV.)
3. When breathing into the mouthpiece, the subject's nose must be securely closed with the noseclip. In addition, the lips must form a tight seal around the mouthpiece to avoid any air leakage.
4. Calibrate the transducer by following the procedures outlined by the software. If, during data acquisition, the chain slips on the scale arm, recalibrate the spirometer before continuing.

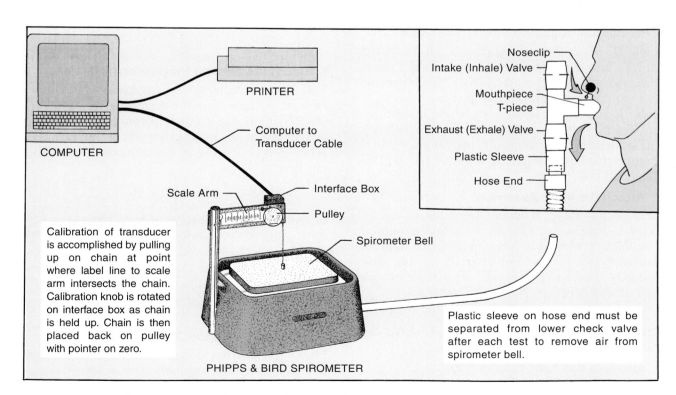

Figure 62.2 The Spirocomp setup for monitoring respiratory volumes.

Groups and Data Files

If there will be an attempt to compare nonsmokers with smokers, nonathletes to athletes, or pre- to post-exercise, your instructor will find it necessary to make specific group assignments so that separate data files can be made. If the class is quite small, the entire class may be a group.

You will be able to add records to any group data file, even if it has been saved to disk; so if there are two or more different lab sections, students from each section may be put into the same group data file. *One thing you don't have to do is separate the groups by gender because the software will do that automatically.*

Experimental Procedure

Now that all components are active, proceed as follows, using the instructions that apply to your particular computer platform.

1. Enter Subject Data.

MS-DOS: Select "Experiment Menu" from the main SPIROCOMP MENU and choose "New" group file. Enter the subject data as prompted. The Spirocomp data acquisition screen will then appear.

Macintosh, Windows 95: Select **"New Database"** from the **File** menu and name the file appropriately. Then select **"Add New Record"** from the **Acquire** menu. The Subject Data Entry window will appear. Enter the subject data and then click the **Volumes** button. The Spirocomp data acquisition window will appear.

Apple II: Select "Run Spirocomp" from the MAIN MENU, and then choose "Run Spirocomp" again. Enter the subject data as prompted. The Spirocomp data acquisition screen will then appear.

Note: If you do not know your height in centimeters, enter it in inches, followed by a quote mark ("). The software will automatically convert the value to centimeters. For example, a value of 69" will be converted to 175 cm.

2. Attach the Valve Assembly.

Remove the valve assembly from the beaker of alcohol, shake it dry, and attach it to the spirometer hose. Next, remove a reusable mouthpiece from the alcohol, shake it dry, and insert it on the valve assembly. Refer to figure 62.3 to make sure that your assembly is correct.

3. Put the Noseclip on the Subject.

Remove the noseclip from the alcohol, shake dry, and place it in position on the subject.

4. Measure the *Tidal Volume (TV).*

Note in figure 62.1 that the tidal volume is the amount of air that moves in and out of the lungs dur-

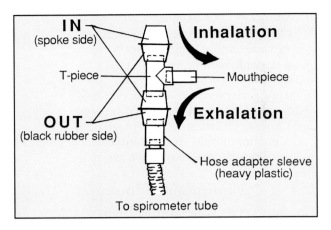

Figure 62.3 Correct method of attaching the valve assembly to the hose and mouthpiece.

ing a single normal respiratory cycle. Although sex, age, and weight determine this value, the average normal tidal volume is around **500 ml.** To get reliable results make certain that the subject is rested, not excited, and not moving.

Have the subject breathe normally into the spirometer for one or two cycles, and then *press the T key during an inhalation.* Follow the instructions on the screen.

The software will read the changes in the spirometer for three cycles, automatically stop, and then graph the average Tidal Volume. Record the TV here:

TV: _____

5. Count the *Resting Respiratory Rate.*

This value pertains to the number of breaths per minute. To determine the *resting respiratory rate (RRR),* simply use a watch to count the number of tidal cycles in one minute. A typical respiratory rate is between 12 and 15 cycles per minute. Record this rate here:

RRR: _____

6. Determine the *Resting Minute Volume.*

This value (RMV) is the amount of air moved in and out of the lungs in one minute. At rest, it is simply the product of the tidal volume and respiratory rate, or RRR × TV. Record here:

RMV: _____

7. Empty Air from the Bell Float.

Empty the air from the bell float of the spirometer by pulling the valve assembly apart as shown in figure 62.4. You can speed up the emptying process by pressing *gently* down on the bell float.

8. Determine the *Expiratory Reserve Volume (ERV).*

As indicated in figure 62.1, the ERV is the amount of air that one can expire beyond the tidal volume.

To measure this volume have the subject breathe normally into the spirometer for one or two cycles, and then *press the E key during an inhalation* and follow the instructions on the screen.

After reading two normal tidal cycles, the computer will instruct you to stop after the next (third) normal exhalation and wait for the next prompt. (This means that after exhaling the third time, the subject should stop all air movement and wait for the next prompt.)

The computer will then prompt you to **exhale maximally.** At this point the subject should empty his or her lungs as forcefully and as completely as possible. *Absolutely no inhaling should take place at this time.*

When the subject is no longer capable of expelling more air, the software will calculate the ERV and graph it on the screen.

If you felt that you made a mistake or did not get good results for some reason, repeat the reading by emptying the bell float and pressing the E key.

9. Determine the *Vital Capacity* and *FEV$_T$*

Note in figure 62.1 that the *vital capacity (VC)* is the sum of the tidal, expiratory reserve, and inspiratory reserve volumes. This value is determined by directing an individual to take as deep a breath as possible and exhaling all the air possible. The average vital capacity for men and women is around **4500 ml.**

While measuring the vital capacity you will also be able to determine *timed vital capacity,* also referred to as the *forced expiratory volume$_T$* (FEV$_T$).

To determine the FEV$_T$ of an individual, the subject takes a deep breath and expels it as fast as possible. An individual with no respiratory impairment should be able to expire 95% of his or her vital capacity within three seconds; however, it is the percentage of vital capacity that is expelled within the first second that is of paramount importance. **An individual with no pulmonary impairment should be able to expel 75% of his or her vital capacity within one second.** Individuals with emphysema and asthma will have a much lower percentage due to air entrapment.

In doing these two tests, both the **volume** and **rate** of air flow will be measured. It is for this reason that the subject must exhale as completely and rapidly as possible. The subject must keep trying to squeeze out every little bit of air until the software stops and graphs the values. Proceed as follows to do these two tests:

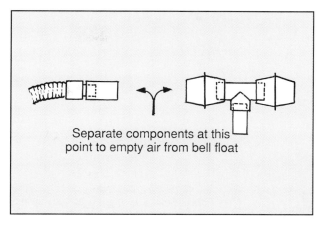

Figure 62.4 When emptying the bell float, separate the valve assembly from the hose as shown.

MS-DOS, Macintosh, Windows 95: Empty the air from the spirometer's bell as described in step 7 and then press the V key. Follow the instructions on the screen. You will be prompted to *inhale maximally* and then *exhale as forcefully, rapidly, and completely as possible* into the spirometer.

The computer will automatically detect the onset of exhalation and compute the forced expiratory volumes FEV$_1$, FEV$_2$, and FEV$_3$.

Apple II: Empty the the spirometer's bell as described in step 7 and then press the V key. Follow the instructions on the screen.

At this point let the subject operate the computer because the computer needs to know *exactly* when the subject begins to exhale.

The subject will be prompted to inhale maximally and then *press the V key as he or she is exhaling maximally. Emphasis must be on exhaling as forcefully, rapidly, and completely as possible.*

10. Repeat for Each Subject.

MS-DOS, Apple II: When you are satisfied with all the measurements for this subject, press the N key and repeat steps 1 through 9 for each subject in the group.

Macintosh, Windows 95: When you are satisfied with all the measurements for this subject from the **Acquire** menu, select "Add New Record," enter the subject data, click the **Volumes** button, and repeat steps 1 through 9 for each subject.

11. View the Group Averages.

MS-DOS: Press ESC to return to the main SPIRO-COMP MENU. Select "Review/Analyze Menu" and then choose "Graph Averages." A stacked bar graph representing the average volumes for all members of the group is displayed. Press the M or F key to view the averages for just the males or

females, respectively. A single record can be viewed by pressing the S key.

Macintosh, Windows 95: View the average volumes for all subjects, only males, or only females by choosing the appropriate items from the **Analyze** menu.

A stacked bar graph representing the average volumes for all members of the group is displayed. An option for viewing any single record is also available from the **Analyze** menu.

Apple II: From the MAIN MENU, select "Data Review (Graph)." You can then choose to view the average volumes for all males, all females, or the entire group. An option for viewing any single record is also available.

12. View the Spreadsheet.

MS-DOS: From the Spirocomp "Review/ Analysis Menu," select "View Spreadsheet." All the records from the entire group are displayed in row/column format. Averages and other summary information are displayed at the end of the listing. If there are more records than will fit on the screen, press the N key to display subsequent pages.

Macintosh, Windows 95: From the **Analyze** menu select "Summary Table." A new window appears with the data from the entire file displayed in row/column format. Averages are displayed at the top of the listing. Resize the window or use the scroll bars to view any portion of the data not visible.

Apple II: From the MAIN MENU, select "Data Review (Spreadsheet)." You can then choose to

view the average volumes for all males, all females, or the entire group.

An option for viewing any single record is also available. Averages and other summary information are displayed at the end of the listing. If there are more records than will fit on the screen, pressing the N key will display subsequent pages.

13. Save Data to Disk.

MS-DOS: Press ESC to return to the SPIROCOMP MENU and select "Data File/Disk Commands." Then choose "SAVE" File and give the file an appropriate name and press the Return (Enter) key.

Macintosh, Windows 95: Each record is automatically saved before a new one is added. To manually save the current record select "Save Record" from the **File** menu.

Apple II: From the MAIN MENU select "Save Current Data" and then select "Catalog/Save Data File." On single floppy drive systems, you will be prompted to insert a data disk. Give the file an appropriate name and press the Return key.

14. Printout.

Press the P key to print out any screens needed by individuals.

Laboratory Report

Since there is no Laboratory Report sheet for this exercise in the back of your manual, your instructor will indicate how this experiment is to be reported.

63

Anatomy of the Digestive System

The study of the alimentary tract, liver, and pancreas will be pursued in this exercise. Cat dissection will provide an in-depth study of the various components.

The Alimentary Canal

Figure 63.1 is a simplified illustration of the digestive system. The alimentary canal, which is about 30 feet long, has been foreshortened in the intestinal area for clarity. The following text, which pertains to this illustration, is related to the passage of food through its entire length.

When food is taken into the oral cavity, it is chewed and mixed with saliva that is secreted by many glands of the mouth. The most prominent of these glands are the parotid, submandibular, and sublingual glands. The **parotid glands** are located in the cheeks in front of the ears, one on each side of the head. The **sublingual glands** are located under the tongue and are the most anterior glands of the three pairs. The **submandibular glands** are situated posterior to the sublinguals, just inside the body of the mandible. (Figure 63.4 illustrates the position of these glands more precisely.)

After the food has been completely mixed with saliva it passes to the **stomach** by way of a long tube, the **esophagus.** The food is moved along the esophagus by wavelike constrictions called *peristaltic waves.* These constrictions originate in the **oropharynx,** which is the cavity at the top of the esophagus, posterior to the tongue. The upper opening of the stomach through which the food enters is the **cardiac sphincter.** The upper rounded portion, or **fundus,** of the stomach holds the bulk of the food to be digested. The lower portion, or **pyloric region,** is smaller in diameter, more active, and accomplishes most of the digestion that occurs in the stomach. That region between the fundus and the pyloric portion is called the **body.** After the food has been acted upon by the various enzymes of the gastric fluid, it is forced into the small intestine through the **pyloric sphincter** of the stomach.

The *small intestine* is approximately 23 feet long and consists of three parts: the duodenum, jejunum, and ileum. The first 10 to 12 inches make up the **duodenum.** The **jejunum** comprises the next 7 or 8 feet, and the last coiled portion is the **ileum.** Complete digestion and absorption of food take place in the small intestine.

Indigestible food and water pass from the ileum into the large intestine, or **colon,** through the **ileocecal valve.** This valve is shown in a cutaway section. The large intestine has four sections: the ascending, transverse, descending, and sigmoid colons. The **ascending colon** is the portion of the large intestine that the ileum empties into. At its lower end the ascending colon has an enlarged compartment, or pouch, called the **cecum.** Note that the cecum has a narrow tube, the **appendix,** extending downward from it.

The ascending colon ascends on the right side of the abdomen until it reaches the undersurface of the liver, where it bends abruptly to the left, becoming the **transverse colon.** The **descending colon** passes down the left side of the abdomen, where it changes direction again, becoming the **sigmoid colon.** The last portion of the alimentary canal is the **rectum,** which is about 5 inches long and terminates with an opening, the **anus.**

Leading downward from the inferior surface of the liver is the **hepatic duct.** This duct joins the **cystic duct,** which connects with the round saclike **gallbladder.** Bile, a secretion of the liver, passes down the hepatic duct and up the cystic duct to the gallbladder, where it is stored until needed. When the gallbladder contracts, bile is forced down the cystic duct into the **common bile duct,** which extends from the juncture of the cystic and hepatic ducts to the intestine.

Between the duodenum and the stomach lies another gland, the **pancreas.** Its duct, the **pancreatic duct,** joins the common bile duct and empties into the duodenum.

Assignment:
Label figure 63.1.

Disassemble a laboratory manikin and identify all of these structures.

Answer the questions on Laboratory Report 63 pertaining to this part of the exercise.

Oral Anatomy

A typical normal mouth is illustrated in figure 63.2. To provide maximum exposure of the oral structures, the lips *(labia)* have been retracted away from the teeth and the cheeks *(buccae)* have been cut.

The lips are flexible folds that meet laterally at the angle of the mouth where they are continuous with the cheeks. The lining of the lips, cheeks, and other oral surfaces consists of mucous membrane, the *mucosa.* Near the median line of the mouth on the inner surface of the lips, the mucosa is thickened to form folds, the **labial frenula,** or *frena.* Of the two frenula, the upper one is usually stronger. The delicate mucosa that covers the neck of each tooth is the **gingiva.**

The hard and soft palates are distinguishable as differently shaded areas on the roof of the mouth. The lighter shaded area adjacent to the teeth is the **hard palate.** Posterior to the hard palate is a darker area, the **soft palate.** The **uvula,** a fingerlike projection seen above the back of the tongue, is a part of the soft palate. It varies considerably in size and shape among individuals.

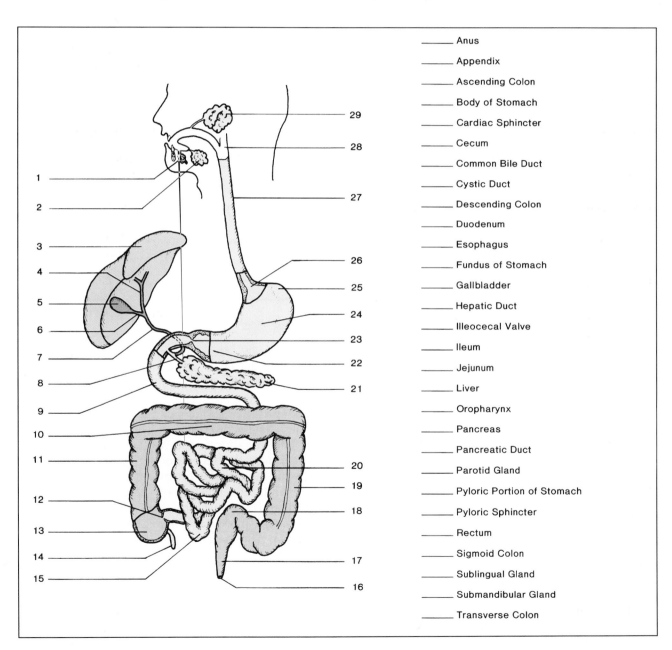

_____ Anus
_____ Appendix
_____ Ascending Colon
_____ Body of Stomach
_____ Cardiac Sphincter
_____ Cecum
_____ Common Bile Duct
_____ Cystic Duct
_____ Descending Colon
_____ Duodenum
_____ Esophagus
_____ Fundus of Stomach
_____ Gallbladder
_____ Hepatic Duct
_____ Illeocecal Valve
_____ Ileum
_____ Jejunum
_____ Liver
_____ Oropharynx
_____ Pancreas
_____ Pancreatic Duct
_____ Parotid Gland
_____ Pyloric Portion of Stomach
_____ Pyloric Sphincter
_____ Rectum
_____ Sigmoid Colon
_____ Sublingual Gland
_____ Submandibular Gland
_____ Transverse Colon

Figure 63.1 The alimentary canal.

On each side of the tongue at the back of the mouth are the **palatine tonsils.** Each tonsil lies in a recess bounded anteriorly by a membrane, the **glossopalatine arch,** and, posteriorly, by a membrane, the **pharyngopalatine arch.**

The tonsils consist of lymphoid tissue covered by epithelium. The epithelial covering of these structures dips inward into the lymphoid tissue forming glandlike pits called **tonsillar crypts** (label 11, figure 63.3). These crypts connect with channels that course through the lymphoid tissue of the tonsil. If they are infected, however, their protective function is impaired and they may actually serve as foci of infection. Inflammation of the palatine tonsils is called *tonsillitis.* Enlargement of the tonsils tends to obstruct the throat cavity and interfere with the passage of air to the lungs.

Assignment:
Label figure 63.2.

The Tongue

Figure 63.3 shows the tongue and adjacent structures. It is a mobile mass of striated muscle completely covered with mucous membrane. It is subdivided into three parts: the apex, body, and root. The **apex** of the tongue is the most anterior tip that rests against the inside surfaces of the anterior teeth. The **body** *(corpus)* is the bulk of the tongue that extends posteriorly from the apex to the root. The body is that part of the tongue that is visible by simple inspection: i.e., without the aid of a mirror. The posterior border of the body is arbitrarily located somewhere anterior to the tonsillar material of the organ. Extending down the median line of the body is a groove, the **central** *(median)* **sulcus.** The **root** of the tongue is the most posterior portion. Its surface is primarily covered by the **lingual tonsil.**

Extending upward from the posterior margin of the root of the tongue is the **epiglottis.** On each side of the tongue are seen the oval **palatine tonsils.**

The dorsum of the tongue is covered with several kinds of projections called *papillae* (see figure HA-20). The cutout section is an enlarged portion of its surface to reveal the anatomical differences between these papillae. The majority of the projections have tapered points and are called **filiform papillae.** These projections are sensitive to touch and give the dorsum a rough texture. This roughness provides friction for the handling of food. Scattered among the filiform papillae are the larger rounded **fungiform papillae.** Two fungiform papillae are

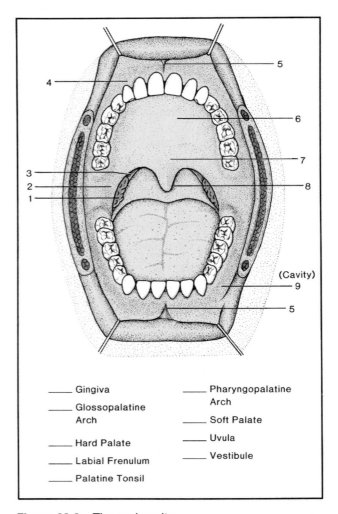

Figure 63.2 The oral cavity.

___ Gingiva	___ Pharyngopalatine Arch
___ Glossopalatine Arch	___ Soft Palate
___ Hard Palate	___ Uvula
___ Labial Frenulum	___ Vestibule
___ Palatine Tonsil	

shown in the sectioned portion. A third type of papilla is the large donut-shaped **vallate** *(circumvallate)* **papilla.** These papillae are arranged in a "V" near the posterior margin of the dorsum. Although the exact number may vary in individuals, eight vallate papillae are shown.

The fungiform and vallate papillae contain **taste buds,** the receptors for taste. Note on the walls of the circular furrow surrounding each vallate papilla are clusters of these round receptors. Comparatively speaking, the vallate papillae contain many more taste buds than the fungiform papillae.

A fourth type of papilla is the **foliate papilla.** These projections exist as vertical rows of folds of mucosa on each side of the tongue, posteriorly. A few taste buds are also scattered among these papillae.

Assignment:
Label figure 63.3.

The Salivary Glands

The salivary glands are grouped according to size. The largest glands, of which there are three pairs, are referred to as the *major salivary glands.* They are the parotids, submandibulars, and sublinguals. The smaller glands, which average only 2–5 mm in diameter, are the *minor salivary glands.*

Major Salivary Glands

Figure 63.4 illustrates the relative positions of the three major salivary glands as seen on the left side of the face. Since all of these glands are paired, it should be kept in mind that the other side of the face has another set of these glands.

The largest gland, which lies under the skin of the cheek in front of the ear is the **parotid gland.** Note that it lies between the skin layer and the *masseter muscle.* Leading from this gland is the **parotid (Stensen's) duct,** which passes over the masseter and through the *buccinator muscle* into the oral cavity. The drainage of this duct usually exits near the upper second molar.

The secretion of the parotid glands is a clear watery fluid that has a cleansing action in the mouth. It contains the digestive enzyme *salivary amylase,* which splits starch molecules into disaccharides (double sugars). The presence of sour (acid) substances in the mouth causes this gland to increase its secretion.

Inside the arch of the mandible lies the **submandibular (submaxillary) gland.** In figure 63.4 the mandible has been cut away to reveal two cut surfaces. Note that the lower margin of the submandibular gland extends down somewhat below the inferior border of the mandible. Also, note that a flat muscle, the *mylohyoid,* extends somewhat into the gland. The secretion of this gland empties into the oral cavity through the **submandibular (Wharton's) duct.** The **opening of the submandibular duct** is located under the tongue near the lingual frenulum. The **lingual frenulum** is a mucosal fold on the median line between the tongue and the floor of the mouth. It is shown in figure 63.4 as a triangular membrane.

The secretion of the submandibulars is quite similar in consistency to that of the parotid glands,

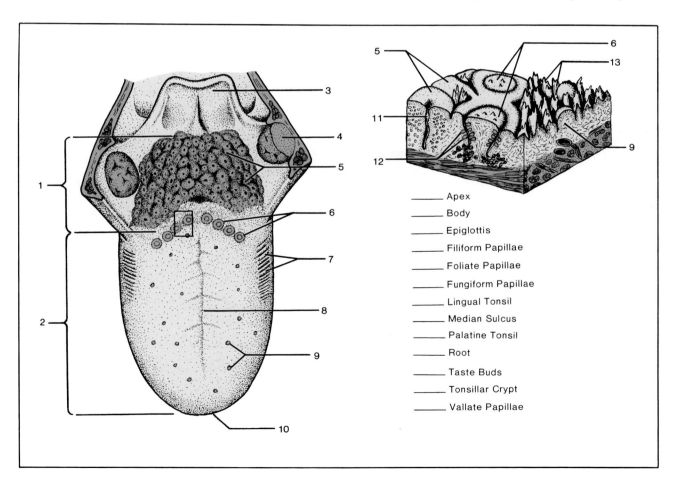

_____ Apex

_____ Body

_____ Epiglottis

_____ Filiform Papillae

_____ Foliate Papillae

_____ Fungiform Papillae

_____ Lingual Tonsil

_____ Median Sulcus

_____ Palatine Tonsil

_____ Root

_____ Taste Buds

_____ Tonsillar Crypt

_____ Vallate Papillae

Figure 63.3   **Tongue anatomy.**

except that it is slightly more viscous due to the presence of some *mucin.* This ingredient of saliva aids in holding the food together in a bolus. Bland substances such as bread and milk stimulate this gland.

The **sublingual gland** is the smallest of the three major salivary glands. As its name would

imply, it is located under the tongue in the floor of the mouth. It is encased in a fold of mucosa, the **sublingual fold,** under the tongue. This fold is shown in figure 63.5 (label 5). Drainage of this gland is through several **lesser sublingual ducts** *(ducts of Rivinus).* The number of openings is

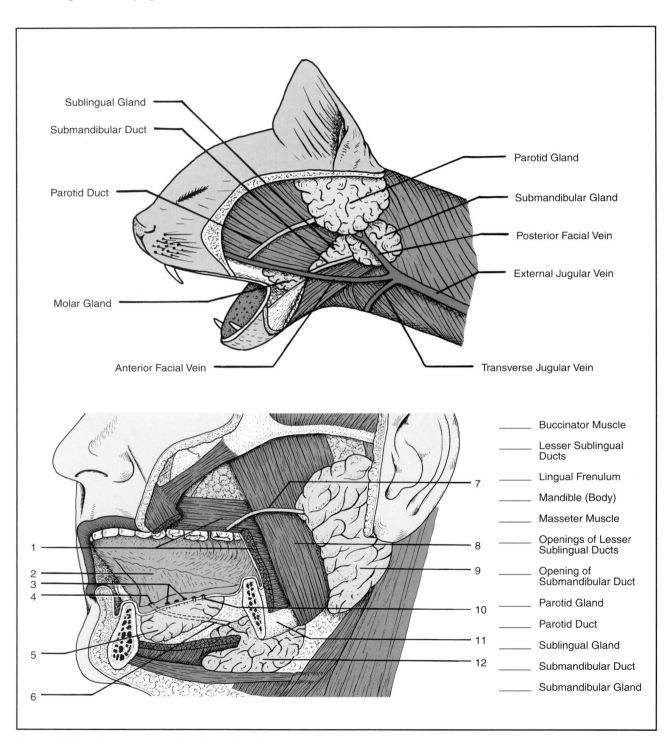

Sublingual Gland

Submandibular Duct

Parotid Duct

Molar Gland

Anterior Facial Vein

Parotid Gland

Submandibular Gland

Posterior Facial Vein

External Jugular Vein

Transverse Jugular Vein

_____ Buccinator Muscle

_____ Lesser Sublingual Ducts

_____ Lingual Frenulum

_____ Mandible (Body)

_____ Masseter Muscle

_____ Openings of Lesser Sublingual Ducts

_____ Opening of Submandibular Duct

_____ Parotid Gland

_____ Parotid Duct

_____ Sublingual Gland

_____ Submandibular Duct

_____ Submandibular Gland

Figure 63.4 Major salivary glands, cat and human.

variable among individuals. The gland differs from the other two in that it is primarily a mucous gland. The high mucin content of its secretion lends a certain degree of ropiness to it.

Minor Salivary Glands

There are four principal groups of these smaller salivary glands in the mouth: the palatine, lingual, buccal, and labial glands. In illustration A, figure 63.5, the palatine mucosa has been partially removed to reveal the closely packed nature of the **palatine glands.** These glands, which are about 2–4 mm in diameter, almost completely cover the roof of the oral cavity. Where the mucosa has been removed near the lower third molar we see that the glands occur even next to the teeth. The glands in this particular area are called **retromolar glands.**

Illustration B in figure 63.5 shows the oral cavity with the tongue pointed upward and the inferior mucosal surface of it partially removed. The gland exposed in this area of the tongue is the **anterior lingual gland.**

The **buccal glands** cover the majority of the inner surface of the cheeks, and the **labial glands** are found under the inner surface of the lips. All of the minor glands except for the palatine glands produce salivary amylase.

Assignment:
Label figures 63.4 and 63.5.

The Teeth

Every individual develops two sets of teeth during the first twenty-one years of life. The first set, which begins to appear at approximately six months of age, are the *deciduous teeth.* Various other names such as *primary, baby,* or *milk teeth* are also used in reference to these teeth.

The second set of teeth, known as *permanent* or *secondary teeth,* begins to appear when the child is about 6 years of age. As a result of normal growth from the sixth year on, the jaws enlarge, the secondary teeth begin to exert pressure on the primary teeth, and *exfoliation,* or shedding, of the deciduous dentition occurs.

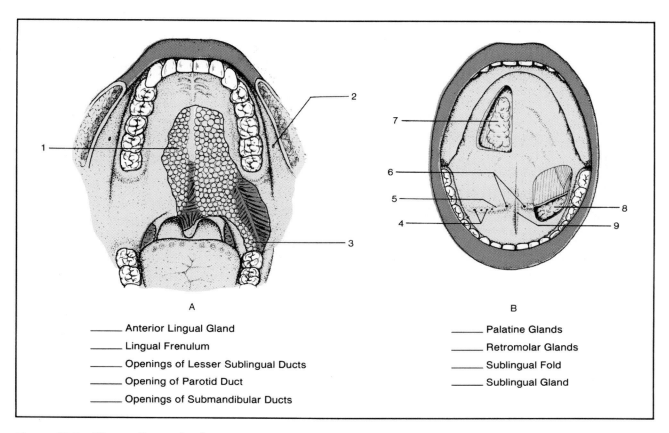

A

_____ Anterior Lingual Gland
_____ Lingual Frenulum
_____ Openings of Lesser Sublingual Ducts
_____ Opening of Parotid Duct
_____ Openings of Submandibular Ducts

B

_____ Palatine Glands
_____ Retromolar Glands
_____ Sublingual Fold
_____ Sublingual Gland

Figure 63.5 Minor salivary glands.

The Deciduous Teeth

The deciduous teeth number twenty in all—five in each quadrant of the jaws. Normally, all twenty have erupted by the time the child is 2 years old. Starting with the first tooth at the median line, they are named as follows: **central incisor, lateral incisor, cuspid** *(canine),* **first molar,** and **second molar.** The naming of the lower teeth follows the same sequence.

Once the 2-year-old child has all of the deciduous teeth, no visible change in the teeth occurs until around the sixth year. At this time the first permanent molars begin to erupt.

The Permanent Teeth

For approximately 5 years (seventh to twelfth year) the child will have a *mixed dentition,* consisting of both deciduous and permanent teeth. As the submerged permanent teeth enlarge in the tissues, the roots of the deciduous teeth undergo *resorption.* This removal of the underpinnings of the deciduous teeth results eventually in exfoliation.

Figure 63.6 illustrates the permanent dentition. Note that there are sixteen teeth in one half of the mouth—thus, a total of thirty-two teeth. Naming them in sequence from the median line of the mouth

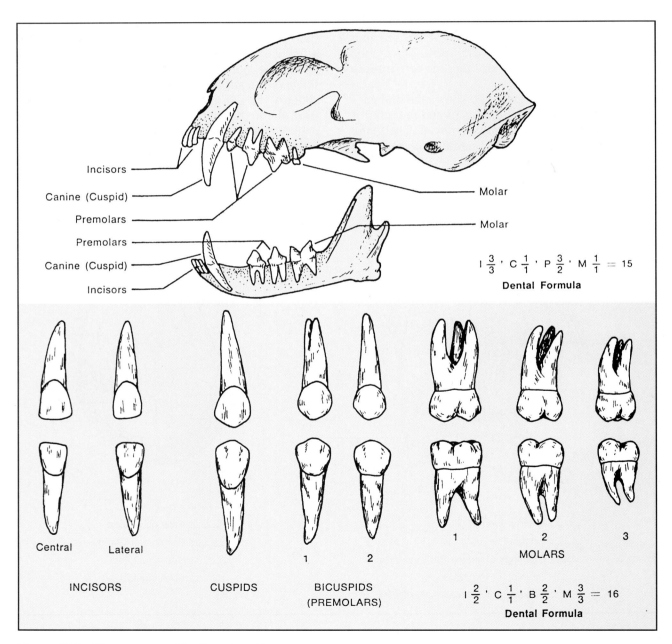

$$I\frac{3}{3} \cdot C\frac{1}{1} \cdot P\frac{3}{2} \cdot M\frac{1}{1} = 15$$
Dental Formula

$$I\frac{2}{2} \cdot C\frac{1}{1} \cdot B\frac{2}{2} \cdot M\frac{3}{3} = 16$$
Dental Formula

Figure 63.6 The permanent teeth, cat and human.

they are **central incisor, lateral incisor, cuspid** *(canine),* **first bicuspid, second bicuspid, first molar, second molar,** and **third molar.** The third molar is also called the *wisdom tooth.*

Comparisons of the teeth in figure 63.6 reveal that the anterior teeth have single roots and the posterior ones have several roots. Of the bicuspids, the only ones that have two roots *(bifurcated)* are the maxillary first bicuspids. Although all of the maxillary molars are shown with three roots *(trifurcated),* there may be considerable variability, particularly with respect to the third molars. The roots of the mandibular molars are generally bifurcated. The longest roots are seen on the maxillary cuspids.

Assignment:
Identify the teeth of skulls that are available in the laboratory.

Tooth Anatomy

The anatomy of an individual tooth is shown in figure 63.7. Longitudinally, it is divided into two portions, the crown and root. The line where these two parts meet is the **cervical line,** or *cemento-enamel juncture.*

The dentist sees the crown from two aspects: the anatomical and clinical crowns. The **anatomical crown** is the portion of the tooth that is covered with enamel. The **clinical crown,** on the other hand, is that portion of the crown that is exposed in the mouth. The structure and physical condition of the soft tissues around the neck of the tooth will determine the size of the clinical crown.

The tooth is composed of four tissues: the enamel, dentin, pulp, and cementum. **Enamel** is the most densely mineralized and hardest material in the body. Ninety-six percent of enamel is mineral. The remaining 4% is a carbohydrate-protein complex. Calcium and phosphorus make up over 50% of the chemical structure of enamel. Microscopically, this tissue is made up of very fine rods or prisms that lie approximately perpendicular to the outer surface of the crown. It has been estimated that the upper molars may have as many as 12 million of these small prisms in each tooth. The hardness of enamel enables the tooth to withstand the abrasive action of one tooth against the other.

Dentin is the material that makes up the bulk of the tooth and lies beneath the enamel. It is not as

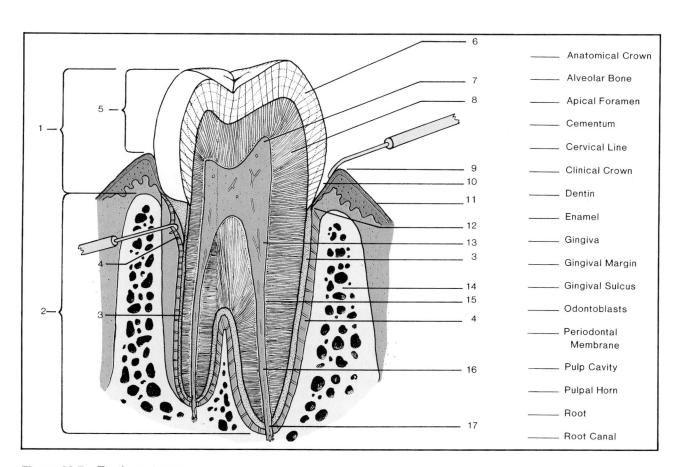

Anatomical Crown

Alveolar Bone

Apical Foramen

Cementum

Cervical Line

Clinical Crown

Dentin

Enamel

Gingiva

Gingival Margin

Gingival Sulcus

Odontoblasts

Periodontal Membrane

Pulp Cavity

Pulpal Horn

Root

Root Canal

Figure 63.7 Tooth anatomy.

hard or brittle as enamel and resembles bone in composition and hardness. This tissue is produced by a layer of **odontoblasts** (label 15) that lies in the outer margin of the pulp cavity.

Cementum is the hard dental tissue that covers the anatomical root of the tooth. It is a modified bone tissue, somewhat resembling osseous tissue. It is produced by cells called *cementocytes,* which are very similar to osteocytes. The primary function of cementum is to provide attachment of the tooth to surrounding tissues in the alveolus.

The cavity in the center of the tooth is called the **pulp cavity.** Extending down into the roots, this cavity becomes the **root canals.** Where this chamber extends up into the cusps of the crown, it forms the **pulpal horns.**

The tissue of the pulp is essentially loose connective tissue. It consists of fibroblasts, intercellular ground substance, and white fibers. Permeating this tissue are blood vessels, lymphatic vessels, and nerve fibers.

The functions of the pulp are to (1) provide nourishment for the living cells of the tooth, (2) provide some sensation to the tooth, and (3) produce dentin. The blood vessels and nerve fibers that enter the pulp cavity do so through openings, the **apical foramina,** in the tips of the roots. Inflammation of the pulp, or *pulpitis,* may result in the destruction of the blood vessels and nerves, producing a dead, or *devitalized,* tooth.

Between the cementum and the alveolar bone lies a vascular layer of connective tissue, the **periodontal membrane.** On the left side of figure 63.7 it is shown pulled away from the root. One of the principal functions of this membrane is tooth retention. The membrane consists of bundles of fibers that extend from the cementum to the alveolar bone, firmly holding the tooth in place. Its presence also acts as a cushion to reduce the trauma of occlusal action. Another very important function of this membrane is to provide sensitivity to touch. This sensitivity is due to the presence of free nerve ending receptors in the membrane. The slightest touch at the surface of the tooth is transmitted to these nerve endings through the medium of the periodontal membrane. Even if apical portions of the membrane are removed, as in root tip resection, the sense of touch is not impaired.

To demonstrate the depth of the **gingival sulcus,** a probe is seen on the right side of figure 63.7. This crevice is frequently a site for bacterial putrefaction and the origin of *gingivitis.* The upper edge of the gingiva is called the **gingival margin.** The portion of the gingiva that lies over the alveolar bone is called the *alveolar mucosa.*

Assignment:
Label figure 63.7.

Cat Dissection

Since the anatomical study of the circulatory, respiratory, and digestive systems of the cat cannot be studied independently of each other, it is very likely that this dissection may be performed simultaneously with the study of the other two systems. It is because of these interrelationships that there will be some repetition of discussion of digestive anatomical structures.

Oral Cavity

If Exercise 56 has already been completed, you will have identified the *oral cavity, nasal cavity, nasopharynx, oropharynx, soft palate, hard palate, palatine tonsils, auditory tube aperture, esophagus, larynx,* and *trachea.* If you have not studied these organs previously refer to page 321 to identify them. It will be necessary for some specimens to be cut along the median line with a bone saw, as described on pages 322–323, to reveal some of these structures.

In addition to the above structures, identify the transverse ridges, or **rugae,** of the hard palate that assist in holding food in the mouth during swallowing. Also identify the **oral vestibule,** which is the space between the teeth and lips.

Lift up the tongue and examine its inferior surface. Do you see any evidence of a **lingual frenulum?** The cat has distinct **filiform** and **fungiform papillae** on the tongue's dorsal surface. The filiform papillae are spinelike and more numerous than the fungiform papillae. In addition to being tactile receptors, the filiform papillae also act as scrapers. Can you locate the **circumvallate papillae** at the back of the tongue? How many are there?

Examine the teeth, referring to figure 63.6 to identify them. How does the number of **incisors** compare with the human's? Note how extremely small they are. The important anterior teeth on the cat are the **canines.** They function to kill small prey and tear away portions of food. Note that there are three upper and two lower **premolars** in the cat. Observe how insignificant the upper **molar** is compared with the lower molar. Since cats are carnivores and swallow chunks of meat without chewing them, the posterior teeth act as shears rather than grinders. No flat occlusal surfaces are seen on the cat's posterior teeth.

Salivary Glands

Remove the skin and platysma muscle from the left side of the head if this has not already been done. Any lymph nodes that obscure the anterior facial vein should be removed.

By referring to figure 63.4, identify the **parotid, submandibular,** and **sublingual glands.** What gland is seen on the cat that is lacking in humans? Can you locate the **parotid** and **submandibular ducts?** A fifth gland, the **infraorbital gland,** exists in the floor of each eye orbit but cannot be seen unless the eye is removed.

Esophagus

Force a probe into the upper end of the esophagus and stretch its walls laterally. Note how much it can be distended. Trace it down to where it penetrates the diaphragm and liver to enter the stomach.

Abdominal Organs

If the circulatory system has already been studied, the abdominal cavity will be open for the remainder of this study. If the abdomen has not been opened yet, make an incision with scissors along the median line from the thoracic region to the symphysis pubis. Also, make additional cuts laterally from the median line in the pubic region so that the muscle of the abdominal wall can be pulled aside to expose all the organs.

The two most prominent structures revealed by removing the abdominal wall are the liver and the greater omentum. Examine the **greater omentum.** It is a double sheet of peritoneum that is attached to the greater curvature of the stomach and to the dorsal body wall. Only a small segment of it is revealed in figure 63.9, because most of it has been cut away to expose the intestines that lie under it.

Lift up the greater omentum and examine both surfaces. Note that the potential space *(omental bursa)* between its two layers of peritoneum is filled with fat. Cut a slit into its surface to confirm its structure. That portion of the greater omentum between the stomach and spleen is called the **gastrosplenic ligament.** Identify this ligament as well as the **spleen** and **stomach.** Cut away and discard most of the greater omentum so that it will not be in the way.

Identify the right medial, right lateral, left medial, and left lateral **lobes of the liver.** Also, locate the soft greenish **gallbladder.** By referring to figure 63.8, identify the falciform and round ligaments. The **falciform ligament** is a sickle-shaped (Latin: *falx, falcis,* sickle), double-layered peritoneal fold that lies in the notch between the right and left medial lobes of the liver. Spread apart these lobes and lift this membrane out with a pair of forceps. Note that its left margin is attached to the diaphragm. Its free margin contains a thickened structure, the **round ligament.** This ligament is a vestige of the umbilical vein, which functioned during prenatal existence.

Identify the **lesser omentum,** which extends from the liver to the lesser curvature of the stomach

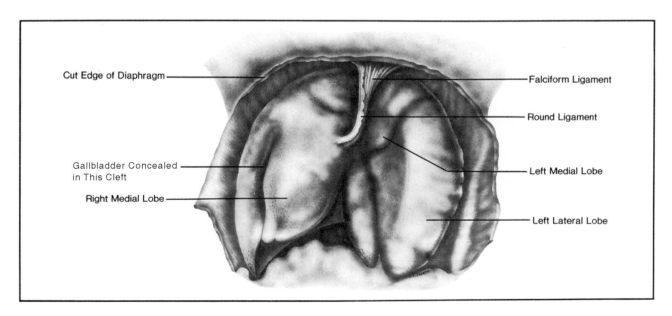

Figure 63.8   Anatomy of cat liver, anterior view.

and a portion of the duodenum. The common bile duct, hepatic artery, and portal vein lie within the lateral border of the lesser omentum. Dissect into this region to see if these latter structures can be seen.

Probe into the space under the right medial lobe of the liver and locate the **cystic duct,** which leads from the gallbladder into the **common bile duct.**

Raise the liver sufficiently to see where the esophagus enters the stomach. Identify the **fundus, body,** and **pylorus** of the stomach. That portion of the stomach where the esophagus joins it is called the **cardiac portion.** Open up the stomach by making the incision along its long axis. Observe that its lining has long folds called **rugae.**

Identify the **duodenum.** Note that it passes caudally about 3 inches and then doubles back on itself for another 3 or 4 inches. Within this duodenal loop lies the duodenal portion of the **pancreas.** The pancreas is quite long in that it extends from the duodenum over to the spleen. The mesentery that supports the duodenum and pancreas in this region is called the *mesoduodenum.*

Expose the **pancreatic duct** by dissecting away some of the pancreatic tissue. This duct lies embedded in the gland. It joins the common bile duct to form a short duct, the **ampulla of Vater,** which empties into the duodenum. A small **accessory pancreatic duct** is often seen entering the duodenum about 2 centimeters caudad to the ampulla of Vater. It is often confused with small arteries in this region.

The remainder of the small intestine is divided, arbitrarily, into a proximal half, the **jejunum,** and a distal half, the **ileum.** Trace the ileum to where it enters the cecum.

Remove a 2-inch section of the ileum or jejunum, slit it open longitudinally, and wash out its interior. Examine its inner lining with a dissecting microscope or hand lens, looking for the minute tubular projections called **villi.**

The large intestine consists of the **cecum, ascending colon, transverse colon,** and **descending colon.** The descending colon empties into the **rectum** and exits through the **anus.** Identify all these structures. Make an incision through the distal end of the ileum and the cecum to reveal the **ileocecal valve.**

Histological Studies

Make a systematic histological study of the organs of the digestive system, using the Histology Atlas for reference.

Materials:
prepared slides of:
  taste buds (H5.120, H5.121)
  salivary glands (H5.131, H5.133, H5.134)
  esophagus (H5.311)
  stomach (H5.415)
  duodenum, jejunum, and ileum (H5.58)
  colon (H5.613)
  appendix (H5.65)
  anal canal (H5.67)
  tooth embryology (H5.1176, H5.1184)

Salivary Glands   Examine a slide of the three major salivary glands, using figure HA-19 to note the histological differences between them.

Taste Buds   Examine slides of vallate or foliate papillae. Identify the structures shown on figure HA-21.

Tooth Embryology   Read the legend on figure HA-22 and examine slides to identify the structures shown in figure HA-22. Note in particular what cells give rise to dentin and enamel.

Esophagus   Identify the four coats that make up the wall of the esophagus. Note in particular the type of epithelial tissue that lines the esophagus. Use figure HA-23 for reference.

Stomach Wall   Examine a slide of the fundus of the stomach and compare it with illustrations C and D of figure HA-23. Study the deep cells within the pits and differentiate the **parietal** and **chief cells.** How do these cells differ in function? How does the mucosa of the stomach differ from the mucosa of the esophagus?

Small Intestine   Study a slide that has all three sections of the small intestine. Use figures HA-24 and HA-25 for reference. Note the differences that exist between the duodenal and intestinal glands. How do these glands differ in function? What is the function of the **lacteal** in a villus? Look for lymphoidal tissue **(Peyer's patches)** in the ileum (see illustration D, figure HA-29). This tissue constitutes the gut-associated lymphoid tissue (GALT) of the digestive tract, an important part of the immune system.

Large Intestine   Examine a slide of the colon and refer to illustrations A and B, figure HA-26. What is lacking on the mucosa of the colon that is seen in the small intestine? Identify all structures.

Appendix   Refer to illustration C, figure HA-26 while studying a slide of the appendix. Compare its structure with the colon. What function would lymph nodules perform here?

Liver and Pancreas When examining slides of the liver and pancreas, read the legend on figure HA-27, which explains the significance of the structures labeled in the four illustrations.

Gallbladder Refer to illustrations A and B, figure HA-28, while studying a slide of the gallbladder.

How does the mucosa of the gallbladder resemble the small intestine? How does it differ?

Anal Canal Determine what kind of tissue makes up the mucosa in this region. Consult illustration D, figure HA-26.

Laboratory Report

Complete Laboratory Report 63 for this exercise.

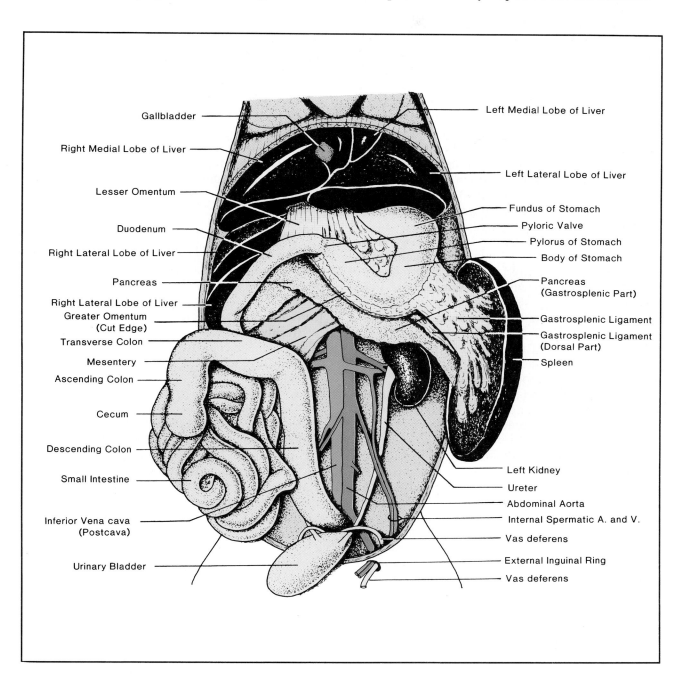

Figure 63.9 Abdominal viscera of the cat.

64

The Chemistry of Hydrolysis

The basic nutrients of the body consist of small quantities of vitamins and minerals plus large amounts of carbohydrates, fats, and proteins. Although vitamins and minerals play a vital role in all physiological activities, it is the carbohydrates, fats, and proteins that provide the energy and raw materials for growth and tissue repair.

Food, as it is normally ingested, consists of large complex molecules of carbohydrates, fats, and proteins that cannot pass through the intestinal wall unless they are converted to smaller molecules by digestion in the stomach and intestines. The conversion of these large molecules to smaller ones is accomplished by digestive enzymes. All digestive enzymes split molecules by hydrolysis.

Hydrolysis is a process whereby the larger molecules are split into smaller units by combining with water. Although hydrolysis without enzymes will occur automatically at body temperature, the process is extremely slow; the catalytic action of digestive enzymes, on the other hand, greatly hastens the process.

The end result of this action is to reduce carbohydrates to monosaccharides, fats to fatty acids and glycerol, and proteins to amino acids. The large size of some food molecules requires that different enzymes work at different levels to produce intermediate-size molecules before the end products are produced.

Digestive enzymes are produced by the salivary glands, the stomach wall, the intestinal wall, and the pancreas. Since the pancreas produces all three of the basic types of enzymes (proteases, lipases, and carbohydrases), we will focus our attention here on this gland.

An enzyme extract from the pancreas of a freshly killed rat will be analyzed for the presence of the three types. Figure 64.1 reveals the steps that will be employed for extracting the pancreatic juice. Since enzymes are relatively unstable and easily denatured by certain adverse environmental conditions, it will be essential that chemical purity, cleanliness, and temperature control be maintained.

Once the pancreatic juice has been extracted, tests will be made for evidence of carbohydrase,
lipase, and protease activity. Small amounts of the pancreatic juice will be added to substrates of starch, fat, and protein to observe the hydrolytic action of the various enzymes. Color changes that occur in each test will indicate the breaking of bonds within the large molecules to produce smaller molecules. Positive and negative test controls will be used for comparisons with the actual tests.

The class will be divided into groups of four students. While two of the students are dissecting the rat, the other two will set up the necessary supplies and equipment for the extraction and assay procedures.

Extraction Procedure

Remove the pancreas from a freshly killed rat and produce the pancreatic extract using the following procedures.

Materials:
 freshly killed rat
 refrigerated centrifuge
 centrifuge tubes
 balance
 Sorval blender
 beaker, 30 ml size
 Erlenmeyer flask, 125 ml size
 graduated cylinder, 100 ml size
 crushed ice
 cold mammalian Ringer's solution
 dissecting pan (wax bottom)
 dissecting instruments
 dissecting pins
 Parafilm

1. Follow the steps outlined in figures 64.2 through 64.5 to open up the abdominal cavity of a freshly killed rat and remove the pancreas.
2. Place the gland into a beaker containing a small amount (about 20 ml) of cold mammalian Ringer's solution.
3. Snip off a small piece of the intestine and add it to the beaker.
4. After washing the tissues in the Ringer's solution, macerate and homogenize them in a

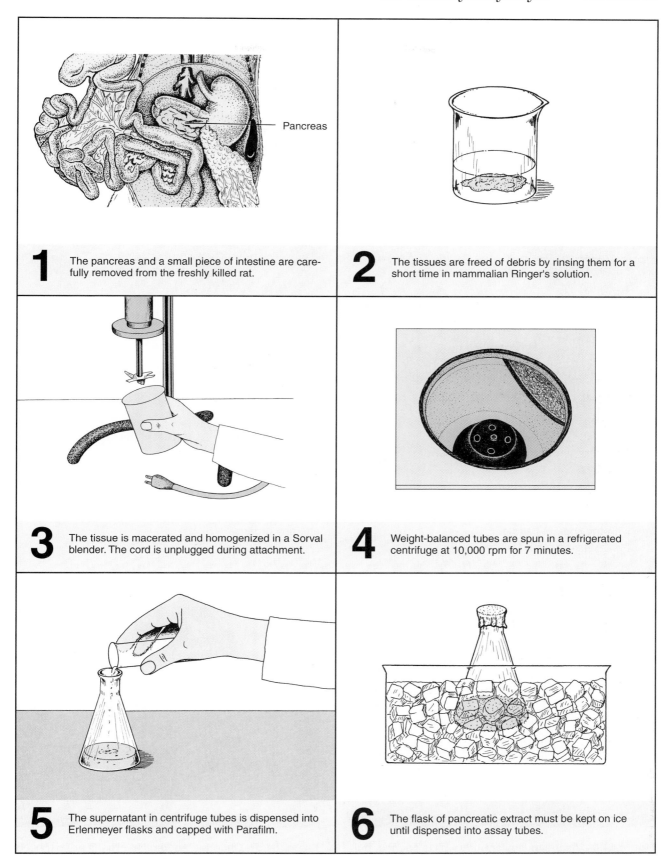

1 The pancreas and a small piece of intestine are carefully removed from the freshly killed rat.

2 The tissues are freed of debris by rinsing them for a short time in mammalian Ringer's solution.

3 The tissue is macerated and homogenized in a Sorval blender. The cord is unplugged during attachment.

4 Weight-balanced tubes are spun in a refrigerated centrifuge at 10,000 rpm for 7 minutes.

5 The supernatant in centrifuge tubes is dispensed into Erlenmeyer flasks and capped with Parafilm.

6 The flask of pancreatic extract must be kept on ice until dispensed into assay tubes.

Figure 64.1 Procedure for extracting pancreatic enzymes.

blender. The tissues of several groups should be blended together to get ample bulk.

5. Distribute the blended mixture into evenly balanced centrifuge tubes and spin the tubes in a refrigerated centrifuge (Beckman or other) for 7 minutes at 10,000 rpm. Tubes must be balanced by weighing carefully and adjusting the contents until they are equal.

6. Decant supernatant from the tubes into Erlenmeyer flasks for each group. The supernatant will be a milky pink mixture of digestive enzymes.

7. Place the flask of pancreatic extract into a container of ice to prevent enzyme deterioration.

Tube Preparation

Figure 64.6 illustrates the number of test tubes that will be set up to assay the pancreatic extract for carbohydrase, protease, and lipase. The detection of the presence of each enzyme will depend on a specific color test after the tubes have been incubated in a 38° C water bath for 1 hour.

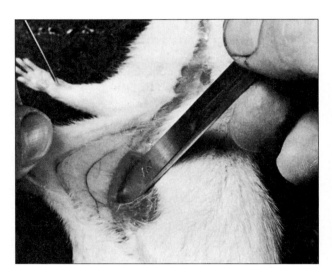

Figure 64.2 After pinning down the feet and cutting through the skin, separate the skin from the musculature with the scalpel handle.

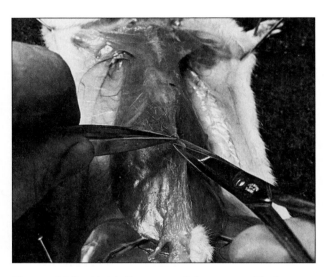

Figure 64.3 Hold the musculature up with forceps while cutting through the wall. Be careful not to damage any organs.

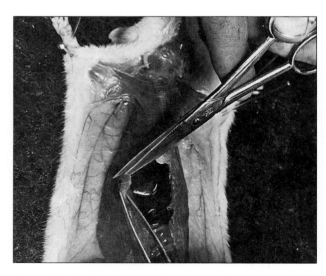

Figure 64.4 As the muscular body wall is opened, pull flaps of muscle wall laterally and pin them down to keep the cavity open.

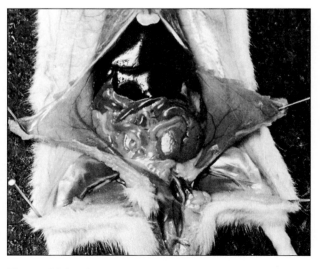

Figure 64.5 Once the abdominal organs are exposed as shown here, lift the liver with forceps to expose the pancreas for removal.

Tube 1 in each series contains pancreatic extract, a substrate, and necessary buffering agents. The last tube in each series is a **positive test control** that contains the enzyme we are testing for in tube 1.

All other tubes in each series should give negative results because of certain deficiencies. Note that tube 2 in each series contains the same ingredients as tube 1 except for two significant variables: (1) the tube is placed on ice while the others are incubated; and (2) the pancreatic extract is not added until after one hour on ice. This tube is designated as our **zero time control.**

All four members of each team will work cooperatively to set up these 16 tubes. After the tubes have been in the water baths for one hour 1 ml of pancreatic juice will be added to the three zero time tubes. Various tests will then be performed to detect the presence of hydrolysis.

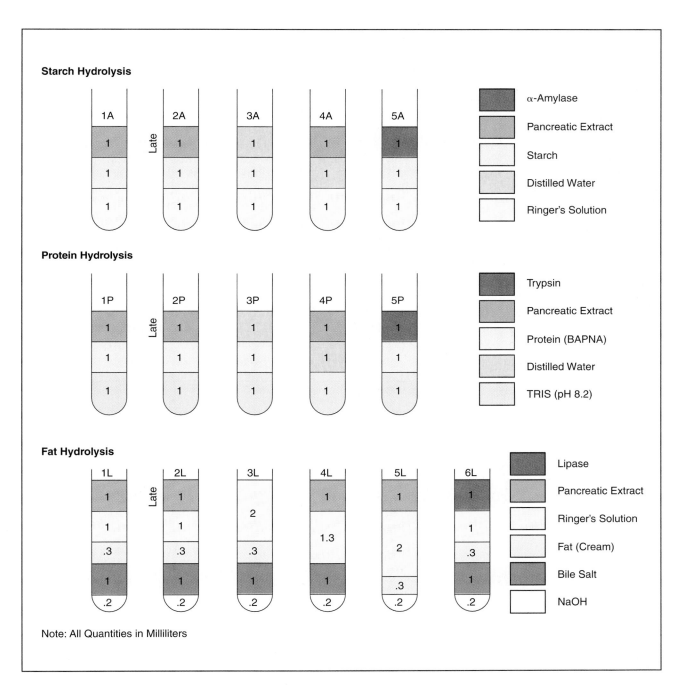

Figure 64.6   Protocol for enzyme assay of pancreatic extract.

Materials:

water baths (38° C)
Vortex mixer
pipetting devices
test-tube rack (wire type)
test tubes (16 mm × 150 mm), 16 per group
hot plates
pipettes, 1 ml, 5 ml, and 10 ml sizes
graduated cylinder (10 ml size)
bromthymol blue pH color standards
china marking pencil
photocolorimeter (Spectronic 20)
spot plates

solutions in dropping bottles

IKI solution	2% sodium taurocholate
Benedict's solution	(a bile salt)
Barfoed's solution	soluble starch, 0.1%
bromthymol blue	cream, unhomogenized
amylase, 1%	albumin, 0.1%
trypsin, 1%	NaCl, 0.2%
lipase, 1%	NaOH, 0.1N
TRIS (pH 8.2)	BAPNA solution

Carbohydrase

Carbohydrate in plants is stored, primarily, in the form of starch. Starch is composed of 98% amylose and 2% amylopectin. Amylose is a straight-chain polysaccharide of glucose units attached by α-**1,4-glucosidic bonds.** Amylopectin is a branching polysaccharide with side chains attached to the main chain by α-**1,6-glucosidic bonds.** These linkages are illustrated in figure 64.7.

The digestion of starch is catalyzed by α-*amylase,* an enzyme present in saliva, pancreatic juice, and intestinal juice. This enzyme randomly attacks the α-1,4 bonds. An α-*1,6-glucosidase* is necessary to break the α-1,6 bond. The action of α-amylase on starch results in the formation of molecules of maltose and isomaltose. These two molecules are converted to glucose by the enzymes maltase and isomaltose in the epithelial cells of the small intestine.

Set up the five tube test series for α-amylase according to the following protocol:

1. With a china marking pencil, label the test tubes 1A through 5A ("A" for amylase).
2. With a 5 ml or 10 ml pipette, deliver 1 ml of **mammalian Ringer's solution** to each of the five tubes. Use a mechanical pipetting device such as the one illustrated in figure 64.10.
3. Flush out the pipette with water and deliver 1 ml of **distilled water** to tubes 3 and 4.
4. Using the same pipette, deliver 1 ml of **starch solution** to tubes 1, 2, 3, and 5. Note that tube 4 does not get any starch.
5. With a fresh 5 ml pipette, deliver 1 ml of **pancreatic juice** to tubes 1 and 4.
6. With a fresh 1 ml pipette, deliver 1 ml of **α-amylase to tube 5.**
7. Mix each tube on a Vortex mixer (figure 64.11) for a few seconds.
8. Remove tube 2 from the series, cover it with Parafilm, and place it in a rack that is immersed in ice water.

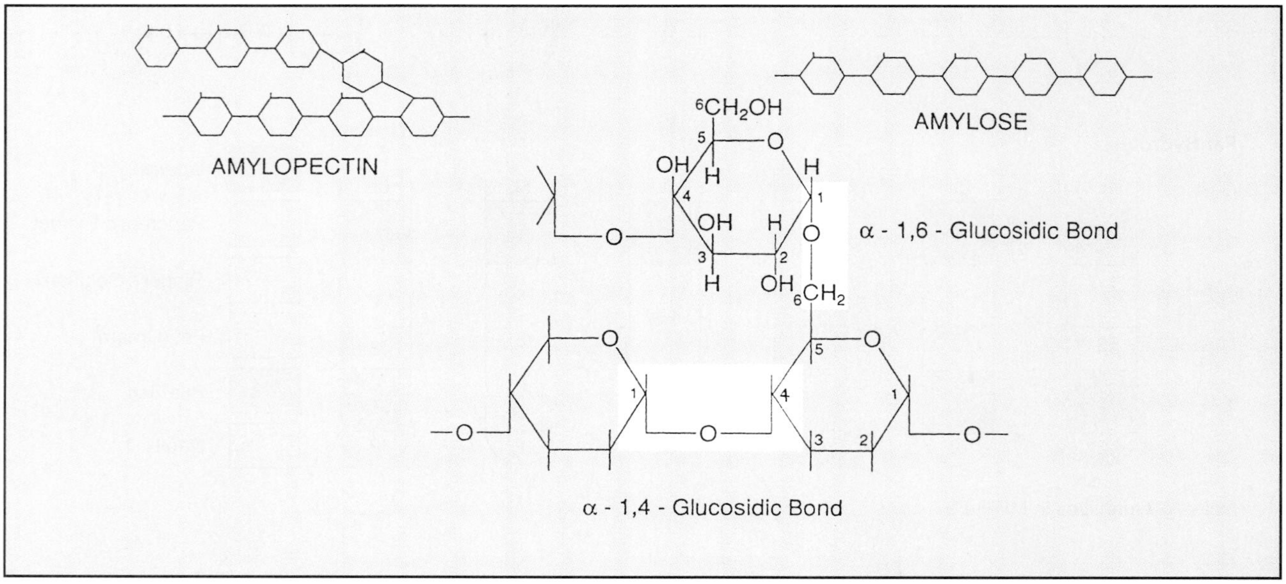

Figure 64.7 Carbohydrate linkage where hydrolysis occurs.

9. Cover the other four tubes with Parafilm and place the test-tube rack with these tubes in a 38° C water bath for **1 hour.**

Protease

The dietary proteins are derived, primarily, from meats and vegetables. These proteins consist of long chains of amino acids that are connected to each other by **peptide linkages.** Note in figure 64.8 that when a peptide linkage is hydrolyzed, the carbon of one amino acid takes on OH^- to form a carboxyl group (COOH), and H^+ is added to the other amino acid to form an amine (NH_2) group.

Protein digestion begins in the stomach with the action of *pepsin* and is completed in the small intestine. Pancreatic juice contains *trypsin, chymotrypsin,* and *carboxypolypeptidase,* which convert proteoses, peptones, and polypeptides to polypeptides and amino acids. *Amino polypeptidase* and *dipeptidase* within the epithelial cells of the small intestine hydrolyze the final peptide linkages as small polypeptides and dipeptides pass through the intestinal mucosa into the portal blood.

Since there are considerable physical differences between the linkages of different amino acids, there need be, and are, a multiplicity of proteolytic enzymes; no single enzyme can possibly digest protein all the way to its amino acids.

In our assay of pancreatic juice, we will use a synthetic peptide substrate that is designated as BAPNA (benzoyl D-L p-arginine nitroaniline). If our pancreatic extract contains trypsin, this protein will be hydrolyzed to release a yellow aniline dye, giving visual evidence of hydrolysis. Trypsin will be used in our positive test control.

1. With a china marking pencil, label the test tubes 1P through 5P ("P" for protease).
2. With a 5 ml or 10 ml pipette, deliver 1 ml of **pH 8.2 buffered solution** to each of the five tubes. Use a mechanical pipetting device such as the one illustrated in figure 64.10.
3. Flush out the pipette with water and deliver 1 ml of **distilled water** to tubes 3 and 4.
4. Using the same pipette, deliver 1 ml of **BAPNA solution** to tubes 1, 2, 3, and 5. Note that tube 4 does not get any BAPNA; it is a negative test control.
5. With a fresh 5 ml pipette, deliver 1 ml of **pancreatic juice** to tubes 1 and 4.
6. With a fresh 1 ml pipette, deliver 1 ml of **trypsin** to tube 5.
7. Mix each tube on a Vortex mixer (figure 64.11) for a few seconds.
8. Remove tube 2 from the series, cover it with Parafilm, and place it in a rack that is immersed in ice water.
9. Cover the other four tubes with Parafilm and place the test-tube rack with these tubes in a 38° C water bath for **one hour.**

Lipase

The hydrolysis of fats to glycerol and fatty acids is catalyzed by *lipase.* Although this enzyme is present

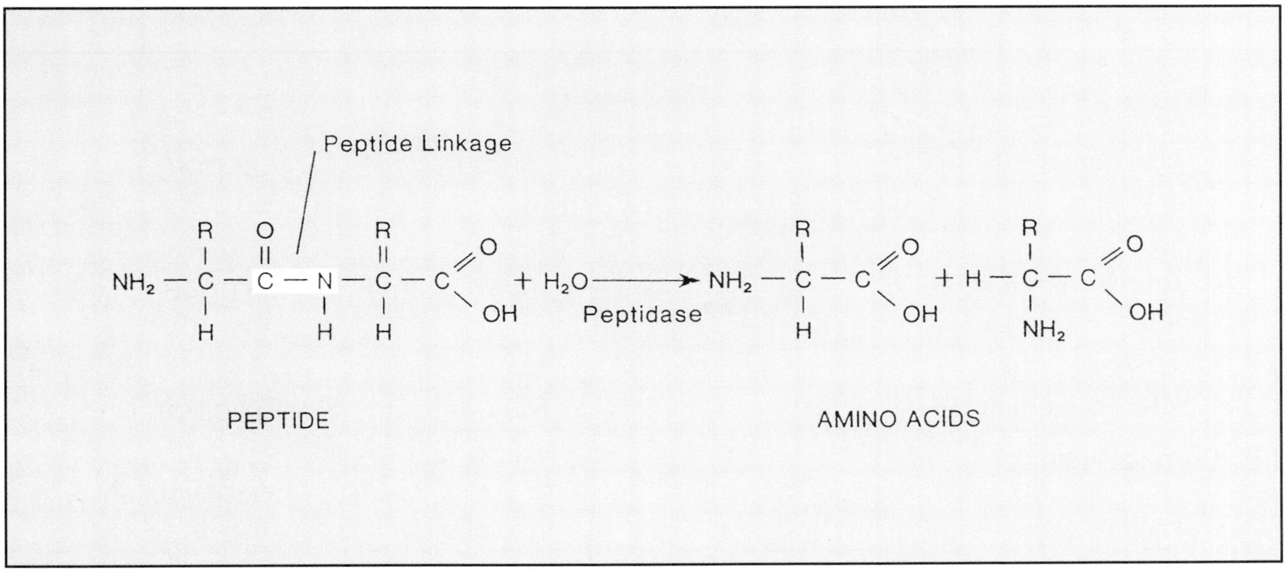

Figure 64.8 Protein hydrolysis.

Exercise 64 • The Chemistry of Hydrolysis

in both gastric and pancreatic secretions, nearly all fat digestion occurs in the small intestine.

Lipase hydrolysis differs from carbohydrase and protease hydrolysis in that it requires an emulsifying agent to hasten the process. *Bile salts,* produced by the liver, act like detergents to break up large fat globules into smaller globules, greatly increasing the surface of fat particles. Since lipases, like all enzymes, are only able to work on the surfaces of their substrate molecules, the increased surface area greatly enhances the speed of fat hydrolysis.

The most common fats in food are neutral fats, or **triglycerides.** As illustrated in figure 64.9, each triglyceride molecule is composed of a glycerol nucleus and three fatty acids. Although the final end products of triglyceride hydrolysis are fatty acids, and glycerol, there are intermediate products of monoglycerides and diglycerides. Diagrammatically, the process looks like this:

E → F → Γ → I + Ξ

TRIGLYCERIDE DIGLYCERIDE MONOGLYCERIDE GLYCEROL FATTY ACIDS

Set up six test tubes to assay pancreatic extract for the presence of lipase:

1. With a china marking pencil, label the test tubes 1L through 6L ("L" for lipase).
2. Using a 1 ml pipette, deliver 0.2 ml of **0.1N NaOH** to each of the tubes. Use a mechanical pipetting device such as the one in figure 64.10.

3. With a 5 ml pipette, deliver 0.2 ml of **mammalian Ringer's solution** to tubes 1, 2, and 6; 2 ml to tubes 3 and 5; and 1.3 ml to tube 4.
4. Flush out the pipette with water and deliver 1 ml of **bile salt solution** to each of the tubes except tube 5.
5. Flush out the pipette with water again and deliver 0.3 ml of **cream** to all tubes except tube 4.
6. With a fresh 5 ml pipette, deliver 1 ml of **pancreatic juice** to tubes 1, 4, and 5.
7. Deliver 1 ml of **lipase** to tube 6, using a fresh 1 ml pipette.
8. Mix each tube on a Vortex mixer (figure 64.11) for a few seconds.
9. Add 2 drops of bromthymol blue to each tube. Check the color against a bromthymol blue standard of pH 7.2.
   If the correct color does not develop, add NaOH, drop by drop, slowly, until the tube reaches the correct color of pH 7.2.
10. Remove tube 2 from the series, cover it with Parafilm, and place it in a rack that is immersed in ice water.
11. Cover the other five tubes with Parafilm and place the test-tube rack with these tubes in a 38° C water bath for **one hour.**

Evaluation of Tests

Remove the tubes from both water baths after one hour and proceed as follows to complete the experiment:

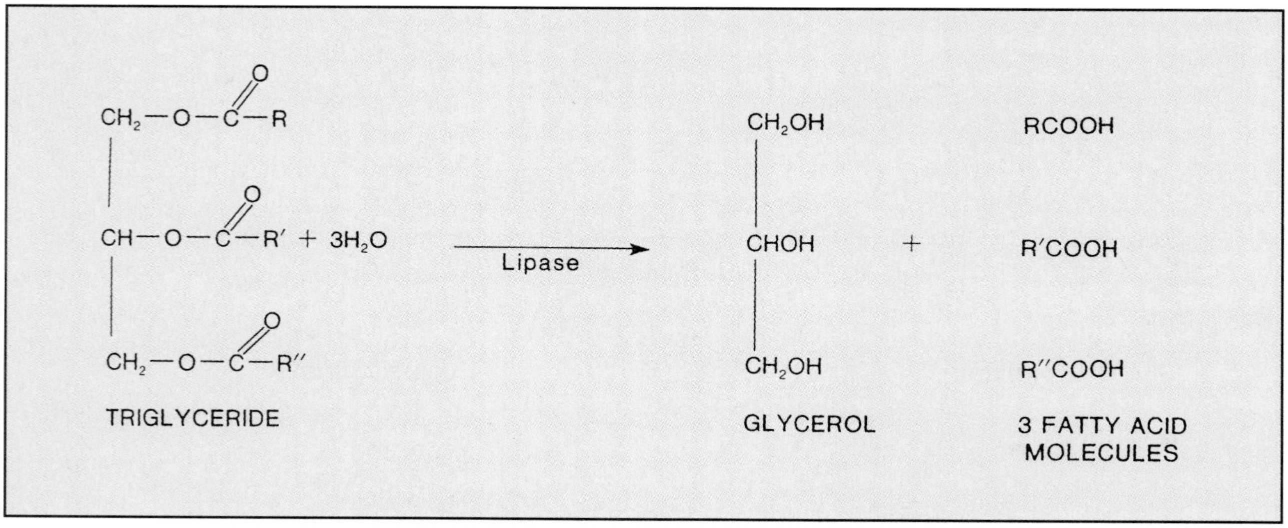

Figure 64.9 Fat hydrolysis.

366

1. To tube 2 of each series add 1 ml of pancreatic juice and mix for a few seconds on a Vortex mixer. Place each # 2 tube in place within its series.

2. **Carbohydrase Test:** Remove a sample of test solution from each of the five tubes (1A through 5A) and place in separate depressions of a spot plate.

    Add 2 drops of IKI solution to each of the solutions on the plate. If starch is present due to lack of hydrolysis, the spot solution will turn blue.

    If hydrolysis has occurred, test the remaining solution in the tube with Barfoed's and Benedict's tests to determine degree of digestion.

    See appendix B for these tests. Record your results on the Laboratory Report.

3. **Protease Test:** Since hydrolysis produces a yellow color in the test tube, record the color of each tube on the Laboratory Report.

    If a photocolorimeter is set up and calibrated at 410 μm, take a reading for each tube and record the optical densities (OD) in the table on the Laboratory Report. Be sure to use appropriate cuvettes for the contents of each tube.

    Note: Before taking any photocolorimeter readings, the photocolorimeter should be standardized with a blank made up of 3 ml of BAPNA and 0.1 ml of 0.001 M HCl.

4. **Lipase Test:** Since fat hydrolysis results in a lowering of pH due to the formation of fatty acids, the bromthymol blue in the tubes will change from blue to yellow.

    Compare the six tubes with a bromthymol blue color standard to determine the pH changes that have occurred.

Laboratory Report

Complete the first portion of combined Laboratory Report 64, 65.

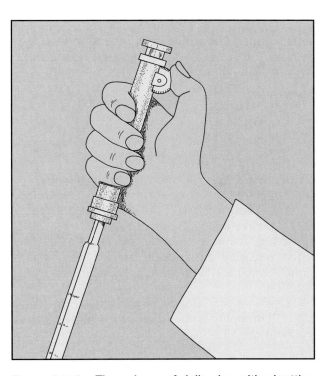

Figure 64.10 The volume of deliveries with pipetting device is controlled with the thumb.

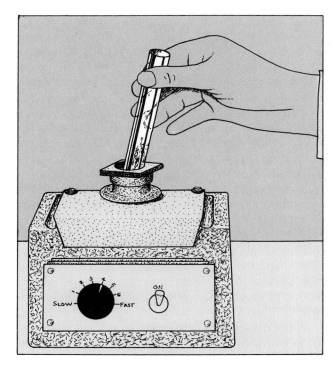

Figure 64.11 Each tube in the assay procedure must be mixed on the Vortex mixer.

65 Factors Affecting Hydrolysis

Enzymes are highly complex protein molecules of somewhat unstable nature due to the presence of weak hydrogen bonds. Many factors influence the rates at which hydrolysis occurs. Temperature and hydrogen ion concentration are probably the most important factors in this respect. It is the influence of these two conditions that will be studied in this exercise.

Salivary amylase (ptyalin) has been selected for this study. This enzyme hydrolyzes the starch amylose to produce soluble starch, maltose, dextrins, achrodextrins, and erythrodextrin. When iodine is used as an indicator of amylase activity, starch and soluble starch produce a blue color and erythrodextrin yields a red color. To detect the presence of maltose, Benedict's solution is used. With heat, maltose is a reducing sugar causing the reduction of soluble cupric to insoluble red cuprous oxide in Benedict's solution.

In this exercise, the saliva of one individual will be used. The saliva will be diluted to 10% with distilled water. To determine the effect of temperature on the rate of hydrolysis, iodine solution (IKI) will be added to a series of ten tubes (figure 65.1) containing starch and saliva. By adding the iodine at one-minute intervals to successive tubes, it is possible to determine, by color, when the hydrolysis of starch has been completed. For temperature comparisons, two water baths will be used. One will be 20° C and the other will be at 37° C. To determine the effects of hydrogen ion concentration on the action of amylase, we will use buffered solutions of pH 5, 7, and 9.

The dispensing of solutions in this exercise may be performed with medicine droppers, graduates, or serological pipettes.

If laboratory time is limited, it might be necessary for the instructor to divide the class into thirds that will perform only a portion of the entire experiment. After each group completes its portion, the results can be tabulated on the chalkboard so that all students can record results for the complete experiment. The best arrangement is for students to work in pairs. The following assignments will work well for a two-hour laboratory period:

GROUP	ASSIGNMENT
A (⅓ of Class)	Controls, 20° C, Boiling
B (⅓ of Class)	Controls, 37° C, Boiling
C (⅓ of Class)	Controls, pH Effects

Saliva Preparation

Before this experiment, the instructor will prepare a 10% saliva solution for the entire class as follows:

Materials:
    50 ml graduate
    250 ml graduated beaker
    paraffin
    12 dropping bottles with labels

Place a small piece (¼″ cube) of paraffin under your tongue and allow it to soften for a few minutes before starting to chew it. As it is chewed, expectorate all saliva into a 50 ml graduate.

When you have collected around 15 to 20 ml of saliva, measure out enough distilled water with a graduate and add it to the saliva, making up a 10% solution.

Dispense the diluted saliva into labeled ("10% saliva") dropping bottles. For a class of 24 students, you will need 12 bottles.

Controls

A set of five control tubes is needed for color comparisons with test results. Tubes 1 and 2 will be used for detection of starch. Tubes 3, 4, and 5 will be used for detection of the presence of maltose. Table 65.1 shows the ingredient in each tube and the significance of test results. Proceed as follows to prepare your set of controls.

Materials:
10% saliva solution (in dropping bottle)
0.1% starch solution
1% maltose solution
Benedict's solution (in dropping bottle)
IKI solution (in dropping bottle)
test tubes (13 mm dia × 100 mm long)
test-tube rack (Wassermann type)
medicine droppers
250 ml beaker
electric hot plate
1 ml sterile pipettes
pipetting device

1. Label five clean test tubes: 1, 2, 3, 4, and 5.
2. Fill the five tubes with the following reagents, *using a different pipette for each reagent.*

    Tube 1: 1 ml starch solution and 2 drops of IKI solution.
    Tube 2: 1 ml distilled water and 2 drops of IKI solution.
    Tube 3: 1 ml starch solution, 2 drops of saliva solution, and 5 drops Benedict's solution.
    Tube 4: 1 ml distilled water, 2 drops saliva solution, and 5 drops Benedict's solution.
    Tube 5: 1 ml maltose solution and 5 drops Benedict's solution.

3. Place tubes 3, 4, and 5 into a beaker of warm water and boil on an electric hot plate for **5 minutes** to bring about the desired color changes in tubes 3 and 5.
4. Set up the five tubes in a test-tube rack to be used for color comparisons. Tube 1 will be your positive starch control. Tubes 3 and 5 will be your positive sugar controls. Tubes 2 and 4 will be your negative controls for starch and sugar.

Effect of Temperature (20° C and 37° C)

In this portion of the experiment we will determine the effect of room temperature (20° C) and body temperature (37° C) on the rate at which amylase acts. Note in figure 65.1 that tubes of starch and saliva are given two drops of IKI solution at one-minute intervals to observe how long it takes for hydrolysis to occur at a given temperature.

When IKI is added to the first tube, the solution will be blue, indicating no hydrolysis. At some point along the way, however, the blue color will disappear, indicating complete hydrolysis. Note that the materials list is for performing the test at only one temperature.

Materials:
10 serological test tubes
(13 mm dia × 100 mm)
test-tube rack (Wassermann type)
water bath (20° C or 37° C)
10% saliva solution
0.1% starch solution
china marking pencil
thermometer (Centigrade scale)
1 ml serological pipettes or 10 ml graduate
mechanical pipetting device

1. Label ten test tubes 1 through 10 with a china marking pencil and arrange them sequentially in the front row of the test-tube rack.
2. With a 1 ml pipette dispense 1 ml of starch solution to each tube.
3. Place the rack of tubes and bottle of saliva in the water bath (20° C or 37° C).
4. Insert a clean thermometer into the bottle of saliva. When the temperature of the saliva has reached the temperature of the water bath (3–5 minutes), proceed to the next step.

Table 65.1 Control tube contents and functions.

TUBE NUMBER	CARBOHYDRATE	SALIVA	REAGENT	TEST RESULTS
1	Starch	—	IKI	Positive for Starch
2	—	—	IKI	Negative for Starch
3	Starch	2 Drops	Benedict's	Positive for Sugar Positive for Hydrolysis
4	—	2 Drops	Benedict's	Negative for Sugar
5	Maltose	—	Benedict's	Positive for Sugar

5. Lift the bottle of saliva from the water bath, **record the time,** and put **1 drop** of saliva into each test tube, starting with tube 1. Leave the rack of tubes in the water bath and be sure that the drop falls directly into the starch solution without touching the sides of the tube.
6. Shake the rack of tubes to mix their contents.
7. **Exactly one minute** after tube 1 has received saliva, add 2 drops of IKI solution to it and mix again. Note the color.
8. **One minute later** add 2 drops of IKI to tube 2 and repeat this process every minute until all ten tubes have received IKI solution.
9. Compare the colors of the tubes with the control tubes and determine the time required to digest the starch.
10. Record your results on the Laboratory Report.
11. Wash the insides of all test tubes with soap and water and rinse thoroughly.

Effect of Boiling

In this portion of this experiment, we will compare the action of amylase that has been boiled for a few seconds with unheated amylase. Figure 65.2 illustrates the procedure.

Materials:
    3 test tubes
    test-tube holder
    Bunsen burner
    10% saliva solution
    0.1 starch solution
    IKI solution

1. Label three test tubes 1, 2, and 3.
2. Dispense 1 ml of saliva solution to tubes 1 and 2, and 1 ml of distilled water to tube 3.
3. Heat tube 1 over a Bunsen burner flame so that it comes to a boil for a few seconds. Use a test-tube holder.
4. Add 1 ml of starch solution to each of the three tubes and place them in a 37° C water bath for 5 minutes.
5. Remove the tubes from the water bath and test for the presence of starch by adding 2 drops of IKI to each tube. Record the results on the Laboratory Report.
6. Wash and rinse the test tubes.

Effect of pH on Enzyme Activity

To determine the effect of different hydrogen ion concentrations on amylase activity, the enzyme, starch, and buffered solutions will be incubated at 37° C. The overall procedure is illustrated in figure 65.3.

Materials:
    test-tube rack (Wassermann type)
    12 test tubes (13 mm dia × 100 mm)
    1 ml pipettes and pipetting device
    10 ml graduate
    water bath (37° C)
    thermometer
    10% saliva solution
    0.1% starch solution
    IKI solution
    pH 5 buffer solution
    pH 7 buffer solution
    pH 9 buffer solution

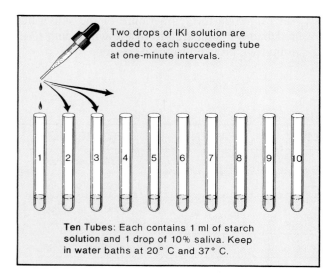

Figure 65.1 Amylase activity at 20° C and 37° C.

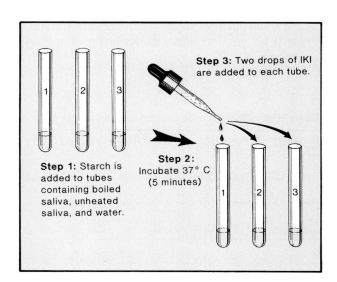

Figure 65.2 Effect of boiling on amylase.

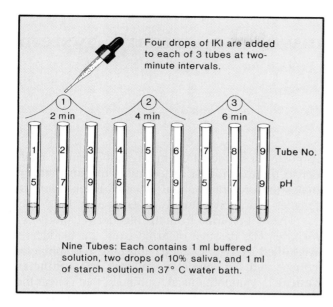

Four drops of IKI are added to each of 3 tubes at two-minute intervals.

Nine Tubes: Each contains 1 ml buffered solution, two drops of 10% saliva, and 1 ml of starch solution in 37° C water bath.

Figure 65.3 Effect of pH on amylase activity.

1. Label nine test tubes according to figure 65.3. Note that the tubes are arranged in three groups. Each group has three tubes of 5, 7, and 9 arranged in sequence.
2. With a 1 ml pipette, deliver 1 ml of pH 5 buffer solution to tubes 1, 4, and 7.
3. With a fresh pipette, deliver 1 ml of pH 7 solution to tubes 2, 5, and 8.
4. With another fresh pipette, deliver 1 ml of pH 9 solution to tubes 3, 6, and 9.

5. Add 2 drops of saliva solution to each of the nine tubes containing buffered solutions.
6. Fill the three other empty tubes about two-thirds full of starch solution (total of approximately 15 ml).
7. Place a clean thermometer in one of the three starch tubes and insert the whole rack of 12 tubes in the 37° C water bath.
8. When the thermometer in the starch tube registers 37° C, pipette 1 ml of warm starch solution to each of the nine tubes. Use the starch from the 3 tubes in the rack.
9. **Record the time** that the starch was added to tube 1.
10. **Two minutes** after tubes 1, 2, and 3 received starch, add 4 drops of IKI solution to each of these three tubes.
11. At **4 minutes** from the original recorded time, add 4 drops of IKI solution to tubes 4, 5, and 6.
12. At **6 minutes** from the original recorded time, add 4 drops of IKI solution to tubes 7, 8, and 9.
13. Remove the tubes from the water bath and compare them with the controls.
14. Record your observations on the Laboratory Report.

Laboratory Report

Complete the last portion of combined Laboratory Report 64, 65.

66 Anatomy of the Urinary System

This study consists of three parts: (1) histological studies, (2) sheep kidney dissection, and (3) cat dissection. Although the cat's kidney will be dissected, a sheep kidney is also included here because of its greater size. A general description of the urinary system in humans will precede these laboratory activities.

Organs of the Urinary System

Figure 66.1 illustrates the components of the urinary system with its principal blood supply. It consists of two kidneys, two ureters, the urinary bladder, and a urethra. The **kidneys** are somewhat bean-shaped, dark brown, and located behind the peritoneum (*retroperitoneal*). The right kidney is usually positioned somewhat lower than the left one, probably because of its displacement by the liver. Each kidney is supplied blood through a **renal artery** that is a branch off the **abdominal aorta.** Blood leaving each kidney drains via a **renal vein** into the **inferior vena cava.**

Urine passes from each kidney to the **urinary bladder** through a **ureter,** entering the bladder on its posterior surface. Leading from the urinary bladder to the exterior is a short tube, the **urethra.** In males the urethra is about 20 centimeters long; in females it is approximately 4 centimeters long.

The exit of urine from the bladder is called *micturition.* The passage of urine from the bladder is controlled by two sphincter muscles, the sphincter vesicae and the sphincter urethrae. The **sphincter vesicae** is a smooth muscle sphincter that is near the exit of the bladder. When approximately 300 ml of urine has accumulated in the bladder, the muscular walls of the bladder are stretched sufficiently to initiate a parasympathetic reflex that causes the bladder wall to contract. These contractions force urine past the sphincter vesicae into the urethra above the **sphincter urethrae.** This second sphincter, which is located approximately 1–3 centimeters below the sphincter vesicae, consists of skeletal muscle fibers and is voluntarily controlled. The presence of urine in the urethra above this sphincter creates a desire to micturate; however, since the valve is under voluntary control, micturition can be inhibited. When both sphincters are relaxed, urine passes from the body.

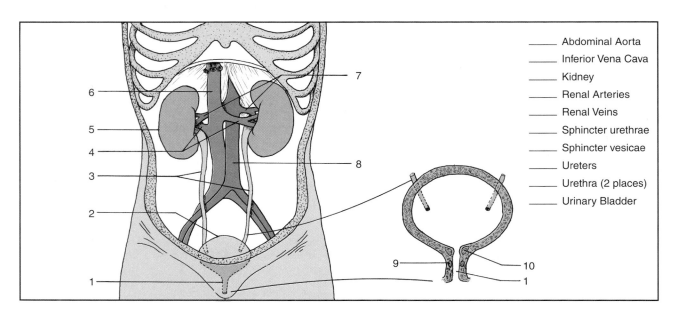

_____ Abdominal Aorta
_____ Inferior Vena Cava
_____ Kidney
_____ Renal Arteries
_____ Renal Veins
_____ Sphincter urethrae
_____ Sphincter vesicae
_____ Ureters
_____ Urethra (2 places)
_____ Urinary Bladder

Figure 66.1 The urinary system.

Assignment:
Label figure 66.1.

Kidney Anatomy

Figure 66.2 reveals a frontal section of a human kidney. Its outer surface is covered with a thin fibrous **renal capsule** (label 8). In addition to this thin covering, the kidney is provided support and protection by a fatty capsule that completely encases it. The latter is not shown in figure 66.1 or 66.2.

Immediately under the capsule is the **cortex** of the kidney. The cortex is reddish brown due to its extensive blood supply. The lighter inner portion is the **medulla.** The medulla is divided into cone-shaped **renal pyramids.** Nine of these pyramids are seen in figure 66.2. Cortical tissue, in the form of **renal columns,** extends down between the pyramids. Each renal pyramid terminates as a **renal papilla,** which projects into a minor calyx. The **minor calyces** (label 9) are short tubes that receive urine from the papillae and empty into a central **major calyx** of the kidney (label 5). The funnel-like **renal pelvis** receives urine from the major calyx and empties it into the **ureter.**

The Nephrons

The basic functioning unit of the kidney is the *nephron.* The enlarged section in figure 66.2 reveals two nephrons. Figure 66.3 illustrates a single nephron in greater detail. It has been estimated that

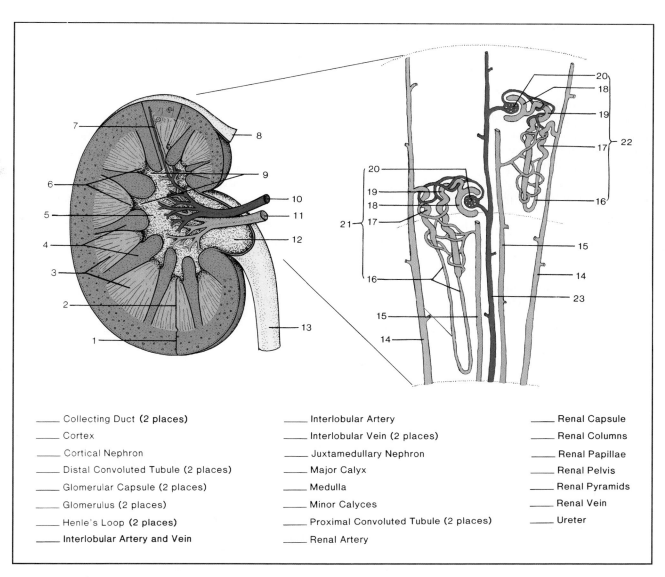

_____ Collecting Duct **(2 places)**
_____ Cortex
_____ Cortical Nephron
_____ Distal Convoluted Tubule **(2 places)**
_____ Glomerular Capsule **(2 places)**
_____ Glomerulus **(2 places)**
_____ Henle's Loop **(2 places)**
_____ Interlobular Artery and Vein

_____ Interlobular Artery
_____ Interlobular Vein **(2 places)**
_____ Juxtamedullary Nephron
_____ Major Calyx
_____ Medulla
_____ Minor Calyces
_____ Proximal Convoluted Tubule **(2 places)**
_____ Renal Artery

_____ Renal Capsule
_____ Renal Columns
_____ Renal Papillae
_____ Renal Pelvis
_____ Renal Pyramids
_____ Renal Vein
_____ Ureter

Figure 66.2 Anatomy of the kidney.

there are approximately one million nephrons in each kidney. Eighty percent of the nephrons are located in the cortex; these are designated as **cortical nephrons.** The remaining 20%, which are located partially in the cortex and partially in the medulla, are called **juxtamedullary nephrons** (label 21, figure 66.2).

The formation of urine by the nephron results from three physiological activities that occur in different regions of the nephron: (1) filtration, (2) reabsorption, and (3) secretion. Because of the differing functions of various regions of the nephron, the fluid that first forms at the beginning of the nephron is quite unlike the urine that enters the calyces of the kidney.

Note that each nephron consists of an enlarged end, the **renal corpuscle,** and a long tubule that empties eventually into the calyx of the kidney. Each renal corpuscle has two parts: an inner tuft of capillaries, the **glomerulus,** and an outer double-walled cuplike structure, the **glomerular** *(Bowman's)* **capsule.** Blood that enters the kidney through the **renal artery** reaches each nephron through an **interlobular artery** (label 2, figure 66.3). A short **afferent arteriole** conveys blood into the glomerulus from the interlobular artery. Blood exits from the glomerulus through the **efferent arteriole,** which has a much smaller diameter than the afferent vessel.

The high intraglomerular blood pressure forces a highly dilute fluid, the *glomerular filtrate,* to pass into the glomerular capsule. This fluid consists of glucose, amino acids, urea, salts, and a great deal of water.

This filtrate passes next through the **proximal convoluted tubule** (label 8, figure 66.3), the **descending limb of Henle's loop** (label 12), the **ascending limb of Henle's loop,** and the **distal convoluted tubule** (label 10) before emptying into the large **collecting duct** on the right side of the diagram. A single collecting duct may have several nephrons emptying into it.

As the filtrate moves through these various regions, water, glucose, amino acids, and other substances are reabsorbed into the blood of the **peritubular capillaries,** which enmesh the entire route. Eighty percent of the water is reabsorbed through the walls of the proximal convoluted tubules. Water reabsorption is facilitated by the *antidiuretic hormone* of the posterior pituitary and *aldosterone* of the adrenal cortex. Cells lining the collecting ducts alter urine composition by secreting ammonia, uric acid, and other substances into the lumen of the duct.

Assignment:
Label figures 66.2 and 66.3.

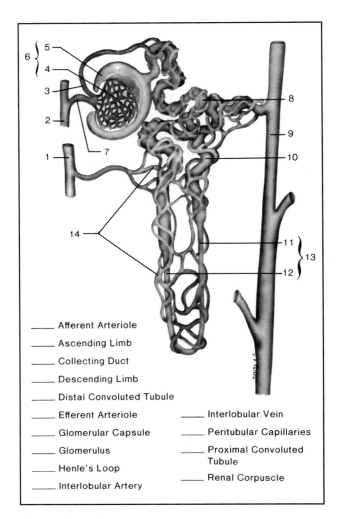

_____ Afferent Arteriole

_____ Ascending Limb

_____ Collecting Duct

_____ Descending Limb

_____ Distal Convoluted Tubule

_____ Efferent Arteriole   _____ Interlobular Vein

_____ Glomerular Capsule   _____ Peritubular Capillaries

_____ Glomerulus   _____ Proximal Convoluted Tubule

_____ Henle's Loop

_____ Interlobular Artery   _____ Renal Corpuscle

Figure 66.3   The nephron.

Histological Study

Materials:
prepared slides of:
    kidney tissue (H9.11, H9.115, or H9.14)
    ureter (H9.16)
    urinary bladder (H9.21)

Examine slides of the **kidney cortex, kidney medulla, urinary bladder,** and **ureter.** Compare the slides with figures HA-32, HA-33, HA-28 (illustrations C and D), and HA-37 (illustrations A and B). Note the differences in kinds of tissue seen in the tubules, calyx, urinary bladder, and ureters.

Sheep Kidney Dissection

Dissection of the sheep kidney may be combined with the cat dissection to provide greater anatomical clarity. Although the injected cat kidney may illustrate better the blood supply, the fresh sheep kidney will reveal more vividly some of the lifelike characteristics of the organ.

Materials:
    dissecting kit, tray, long knife
    fresh sheep kidneys

1. If the kidneys are still encased in fat, peel it off carefully. As you lift the fat away from the kidney, look carefully for the **adrenal gland,** which should be embedded in the fat near one end of the kidney.
2. Remove the adrenal gland from the fat and cut it in half. Note that the gland has a distinct outer **cortex** and an inner **medulla.**
3. Probe into the surface of the kidney with a sharp dissecting needle to see if you can differentiate the **capsule** from the underlying tissue.
4. With a long (butcher) knife, slice the kidney longitudinally to produce a frontal section similar to figure 66.2. Wash out the cut halves with running water.
5. Identify all the structures seen in the kidney section of figure 66.2.

Cat Dissection

The close association of the urinary and reproductive organs necessitates the study of the organs of these two systems together. Most emphasis, however, will be placed here on the urinary organs.

As you proceed through the following dissection, refer to figures 66.4 and 66.5 to identify the various organs. A more detailed study of the reproductive organs will be made in Exercise 69.

1. To expose the urogenital system, remove the liver, spleen, and stomach.
2. With forceps and scalpel, clear away excess fat, the peritoneum, and other connective tissue in the area.
    Take care not to injure the internal spermatic (or ovarian) blood vessels and the ureters, which are embedded in fat.
3. Note that the **kidneys** are *retroperitoneal* in that they lie between the peritoneum and the body wall. Note also that the left kidney is somewhat caudad of the right one.

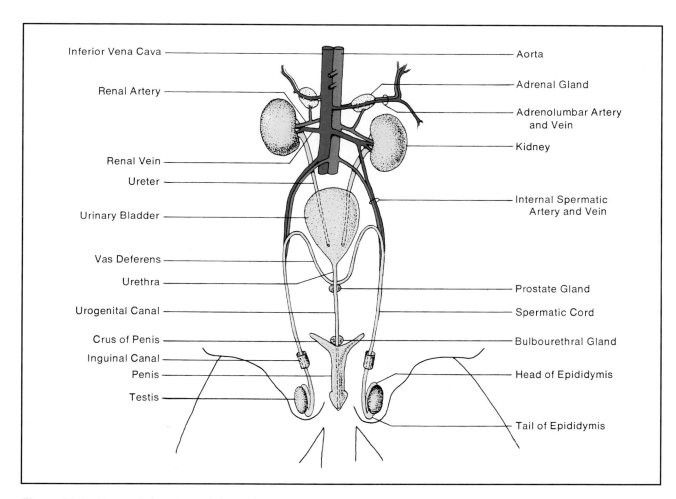

Figure 66.4 **Urogenital system of the male cat.**

4. Identify the two **adrenal glands,** which are embedded in connective tissue between the anterior ends of the kidneys and the large abdominal blood vessels (aorta and inferior vena cava).
5. Remove both kidneys. With a razor blade or sharp scalpel slice one longitudinally to produce a series of thin frontal sections revealing its internal structure.
6. Slice the other kidney to produce a series of thin transverse sections.
7. Identify the **cortex, medulla,** and **papilla** on the sections. Observe that there is only one renal papilla in the cat kidney emptying into the pelvis of the ureter. In the human there are approximately twelve papillae.
8. Trace the **ureter** of each kidney to the bladder where it enters the dorsal surface of the bladder.
9. Note that the **urinary bladder** is also retroperitoneal. It is a musculomembranous sac attached to the abdominopelvic walls by ligaments formed from folds of peritoneum.

The ventral side of the bladder is attached to the linea alba by the **medial ligament.** Two **lateral ligaments,** which contain sizable fat deposits, connect the sides of the bladder to the dorsal body wall.

10. Make a linear ventral incision in the bladder with a scalpel and examine the inside wall.

Locate the openings of the two ureters and urethra. Also, cut into the urethra where it joins the bladder. Do you see any evidence of **sphincter muscles** in this region?
11. Observe that the urethra of the female cat empties into a **urogenital sinus.** In the male it passes through the penis.
12. Trade specimens with students who have the opposite sex of your specimen so that you can become familiar with the anatomy of both sexes.

Laboratory Report

Complete Laboratory Report 66 for this exercise.

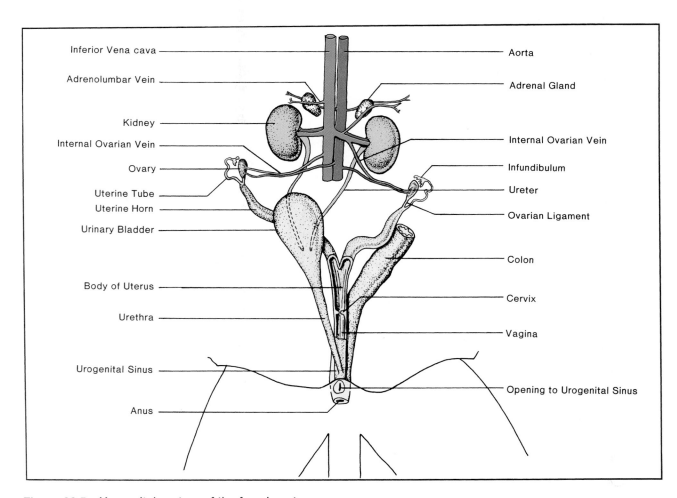

Figure 66.5 Urogenital system of the female cat.

Urine: Composition and Tests

Urine is a highly complex aqueous solution of organic and inorganic substances. Urea, uric acid, creatinine, sodium chloride, ammonia, and water are the principal ingredients. Most substances are either waste products of cellular metabolism or products derived directly from certain foods that are eaten.

Our principal focus in this laboratory session will be with the detection of abnormal substances in urine that may have pathological significance. Some tests will be of physical properties; most tests will be for chemical substances. A total of ten specific tests will be made.

The chemical tests will be of two types: (1) classical *in vitro* methods, and (2) test strip methods. In most laboratories the test strip method will be preferred because of time saved and test reliability. Proceed as follows to test a urine sample:

Materials:

test strips: *Albustix, Clinistix, Ketostix, Hemastix, Bili-Labstix*
other items: listed under each test

Collection of Specimen

Urine should be collected in a clean container and stored in a cool place until tested. If the sample is to be collected just prior to lab testing, plastic disposable containers will be made available on the supply table. The best time for collecting a urine sample is 3 hours after a meal. First samples taken in the morning are least likely to reveal abnormal substances in the urine.

Appearance of Test Specimen

The visual appearance of a urine sample is the first important thing to consider. Its color and turbidity can provide clues as to evidence of pathology. Evaluate its color and transparency according to the following standards:

Color

Normal urine will vary from light straw to amber color. The color of urine is due to a pigment called *urochrome* that is the end product of hemoglobin breakdown:

$$\text{Hemoglobin} \rightarrow \text{Hematin} \rightarrow \text{Bilirubin}$$
$$\text{Urochrome} \leftarrow \text{Urochromogen} \quad \swarrow$$

The following are deviations from normal color that have pathological implications:

Milky: pus, bacteria, fat or chyle.

Reddish amber: urobilinogen or porphyrin. Urobilinogen is produced in the intestine by the action of bacteria on bile pigment.
Porphyrin may be evidence of liver cirrhosis, jaundice, Addison's disease, and other conditions.

Brownish yellow or green: bile pigments. Yellow foam is definite evidence of bile pigments.

Red to smoky brown: blood and blood pigments.

Carrots, beets, rhubarb, and certain drugs may color the urine, yet have no pathological significance. Carrots may cause increased yellow color due to carotene; beets cause reddening, and rhubarb may cause urine to become brown.

Evaluate your urine sample according to the above criteria and record the information on the Laboratory Report.

Transparency (Cloudiness)

A fresh sample of normal urine should be clear but may become cloudy after standing for a while. Cloudy urine may be evidence of phosphates, urates, pus, mucus, bacteria, epithelial cells, fat, and chyle.

Phosphates disappear with the addition of dilute acetic acid, and urates will dissipate with heat. Other causes of turbidity can be analyzed by microscopic examination.

After shaking your sample, determine the degree of cloudiness and record it on the Laboratory Report.

Specific Gravity

The specific gravity of a 24-hour specimen of normal urine will be between 1.015 and 1.025. Single urine specimens may range from 1.002 to 1.030. The more solids in solution, the higher will be the specific gravity. The greater the volume of urine in a 24-hour specimen, the lower will be its specific gravity. A low specific gravity will be present in chronic nephritis and *diabetes insipidus.* A high specific gravity may indicate *diabetes mellitus,* fever, and acute nephritis.

Materials:
urinometer (cylinder and hydrometer)
thermometer (Centigrade scale)
filter paper

1. Fill the urinometer cylinder three-fourths full of well-mixed urine (figure 67.1). Remove any foam on the surface with a piece of filter paper.
2. Insert the hydrometer into the urine and read the graduation on the stem at the level of the bottom of the meniscus.
3. Take the temperature of the urine and compensate for the temperature by adding 0.001 to the specific gravity for each 3° C above 25° C, and subtracting the same amount for each degree below 25° C. This value is the **adjusted specific gravity.**
4. Record the adjusted specific gravity on the Laboratory Report.
5. Wash the cylinder and hydrometer with soap and water after completing the test.

Hydrogen Ion Concentration

Although freshly voided urine is usually acid (around pH 6), the normal range is between 4.8 and 7.5. The pH will vary with the time of day and diet. Twenty-four hour specimens are less acid than fresh specimens and may become alkaline after standing due to bacterial decomposition of urea.

High acidity is present in acidosis, fevers, and high-protein diets. Excess alkalinity may be due to urine retention in the bladder, chronic cystitis, anemia, obstructing gastric ulcers, and alkaline therapy. The simplest way to determine pH is to use pH indicator paper strips as shown in figure 67.2.

Materials:
pH indicator paper strip (*pHydrion* or *nitrazine* papers)

1. Dip a strip of pH paper into the urine three consecutive times and shake off the excess liquid.
2. **After 1 minute,** compare the color with color chart. Record observations on the Laboratory Report.

Protein Detection (Exton's Method)

Although the large size of protein molecules normally prohibits their appearance in normal urine, certain conditions can allow them to filter through. Excessive muscular exertion, prolonged cold baths, and excessive protein ingestion may result in what is termed *physiologic albuminuria.*

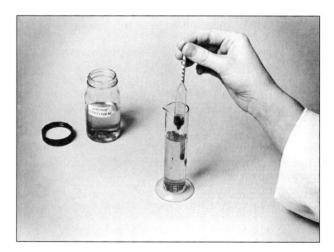

Figure 67.1 Specific gravity. The hydrometer is lowered into the urine after the foam has been removed with a piece of filter paper. The reading is taken from the bottom of the meniscus.

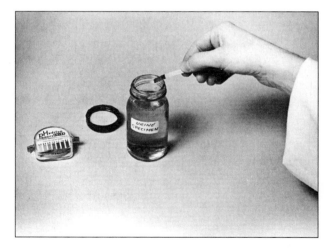

Figure 67.2 Determination of pH. Only fresh urine samples should be tested because pH tends to rise with aging of urine. The pH should be read one minute after the paper has been dipped.

Pathologic albuminuria, on the other hand, exists when albumin in the urine results from kidney congestion, toxemia of pregnancy, febrile diseases, and anemias.

Detection of albumin may be accomplished by precipitation with chemicals, heat, or both. In our test here we will use *Exton's method,* which employs sulfasalicylic acid and heat.

Materials:
    Exton's reagent
    test tubes, one for each urine sample
    test-tube holder
    Bunsen burner
    pipettes (5 ml size) or graduates

1. Pour or pipette equal volumes of urine and Exton's reagent into a test tube. Approximately 3 ml of each should suffice.
2. Shake the tube from side to side, or roll it between the palms of your hands to mix thoroughly. If precipitate occurs, albumin is present.
3. Heat tube in Bunsen flame (figure 67.4). If precipitate persists, albumin is definitely present in the urine.
4. Record your results on the Laboratory Report.

Test Strip Method

Shake the sample of urine and dip the test portion of an *Albustix* test strip into the urine. Touch the tip of the strip against the edge of the urine container to remove excess urine. Immediately compare the test area with the color chart on the bottle. Note that the color scale runs from yellow (negative) to turquoise (++++). Record your results on the Laboratory Report.

Detection of Mucin
(Using Glacial Acetic Acid)

Inflammation of the mucous membranes of the urinary tract and vagina may result in large amounts of mucin being present in urine. Since mucin can be confused with albumin, its presence should be confirmed. Mucin and mucoid masses are glycoproteins that will reduce Benedict's reagent in the presence of acid or alkali. Glacial acetic acid will be used here for the detection of mucin.

Materials:
    test tubes and test-tube holder
    Bunsen burner
    acetic acid (glacial)
    sodium hydroxide (10%)
    pipettes (5 ml size)
    filter paper
    funnel
    ring stand
    small graduate

1. If urine was positive for albumin, remove the albumin by boiling 5 ml of urine for a few minutes and filtering while hot. See figure 67.5.
2. After the urine has cooled, add 6 ml of **distilled water** to 2 ml of the urine to prevent precipitation of urates.
3. Add a few drops of **glacial acetic acid.** If mucin is present, the urine will become turbid.
4. To further prove that the precipitate is mucin, add a few drops of **sodium hydroxide.** The precipitate will disappear if it is mucin.
5. Record the results of this test on your Laboratory Report.

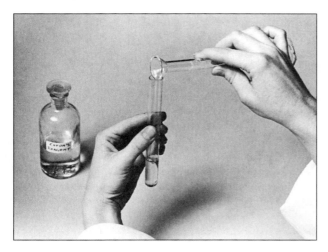

Figure 67.3 Albumin test. If albumin is present in the urine, the addition of Exton's reagent to it will cause a precipitate to form.

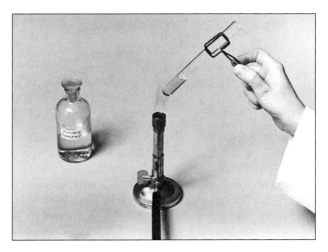

Figure 67.4 Albumin confirmation. If precipitate formed from the addition of Exton's reagent persists after heating, the presence of albumin is confirmed.

Detection of Glucose
(Using Benedict's Reagent)

Normally, urine will have no more than 0.03 gm of glucose per 100 ml of urine. When glucose exists in urine in amounts greater than this *glycosuria* exists. This usually indicates that *diabetes mellitus* is present.

The renal threshold of glucose is around 160 mg per 100 ml. Glycosuria indicates that blood levels of glucose exceed this amount and the kidneys are unable to accomplish 100% reabsorption of the carbohydrate.

The detection of glucose in urine is usually performed with *Benedict's reagent.* In the presence of glucose a precipitate ranging from yellowish green to red will form. The color of the precipitate will indicate the approximate percentage of glucose that is present.

Materials:

2 test tubes
electric hot plate
beaker (250 ml size)
Benedict's qualitative reagent
pipettes (5 ml size) or graduates
urine sample to be tested
urine sample (positive for glucose)

1. Label one tube for each sample to be tested. One tube should be for a known positive sample, if available.
2. Pour 5 ml of **Benedict's reagent** into each tube.
3. Add 8 drops (or 0.5 ml) of urine to each tube of Benedict's reagent. Use separate clean pipettes for each urine sample.
4. Place the tubes in a beaker of warm water and bring to boil for 5 minutes (figure 67.6).

5. Determine the amount of glucose present by color:

Negative = clear blue to cloudy green
+ = yellowish green
(0.5 to 1.0 gm %)
+ + = greenish yellow
(1.0 to 1.5 gm %)
+ + + = yellow (1.5 to 2.5 gm %)
+ + + + = orange (2.5 to 4 gm %)
red (4 gm % and over)

Test Strip Method

Shake the sample of urine and dip the test portion of a Clinistix test strip into the urine. Ten seconds after wetting, compare the color of the test area with the color chart on the label of the bottle. Note that there are three degrees of positivity.

The light intensity generally indicates 0.25% or less glucose. The dark intensity indicates 0.5% or more glucose. The medium intensity has no quantitative significance. Record your results on the Laboratory Report.

Detection of Ketones
(Using Rothera's Test)

Normal catabolism of fats produces carbon dioxide and water as final end products. When there is inadequate carbohydrate in the diet, or when there is a defect in carbohydrate metabolism, the body begins to utilize an increasing amount of fatty acids. When this increased fat metabolism reaches a certain point, fatty acid utilization becomes incomplete, and intermediary products of fat metabolism occur

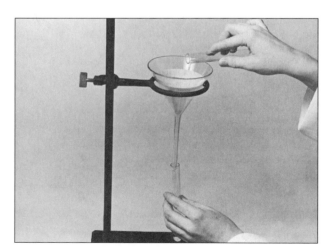

Figure 67.5 **Mucin test.** Before testing for mucin it is necessary to remove albumin by filtering boiled urine. Acetic acid causes mucin to precipitate.

Figure 67.6 **Glucose test.** To detect glucose in urine it is necessary to boil 0.5 ml of urine in 5 ml of Benedict's reagent for 5 minutes to get a color change.

in the blood and urine. These intermediary substances are the ketones acetoacetic acid, acetone, and beta hydroxybutyric acid. The presence of these substances in urine is called *ketonuria*.

Diabetes mellitus is the most important disorder in which ketonuria occurs. Progressive diabetic ketosis is the cause of diabetic acidosis, which can eventually lead to coma or death. It is for this reason that the detection of ketonuria in diabetics is of great significance.

The test method used here is called *Rothera's test*. The active ingredients in Rothera's reagent are sodium nitroprusside and ammonium sulfate. In the presence of ketones and ammonium hydroxide, this reagent produces a pinkish purple color.

Materials:
Rothera's reagent
test tubes
ammonium hydroxide (concentrated)

1. Add about 1 gm of **Rothera's reagent** to 5 ml of urine in a test tube.
2. Layer over the urine 1 to 2 ml of concentrated **ammonium hydroxide** by allowing it to flow gently down the side of the inclined test tube.
3. If a pink-purple ring develops at the interface, ketones are present. No ring, or a brown ring, is negative.
4. Record your results on the Laboratory Report.

Test Strip Method

As with the previous rapid methods, dip the test portion of a Ketostix test strip into the urine sample and tap it dry on the edge of the urine container. Fifteen seconds after wetting, compare the color of the test strip with the color chart on the label of the bottle.

Record your results on the Laboratory Report.

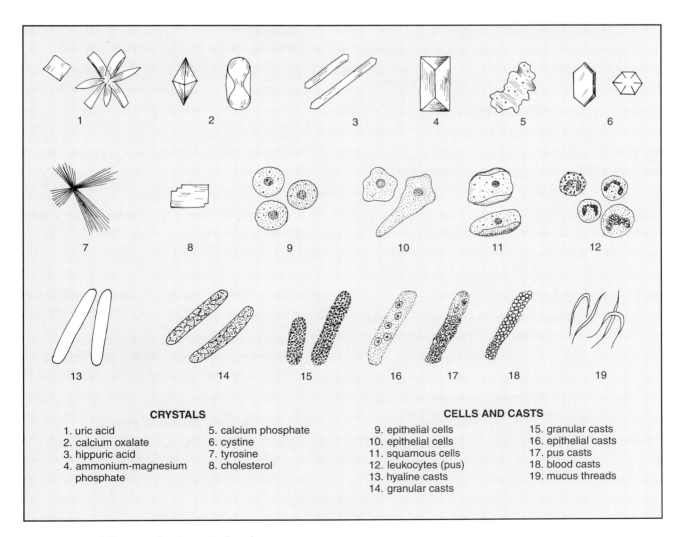

CRYSTALS

1. uric acid
2. calcium oxalate
3. hippuric acid
4. ammonium-magnesium phosphate
5. calcium phosphate
6. cystine
7. tyrosine
8. cholesterol

CELLS AND CASTS

9. epithelial cells
10. epithelial cells
11. squamous cells
12. leukocytes (pus)
13. hyaline casts
14. granular casts
15. granular casts
16. epithelial casts
17. pus casts
18. blood casts
19. mucus threads

Figure 67.7 Microscopic elements in urine.

Detection of Hemoglobin
(Using Hemastix Test Strips)

When red blood cells disintegrate (hemolyze) in the body, hemoglobin is released into the surrounding fluid. If the hemolysis occurs in the blood vessels, the hemoglobin becomes a constituent of the plasma and some of it will be excreted by the kidneys into the urine. If red blood cells enter the urinary tract due to disease or trauma to parts of the urinary system, the cells will hemolyze in the urine. Regardless of the cause of hemoglobin in urine, the condition is called *hemoglobinuria.*

Hemoglobinuria may be evidence of hemolytic anemia, transfusion reactions, yellow fever, smallpox, malaria, hepatitis, mushroom poisoning, renal infarction, burns, etc. The simplest way to test for hemoglobin is to use *Hemastix* test strips as follows:

1. Shake the sample of urine and dip the test portion of a Hemastix test strip into the urine.
2. Tap the edge of the strip against the edge of the urine container and let dry for **30 seconds.**
3. Compare the color of the test strip with the color chart on the bottle.
4. Record your results on the Laboratory Report.

Detection of Bilirubin
(Using Bili-Labstix)

We have seen that bilirubin is formed in the second stage of hemoglobin degradation from hematin. It is normally present in urine in very small quantities. When present in large amounts, however, it usually indicates damage to the liver by hepatoxic agents (chemical or biological). The presence of significant amounts of bilirubin is designated as *bilirubinuria.* The simplest way to test for this constituent in urine is to use *Bili-Labstix.* The procedure is as follows:

1. Shake the sample of urine and dip the test portion of a Bili-Labstix test strip into the urine.
2. Tap the edge of the strip against the ridge of the urine container and let dry for **20 seconds.**
3. Compare the color of the test strip with the color chart on the bottle. The results are interpreted as negative, small (+), moderate (++), and large (+++) amounts of bilirubin.
4. Record your results on the Laboratory Report.

Microscopic Study

Microscopic examination of elements in urine is the most important phase of urine analysis. However, the scope of this phase goes considerably beyond the limited scope of this course. A complete microscopic examination includes not only visual analysis of the sediment but also bacteriological differentiation.

Normal urine will contain an occasional leukocyte, some epithelial cells, mucin, bacteria, and crystals of various kinds. The experienced technologist has to determine when these substances exist in excess amounts and must be able to differentiate the various types of casts, cells, and crystals that predominate.

Figure 67.7 illustrates only a few of the elements that might be encountered in urine. No attempt shall be made here to identify all particulate matter in urine.

Materials:
    microscope slides and cover glasses
    Pasteur pipettes (optional)
    centrifuge
    centrifuge tubes (conical tipped)
    pipettes (5 ml size) or graduates
    mechanical pipetting device
    wire loop

1. Measure 5 ml of urine into a centrifuge tube after shaking the urine sample to resuspend the sediment. Use a mechanical pipetting device or a graduate.

    Be sure to balance the centrifuge with an even number of loaded tubes.
2. Centrifuge the tubes for **5 minutes** at a slow speed (1500 rpm).
3. Pour off all urine and allow the remaining sediment in the tube to settle down into the bottom of the tube.
4. With a Pasteur pipette or flamed wire loop, transfer a small amount of the sediment to a microscope slide and cover with a cover glass.
5. Examine under the microscope with low- and high-power objectives. Reduce the lighting by adjusting the diaphragm. Refer to figure 67.7 to identify structures.

Laboratory Report

Complete Laboratory Report 67 for this exercise.

The Endocrine Glands

68

In various dissections of previous exercises, endocrine glands have been encountered and discussed briefly. In this exercise all of the glands will be studied in a unified manner to review what has been stated previously and to explore more in depth the histology of each gland. By studying stained microscope slides of the various glands, and by comparing them with photomicrographs, you should have no difficulty identifying the specific cells in most glands that produce the various hormones.

This laboratory experience should help to dispel some of the abstractness one usually encounters when attempting to assimilate a mass of endocrinological facts.

The Thyroid Gland

The thyroid gland consists of two lobes joined by a connecting isthmus. A posterior view of this gland is seen in figure 68.1. Note in figure 68.2 that, microscopically, it consists of large numbers of spherical sacs called **follicles.** These follicles are filled with a colloidal suspension of a glycoprotein, **thyroglobulin.** The principal hormones of this gland are the *thyroid hormone* and *calcitonin.*

The Thyroid Hormone

Thyroid hormone consists of thyroxine, triiodothyronine, and a small quantity of closely related iodinated hormones. They form within the thyroglobulin molecule, emerge from the molecule, and enter blood vessels in the gland to be transported to all tissues of the body for utilization. Excess hormones remain stored in the thyroglobulin.

Action   The combined activity of thyroxine and triiodothyronine increases the metabolic activity of most tissues of the body. Bone growth in children is accelerated; carbohydrate metabolism is enhanced; cardiac output and heart rate are increased; respiratory rate is increased; and mental activity is increased.

Deficiency Symptoms   In children, hypothyroidism may result in *cretinism,* a condition characterized by physical and mental retardation. In

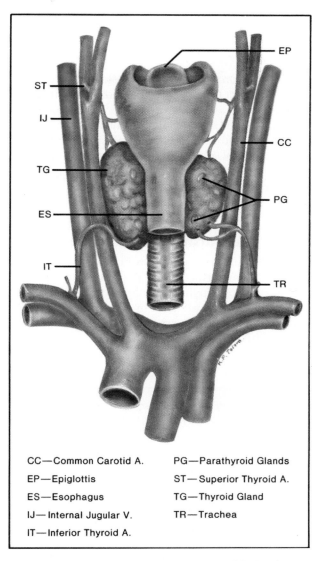

CC—Common Carotid A.   PG—Parathyroid Glands
EP—Epiglottis   ST—Superior Thyroid A.
ES—Esophagus   TG—Thyroid Gland
IJ—Internal Jugular V.   TR—Trachea
IT—Inferior Thyroid A.

Figure 68.1   The thyroid and parathyroid glands.

adults, severe long-term deficiency causes *myxedema,* which is characterized by obesity, slow pulse, lack of energy, and mental depression.

Oversecretion Symptoms   Hyperthyroidism is characterized by weight loss, rapid pulse, intolerance to heat, nervousness, and inability to sleep. *Exophthalmia,* or protrusion of the eyes, is often present.

383

Goiters Enlarged thyroid glands ("goiters") may be present in both hypo- and hyperthyroidism. Hypothyroid goiters may be due to iodine deficiency *(endemic goiter)* or some other unknown cause *(idiopathic nontoxic goiter)*. Hyperthyroid goiters may or may not be due to malignancy of the gland.

Regulation Thyrotropin, which is produced by the anterior pituitary gland, stimulates the thyroid to produce the thyroid hormone.

Calcitonin

Calcitonin is produced by the parafollicular cells in the interstitium of the thyroid gland. When injected experimentally, calcitonin causes a decrease in blood calcium. This is due, presumably, to increased osteoblastic and decreased osteoclastic activity.

It probably plays an important role in bone remodeling during growth in children, but in adults it appears to have very little, if any, effect on blood calcium levels over the long term.

The Parathyroid Glands

There are four small parathyroid glands embedded in the posterior surface of the thyroid gland. See figure 68.1. Occasionally, parathyroid tissue is seen outside of the thyroid gland in the neck region.

Two kinds of cells are seen in this gland: **chief** and **oxyphil cells.** Both types of cells are shown in

figure 68.2. Note that the chief cells are more numerous, smaller, and arranged in cords. They produce the only hormone of this gland, which has been designated, simply, as the *parathyroid hormone (PTH)*. The function of the larger oxyphil cells is unknown at this time.

Action PTH raises blood calcium levels and lowers the phosphorus levels in the blood. This is accomplished by activating osteoclasts in bone tissue. To prevent calcium loss in the kidneys, the hormone also promotes the reabsorption of calcium in the renal tubules. Removal of all parathyroid tissue results in *tetany* and death.

Regulation The plasma level of calcium in humans is remarkably stable: about 10 mg/100 ml. Production of PTH appears to be regulated by the level of calcium in the blood. Excess blood calcium levels inhibit hormone production, and lowered blood calcium levels cause the glands to produce more of the hormone.

The Pancreas

Figure 68.3 illustrates the histology of the pancreas. The endocrine-secreting portion of the pancreas is performed by clusters of cells called **pancreatic islets** (islets of Langerhans). These cells are located between the saclike glands *(acini)* that produce the pancreatic digestive enzymes. Insulin, glucagon,

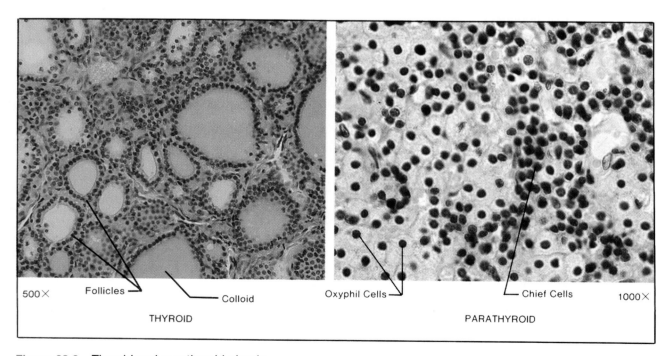

500× Follicles — Colloid THYROID Oxyphil Cells — Chief Cells 1000× PARATHYROID

Figure 68.2 Thyroid and parathyroid glands.

and somatostatin are produced by different types of cells in the islets.

Insulin

Insulin is an anabolic hormone that is produced by the **beta cells** of the pancreatic islets. These cells can be differentiated from the other islet cells by the presence of many granules in the cytoplasm.

Insulin promotes the storage of glucose by converting it to glycogen. It also promotes the storage of fatty acids and amino acids. Insulin deficiency results in *diabetes mellitus,* which is characterized by *hyperglycemia* (high blood sugar).

Regulation   Although a variety of stimulatory and inhibitory factors affect insulin production, feedback control by blood glucose levels on the beta cells is the major controlling factor. As glucose levels become elevated, insulin production is stepped up; when the glucose level is normal or low, the rate of insulin production is low.

Glucagon

Glucagon is a catabolic hormone produced by the **alpha cells** of the pancreatic islets. Its action is just the opposite of insulin in that it mobilizes glucose, fatty acids, and amino acids in tissues. Deficiency of this hormone will result in hypoglycemia. Glucagon also stimulates the production of the growth hormone, insulin, and pancreatic somatostatin.

Somatostatin

A third type of islet cells, called the **delta cells,** produce somatostatin, a growth-inhibiting hormone. With Mallory's stain the cytoplasm of these cells stains blue.

This hormone is growth inhibiting in that it inhibits the release of the growth hormone by the anterior pituitary gland. In addition, it inhibits the production of insulin and glucagon. Tumors involving the delta cells result in hyperglycemia and other diabetes-like symptoms. Removal of the tumors causes the symptoms to disappear.

The Adrenal Glands

Each adrenal gland has an outer *cortex* and an inner *medulla.* Surrounding the entire gland is a capsule of fibrous connective tissue.

The embryological origins of the cortex and medulla explain their differences in function. The cells of the medulla originate from the neural crest of the embryo. This embryonic tissue also gives rise to ganglionic cells of the sympathetic nervous system; thus, we can expect a close kinship in function of the adrenal medulla and the sympathetic nervous system.

Cells of the cortex, on the other hand arise from embryonic tissue associated with the gonads.

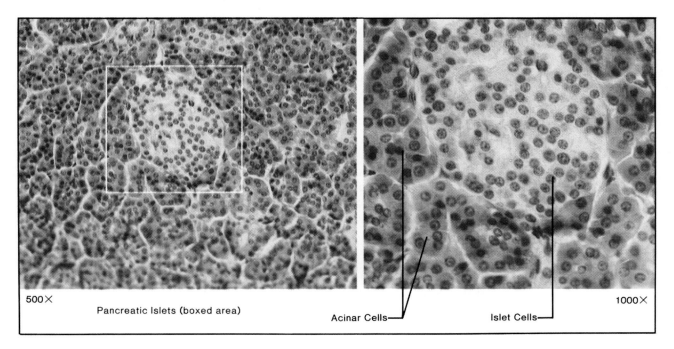

500×

Pancreatic Islets (boxed area)

Acinar Cells

Islet Cells

1000×

Figure 68.3   **Pancreas histology.**

Hormones having some relationship to the reproductive organs might, thus, be expected from the cortex.

The Adrenal Cortex

Figures 68.4 and 68.5 reveal the three layers of the cortex. Note that immediately under the capsule lies a layer called the **zona glomerulosa.** The innermost layer of cells, which interfaces with the medulla, is the **zona reticularis.** The middle layer, which is the thickest, is the **zona fasciculata.**

In all layers are seen vascular channels, called **sinusoids,** which are larger in diameter than capillaries but resemble capillaries in that their walls are one cell thick. It is in these channels that the hormones collect before passing directly into the general circulation.

The adrenal cortex secretes many different hormones, all of which belong to a group of substances called *steroids.* More specifically, they are referred to as **corticosteroids.** All of them are synthesized from cholesterol and have similar molecular structure. Whereas destruction of the adrenal medulla is more or less inconsequential to survival, obliteration of the adrenal cortex results in death.

The corticosteroids fall into three groups: the mineralocorticoids, the glucocorticoids, and the androgenic hormones. The *mineralocorticoids* are hormones that control the excretion of Na^+ and K^+, which affects the electrolyte balance. The *glucocorticoids* affect primarily the metabolism of glucose and protein. The *androgenic hormones* produce some of the same effects on the body as the male sex hormone (testosterone). Of the 30 or more corticosteroids that are produced by the adrenal cortex, the most important two are aldosterone and cortisol. Descriptions of these two hormones follow.

Aldosterone This mineralocorticoid is produced by cells of the zona glomerulosa. The most important function of this hormone is to increase the rate of renal tubular absorption of sodium. Any condition that impairs the production of aldosterone will result in death within two weeks if salt replacement or mineralocorticoid therapy is not provided.

In the absence of aldosterone, the K^+ concentration in extracellular fluids rises, Na^+ and Cl^- concentrations decrease, and the total volume of body fluids becomes greatly reduced. These conditions cause reduced cardiac output, shock, and death.

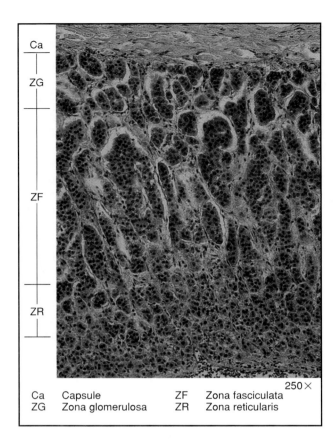

Ca ZG ZF ZR

250×

| Ca | Capsule | ZF | Zona fasciculata |
| ZG | Zona glomerulosa | ZR | Zona reticularis |

Figure 68.4 **The adrenal cortex.**

Factors that affect aldosterone production are ACTH, K^+ concentration, Na^+ concentration, and the renin-angiotensin system.

Cortisol (Hydrocortisone, Compound F)

Approximately 95% of the glucocorticoid activity results from the action of cortisol. This hormone is produced primarily by cells in the zona fasciculata; some of it is also produced by the zona reticularis.

When an excess of cortisol exists, the following physiological activities take place in the body: (1) the rate of glucogenesis is stepped up; (2) glucose utilization by cells is decreased; (3) protein catabolism in all cells except the liver is increased; (4) amino acid transport to muscle cells is decreased; and (5) fatty acids are mobilized from adipose tissue.

Cushing's syndrome is a condition in which many of the above reactions occur. It is caused by an *excess of glucocorticoids,* induced by a pituitary tumor. Protein depletion in these patients causes them to have poorly developed muscles, weak bones (due to bone dissolution), slow healing of wounds, hyperglycemia, and hair that is thin and scraggly.

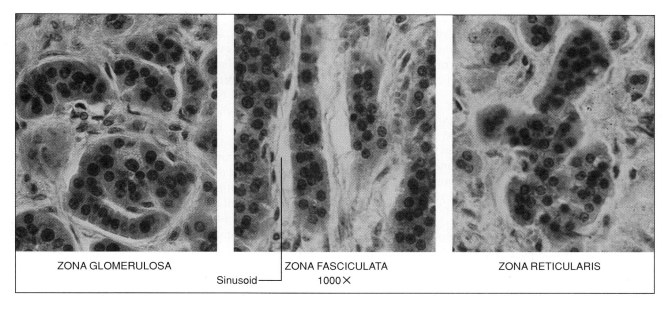

ZONA GLOMERULOSA   ZONA FASCICULATA   ZONA RETICULARIS

Sinusoid —|

1000×

Figure 68.5   Histology of the three layers of the adrenal cortex.

Addison's disease is a condition in which total glucocorticoid insufficiency exists. It can be caused by destruction of the adrenal cortex by cancer, tuberculosis, or some other infectious agent.

Although several substances such as vasopressin, serotonin, and angiotensin II stimulate the adrenal cortex, it is primarily the action of ACTH that induces the cells of the zona fasciculata and zona reticularis to produce cortisol.

ACTH production by the anterior pituitary is partially initiated by stress through the hypothalamus. Stress induces the hypothalamus to produce CRF (corticotropin-releasing factor), which passes to the anterior pituitary, causing the latter to produce ACTH.

The Adrenal Medulla

This portion of the adrenal gland produces the two *catecholamines,* norepinephrine and epinephrine. Although different cells in the medulla produce each of these hormones, cell differences are difficult to detect microscopically.

Approximately 80% of the medullary secretion is epinephrine. Norepinephrine makes up the remaining 20%. Norepinephrine is also produced by certain nerve endings of the autonomic nervous system.

Action   The effects of norepinephrine and epinephrine in different tissues depend on the types of receptors that exist in the tissues. There are two classes of adrenergic receptors: alpha and beta. In addition, there are two types of alpha receptors (α_1 and α_2) and two types of beta receptors (β_1 and β_2).

Both norepinephrine and epinephrine increase the force and rate of contraction of the isolated heart. These reactions are mediated by the β_1 receptors. Norepinephrine causes vasoconstriction in all organs through α_1 receptors.

Epinephrine causes vasoconstriction everywhere except in the muscles and liver, where beta receptors bring about vasodilation. Increased mental alertness is also induced by both catecholamines, which may be induced by the increased blood pressure.

Blood glucose levels are increased by both hormones. This is accomplished by glycogenolysis (glycogen → glucose) in the liver. The β_2 receptors account for this reaction.

Norepinephrine and epinephrine are equally potent in mobilizing free fatty acids through beta receptors. The metabolic rate is also increased by these hormones; how this takes place is not precisely understood at this time.

Regulation   Production of both hormones is initiated when the adrenal medulla is stimulated by sympathetic nerve fibers.

The Thymus Gland

The thymus consists of two long lobes joined by connective tissue. It lies in the upper chest region above the heart. It is most highly developed before birth and during the growing years. After puberty, it begins to regress and continues to do so throughout life.

Histologically, this gland consists of a cortex and medulla. Figure 68.6 reveals its microscopic appearance. The **cortex** consists of lymphoidlike tissue filled with a large number of lymphocytes. The **medulla** contains a smaller number of lymphocytes and structures called **thymic corpuscles** (Hassall's bodies). The function of the latter structures is unknown.

Function It appears that the principal role of the thymus is to process lymphocytes into T-lymphocytes, as described on page 319. It is speculated that this gland produces one or more hormones that function in T-lymphocyte formation. Thymosin and several other extracts from this gland have been under study.

The Pineal Gland

The pineal gland is situated on the roof of the third ventricle under the posterior end of the corpus callosum. It is supported by a stalk that contains postganglionic sympathetic nerve fibers that do not seem to extend into the gland. The gland consists of neuroglial and parenchymal cells that suggest a secretory function. Figure 68.6, right-hand illustration, reveals its histological appearance.

In young animals and infants the gland is large and more glandlike; cells tend to be arranged in alveoli. Just before puberty, however, the gland begins to regress, and small concretions of calcium carbonate form that are called **pineal sand.** There is

some evidence, though not conclusive, that the gland contains some gonadotropin peptides. Most attention, however, is on the production of melatonin by the gland. **Melatonin** is an indole that is synthesized from serotonin. Its production appears to be regulated by daylight (circadian rhythm).

During daylight the production of melatonin is suppressed. At night the gland becomes active and produces considerable quantities of melatonin. Regulation occurs through the eyes. Light striking the retina sends messages to the pineal gland through a retina-hypothalmic pathway. Norepinephrine reaches the cells from postganglionic sympathetic nerve endings in the pineal stalk. Beta adrenergic receptors to the cells are mediators in the inhibition of melatonin production.

Although considerable speculation exists that melatonin inhibits the estrus (menstrual) cycle in humans, as it probably does in lower animals, there is no definitive proof at this time that this is the case.

The Testes

A section through the testis (figure 68.7) reveals that it consists of coils of **seminiferous tubules** where spermatozoa are produced, and **interstitial cells** where the male sex hormone, **testosterone,** is secreted. Note that the interstitial cells lie in the spaces between the seminiferous tubules.

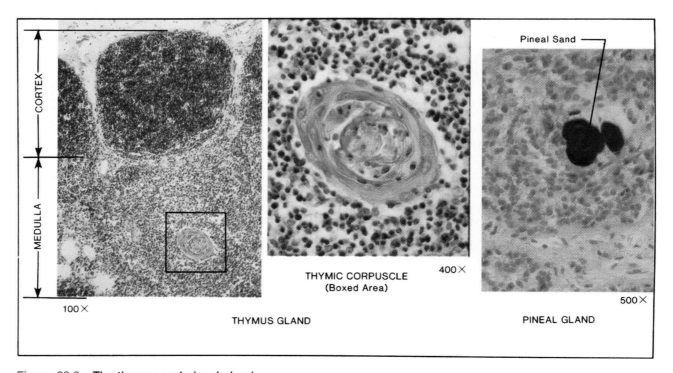

Figure 68.6 The thymus and pineal glands.

Testosterone is produced in large amounts during embryological development, but from early childhood to puberty very little of the hormone is produced. With the onset of puberty at the age of 10 or 11 years the testes begin to produce large quantities of testosterone as the development of sexual maturity takes place.

During embryological development, testosterone is responsible for the development of the male sex organs. If the testes are removed from a fetus at an early stage, the fetus will develop a clitoris and vagina instead of a penis and scrotum, even though the fetus is a male. The hormone also controls the development of the prostate gland, seminal vesicles, and male ducts while suppressing the formation of female genitals.

Embryological development of the testes occurs within the body cavity. During the last two months of development the testes descend into the scrotum through the inguinal canals. The descent of the testes is controlled by testosterone.

During puberty the testes become larger and produce a great deal of testosterone, which causes considerable enlargement of the penis and scrotum. At the same time, the male secondary sexual characteristics develop. These characteristics are (1) the appearance of hair on the face, axillae, chest, and pubic region; (2) the enlargement of the larynx accompanied by a voice change; (3) increased skin thickness; (4) increased muscular development; (5) increased bone thickness and roughness; and (6) baldness.

During embryological development the production of testosterone is regulated by **chorionic gonadotropin,** a hormone produced by the placenta. At the onset of puberty the testes are stimulated to produce testosterone by the luteinizing hormone (LH), which is produced by the anterior lobe of the pituitary gland. This hormone stimulates the interstitial cells to produce testosterone. Maturation of spermatozoa in the testes is controlled by the follicle-stimulating hormone (FSH), which is also produced by the anterior pituitary gland. Testosterone assists FSH in this process.

The Ovaries

During fetal development small groups of cells move inward from the germinal epithelia of each ovary and develop into primordial follicles. As seen in figure 68.8, the **germinal epithelium** is a layer of cuboidal cells near the surface of the ovary, and the **primordial follicles** are the small round bodies that contain **ova.**

The ovaries of a young female at the onset of puberty are believed to contain between 100,000 and 400,000 immature follicles. During all the reproductive years of a woman, only about 400 of these follicles will reach maturity and expel their ova.

During puberty, selected primordial follicles enlarge to form **Graafian follicles,** and every 28 days a maturing Graafian follicle expels an ovum in the ovulation process. The development of these follicles and the subsequent corpus luteum results in

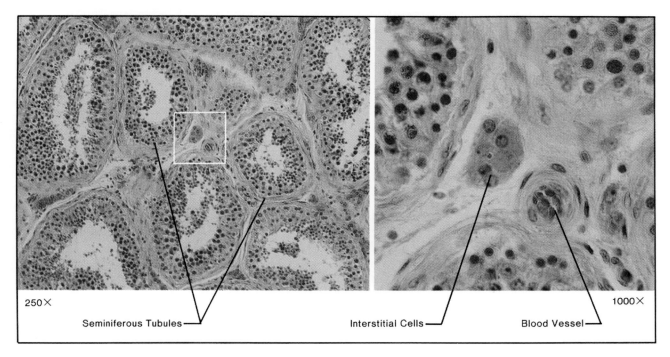

Figure 68.7 Testicular tissue.

the production of **estrogens** and **progesterone,** the principal female sex hormones. A discussion of each hormone follows.

Estrogens

The principal functions of estrogens is to promote the growth of specific cells in the body and to control the development of female secondary sex characteristics. In addition to being produced in Graafian follicles, estrogens are also produced by the corpus luteum, placenta, adrenal cortex, and testes. The production of estrogens at puberty is initiated by the secretion of FSH by the anterior lobe of the pituitary gland.

Of the six or seven estrogens that have been isolated from the plasma of women, **β-estradiol, estrone,** and **estriol** are the most abundant ones produced. The most potent of all the estrogens is β-estradiol.

During puberty when the estrogens are produced in large quantities the following changes occur: (1) the female sex organs become fully developed; (2) the vaginal epithelium changes from cuboidal to stratified epithelium; (3) the uterine lining (endometrium) becomes more glandular in preparation for implantation of the fertilized ovum; (4) the breasts form; (5) osteoblastic activity increases with more rapid bone growth; (6) calcification of the epiphyses in long bones is hastened; (7) pelvic bones enlarge and change shape, increasing the size of the pelvic outlet; (8) fat deposition under the skin, in the hips, buttocks, and thighs is increased; and (9) skin vascularization is increased.

Progesterone

Once ovulation takes place in the ovary the Graafian follicle is replaced by a **corpus luteum.** Progesterone is the principal hormone produced by this body. The ovary in figure 68.13 shows the development of Graafian follicles and the corpus luteum.

The most important function of this hormone is to promote secretory changes in the endometrium in preparation for implantation of the fertilized ovum. In addition, the hormone causes mucosal changes in the uterine tubes and promotes the proliferation of alveolar cells in the breasts in preparation for milk production.

When the plasma level of progesterone falls, due to regression of the corpus luteum, deterioration of the endometrium takes place and menstruation occurs. If pregnancy occurs the corpus luteum enlarges and produces additional quantities of progesterone. The conversion of a Graafian follicle into a corpus luteum is completely dependent on LH production.

The Pituitary Gland

The pituitary gland, or hypophysis, consists of two lobes: anterior and posterior. The anterior lobe, or **adenohypophysis,** develops from the roof of the oral

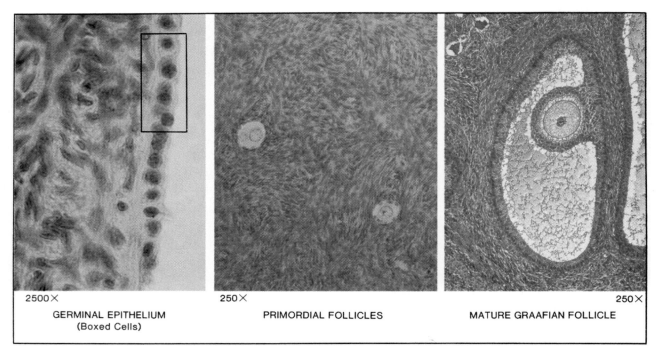

2500×	250×	250×
GERMINAL EPITHELIUM (Boxed Cells)	PRIMORDIAL FOLLICLES	MATURE GRAAFIAN FOLLICLE

Figure 68.8 Ovarian tissue.

cavity of the embryo. This part of the hypophysis is made up of glandular cells. The posterior lobe, or **neurohypophysis,** develops as an outgrowth of the floor of the brain during embryological development. Unlike the adenohypophysis, the cells of this lobe are nonsecretory and resemble neuroglial tissue. The entire gland is invested by an extension of the dura mater.

Figure 68.9 reveals portions of the two lobes. Note that the neurohypophysis in this region consists of two parts: the **pars intermedia** and the **pars nervosa.**

The Adenohypophysis

Two basic types of glandular cells are seen in this portion of the hypophysis: chromophobes and chromophils. Both types of cells can be seen in figure 68.10.

Chromophobes are cells that lack an affinity for routine dyes used in staining tissues. **Chromophils** stain readily and are of two types: acidophils and basophils. **Acidophils** take on the pink stain of eosin. **Basophils,** on the other hand, stain readily with basic stains, such as methylene blue and crystal violet.

Six hormones are produced by these various cells in the anterior lobe. Except for somatotropin, all these hormones are targeted specifically for glands. Production of the hormones is regulated by the hypothalamus. Between the hypothalamus and adenohypophysis is a vascular connection, the *hypophyseal portal system,* that transports releasing factors from the hypothalamus to the secretory cells. The releasing factors are secreted by cells in the hypothalamus. The six hormones of the adenohypophysis are as follows:

Somatotropin (SH)   This hormone, also called the **growth hormone** (**GH**), is produced by **somatotrophs,** a type of acidophil. In males and nonpregnant females most of the acidophils are of this type.

Somatotropin increases the growth rate of all cells in the body by enhancing amino acid uptake and protein synthesis. Excess production of the hormone during the growing years causes **gigantism** due to the stimulation of growth in the epiphyses of the long bones. An adult with excess SH production develops **acromegaly,** which is characterized by enlargement of the mandible and forehead as well as of the small bones of the hands and feet. Deficiency of the hormone can cause **dwarfism.**

Prolactin (*Luteotropic Hormone, LTH*)   This hormone is produced by another type of acidophil, called a **mammotroph.** With suitable staining methods, it is possible to differentiate these cells from the somatotrophs.

Prolactin promotes the production of milk in the breasts after childbirth. The release of this hormone prior to childbirth is inhibited by *PIF (prolactin-inhibiting factor),* which is produced in the hypothalamus. The presence of large amounts of estrogens and progesterone prior to childbirth causes the hypothalamus to produce the inhibiting factor.

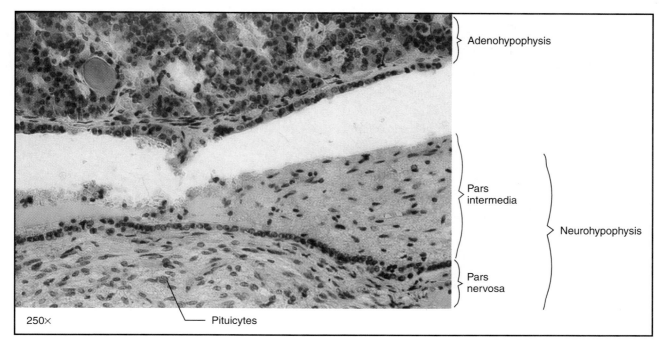

250×   Pituicytes

Adenohypophysis

Pars intermedia

Pars nervosa

Neurohypophysis

Figure 68.9   The pituitary gland.

Thyrotropin (TSH) This hormone is produced by large irregular basophilic cells called **thyrotrophs.** TSH regulates the rate of iodine uptake and the synthesis of the thyroid hormone by the thyroid gland. Excess amounts of TSH will cause hyperthyroidism. Hypothyroidism results from a deficiency of the hormone.

TSH production is regulated by the hypothalamic *thyrotropin releasing factor (TRF)*. TRF is secreted by nerve endings in the hypothalamus and passes to the adenohypophysis via the hypophyseal portal system. Thyrotropin production is also regulated by a thyroid hormone feedback mechanism.

Adrenocorticotropin (ACTH) This hormone is probably produced by a type of chromophobe. The suspected cell is a large chromophobe with large cytoplasmic granules, peripherally located. ACTH controls glucocorticoid production of the adrenal cortex.

As stated previously, ACTH production is initiated by CRF (corticotropin-releasing factor). It appears that stress causes cells in the hypothalamus to release CRF into the hypophyseal portal system, activating the adenohypophysis to produce ACTH. In addition to CRF the gland is regulated by a glucocorticoid feedback mechanism.

Follicle-Stimulating Hormone (FSH) This hormone is produced by basophilic **gonadotrophs.** FSH promotes the development of Graafian folli-

cles in the ovaries and the maturation of spermatozoa in the testes. FSH production by the adenohypophysis is regulated by estrogen and testosterone feedback mechanisms.

Luteinizing Hormone (LH) This hormone is produced by basophilic **gonadotrophs** that are somewhat larger than the FSH gonadotrophs. Whereas the FSH cells tend to be located near the periphery of the gland, LH-producing cells are more generally distributed throughout the gland.

Maturation of the Graafian follicles and ovulation during the menstrual cycle are dependent on this hormone. Although FSH contributes much to early follicle development, ovulation is entirely dependent on LH. The production of estrogens and progesterone is dependent on LH also.

In addition, this hormone stimulates the interstitial cells of the testes to produce testosterone; thus, it has been referred to as the **interstitial cell-stimulating hormone (ICSH).** LH production by the adenohypophysis is regulated by estrogen and progesterone feedback mechanisms.

The Neurohypophysis

The neuroglial-like cells in the neurohypophysis are called **pituicytes.** These cells are small and have numerous processes. Unlike neuroglia, pituicytes contain fat and pigment granules in their cytoplasm. Figure 68.11 is representative of these cells.

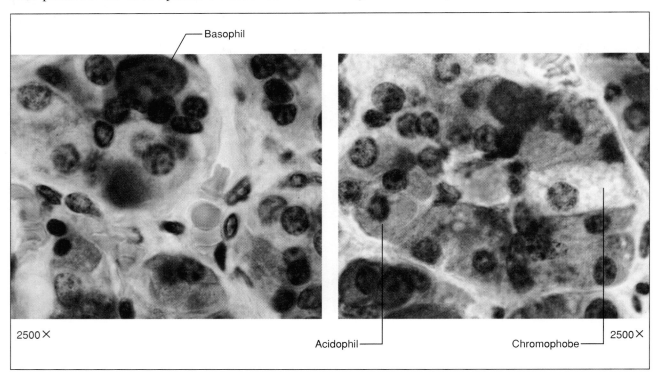

2500×   Basophil   Acidophil   Chromophobe   2500×

Figure 68.10 Cells of the adenohypophysis.

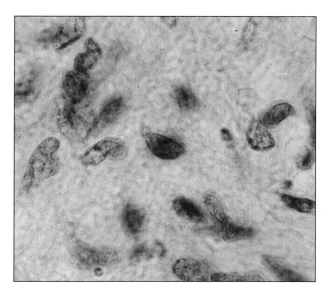

Figure 68.11 Pituicytes of the neurohypophysis (2500x).

Pituicytes are nonsecretory. Their function is to provide support for nerve fibers from tracts that originate in the hypothalamus. Two hormones, the **antidiuretic hormone** and **oxytocin,** are liberated by the ends of these nerve tracts. The hormones are actually synthesized in the cell bodies of **neurose-cretory neurons** of the hypothalamus and pass down the fibers into the neurohypophysis where they are released.

Antidiuretic Hormone (ADH) This hormone controls the permeability of collecting tubules in nephrons to water absorption. When ADH is present in even minute amounts, water is easily reabsorbed back into the blood through the walls of the collecting tubules. When ADH is lacking, rapid water loss through the kidneys takes place. *Diabetes insipidus* is a disease in which the neurohypophysis fails to provide enough ADH. Without this hormone the osmotic balance of body fluids cannot remain stable very long.

Another function of ADH is to increase the arterial blood pressure when a severe blood loss has reduced blood volume. A 25% loss of blood will increase ADH production by 25 to 50 times normal. ADH increases blood pressure by constricting the arterioles. Because ADH has this potent pressor capability, it is also called **vasopressin.** The mechanism for increased ADH production during blood loss lies in baroreceptors that are located in the walls of the carotid, aortic, and pulmonary arteries.

Oxytocin Any substance that will cause uterine contractions during pregnancy is designated as

being "oxytocic." During childbirth, pressure of the unborn child on the uterine cervix causes a neurogenic reflex to the neurohypophysis, stimulating it to produce the hormone oxytocin. This hormone, being oxytocic, increases the strength of uterine contractions to facilitate childbirth. Oxytocin also activates the myoepithelial cells of the mammary gland, resulting in the flow of milk.

Laboratory Assignment
Microscopic Studies
Do a systematic study of the various endocrine glands using the slides that are available. Make drawings, if required.

Materials:
prepared slides of the following glands:
thyroid (H14.11, H14.12, or H14.13)
parathyroid (H14.16)
thymus (H14.21 or H14.22)
adrenal (H14.26)
ovary (H9.310 or H9.3181)
testis (H9.415)
pituitary (H14.31)
pineal (H14.36)

While examining these slides, refer to the various photomicrographs in this exercise to identify the various kinds of cells that are mentioned in the text. Use low power for scanning and oil immersion wherever necessary.

Labeling
Figures 68.12 and 68.13 summarize many of the endocrinological facts learned in this exercise. Figure 68.12 pertains primarily to the various tissues that produce the hormones. Figure 68.13 depicts most of the hormones produced by the pituitary gland that regulate growth and glands of the body. It also includes many of the hormones produced by the glands that are regulated by the pituitary gland.

It will probably be necessary for you to refer back to various portions of the text in this exercise to seek out the various hormones. Once both of these diagrams are completed, you should have a comprehensive picture of the endocrine system.

Laboratory Report
Complete Laboratory Report 68 for this exercise.

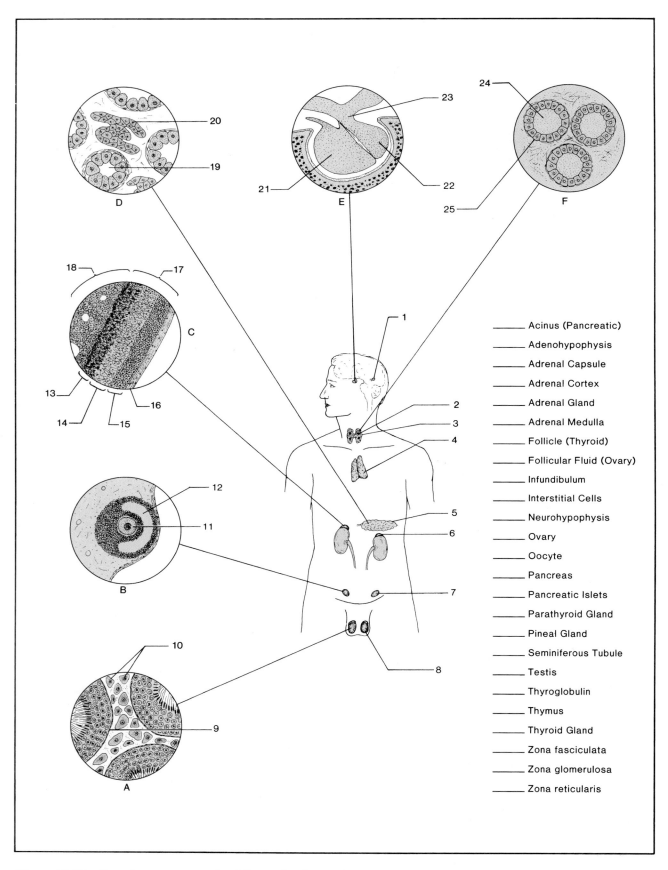

Acinus (Pancreatic)
Adenohypophysis
Adrenal Capsule
Adrenal Cortex
Adrenal Gland
Adrenal Medulla
Follicle (Thyroid)
Follicular Fluid (Ovary)
Infundibulum
Interstitial Cells
Neurohypophysis
Ovary
Oocyte
Pancreas
Pancreatic Islets
Parathyroid Gland
Pineal Gland
Seminiferous Tubule
Testis
Thyroglobulin
Thymus
Thyroid Gland
Zona fasciculata
Zona glomerulosa
Zona reticularis

Figure 68.12   Secretory components of the endocrine system.

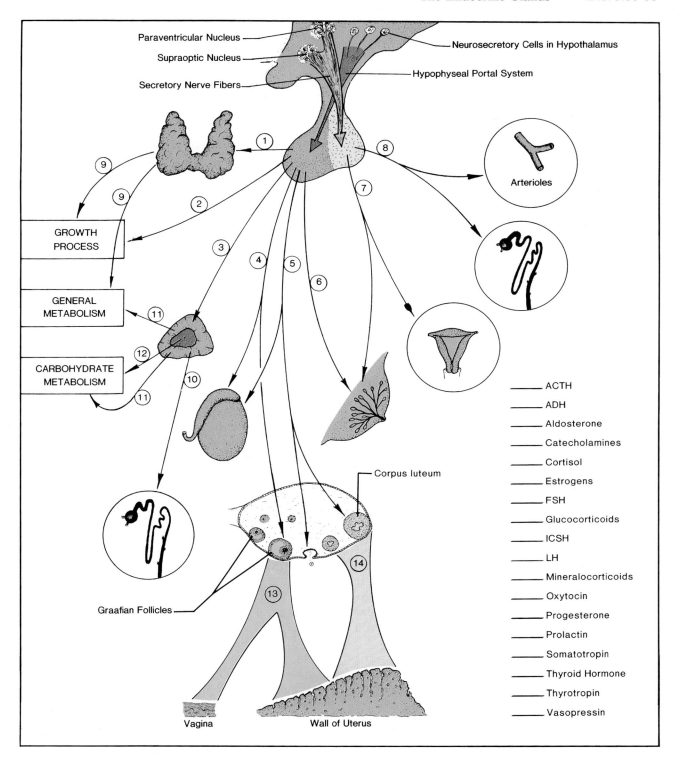

Figure 68.13   The pituitary regulatory mechanism.

Histology Self-Quiz No. 3

If at this point you have completed all the histology assignments in Exercises 40, 56, and 63, you should be ready to test your knowledge of the histology of these areas by taking self-quiz no. 3 on page 415. After recording your answers on a sheet of paper, check your answers against the key that you find on page 298.

69

The Reproductive System

This exercise pertains to the human reproductive system, using cat dissection for anatomical studies. The histology of the various organs, as well as microscopic studies of spermatogenesis and oogenesis, are also included.

Materials:
models of human reproductive organs
prepared slides of human testis
ocular micrometer
cat and dissection instruments

The Male Organs

Figure 69.1 is a sagittal section of the male reproductive organs. The primary sex organs are the paired oval **testes** *(testicles),* which lie enclosed in a sac, the **scrotum.** Partially surrounding each testis is a double-layered serous membrane, the **tunica vaginalis** (label 1), which is a downward extension of the peritoneum into the scrotal sac. It is formed during the descent of the testes from the body cavity prior to birth.

Between the tunica vaginalis and the outer skin of the scrotum is a superficial fascia (not shown) and the cremaster muscle. Only a small portion of the **cremaster muscle** (label 6) is shown on the spermatic cord in figure 69.1. Since this muscle is temperature sensitive, it is capable of raising or lowering the testis to maintain an internal testicular temperature of 94°–95° F. Higher temperatures tend to inhibit spermatozoa maturation.

Note in the enlarged sectional view of the testis that it is divided into several chambers by **testicular septa.** Within each chamber lie several coiled-up **seminiferous tubules** where the spermatozoa are produced. Note also that the seminiferous tubules unite to form a plexus of canals called the **rete testis,** and leading from this plexus are several ducts called **vasa efferentia.**

Mature spermatozoa pass from the seminiferous tubules through the rete testis and vasa efferentia to storage ducts within the epididymis. The **epididymis** (label 26) consists of a mass of coiled-up tubules on the posterior, exterior surface of the testis.

In the act of ejaculation the spermatozoa leave the epididymis by way of a tube, the **vas deferens** *(ductus deferens),* which is the central canal of the **spermatic cord** (label 5). In addition to the vas deferens, the spermatic cord also contains nerves and blood vessels that supply the testis.

Tracing the vas deferens upward, we see that it passes through the inguinal canal, over the pubic bone and bladder into the pelvic cavity. The terminus of the vas deferens is enlarged to form the **ampulla of the vas deferens.**

Ducts from a pair of glands, the **seminal vesicles** (color code, blue) and the ampulla join to form a **common ejaculatory duct** that passes through the prostate gland to join the **prostatic urethra.** From here the urethra passes through the **urogenital diaphragm** (label 27) and into the penis, where it becomes the **penile urethra.**

The prostate gland, seminal vesicles, and bulbourethral glands contribute alkaline secretions to the seminal fluid. This alkalinity stimulates sperm motility. The **prostate gland,** which is positioned inferior to the urinary bladder, is the largest of these secondary sex glands.

The **bulbourethral** *(Cowper's)* **glands** (color code, blue) are the smallest of the three glands. These pea-sized glands have ducts about one inch long that empty into the urethra at the base of the penis. The secretion of these latter glands is a clear mucoid fluid that lubricates the end of the penis and prepares the urethra for seminal fluid.

The **penis** consists of three cylinders of erectile tissue: one corpus spongiosum and two corpora cavernosa. The **corpus spongiosum** is a cylinder of tissue that surrounds the penile urethra. Both ends of this body are enlarged: its proximal end forms the **bulb of the penis,** and its distal enlarged end is the **glans penis.** The enlarged portion of the urethra within the glans is the **navicular fossa.**

The two **corpora cavernosa** are located in the dorsal part of the organ and are separated on the midline by a **septum penis.** The cross section of the penis in figure 69.1 shows best the relationship of the three corpora to the penile urethra and septum. This view also reveals that a **tunica albuginea** ensheathes the corpora cavernosa and a **tunica**

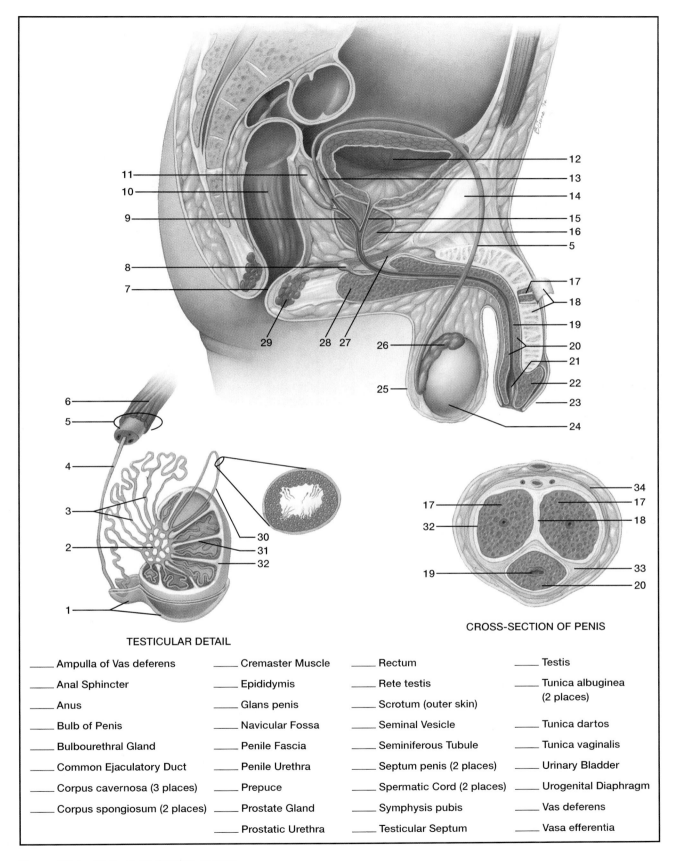

TESTICULAR DETAIL

CROSS-SECTION OF PENIS

_____ Ampulla of Vas deferens

_____ Anal Sphincter

_____ Anus

_____ Bulb of Penis

_____ Bulbourethral Gland

_____ Common Ejaculatory Duct

_____ Corpus cavernosa (3 places)

_____ Corpus spongiosum (2 places)

_____ Cremaster Muscle

_____ Epididymis

_____ Glans penis

_____ Navicular Fossa

_____ Penile Fascia

_____ Penile Urethra

_____ Prepuce

_____ Prostate Gland

_____ Prostatic Urethra

_____ Rectum

_____ Rete testis

_____ Scrotum (outer skin)

_____ Seminal Vesicle

_____ Seminiferous Tubule

_____ Septum penis (2 places)

_____ Spermatic Cord (2 places)

_____ Symphysis pubis

_____ Testicular Septum

_____ Testis

_____ Tunica albuginea
      (2 places)

_____ Tunica dartos

_____ Tunica vaginalis

_____ Urinary Bladder

_____ Urogenital Diaphragm

_____ Vas deferens

_____ Vasa efferentia

Figure 69.1 Male reproductive organs.

dartos surrounds all three of the corpora. Between the tunica dartos and tunica albuginea is a layer of connective tissue called the **penile fascia,** or *Buck's fascia*. It is during sexual arousal that erection of the penis is achieved by filling of these three corpora with blood.

Over the end of the glans penis lies a circular fold of skin, the **prepuce.** Around the neck of the glans, and on the inner surface of the prepuce, are scattered small *preputial glands*. These sebaceous glands produce a secretion that readily undergoes decomposition to form a whitish substance called *smegma*. *Circumcision* is a surgical procedure that involves the removal of the prepuce to facilitate sanitation.

Assignment:
Label figure 69.1.

The Female Organs

Figures 69.2, 69.3, and 69.4 illustrate the female reproductive organs. Models and wall charts will be helpful in identifying all the structures.

External Genitalia

The **vulva** of the external female reproductive organs includes the mons pubis, labia majora, labia minora, and hymen. All of these structures are shown in figure 69.2. The **mons pubis** *(mons veneris)* is the most anterior portion and consists of a firm cushionlike elevation over the symphysis pubis. It is covered with hair.

Two folds of skin on each side of the **vaginal orifice** (label 11) lie over the opening. The larger exterior folds are the **labia majora.** Their exterior surfaces are covered with hair; their inner surfaces are smooth and moist. The labia majora are homologous (of similar embryological origin) to the scrotum of the male.

Medial to the labia majora are the smaller **labia minora.** These folds meet anteriorly on the median line to form a fold of skin, the **prepuce of the clitoris.** The **clitoris** is a small protuberance of erectile tissue under the prepuce that is homologous to the penis in males. It is highly sensitive to sexual excitation. The fold of skin that extends from the clitoris to each labium minus is called the **frenulum of the clitoris.** Posteriorly, the labia minora join to form a transverse fold of skin, the **posterior commissure,** or *fourchette*. Between the anus and the posterior commissure is the **central tendinous point of the perineum.** The **perineum** corresponds to the outlet of the pelvis.

The *vestibule* is the area between the labia minora that extends from the clitoris to the vaginal orifice, hymen, urethral orifice, and fourchette. Situated within the vestibule are the vaginal orifice,

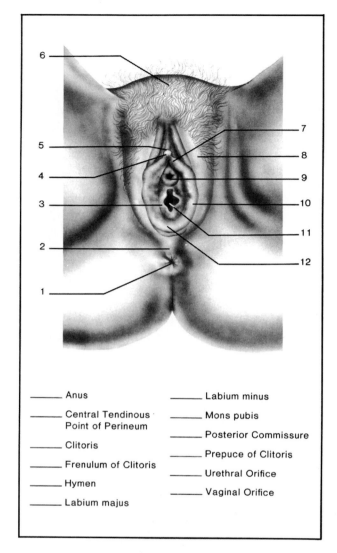

_____ Anus	_____ Labium minus
_____ Central Tendinous Point of Perineum	_____ Mons pubis
_____ Clitoris	_____ Posterior Commissure
_____ Frenulum of Clitoris	_____ Prepuce of Clitoris
_____ Hymen	_____ Urethral Orifice
_____ Labium majus	_____ Vaginal Orifice

Figure 69.2 Female genitalia.

hymen, urethral orifice, and openings of the vestibular glands.

The **urethral orifice** lies about 2–3 centimeters posterior to the clitoris. Many small **paraurethral glands** surround this opening. These glands are homologous to the prostate gland of the male.

On either side of the vaginal orifice are openings from the two **greater vestibular** *(Bartholin's)* **glands.** These glands are homologous to the bulbourethral glands in males. They provide mucous secretions for vaginal lubrication.

The **hymen** is a thin fold of mucous membrane that separates the vagina from the vestibule. It may be completely absent or cover the vaginal orifice partially or completely. Its condition or absence is not a determinant of virginity.

Assignment:
Label figure 69.2.

Internal Organs

Figure 69.3 is a posterior view of the female reproductive organs that shows the relationship of the vagina, uterus, ovaries, and uterine tubes. It also reveals many of the principal supporting ligaments.

The Uterus   The uterus is a pear-shaped, thick-walled, hollow organ that consists of a fundus, corpus, and cervix. The **fundus** is the dome-shaped portion at the large end. The body, or **corpus,** extends from the fundus to the cervix. The **cervix,** or neck, of the uterus is the narrowest portion; it is about one inch long. Within the cervix is an **endocervical canal** that has an outer opening, the **external os,** that opens into the **vagina.** An inner opening, the **internal os,** opens into the cavity of the corpus.

The thick muscular portion of the uterine wall is called the **myometrium.** The inner lining, or **endometrium,** consists of mucosal tissue. The ante-rior and posterior surfaces of the corpus, as well as the fundus, are covered with peritoneum.

The Uterine *(Fallopian)* **Tubes**   Extending laterally from each side of the uterus is a *uterine tube.* The distal end of each tube is enlarged to form a funnel-shaped fimbriated **infundibulum** that surrounds an **ovary.** Although the infundibulum doesn't usually contact the ovary, one or more of the fingerlike **fimbriae** on the edge of the infundibulum usually do contact it. The infundibula and fimbriae receive oocytes produced by the ovaries.

The uterine tubes are lined with ciliated columnar cells that assist in transporting egg cells from the ovary into the uterus. The tubes also have a muscular wall, which produces peristaltic movements that help propel the egg cells. Fertilization usually occurs somewhere along the uterine tube. Note that each uterine tube narrows down to form an **isthmus** near the uterus.

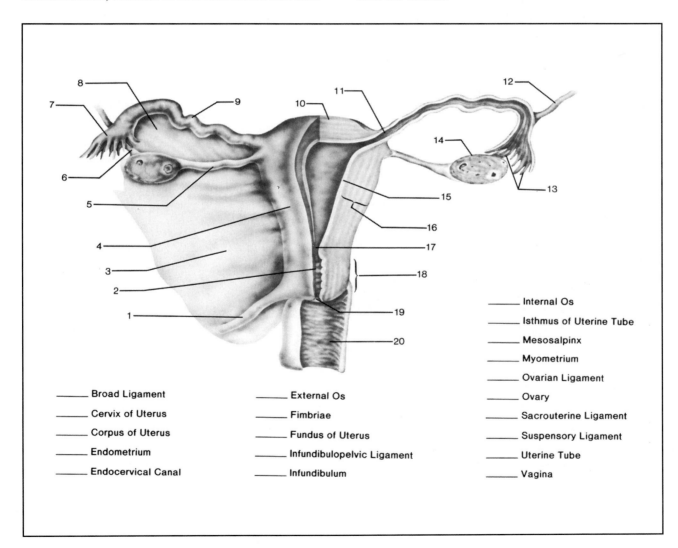

_____ Broad Ligament

_____ Cervix of Uterus

_____ Corpus of Uterus

_____ Endometrium

_____ Endocervical Canal

_____ External Os

_____ Fimbriae

_____ Fundus of Uterus

_____ Infundibulopelvic Ligament

_____ Infundibulum

_____ Internal Os

_____ Isthmus of Uterine Tube

_____ Mesosalpinx

_____ Myometrium

_____ Ovarian Ligament

_____ Ovary

_____ Sacrouterine Ligament

_____ Suspensory Ligament

_____ Uterine Tube

_____ Vagina

Figure 69.3   Posterior view of female reproductive organs.

The Urinary Bladder   The sagittal section in figure 69.4 reveals the relationship of the *urinary bladder* to the reproductive organs. Note that it lies between the uterus and the **symphysis pubis** (label 5). Observe, also, that the bladder is in direct contact with the anterior wall of the vagina. The short **urethra** passes from the bladder through the **urogenital diaphragm** (label 25) to the *urethral orifice* within the vulva.

The superior surface of the bladder is covered with peritoneum that is continuous with the peritoneum on the anterior face of the uterus. The small cavity lined with peritoneum between the bladder and uterus is known as the **anterior cul-de-sac.** The space between the uterus and rectum (label 18) is the **rectouterine pouch,** or *pouch of Douglas.*

Ligaments   The uterus, ovaries, and uterine tubes are held in place and supported by various ligaments formed from the muscle and connective tissue encased in peritoneum, or simply folds of peritoneum. Figure 69.3 reveals the majority of them. A few are shown in figure 69.4.

The largest supporting structure is the **broad ligament** (label 3, figure 69.3). It is an extension of the peritoneal layers that covers the fundus and corpus of the uterus. Extending up over the uterine tubes, this ligament forms a mesentery, the **mesosalpinx,** on each side of the uterus. This portion of the broad ligament, which is shown in figure 69.3 between the ovary and uterine tube, contains blood vessels that supply nutrients to the uterine tube.

Attached to the cervical region of the uterus are two **sacrouterine ligaments** that anchor the uterus to the sacral wall of the pelvic cavity. A **round ligament** (label 16, figure 69.4) extends from each side of the uterus to the body wall. This ligament is not shown in figure 69.3 because it is anterior in position.

Each ovary is held in place by ovarian and suspensory ligaments. The **ovarian ligament** extends from the medial surface of the ovary to the uterus. The **suspensory ligament** is a peritoneal fold on the other side of the ovary that attaches the ovary to the uterine tube.

Each uterine tube is held in place by an **infundibulopelvic ligament** that is attached to the posterior surface of the infundibulum.

Assignment:
Label figures 69.3 and 69.4.

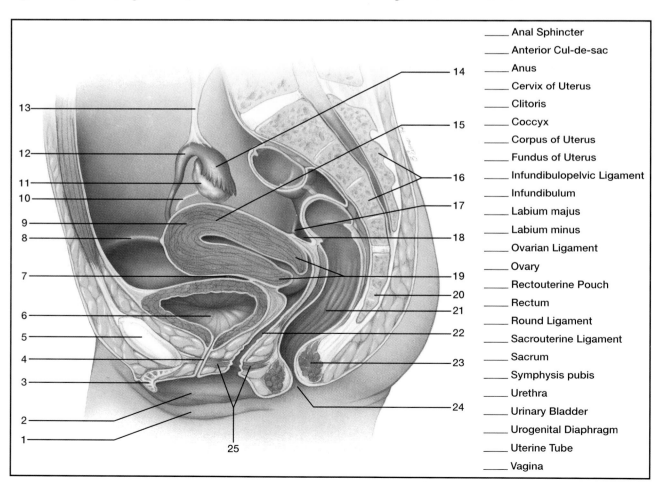

_____ Anal Sphincter
_____ Anterior Cul-de-sac
_____ Anus
_____ Cervix of Uterus
_____ Clitoris
_____ Coccyx
_____ Corpus of Uterus
_____ Fundus of Uterus
_____ Infundibulopelvic Ligament
_____ Infundibulum
_____ Labium majus
_____ Labium minus
_____ Ovarian Ligament
_____ Ovary
_____ Rectouterine Pouch
_____ Rectum
_____ Round Ligament
_____ Sacrouterine Ligament
_____ Sacrum
_____ Symphysis pubis
_____ Urethra
_____ Urinary Bladder
_____ Urogenital Diaphragm
_____ Uterine Tube
_____ Vagina

Figure 69.4   Midsagittal section of female reproductive organs.

Gametogenesis

When the ovum is fertilized by the sperm, 23 chromosomes from each gamete unite to form a zygote of 46 chromosomes. When the gametes are produced in the testes and ovaries from primordial germ cells, the reduction of chromosome numbers is achieved by a series of cell divisions. In the testes this process is called **spermatogenesis.** In the ovaries, a similar process is called **oogenesis.** Figures 69.5 and 69.6 illustrate these two processes.

Spermatogenesis

The production of spermatozoa begins at puberty and is continuous without interruption throughout life. Figure 69.5 illustrates a section of a seminiferous tubule and a diagram of the various stages in the development of mature spermatozoa. The formation of these germ cells takes place as the result of both mitosis and meiosis.

In this study of spermatogenesis, slides of human or animal testes will be examined under the microscope. Before examining the slides, however, familiarize yourself with the characteristic differences between meiosis and mitosis.

Materials:
prepared stained slides of:
    human testis (H9.413 or H9.42)
    rat testis (E17.12)
    rabbit testis (E17.22)

Mitosis

All spermatozoa originate from **spermatogonia** of the **primary germinal epithelium.** This layer of cells is located at the periphery of the seminiferous tubule. These cells contain the same number of chromosomes as other body cells (i.e., 23 pairs) and are said to be *diploid.* Mitotic division of these cells occurs constantly, producing other spermatogonia.

As the spermatogonia move toward the center of the tubule, they enlarge to become **primary spermatocytes.** In this stage the homologous chromosomes unite to form chromosomal units called *tetrads.* This union of homologous chromosomes is called *synapsis.*

Meiosis

Each primary spermatocyte divides further to produce two **secondary spermatocytes.** This division occurs by meiosis instead of mitosis. *Meiosis,* or reduction division, results in the distribution of a *haploid* (half) number of chromosomes to each secondary spermatocyte. The result of this process is

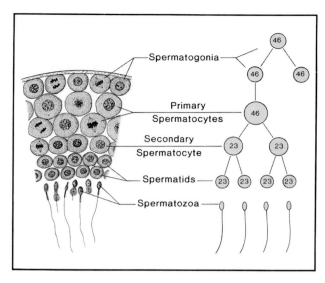

Figure 69.5 **Spermatogenesis.**

that each secondary spermatocyte has only twenty-three chromosomes instead of forty-six. The individual chromosomes of the secondary spermatocytes are formed by splitting of the tetrads along the line of previous conjugation to form chromosomes called *dyads.*

A second meiotic division occurs when each secondary spermatocyte divides to produce two haploid **spermatids.** In this division the dyads split to form single chromosomes, or *monads.* Each spermatid metamorphoses directly into a mature sperm cell. Note that four **spermatozoa** form from each spermatogonium.

Assignment:
Examine slides of human, rat, or rabbit testicular tissue to identify the various stages of spermatogenesis. Of the three listed in the materials list, above, the rabbit may be best to use.

Refer to figure HA-40 of the Histology Atlas to identify the various types of cells.

Draw a representative section of a tubule, labeling all cell types.

Oogenesis

The production of egg cells by the ovaries also begins at puberty. It was pointed out on page 389 that as many as 400,000 immature follicles develop in the ovaries from the germinal epithelium. At the onset of puberty all the potential ova in these follicles contain 46 chromosomes and are designated as **primary oocytes.** Only about 400 of these oocytes will mature during the reproductive lifetime of a woman.

Note in figure 69.6 that the primary oocyte, which forms from an oogonium of the germinal epithelium during embryological development, has

46 chromosomes. In the prophase stage of a dividing oocyte double-stranded chromosomes *(dyads)* unite to form 23 homologous pairs of chromosomes during synapsis. Each pair is made up of four chromatids; thus, it is called a *tetrad.* When the primary oocyte divides in the first meiotic division each dyad member of each homologous pair enters a different cell. The first meiotic division produces a **secondary oocyte** with all the yolk of the primary oocyte and a **polar body** with no yolk. The secondary oocyte and polar body each contain 23 dyads.

The second meiotic division produces a **mature ovum** with 23 chromosomes that are *monads;* that is, each chromosome consists of a single chromatid. Another polar body is also formed in this division. The first polar body divides also to produce two new polar bodies with monads. The end result is that one primary oocyte produces one mature ovum and three polar bodies. The polar bodies never serve any direct function in the fertilization process.

When ovulation takes place the "ovum" is a secondary oocyte. Usually, it is necessary for the sperm to penetrate the cell to trigger the second meiotic division. When the 23 chromosomes in the sperm unite with the chromosomes of the mature ovum, a zygote of 46 chromosomes is formed.

Assignment:
Study figures 69.6 and 68.8 (ovarian tissue) thoroughly to familiarize yourself with the terminology and kinds of divisions. Several questions on the Laboratory Report pertain to this phenomenon.

Cat Dissection

In Exercise 66 the study of the urinary system necessitated a cursory examination of the reproductive organs in the cat. It will now be necessary to study these organs in greater detail. You may find it necessary to refer to figures 66.4 and 66.5 in some instances in this study as well as to figures 69.7 and 69.8 on these pages when identifying the various organs in this dissection.

The Male Reproductive System

This dissection will begin with an examination of the penis, which will necessitate the eventual removal of the right hind leg to reveal the inner structures. Proceed as follows:

1. Locate the **penis.** The skin that forms the sheath around the organ is called the **prepuce.** Pull back the prepuce to expose the enlarged end, which is the **glans.** The glans in the cat is covered with minute horny papillae.
2. Remove the **scrotum,** which is the skin that encloses the two **testes.** Note that each testis is covered by a *fascial sac.* Do not open this sac at this time.
3. Trace the **spermatic cord** from one of the testes to the body wall. It is composed of the **vas deferens** *(ductus deferens),* blood vessels, nerves and lymphatics that supply and drain the testis. Note that the covering of the spermatic cord is continuous with the fascial sac of the testis.

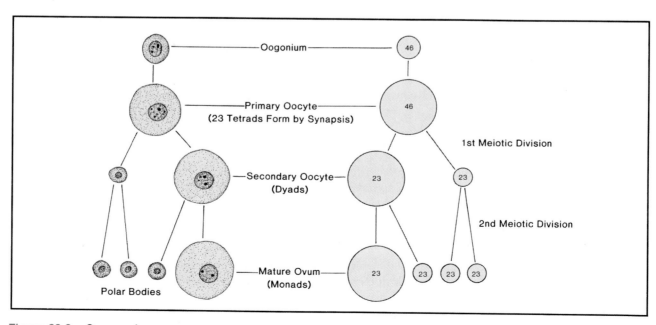

Figure 69.6 Oogenesis.

4. The vas deferens and blood vessels pass through the abdominal wall via the *inguinal canal.* From the inguinal canal the vas deferens passes up over the ureter and enters the penis near the prostate gland.

To be able to study the structures shown in figure 69.7, remove the right leg near the hip joint. To do this it will be necessary to cut through the symphysis pubis.

It will also be necessary to cut through the hipbone above the acetabulum. Clear away the pelvic muscles to expose the structures as shown in figure 69.7.

5. Locate the **prostate gland** where the vas deferens joins the urethra. The tube leading from the prostate through the penis is the **urogenital canal** (penile urethra). At the base of the penis will be seen a pair of **bulbourethral glands.**

6. Cut through the penis to produce a transverse section. Identify the two **corpora cavernosa** that cause erection. Near the symphysis pubis the two corpora diverge forming the **right** and **left crura** (*crus,* singular). Each crus is attached

to a part of the ischium. Refer to figure 66.4 to see both crura.

7. Cut through the fascial sac, exposing the testis. The space between this sac and the testis is called the **vaginal sac.** This cavity is a downward extension of the peritoneal cavity.

8. Trim away the fascial sac and trace the vas deferens toward the testis, noting its convoluted nature. Identify the **epididymis,** which lies on the dorsal side of the testis and is continuous with the vas deferens.

9. With a sharp scalpel or razor blade cut through the testis and the epididymis to produce a section similar to the lower left illustration in figure 69.1. Identify the structures labeled in that diagram.

The Female Reproductive System

Examination of the female reproductive organs will begin with the ovaries. Figure 66.5 will be useful for this portion of the study. Examination of the lower portion of the reproductive tract will necessitate removal of the right leg. Proceed as follows:

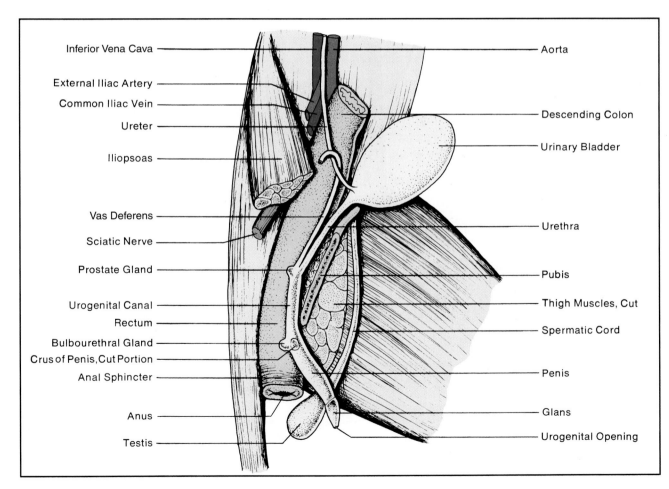

Figure 69.7 Lateral view of the urogenital system of the male cat.

1. Identify the small, light-colored oval ovaries, which lie posterior to the kidneys. Note that they are held by a fold of tissue to the dorsal body wall. This peritoneal fold is called the **mesovarium.** Note, also, that the ovary is attached to the uterine horn by the **ovarian ligament.**

2. Identify the **uterine tube** (oviduct) with its enlarged funnel-like terminis, the **infundibulum** (abdominal ostium). Note that the perimeter of the infundibulum has many irregular projections called **fimbriae.**

   When ova leave the ovary they are drawn into the uterine tube through the infundibulum by ciliated epithelium on the tissues of the area.

3. Observe that the uterus is Y-shaped, having two horns and a body. The **uterine horns** lie along each side of the abdominal cavity. The **uterine body** lies between the urinary bladder and the rectum (see figure 69.8).

4. To be able to observe the relationship of the vagina to other structures, as shown in figure 69.8, remove the right leg near the hip joint. To do this, it will be necessary to cut through the hipbone above the acetabulum. Remove the cut portion of the hipbone and clear away the pelvic muscles to expose the structures, as shown in figure 69.8. Preserve the blood vessels as much as possible.

5. Identify the general areas of the uterine body, vagina, and urogenital sinus.

6. Remove the entire female reproductive tract and open it to reveal the inner structures shown in figure 66.5, which include the **cervix, vagina,** and **urogenital sinus.**

Histological Study

Assuming that spermatogenesis slides have been previously studied, we will focus here, primarily, on a histological study of the various reproductive organs of both sexes. Appropriate pages of the Histology Atlas and some of the illustrations in Exercise 68 (Endocrine Glands) will be used for reference.

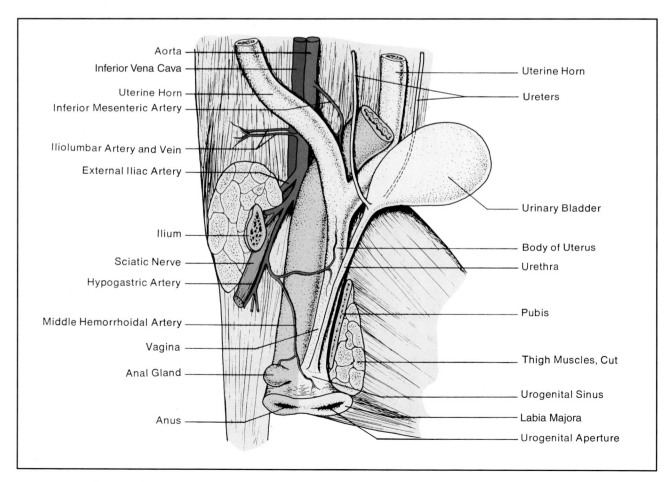

Figure 69.8 Lateral view of the urogenital system of the female cat.

Materials:
prepared slides:
ovary (H9.310 or H9.3181)
uterus (H9.331)
vagina (H9.341)
uterine tube (H9.321)
vas deferens (H9.432)
penis (H9.462)
prostate gland (H9.443)
seminal vesicle (H9.442)

Ovary Study various slides of the ovary to identify **primordial follicles,** the **germinal epithelium,** and **Graafian follicles.** Since the ovaries are not included in the Histology Atlas, use figure 68.8, page 390, for reference.

Note that the developing ova in primordial follicles are surrounded by an incomplete layer of low cuboidal or flattened epithelium.

Mature Graafian follicles are not always seen on some slides. If sectioned properly, you should see a large space surrounding the oocyte that contains large quantities of estrogen produced by the follicular cells.

Uterus Differentiate the **endometrium** from the **myometrium.** Locate the **uterine glands** and **endometrial epithelium,** which consists of **columnar cells.** Refer to figure HA-34.

Uterine Tube Scan a slide of the uterine tube with low-power objective to find the lumen of the tube. Refer to figure HA-35. Note the extremely irregular outline of the mucosa, which consists of a mixture of ciliated and nonciliated columnar epithelium. How many layers of smooth muscle tissue do you find in the muscularis?

Vagina Examine a slide of the mucosa of the vagina and compare it with illustration C, figure HA-36. How does the vaginal mucosa differ from the skin? What function do the lymphocytes perform in the lamina propria?

Uterine Cervix Study a longitudinal section through the cervical area of the uterus that shows where the vaginal lining meets the cervical region.

Consult illustration A, figure HA-36. Note the large number of spaces, or **cervical crypts,** that lie deep in the cervical tissue. Read the part of the legend in figure HA-36 that pertains to the significance of these crypts.

Vas Deferens Examine a cross-sectional slide of the vas deferens, using minimum magnification to study the entire structure. Compare your slide with figure HA-38.

Note that between its inner **epithelium** and outer **adventitia** are three layers of smooth muscle tissue: an inner longitudinal layer, an outer longitudinal layer, and a middle layer of circular fibers. Read what the legend states about the **stereocilia** that are seen on the epithelium.

Penis Examine a cross-sectional preparation of the penis. The photomicrographs in illustrations C and D, figure HA-37, were made from an H9.462 Turtox slide.

Examine the epithelium under high-dry magnification. Does the epithelium on your slide look like illustration D? If not, why not? Read the comments in the legend that pertain to epithelial differences.

Seminal Vesicle A cross section of a portion of a seminal vesicle is seen in illustrations A and B, figure HA-39. Examine a slide first with low-power magnification and note the striking way in which the mucosa is folded upon itself to produce a maze of pockets.

Examine the epithelium under high-dry or oil immersion. What is the nature and function of the seminal fluid?

Prostate Gland Examine a slide of this gland under low power first to identify the structures shown in illustration C, figure HA-39. Note the large number of **tubulo-alveolar glands** that secrete the prostatic fluid. Why does the prostate gland tend to become hardened and restrict micturition in older men?

Laboratory Report
Complete Laboratory Report 69 for this exercise.

Histology Self-Quiz No. 4
If at this point you have completed all the histology assignments in Exercises 66, 68, and 69, you should be ready to test your knowledge of the histology of these areas by taking the last self-quiz on page 419. After recording your answers on a sheet of paper, check your answers against the key that you will find on page 298.

Histology Self-Quiz No. 1

Once you have completed all the exercises in parts 1–4 and studied the histology slides depicted in HA-1 through HA-12 of the Histology Atlas, test your knowledge of this material by taking this test. The answers to this quiz are on page 196. This will help to prepare you for a laboratory practical that may be given.

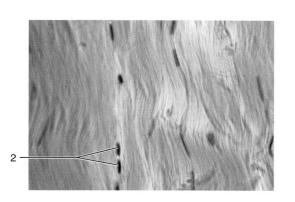

1. What kind of tissue is this?
2. What are these dark stained oval bodies?
3. What do the fine lines represent?

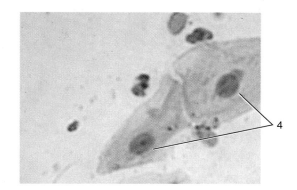

4. What kind of cells are these?
5. Where might you find these cells?

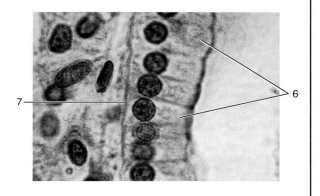

6. What kind of epithelial tissue is this?
7. What is the name of this structure?

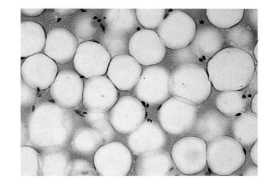

8. What kind of connective tissue is this?
9. Give one function of this type of tissue.

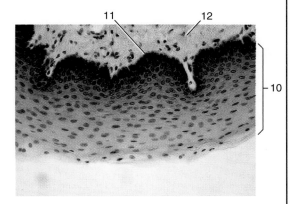

10. What kind of epithelial tissue is this?
11. What is the name of this layer of cells?
12. What is this region called?

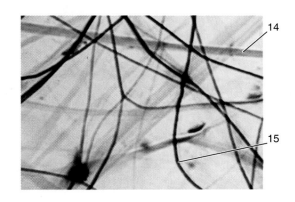

13. What kind of tissue is this?
14. What kind of fiber is this?
15. What kind of fiber is this?

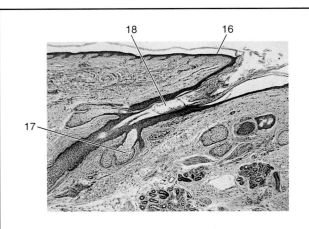

16. Name this layer.
17. Identify this structure.
18. What is this tube called?

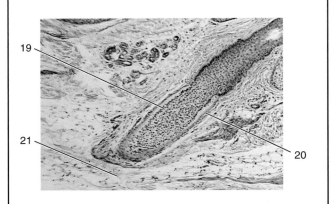

19. Identify this layer of cells.
20. Identify this layer of cells.
21. What is this deep tissue called?

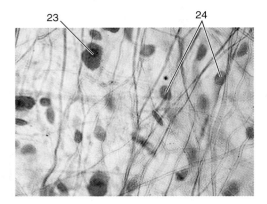

22. What kind of tissue is this?
23. What might this large cell be?
24. What are these smaller cells?

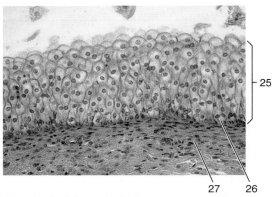

25. What kind of tissue is this?
26. What is this line called?
27. What is this area called?

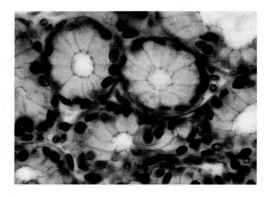

28. What kind of tissue forms the circular sections?
29. What organs in the body have this type of tissue?

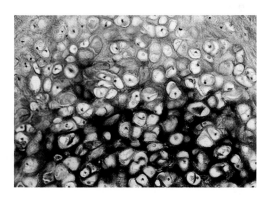

30. What kind of tissue is this?

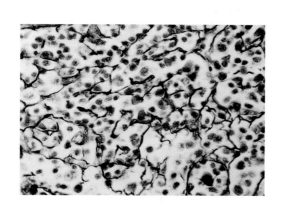

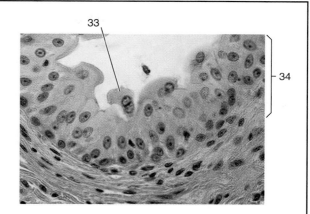

31. What kind of tissue is this?
32. In what organs might one find this tissue?

33. What kind of a cell is this?
34. What kind of tissue is this?

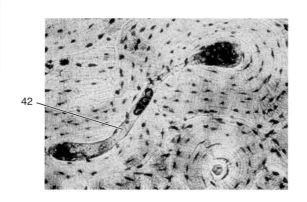

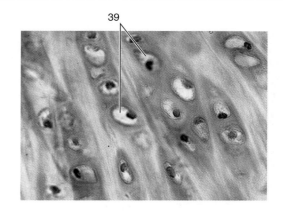

35. What kind of tissue is this?
36. What kind of specialized cell is this?
37. Identify this structure.

38. What kind of tissue is this?
39. What are these cells called?
40. What type of fibers are present in the matrix?

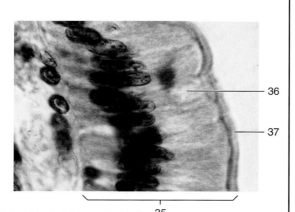

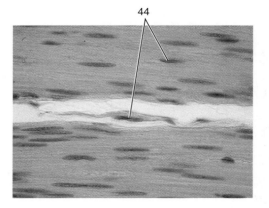

41. What kind of tissue is this?
42. Name this structure.

43. What kind of tissue is this?
44. What are these oval darkly stained bodies?

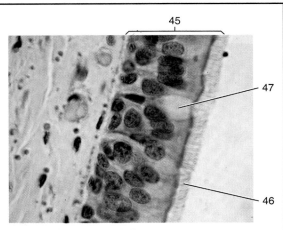

45. What kind of tissue is this?
46. Identify this area.
47. What name is given to this specialized cell?

48. What kind of tissue is this?

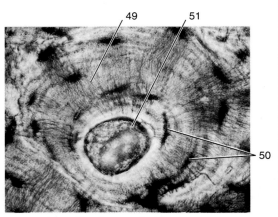

49. What are these fine lines called?
50. What kind of cells are these?
51. What is this circular structure called?

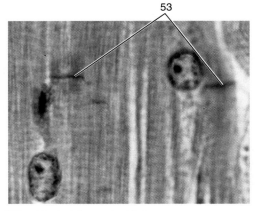

52. What kind of tissue is this?
53. What are these structures called?
54. Name two structures where this tissue is found.

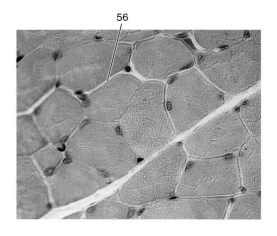

55. What kind of tissue is this?
56. Identify this structure.

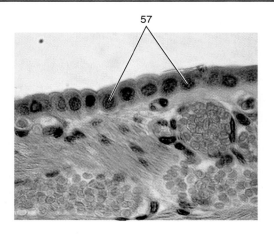

57. What kind of tissue is this?

HQ 1-4

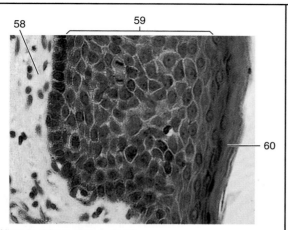

58. What is this region called?
59. What is this layer called?
60. What process occurs to form this layer?

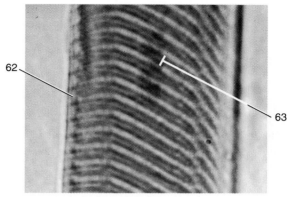

61. What kind of tissue is this?
62. What is this outer membrane called?
63. What is this band called?

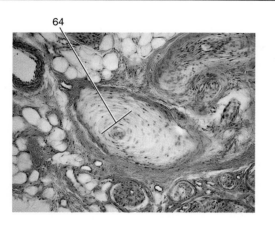

64. What is this oval structure called?
65. What is its function?

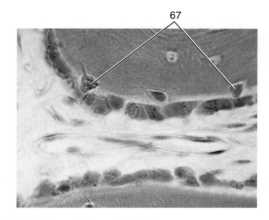

66. What type of tissue is this?
67. What kind of cells are these?
68. What is the function of these cells?

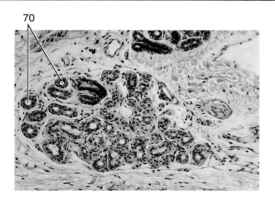

69. What is the name of this large complex structure?
70. Identify these round structures.
71. In what portion of the skin is #69 located?

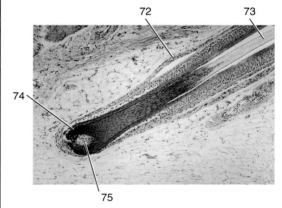

72. Identify this layer of cells.
73. Identify this structure.
74. What is this enlarged region called?
75. What is this clear central portion called?

Histology Self-Quiz No. 2

Once you have completed all the exercises that pertain to nerve tissue, the eye, and the ear, and have studied the histology slides depicted in HA-13 through HA-18 of the Histology Atlas, test your knowledge of this material by taking this test. The answers to this quiz are on page 196. This will help to prepare you for a laboratory practical that may be given.

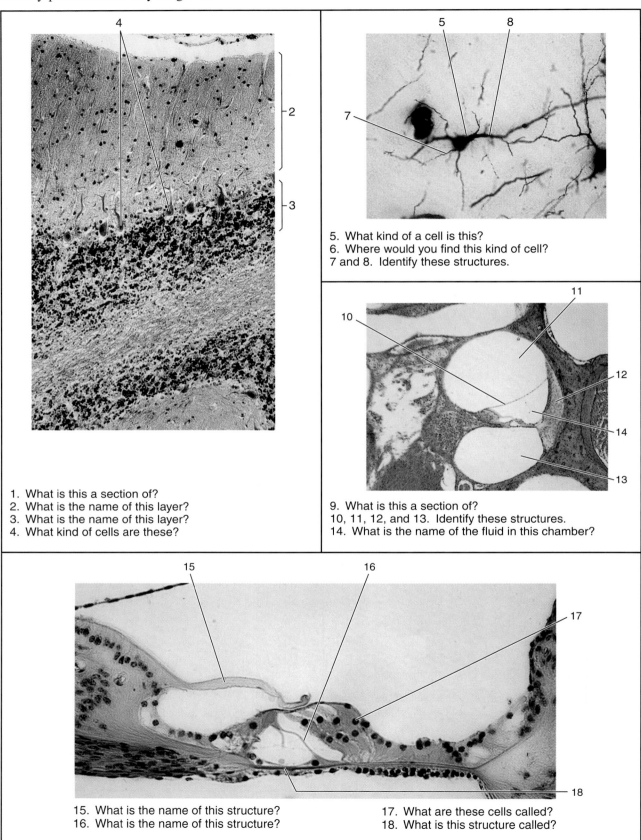

1. What is this a section of?
2. What is the name of this layer?
3. What is the name of this layer?
4. What kind of cells are these?

5. What kind of a cell is this?
6. Where would you find this kind of cell?
7 and 8. Identify these structures.

9. What is this a section of?
10, 11, 12, and 13. Identify these structures.
14. What is the name of the fluid in this chamber?

15. What is the name of this structure?
16. What is the name of this structure?
17. What are these cells called?
18. What is this structure called?

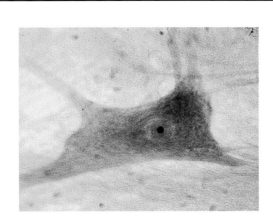

19. What kind of a cell is this?
20. Where are these cells located?

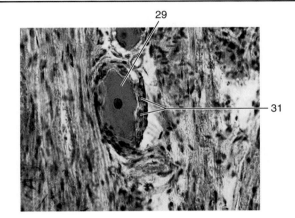

21. What kind of cell is this?
22. What are these dark stained bodies called?
23. Identify this fiber.
24. Identify this fiber.

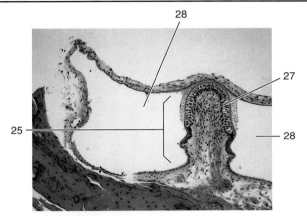

25. Identify this structure.
26. Where is this structure located?
27. What are these cells called?
28. What is the fluid called in this area?

29. What kind of cell is this?
30. Where is #29 located?
31. What are these small cells called?

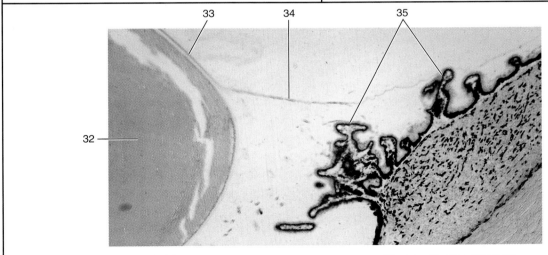

32. Identify this structure.
33. Identify this structure.
34. Identify this structure.
35. Identify this structure.

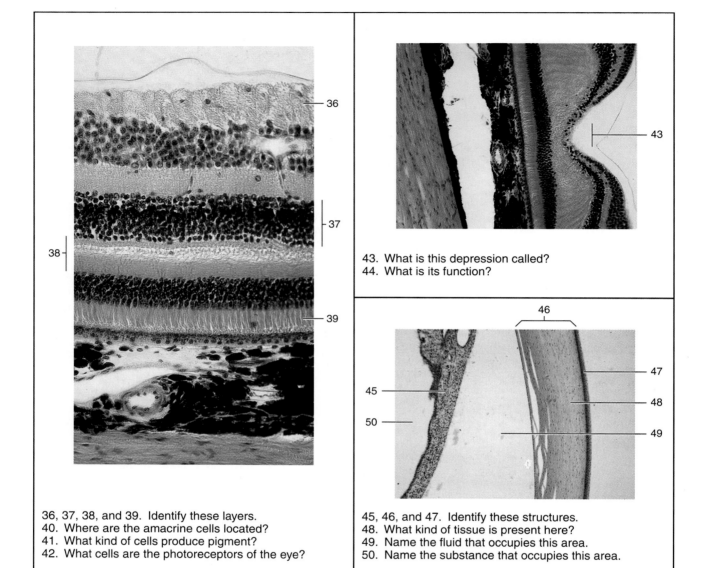

36, 37, 38, and 39. Identify these layers.
40. Where are the amacrine cells located?
41. What kind of cells produce pigment?
42. What cells are the photoreceptors of the eye?

43. What is this depression called?
44. What is its function?

45, 46, and 47. Identify these structures.
48. What kind of tissue is present here?
49. Name the fluid that occupies this area.
50. Name the substance that occupies this area.

HQ 2-3

Histology Self-Quiz No. 3

Once you have completed the exercises related to blood, respiration, and digestion and studied the histology slides depicted in HA-19 through HA-28 of the Histology Atlas, test your knowledge of this material by taking this test. The answers to this quiz are on page 298. This will help to prepare you for a laboratory practical that may be given.

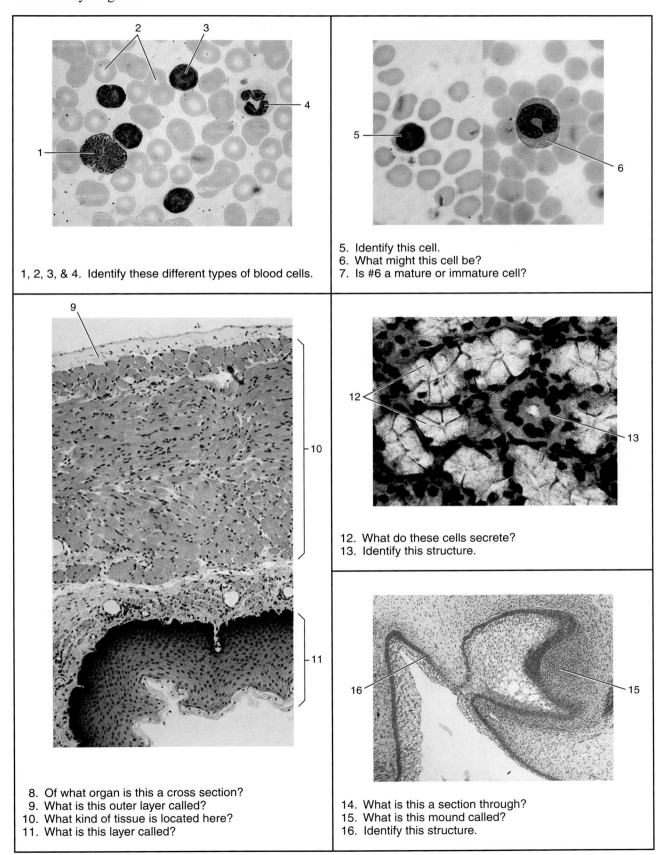

1, 2, 3, & 4. Identify these different types of blood cells.

5. Identify this cell.
6. What might this cell be?
7. Is #6 a mature or immature cell?

8. Of what organ is this a cross section?
9. What is this outer layer called?
10. What kind of tissue is located here?
11. What is this layer called?

12. What do these cells secrete?
13. Identify this structure.

14. What is this a section through?
15. What is this mound called?
16. Identify this structure.

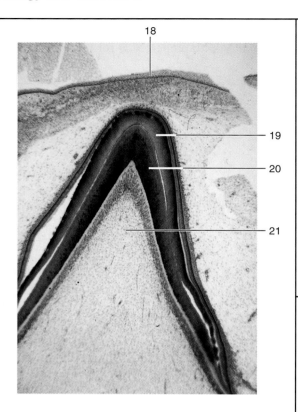

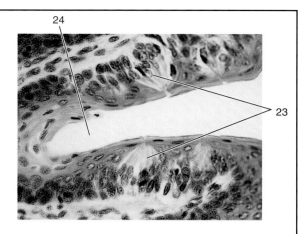

22. Of what organ is this a section?
23. Identify these structures.
24. What is this space called?

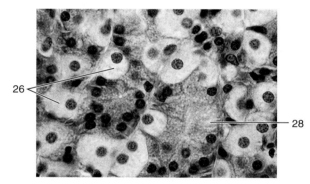

17. What is this entire structure?
18. Identify this layer of tissue.
19. Identify this layer.
20. Identify this layer.
21. What is this area called?

25. Where is this tissue located?
26. What are these pale cells called?
27. What is the function of #26 cells?
28. What are these cells called?
29. What is the function of #28 cells?

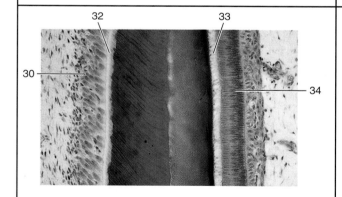

30. What are these cells called?
31. What do #30 cells secrete?
32 and 33. Identify these layers.
34. What are these cells called?
35. What do #34 cells produce?

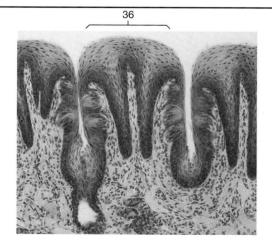

36. Identify this mound.
37. Where are these mounds located?

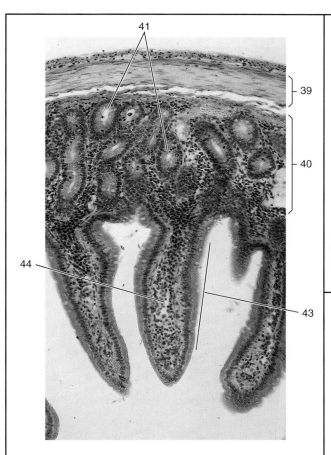

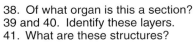

38. Of what organ is this a section?
39 and 40. Identify these layers.
41. What are these structures?
42. What is the function of #41?
43. What is this long structure called?
44. What does this clear area represent?
45. What is the function of #44?

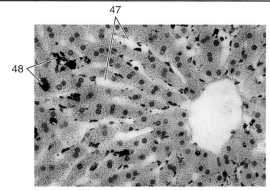

46. Of what organ is this a section?
47. What are these clear areas called?
48. Identify these dark cells.
49. What is the function of #48?

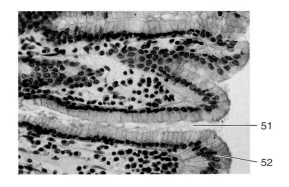

50. Of what organ is this a section?
51. What is this depression called?
52. What kind of tissue is this on the edges?

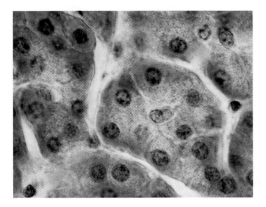

53. Of what organ is this a section?
54. What do the cells in this view produce?

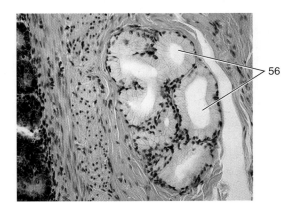

55. Of what organ is this a section?
56. What are these structures?
57. What is the function of #56?

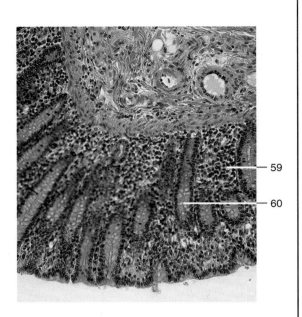

58. Of what organ is this a section?
59. Identify this region.
60. Identify this structure.

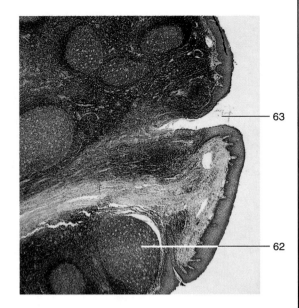

61. Of what organ is this a section?
62. Identify this oval structure.
63. What is this depression called?

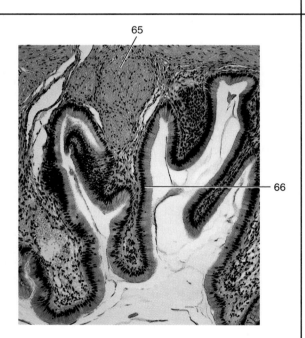

64. Of what organ is this a section?
65. What type of tissue is present in this region?
66. What type of tissue is present here?

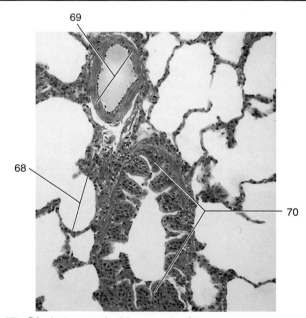

67. Of what organ is this a section?
68. What is this saclike structure called?
69. Identify this thick-walled oval structure?
70. What is this large structure called?

HQ 3-4

Histology Self-Quiz No. 4

Once you have completed all the exercises pertaining to excretion, endocrines, and reproduction and studied the histology slides depicted in HA-29 through HA-40 of the Histology Atlas, test your knowledge of this material by taking this test. The answers to this quiz are on page 298.

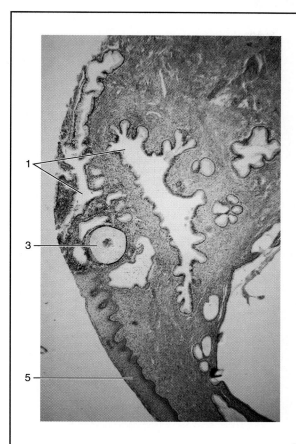

1. What are these spaces called?
2. What kind of fluid accumulates in #1 spaces?
3. Identify this structure.
4. Why did it form in this tissue?
5. Identify this layer of tissue.

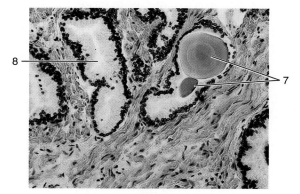

6. Of what organ is this a section?
7. Identify these two structures.
8. What fluid fills this space?

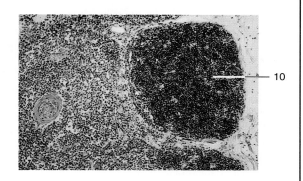

9. Of what endocrine gland is this a section?
10. What portion of the gland is this dark area?
11. What types of blood cells are produced in #10?

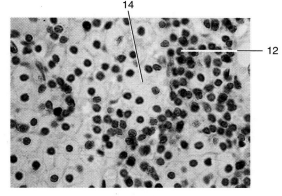

12. What are these dense cells of the parathyroid gland called?
13. What do they secrete?
14. What are these pale cells called?
15. What is the known function of #14 cells?

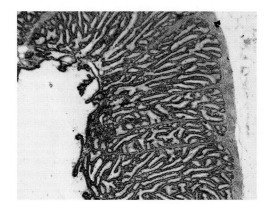

16. Of what gland is this a section?
17. How many of these glands are present in the body?
18. Give one function of the secretion of this gland.

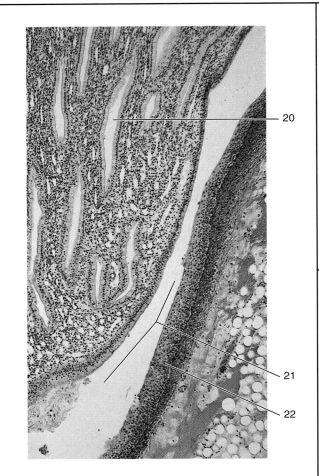

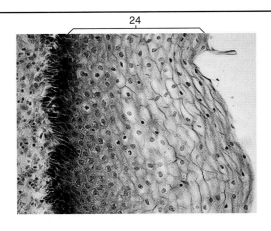

23. Of what organ is this the lining?
24. What type of tissue is this?

19. Of what organ is this a section?
20. Identify this structure.
21. What is the name of this large body?
22. Of what part of #19 is this the wall?

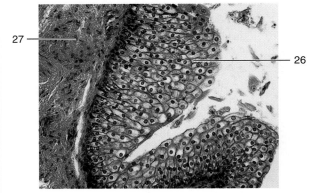

25. Of what duct is this a section?
26. What type of tissue is this?
27. Identify this area.

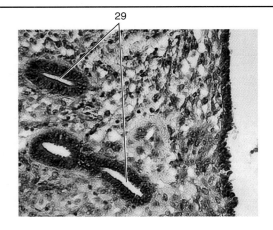

28. Of what organ is this a section?
29. Identify these structures.

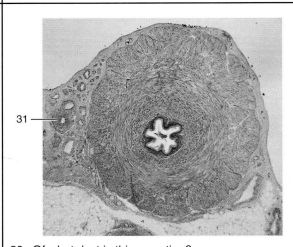

30. Of what duct is this a section?
31. Identify this structure.

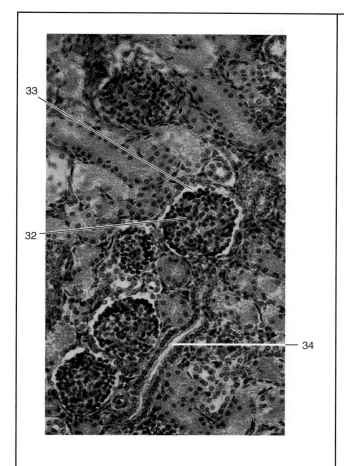

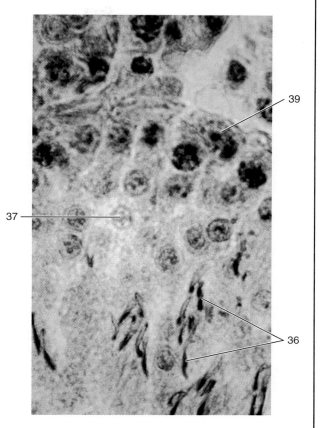

32. Identify this large round body.
33. Identify the clear area around #32.
34. What is this duct called?
35. In what portion of the kidney is this section?

36. Identify these structures.
37. Identify this small cell.
38. Is #37 diploid or haploid?
39. Identify this large cell.
40. Is #39 diploid or haploid?

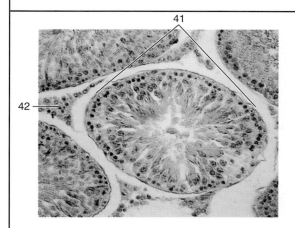

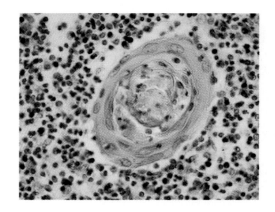

41. What is this large ovoid structure called?
42. Identify this group of cells.
43. What is the function of #42 cells?

44. In what endocrine gland is this oval structure found?
45. Does it have a known function?
46. What is the structure called?

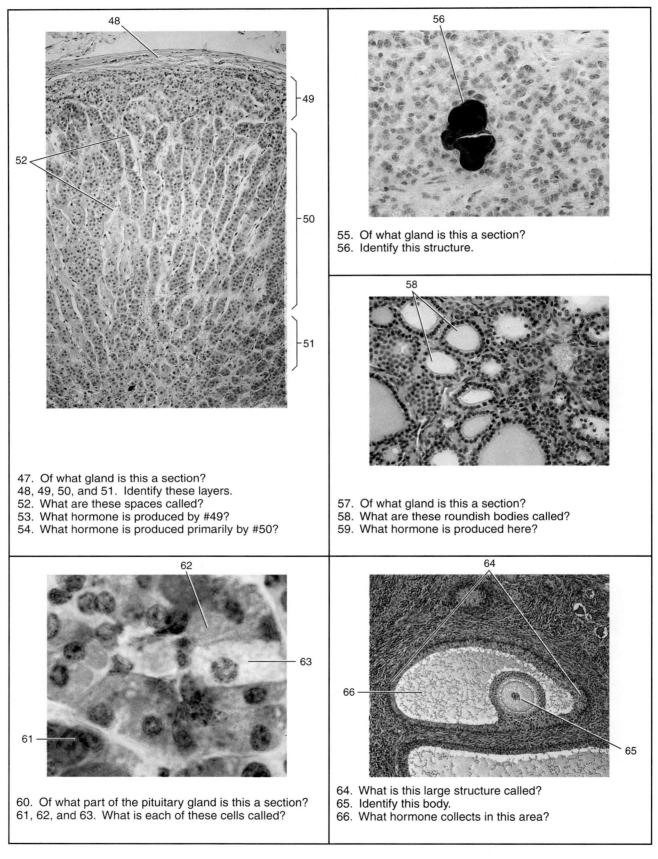

47. Of what gland is this a section?
48, 49, 50, and 51. Identify these layers.
52. What are these spaces called?
53. What hormone is produced by #49?
54. What hormone is produced primarily by #50?

55. Of what gland is this a section?
56. Identify this structure.

57. Of what gland is this a section?
58. What are these roundish bodies called?
59. What hormone is produced here?

60. Of what part of the pituitary gland is this a section?
61, 62, and 63. What is each of these cells called?

64. What is this large structure called?
65. Identify this body.
66. What hormone collects in this area?

LABORATORY REPORT
1,2

Student: _____

Desk No.: _____ Section: _____

Anatomical Terminology, Body Cavities, and Membranes

A. Illustration Labels
Record the label numbers for all the illustrations in the appropriate places in the answer columns.

B. Terminology
From the list of positions, sections, and membranes, select those that are applicable to the following statements. In some cases more than one answer may apply.

Relative Positions		Sections	Membranes
anterior—1	lateral—7	frontal—13	mesentery—17
caudal—2	medial—8	midsagittal—14	pericardium, parietal—18
cranial—3	posterior—9	sagittal—15	pericardium, visceral—19
distal—4	proximal—10	transverse—16	peritoneum, parietal—20
dorsal—5	superior—11		peritoneum, visceral—21
inferior—6	ventral—12		pleura, parietal—22
			pleura, pulmonary—23

Sections:
1. Divides body into front and back portions.
2. Divides body into unequal right and left sides.
3. Three sections that are longitudinal.
4. Divides body into equal right and left sides.
5. Two sections that expose both lungs and heart in each section.

Positions:
6. The tongue is _____ to the palate.
7. The shoulder of a dog is _____ to its hip.
8. A hat is worn on the _____ surface of the head.
9. The cheeks are _____ to the tongue.
10. The fingertips are _____ to all structures of the hand.
11. The human shoulder is _____ to the hip.
12. The shoulder is the _____ (proximal, distal) portion of the arm.

Membranes:
13. Serous membrane attached to lung surface.
14. Serous membrane attached to thoracic wall.
15. Double-layered membrane that holds abdominal organs in place.
16. Double-layered membrane that surrounds the heart.

Select the surfaces on which the following are located:
17. Ear
18. Adam's apple
19. Kneecap
20. Palm of hand

Answers

Terms	Fig. 1.1
1. _____	_____
2. _____	_____
3. _____	_____
4. _____	_____
5. _____	_____
6. _____	_____
7. _____	_____
8. _____	_____
9. _____	_____
10. _____	_____
11. _____	_____
12. _____	_____
13. _____	_____
14. _____	_____
15. _____	_____
16. _____	_____
17. _____	_____
18. _____	_____
19. _____	_____
20. _____	_____

Fig. 2.1	Fig. 1.3

423

C. Localized Areas

Identify the specific areas of the body described by the following statements:

antebrachium—1	cubital—7	hypogastric—13	perineum—19
antecubital—2	epigastric—8	iliac—14	plantar—20
axilla—3	flank—9	inguinal—15	popliteal—21
brachium—4	gluteal—10	leg—16	sternal—22
carpus—5	groin—11	lumbar—17	tarsus—23
costal—6	hypochondriac—12	pectoral—18	thigh—24
			umbilical—25

1. The elbow area.
2. The "rump" area.
3. The underarm area.
4. The sole of the foot.
5. The upper arm.
6. The forearm.
7. Area synonymous with the mammary area.
8. A long, narrow vertical region in the center of the chest area between the mammary areas.
9. Depression on back of leg behind the knee.
10. Side of the abdomen between the lower edge of rib cage and upper edge of hipbone.
11. Area over the ribs on the dorsum.
12. Area between the anus and the external genitalia.
13. The lower portion of the lower extremity.
14. The upper portion of the lower extremity.
15. The ankle.
16. The wrist.
17. Term synonymous with the groin area.
18. Anterior surface of the elbow.
19. Abdominal area that surrounds the navel.
20. Abdominal area that is lateral to the pubic region.
21. Area of the abdominal wall that covers the stomach.
22. Abdominal area that is lateral to umbilical area.
23. Abdominal areas between the transpyloric and transtubercular planes. (Two answers)
24. Areas located on each side of the epigastric area. (One answer)

D. Body Cavities

From the following list of cavities select those applicable to the following statements:

abdominal—1	dorsal—4	spinal—8
abdominopelvic—2	pelvic—5	thoracic—9
cranial—3	pericardial—6	ventral—10
	pleural—7	

1. Cavity that consists of cranial and spinal cavities.
2. Most inferior portion of the abdominopelvic cavity.
3. Cavity separated into two major divisions by the diaphragm.
4. Cavity inferior to the diaphragm that consists of two parts.

Select the cavity in which the following organs are located:

5. Kidneys	9. Rectum	13. Spleen
6. Duodenum	10. Pancreas	14. Spinal cord
7. Brain	11. Lungs and heart	15. Urinary bladder
8. Heart	12. Liver	16. Stomach

Answers

Areas	Fig. 1.2	Fig 2.2
1. _____	_____	_____
2. _____	_____	_____
3. _____	_____	_____
4. _____	_____	_____
5. _____		
6. _____	_____	_____
7. _____	_____	_____
8. _____	_____	_____
9. _____	_____	_____
10. _____	_____	_____
11. _____	_____	
12. _____	_____	_____
13. _____	_____	Fig. 2.3
14. _____	_____	
15. _____	_____	_____
16. _____	_____	_____
17. _____	_____	_____
18. _____	_____	_____
19. _____	_____	_____
20. _____	_____	
21. _____	_____	
22. _____	_____	
23. _____	_____	
24. _____	_____	

Cavities	
1. _____	_____
2. _____	_____
3. _____	_____
4. _____	_____
5. _____	_____
6. _____	_____
7. _____	_____
8. _____	_____
9. _____	_____
10. _____	_____
11. _____	_____
12. _____	_____
13. _____	_____
14. _____	_____
15. _____	
16. _____	

LABORATORY REPORT

3

Student: _____

Desk No.: _____ Section: _____

Organ Systems:
Rat Dissection

A. System Functions

Select the system (or systems) that perform the following functions in the body. Record the numbers in the answer column. More than one answer may apply.

cardiovascular—1   lymphatic—5   respiratory—9
digestive—2   muscular—6   reticuloendothelial—10
endocrine—3   nervous—7   skeletal—11
integumentary—4   reproductive—8 urinary—12

1. Removes carbon dioxide from the blood.
2. Provides a shield of protection for vital organs such as the brain and heart.
3. Movement of ribs in breathing.
4. Carries heat from the muscles to the surface of the body for dissipation.
5. Provides rigid support for the attachment of muscles.
6. Transportation of food from the intestines to all parts of the body.
7. Disposes of urea, uric acid, and other cellular wastes.
8. Movement of arms and legs.
9. Helps to maintain normal body temperature by dissipating excess heat.
10. Destroys microorganisms once they break through the skin.
11. Phagocytic cells of this system remove bacteria from lymph.
12. Carries messages from receptors to centers of interpretation.
13. Controls the rate of growth.
14. Returns tissue fluid to the blood.
15. Produces various kinds of blood cells.
16. Enables the body to adjust to changing internal and external environmental conditions.
17. Ensures continuity of the species.
18. Eliminates excess water from the body.
19. Carries hormones from glands to all tissues.
20. Acts as a shield against invasion by bacteria.
21. Carries fats from the intestines to the blood.
22. Coordination of all body movements.
23. Breaks food particles down into small molecules.
24. Controls the development of the ovaries and testes.
25. Production of spermatozoa and ova.

Answers
System Functions
1. _____
2. _____
3. _____
4. _____
5. _____
6. _____
7. _____
8. _____
9. _____
10. _____
11. _____
12. _____
13. _____
14. _____
15. _____
16. _____
17. _____
18. _____
19. _____
20. _____
21. _____
22. _____
23. _____
24. _____
25. _____

B. Organ Placement

Select the system at the right that includes the following organs.

1. Bones	cardiovascular—1	
2. Brain	digestive—2	
3. Bronchi	endocrine—3	
4. Dermis	integumentary—4	
5. Esophagus	lymphatic—5	
6. Hair	muscular—6	
7. Heart	nervous—7	
8. Kidneys	reproductive—8	
9. Lungs	respiratory—9	
10. Pancreas	skeletal—10	
11. Spleen	urinary—11	
12. Stomach		
13. Teeth		
14. Trachea		
15. Toenails		
16. Tonsils		
17. Thyroid		
18. Ureters		
19. Uterus		
20. Veins		

Answers

Organ Placement

1. _____
2. _____
3. _____
4. _____
5. _____
6. _____
7. _____
8. _____
9. _____
10. _____
11. _____
12. _____
13. _____
14. _____
15. _____
16. _____
17. _____
18. _____
19. _____
20. _____

LABORATORY REPORT

4

Student: _____

Desk No.: _____ Section: _____

Microscopy

A. Completion Questions

Record the answers to the following questions in the column at the right.

1. List three fluids that may be used for cleaning lenses.
2. How can one greatly increase the bulb life on a microscope lamp if the voltage is variable?
3. What characteristic of a microscope enables one to switch from one objective to another without altering the focus?
4. What effect *(increase* or *decrease)* does closing the diaphragm have on the following?
   a. Image brightness
   b. Image contrast
   c. Resolution
5. In general, at what position should the condenser be kept?
6. Express the maximum resolution of the compound microscope in terms of micrometers (μm).
7. If you are getting 225$\times$ magnification with a 45$\times$ high-dry objective, what is the power of the eyepiece?
8. What is the magnification of objects observed through a 100$\times$ oil immersion objective with a 7.5$\times$ eyepiece?
9. Immersion oil must have the same refractive index as _____ to be of any value.
10. Substage filters should be of a _____ color to get the maximum resolution of the optical system.

B. True–False

Record these statements as True or False in the answer column.

1. Eyepieces are of such simple construction that almost anyone can safely disassemble them for cleaning.
2. Lenses can be safely cleaned with almost any kind of tissue or cloth.
3. When swinging the oil immersion objective into position after using high-dry, one should always increase the distance between the lens and slide to keep from damaging the oil immersion lens.
4. Instead of starting first with the oil immersion lens, it is best to use one of the lower magnifications first, and then swing the oil immersion into position.
5. The 45$\times$ and 100$\times$ objectives have shorter working distances than the 10$\times$ objective.

Answers

Completion

1a. _____
b. _____
c. _____
2. _____
3. _____
4a. _____
b. _____
c. _____
5. _____
6. _____
7. _____
8. _____
9. _____
10. _____

True–False

1. _____
2. _____
3. _____
4. _____
5. _____

C. Multiple Choice

Select the best answer for the following statements:

1. The resolution of a microscope is increased by
   1. using blue light.
   2. stopping down the diaphragm.
   3. lowering the condenser.
   4. raising the condenser to its highest point.
   5. Both 1 and 4 are correct.

2. The magnification of an object seen through the 10× objective with a 10× ocular is
   1. ten times.
   2. twenty times.
   3. one thousand times.
   4. None of these are correct.

3. The most commonly used ocular is
   1. 5×.
   2. 10×.
   3. 15×.
   4. 20×.

4. Microscope lenses may be cleaned with
   1. lens tissue.
   2. a soft linen handkerchief.
   3. an air syringe.
   4. Both 1 and 3 are correct.
   5. 1, 2, and 3 are correct.

5. When changing from low power to high power, it is generally necessary to
   1. lower the condenser.
   2. open the diaphragm.
   3. close the diaphragm.
   4. Both 1 and 2 are correct.
   5. Both 1 and 3 are correct.

Answers
Multiple Choice
1. _____
2. _____
3. _____
4. _____
5. _____

LABORATORY REPORT

5,6

Student: _____

Desk No.: _____ Section: _____

Cytology: Basic Cell Structure and Mitosis

A. Figure 5.1

Record the labels for figure 5.1 in the answer column.

B. Cell Drawings

Sketch in the space below one or two epithelial cells as seen under high-dry magnification. Label the **nucleus, cytoplasm,** and **cell membrane.** If mitosis drawings are to be made, put them on a separate sheet of paper.

C. Cell Structure

From the list of structures below, select those that are described by the following statements. More than one structure may apply to some of the statements.

centrosome—1	lysosome—8	nucleus—15
chromatin granules—2	microvilli—9	plasma membrane—16
cilia—3	mitochondria—10	polyribosomes—17
condensing vacuoles—4	microfilaments—11	RER—18
cytoplasm—5	microtubules—12	ribosomes—19
flagellum—6	nuclear envelope—13	secretory granules—20
Golgi apparatus—7	nucleolus—14	SER—21

1. Prominent body in nucleus that produces ribosomes.
2. Trilaminar unit membrane structure.
3. Outer membrane of the cell.
4. A nonmembranous organelle consisting of two bundles of microtubules.
5. Bodies within the nucleus that represent chromosomal material.
6. ER that lacks ribosomes.
7. Double-layer unit membrane structure.
8. Small bodies of RNA and protein attached to surfaces of the RER.
9. Extensive system of tubules, vesicles, and sacs within the cytoplasm.
10. All protoplasmic material between the plasma membrane and nucleus.
11. Sacs in the cytoplasm that contain digestive enzymes.
12. A unit membrane organelle in the cytoplasm that contains cristae.
13. Short, hairlike appendages on the surface of some cells.
14. ER that is covered by ribosomes.
15. Chains and rosettes of small bodies scattered throughout the cytoplasm.
16. Unit membrane structure that is continuous with the plasma and nuclear membranes.
17. Extranuclear bodies that possess DNA and are unable to replicate themselves.
18. A layered concave structure in the cytoplasm that is similar to the basic structure of SER.
19. Surface protuberances that form a delicate brush border.
20. Bodies produced by Golgi and exocytosed by the cell.

Answers

Cell Structure	Fig. 5.1
1._____	_____
2._____	_____
3._____	_____
4._____	_____
5._____	_____
6._____	_____
7._____	_____
8._____	_____
9._____	_____
10._____	_____
11._____	_____
12._____	_____
13._____	_____
14._____	_____
15._____	_____
16._____	_____
17._____	
18._____	
19._____	
20._____	

D. Organelle Functions

Select the organelles that perform the following functions. More than one structure may apply to some statements.

centrosome—1 microvilli—5 plasma membrane—9
flagellum—2 mitochondrion—6 polyribosomes—10
Golgi apparatus—3 nucleolus—7 RER—11
lysosome—4 nucleus—8 SER—12

1. Forms asters during mitosis.
2. Produces ribonucleoprotein for ribosomes.
3. The microcirculatory system of the cell.
4. Contains genetic code of the cell.
5. Place where oxidative respiration occurs.
6. Synthesize protein for endogenous use.
7. Brings about rapid hydrolysis (digestion) of cell after death.
8. Place where ATP is synthesized and stored.
9. Regulates the flow of materials into and out of cell.
10. Synthesizes protein for extracellular distribution.
11. Disposal organelles in cytoplasm that digest protein.
12. Increases absorptive area of the cell.
13. Synthesis, packaging, and transportation of substances.
14. Plays a role in flagellar development.
15. Accomplishes some motile function for some cells.

E. True–False

Evaluate the validity of the following statements concerning cell structure and function.

1. Secretory cells have well-developed Golgi apparatuses.
2. Mitochondria are able to replicate themselves.
3. Practically all molecules that pass through the plasma membrane are actively assisted by the membrane.
4. Ribosomes are found in the cytoplasm, mitochondria, RER, and SER.
5. The inner and outer layers of the plasma membrane are essentially lipoidal in nature with scattered molecules of protein and glycoprotein.
6. A "brush border" consists of microvilli.
7. Spermatozoa move with cilia.
8. Ribosomes are primarily involved in oxidative respiration.
9. Cilia form from basal bodies derived from centrioles.
10. Asters develop from mitochondria.

F. Mitosis (Ex. 6)

Select the phase in which the following events occur.

1. Nuclear membrane disappears.
2. Cleavage furrow forms.
3. Spindle fibers begin to form.
4. New nuclear membrane forms.
5. Asters form from centrioles.
6. Centriole replication occurs.
7. Replication of DNA occurs.
8. Sister chromosomes separate and pass to opposite poles.
9. Formed after telophase is completed.
10. Chromosomes become oriented on equatorial plane.

interphase—1
prophase—2
metaphase—3
anaphase—4
telophase—5
daughter cells—6

Answers

Functions
1. ___ 2. ___ 3. ___ 4. ___ 5. ___ 6. ___ 7. ___ 8. ___ 9. ___ 10. ___ 11. ___ 12. ___ 13. ___ 14. ___ 15. ___

True–False
1. ___ 2. ___ 3. ___ 4. ___ 5. ___ 6. ___ 7. ___ 8. ___ 9. ___ 10. ___

Mitosis
1. ___ 2. ___ 3. ___ 4. ___ 5. ___ 6. ___ 7. ___ 8. ___ 9. ___ 10. ___

LABORATORY REPORT

7

Student: _____

Desk No.: _____ Section: _____

Osmosis and Cell Membrane Integrity

A. Molecular Movement

1. Brownian Movement

    a. Were you able to see any movement of carbon particles under the microscope? _____

    b. What forces propel the movement of these particles? _____

    _____

    _____

    _____

2. Diffusion Plot the distance of diffusion of molecules for the two crystals on the following graphs.

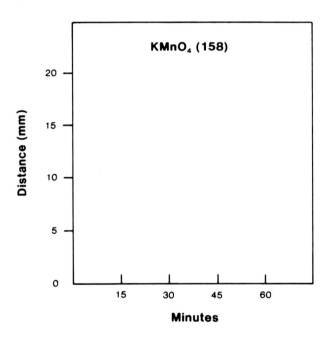

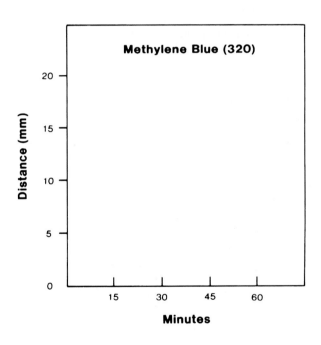

3. Conclusion What is the relationship of molecular weight to the rate of diffusion? _____

_____

_____

_____

_____

B. Osmotic Effects

As you examine each tube macroscopically and microscopically, record your results in the following table:

SOLUTION	MACROSCOPIC APPEARANCE (lysis or no lysis)	MICROSCOPIC APPEARANCE (crenation, lysis, or isotonic)
0.15M Sodium Chloride		
0.30M Sodium Chloride		
0.28M Glucose $C_6H_{12}O_6$		
0.30M Glycerine $C_3H_5(OH)_3$		
0.30 Urea $CO(NH_2)_2$		

1. Are any of the above solutions isotonic for red blood cells? _____

   If so, which ones? _____

2. With the aid of the atomic weight table in appendix A, calculate the percentages of solutes in those solutions that proved to be isotonic. _____

   _____

   _____

LABORATORY REPORT
8,9

Student: _____

Desk No.: _____ Section: _____

Epithelial and Connective Tissues

A. Microscopic Study

If any drawings of tissue are to be made, make them on separate sheets of paper and label all identifiable structures.

B. Figure 9.5

Record the labels for this illustration in the answer column.

C. Epithelial Tissue Characteristics

From the list of tissues select those that are described by the following statements. More than one tissue may apply.

1. Elongated cells, single layer.	columnar
2. Cells with blocklike cross section.	pseudostratified—1
3. Tissue with goblet cells.	simple—2
4. Some cells are binucleate.	stratified—3
5. Have basal lamina.	cuboidal—4
6. Lack intercellular matrix.	squamous
7. Flattened cells, single layer.	simple—5
8. Flattened cells, several layers.	stratified—6
9. Cells are layered and loosely packed.	transitional—7
10. Cells possess brush border.	all of above—8
11. Cells near surface are dome-shaped.	
12. Produce mucus on free surface.	
13. Elongated cells, with some cells not extending up to free surface.	
14. Cells have microvilli on free surface.	

D. Epithelial Tissue Locations

Identify the kind of epithelial tissues found in the following structures.

columnar, ciliated—1	pseudostratified, plain—6
columnar, plain—2	squamous, simple—7
columnar, stratified—3	squamous, stratified—8
cuboidal—4	transitional—9
pseudostratified, ciliated—5	none of above—10

1. Epidermis of skin	11. Intestinal lining
2. Lining of kidney calyces	12. Lining of trachea
3. Peritoneum	13. Lining of blood vessels
4. Auditory tube lining	14. Lining of urinary bladder
5. Cornea of the eye	15. Lining of stomach
6. Capillary walls	16. Lining of male urethra
7. Capsule of lens of the eye	17. The conjunctiva
8. Lining of the pharynx	18. Lining of vagina
9. Lining of the mouth	19. Anal lining
10. Pleural membranes	20. Parotid gland

Answers

Characteristics	Fig. 9.5
1. _____	_____
2. _____	_____
3. _____	_____
4. _____	_____
5. _____	_____
6. _____	_____
7. _____	_____
8. _____	_____
9. _____	_____
10. _____	_____
11. _____	_____
12. _____	_____
13. _____	_____
14. _____	_____

Locations

1. _____	11. _____
2. _____	12. _____
3. _____	13. _____
4. _____	14. _____
5. _____	15. _____
6. _____	16. _____
7. _____	17. _____
8. _____	18. _____
9. _____	19. _____
10. _____	20. _____

E. Connective Tissue Characteristics

From the list of tissues, select those that are described by the following statements. More than one tissue may apply.

1. Contains all three types of fibers.
2. Reinforced mostly with collagenous fibers.
3. Cells are contained in lacunae.
4. Cells have large fat vacuoles.
5. Contain mast cells.
6. Nonyielding supporting tissue.
7. Devoid of blood supply.
8. Contain macrophages.
9. Most flexible of supporting tissues.
10. Converted to bone during growing years.
11. Matrix impregnated with calcium and phosphorous salts.
12. Toughest ordinary tissue (for stretching).
13. Cells have signet ring-like appearance.
14. Reinforced mostly with yellow fibers.
15. Has smooth, glassy appearance.

adipose—1
areolar—2
bone—3
cartilage
   elastic—4
   fibro—5
   hyaline—6
dense irregular—7
dense regular—8
reticular—9

F. Connective Tissue Locations

From the above list of tissues, select the connective tissue that is found in the following:

1. Deep fasciae
2. Tendon sheaths
3. Intervertebral disks
4. Superficial fasciae
5. Epiglottis
6. Basal lamina
7. Lymph nodes
8. Aponeuroses
9. Nerve sheaths
10. Symphysis pubis
11. Auditory tube
12. Skeletal components
13. Periosteum (outer layer)
14. Sharpey's fiber
15. Ligaments and tendons
16. Hemopoietic tissue
17. Tracheal rings
18. Pinna of outer ear

G. Terminology

Select the terms on the right that are described by the following statements:

1. Protein in white fibers.
2. Cell that secretes mucus.
3. Material between fibroblasts.
4. Cartilage-producing cells.
5. Inner layer of periosteum.
6. Cell that produces fibers.
7. Synonym for Haversian system.
8. Protein in reticular fibers.
9. Bony plates in spongy bones.
10. Anchors periosteum to bone tissue.
11. Layers of bony matrix around central (Haversian) canal.
12. Layer of reticular fibers at base of epithelial tissues.
13. Cell found in loose fibrous tissue.
14. Minute canals radiating out from lacunae in bone tissue.
15. Canal in bone that contains capillaries.
16. Horizontal canals that connect adjacent central canals.

basal lamina—1
canaliculi—2
central canal—3
chondrocytes—4
collagen—5
fibroblast—6
goblet cell—7
lamellae—8
macrophage—9
mast cell—10
matrix—11
osteocyte—12
osteogenic layer—13
osteon—14
perforating canal—15
Sharpey's fibers—16
trabeculae—17

Answers

Characteristics	Terms
1. _____	1. _____
2. _____	2. _____
3. _____	3. _____
4. _____	4. _____
5. _____	5. _____
6. _____	6. _____
7. _____	7. _____
8. _____	8. _____
9. _____	9. _____
10. _____	10. _____
11. _____	11. _____
12. _____	12. _____
13. _____	13. _____
14. _____	14. _____
15. _____	15. _____
	16. _____

Locations

1. _____
2. _____
3. _____
4. _____
5. _____
6. _____
7. _____
8. _____
9. _____
10. _____
11. _____
12. _____
13. _____
14. _____
15. _____
16. _____
17. _____
18. _____

LABORATORY REPORT

10

Student: _____

Desk No.: _____ Section: _____

The Integument

A. Figures 10.1 and 10.2

Record the labels for these illustrations in the answer column.

B. Microscopic Study

If the space provided below is adequate for your histology drawings, use it. Use a separate sheet of paper if the space is inadequate.

C. Questions

Select the answer that completes the following statements. Only one answer applies for each question.

1. Granules of the stratum basale consist of
   1. melanin.   2. keratin.   3. eleidin.

2. The papillary layer is a part of the
   1. dermis.   2. epidermis.   3. stratum spinosum.

3. The epidermis consists of the following number of distinct layers:
   1. three.   2. four.   3. five.   4. six.

4. Meissner's corpuscles are sensitive to
   1. temperature.   2. pressure.   3. touch.

5. Pacinian corpuscles are sensitive to
   1. temperature.   2. pressure.   3. touch.

6. The outermost layer of the epidermis is the
   1. stratum corneum.   2. stratum basale.
   3. stratum lucidum.   4. stratum spinosum.

7. New cells of the epidermis originate in the
   1. dermis.   2. stratum basale.
   3. stratum corneum.   4. stratum lucidum.

8. Melanin granules are produced by the
   1. stratum basale.   2. melanocytes.
   3. stratum lucidum.   4. stratum corneum.

Answers

Questions	Fig. 10.1
1._____	_____
2._____	_____
3._____	_____
4._____	_____
5._____	_____
6._____	_____
7._____	_____
8._____	_____
	_____
	_____
	_____
	_____
	_____
	_____
	_____
	_____
	_____

Fig. 10.2

_____
_____
_____
_____
_____
_____

9. The secretions of the apocrine glands usually empty
   1. into a hair follicle.
   2. directly out through the skin surface.
   3. Both 1 and 2.
   4. None of these is correct.

10. Sebum is secreted by
    1. sweat glands.
    2. eccrine glands.
    3. apocrine glands.
    4. None of these is correct.

11. Keratin's function is primarily to
    1. destroy bacteria.
    2. provide waterproofing.
    3. cool the body.

12. Meissner's corpuscles are located in the
    1. papillary layer.
    2. reticular layer.
    3. subcutaneous layer.

13. Arrector pili muscles assist in
    1. maintaining skin tonus.
    2. limiting excessive sweating.
    3. forcing out sebum.
    4. None of these is correct.

14. The hair follicle is
    1. a shaft of hair.
    2. the root of a hair.
    3. a tube in the skin.
    4. a glandular structure.

15. Pacinian corpuscles are located in the
    1. papillary layer.
    2. reticular layer.
    3. subcutaneous layer.

16. Sweat is produced by
    1. sebaceous glands.
    2. eccrine glands.
    3. ceruminous glands.
    4. None of these is correct.

17. The coiled-up portion of an apocrine gland is located in the
    1. epidermis.
    2. dermis.
    3. subcutaneous layer.
    4. hair follicle.

18. Apocrine glands are located
    1. in the axillae.
    2. on the scrotum.
    2. in the ear canal.
    4. All of these are correct.

Answers

Questions

9. _____
10. _____
11. _____
12. _____
13. _____
14. _____
15. _____
16. _____
17. _____
18. _____

11

The Skeletal Plan

A. Illustrations

Record the labels for figures 11.1, 11.2, and 11.3 in the answer columns.

B. Long Bone Structure

Identify the terms described by the following statements:

1. Shaft portion of bone.
2. Hollow chamber in bone shaft.
3. Type of marrow in medullary canal.
4. Enlarged end of a bone.
5. Type of bone in diaphysis.
6. Fibrous covering of bone shaft.
7. Linear growth area of long bone.
8. Smooth gristle covering bone end.
9. Type of bone marrow in bone ends.
10. Lining of medullary canal.
11. Type of bone tissue in bone ends.
12. Lines the central canals.

articular cartilage—1
cancellous bone—2
compact bone—3
diaphysis—4
endosteum—5
epiphyseal disk—6
epiphysis—7
medullary canal—8
periosteum—9
red marrow—10
yellow marrow—11

C. Bone Identification

Select the structures on the right that match the statements on the left.

1. Shoulder blade
2. Collarbone
3. Breastbone
4. Shinbone
5. Kneecap
6. Upper arm bone
7. Bones of spine
8. Thighbone
9. Lateral bone of forearm
10. Joint between ossa coxae
11. Horseshoe-shaped bone
12. Bones of shoulder girdle
13. One half of pelvic girdle
14. Medial bone of forearm
15. Thin bone paralleling tibia (calf bone)

clavicle—1
femur—2
fibula—3
humerus—4
hyoid—5
os coxa—6
patella—7
radius—8
scapula—9
sternum—10
symphysis pubis—11
tibia—12
ulna—13
vertebrae—14

Answers

Structure	Bones
1. _____	1. _____
2. _____	2. _____
3. _____	3. _____
4. _____	4. _____
5. _____	5. _____
6. _____	6. _____
7. _____	7. _____
8. _____	8. _____
9. _____	9. _____
10. _____	10. _____
11. _____	11. _____
12. _____	12. _____
	13. _____
Fig. 11.2	14. _____
_____	15. _____
_____	
_____	**Fig. 11.1**
_____	_____
_____	_____
_____	_____
_____	_____
_____	_____
_____	_____
_____	_____
_____	_____
_____	_____
_____	_____
_____	

D. Medical

Select the condition that is described by the following statements. Since not all conditions are described in this manual, it will be necessary for you to consult your lecture text or medical dictionary for some of the terminology.

closed reduction—1   fissured fracture—6   osteoporosis—11
Colles' fracture—2   greenstick fracture—7   pathological fracture—12
comminuted fracture—3   open reduction—8   Pott's fracture—13
compacted fracture—4   osteomalacia—9   rickets—14
compound fracture—5   osteomyelitis—10   simple fracture—15

1. Fracture due to weak bone structure, not trauma.
2. Fracture in which the skin is not broken.
3. Fracture characterized by bone ends penetrating the skin.
4. Procedure used to set broken bones without using surgery.
5. Fracture caused by severe vertical forces.
6. Skeletal softness in adults.
7. Bone is split longitudinally.
8. Fracture characterized by two or more fragments.
9. Term applied to bone setting with the aid of surgery.
10. Skeletal softness in children due to vitamin D deficiency.
11. Infection of bone marrow.
12. Bone fracture extends only partially through a bone; incomplete fracture.
13. Skeletal deformation due to vitamin D deficiency.
14. Bone condition in which increased porosity occurs due to widening of central canals.
15. Outward displacement of foot due to fracture of lower part of fibula and malleolus.
16. Incomplete bone fracture in which fracture is apparent only on convex surface.
17. Displacement of hand backward and outward due to fracture of lower end of radius.

E. Terminology

The following statements decribe various processes, depressions, openings, and cavities that are seen on various bones of the skeleton. Identify the term at the right that best matches each of the following descriptive statements:

1. A long, tubelike passageway.   condyle—1
2. A rounded, knucklelike process that   fissure—2
   articulates with another bone.   foramen—3
3. A hole through which nerves pass.   fossa—4
4. A narrow slit.   meatus—5
5. A hollow cavity within a bone.   sinus—6
6. A small, rounded process.   spine—7
7. A sharp or long, slender process.   trochanter—8
8. A depression in a bone.   tubercle—9
9. A very large process on a bone.   tuberosity—10
10. A large, roughened process on a bone
    to which a muscle is attached.

Answers

Medical

1. _____
2. _____
3. _____
4. _____
5. _____
6. _____
7. _____
8. _____
9. _____
10. _____
11. _____
12. _____
13. _____
14. _____
15. _____
16. _____
17. _____

Fig. 11.3

_____
_____
_____
_____
_____
_____
_____
_____
_____
_____

Terms

1. _____
2. _____
3. _____
4. _____
5. _____
6. _____
7. _____
8. _____
9. _____
10. _____

LABORATORY REPORT

12

Student: _____

Desk No.: _____ Section: _____

The Skull

A. Illustrations

Record the labels for all the illustrations of this exercise in the answer columns.

B. Bone Names

Select the bones that are described by the following statements. More than one answer may apply.

1. Forehead bone
2. Upper jaw
3. Cheekbone
4. Lower jaw
5. Sides of skull
6. On walls of nasal fossa
7. Unpaired bones (usually)
8. Bones of hard palate (2 pairs)
9. Back and bottom of skull
10. Contain sinuses (4 bones)
11. External bridge of nose
12. On median line in nasal cavity

ethmoid—1
frontal—2
lacrimal—3
mandible—4
maxilla—5
nasal—6
nasal conchae—7
occipital—8
palatine—9
parietal—10
sphenoid—11
temporal—12
vomer—13
zygomatic—14

C. Terminology

From the above list of bones select those that have the following structures:

1. Alveolar process
2. Angle
3. Antrum of Highmore
4. Chiasmatic groove
5. Coronoid process
6. Cribriform plate
7. Crista galli
8. External acoustic meatus
9. Incisive fossa
10. Inferior temporal line
11. Internal acoustic meatus
12. Mandibular condyle
13. Mandibular fossa
14. Mastoid process
15. Mylohyoid line
16. Occipital condyles
17. Ramus
18. Sella turcica
19. Styloid process
20. Superior nasal concha
21. Superior temporal line
22. Symphysis
23. Tear duct
24. Zygomatic arch

D. Sutures

Identify the sutures described by the following statements:

1. Between parietal and occipital.
2. Between two maxillae on median line.
3. Between the two parietals on the median line.
4. Between frontal and parietal.
5. Between parietal and temporal.
6. Between maxillary and palatine.

coronal—1
lambdoidal—2
median palatine—3
sagittal—4
squamosal—5
transverse palatine—6

Answers

Bones	Fig. 12.1	Fig. 12.2
1._____	_____	_____
2._____	_____	_____
3._____	_____	_____
4._____	_____	_____
5._____	_____	_____
6._____	_____	_____
7._____	_____	_____
8._____	_____	_____
9._____	_____	_____
10._____	_____	_____
11._____	_____	_____
12._____	_____	_____

Terms		
1._____	_____	_____
2._____	_____	_____
3._____	_____	_____
4._____	_____	_____
5._____	_____	_____
6._____	_____	_____
7._____	_____	_____
8._____	_____	_____
9._____	_____	_____
10._____	Fig. 12.3	_____
11._____	_____	_____
12._____	_____	_____
13._____	_____	_____
14._____	_____	_____
15._____	_____	_____
16._____	_____	_____
17._____	_____	Sutures
18._____	_____	1._____
19._____	_____	2._____
20._____	_____	3._____
21._____	_____	4._____
22._____		5._____
23._____		6._____
24._____		

E. *Foramina Locations*

By referring to the illustrations, select the bones on which the following foramina or canals are located:

1. Anterior palatine foramen
2. Carotid canal
3. Foramen magnum
4. Foramen ovale
5. Hypoglossal canal
6. Infraorbital foramen
7. Internal acoustic meatus
8. Mandibular foramen
9. Mental foramen
10. Optic canal
11. Supraorbital foramen
12. Zygomaticofacial foramen

frontal—1
mandible—2
maxilla—3
occipital—4
palatine—5
parietal—6
sphenoid—7
temporal—8
zygomatic—9

F. *Foramen Function*

Select the cranial nerves and other structures that pass through the following foramina, canals, and fissures. Note that the number following each cranial nerve corresponds to the number designation of the particular cranial nerve. (Example: the abducens nerve is also known as the sixth cranial nerve.) Also, note that the fifth cranial nerve (trigeminal) has three divisions.

1. Carotid canal
2. Cribriform plate foramina
3. Foramen ovale
4. Foramen magnum
5. Hypoglossal canal
6. Internal acoustic meatus
7. Optic canal

Cranial Nerves
olfactory—1
optic—2
oculomotor—3
trochlear—4
trigeminal branches
   ophthalmic—5.1
   maxillary—5.2
   mandibular—5.3
abducens—6
facial—7
vestibulocochlear—8
spinal accessory—11
hypoglossal—12
brain stem—13
internal carotid a.—14

G. *Completion*

Provide the names of the structures described by the following statements:

1. Lighten the skull.
2. Allow compression of skull bones during birth.
3. Provides attachment site for falx cerebri.
4. Allow sound waves to enter the skull.
5. Provides anchorage for sternocleidomastoideus muscle.
6. Point of attachment for some tongue muscles.
7. Point of articulation between skull and first vertebra.
8. Points of articulation between mandible and skull.

Answers

Locations	Fig. 12.5	Fig. 12.6
1. ____	_____	_____
2. ____	_____	_____
3. ____	_____	_____
4. ____	_____	_____
5. ____	_____	_____
6. ____	_____	_____
7. ____	_____	_____
8. ____	_____	_____
9. ____	_____	_____
10. ____	_____	_____
11. ____	_____	_____
12. ____	_____	_____

Function		
1. ____	_____	_____
2. ____	_____	_____
3. ____	_____	_____
4. ____	_____	_____
5. ____	_____	_____
6. ____	_____	_____
7. ____	_____	_____

Fig. 12.4	Fig.12.7
_____	_____
_____	_____
_____	_____
_____	_____
	_____
	_____

Completion	Fig. 12.8
1. _____	_____
2. _____	_____
3. _____	_____
4. _____	_____
5. _____	_____
6. _____	_____
7. _____	_____
8. _____	_____

LABORATORY REPORT

13

Student: _____

Desk No.: _____ Section: _____

The Vertebral Column and Thorax

A. Illustrations

Record the labels for the four figures in the answer columns.

B. Bone Names

Select the proper anatomical terminology for the following bones:

1. Tailbone
2. Breastbone
3. False rib
4. Floating rib
5. True rib
6. Second cervical vertebra
7. First cervical vertebra
8. Inferior portion of breastbone
9. Midportion of breastbone
10. Superior portion of breastbone

atlas—1
axis—2
coccyx—3
gladiolus—4
manubrium—5
sacrum—6
sternum—7
vertebral ribs—8
vertebrochondral
  ribs—9
vertebrosternal
  ribs—10
xiphoid—11

C. Structures

Select the structures or foramina that perform the following functions:

body—1
intervertebral disks—2
intervertebral foramina—3
odontoid process—4
rib cage—5

spinal curves—6
spinous process—7
sternum—8
transverse process—9
transverse foramina—10

1. Provides anterior attachment for vertebrosternal ribs.
2. Openings through which spinal nerves exit.
3. Portion of vertebra that supports body weight.
4. Protects vital thoracic organs.
5. Impart springiness to vertebral column (two answers).
6. Part of thoracic vertebra that provides attachment for rib.
7. Acts as a pivot for the atlas to move around.
8. Openings through which vertebral artery and vein pass.

Answers

Bone Names	Fig. 13.1
1. _____	_____
2. _____	_____
3. _____	_____
4. _____	_____
5. _____	_____
6. _____	_____
7. _____	_____
8. _____	_____
9. _____	_____
10. _____	_____

Structures	
1. _____	_____
2. _____	_____
3. _____	_____
4. _____	_____
5. _____	_____
6. _____	_____
7. _____	_____
8. _____	_____

Fig. 13.2	
_____	_____
_____	_____
_____	_____
_____	_____
_____	_____
_____	_____
_____	_____
_____	_____

D. Vertebral Differences

Select the type of vertebrae from the column on the right that is unique for the following characteristics:

1. Transverse foramina   atlas—1
2. Large, massive body   axis—2
3. Odontoid process   cervical vertebrae—3
4. Downward-sloping spinous process   thoracic vertebrae—4
5. Rib facets on transverse process   lumbar vertebrae—5

E. Numbers

Indicate the number of components that constitute the following:

1. Cervical vertebrae   7. Spinal curvatures
2. Thoracic vertebrae   8. Pairs of vertebrochondral
3. Lumbar vertebrae   ribs
4. Sacrum (fused vertebrae)   9. Pairs of vertebrosternal
5. Coccyx (rudimentary vertebrae)   ribs
6. Pairs of ribs   10. Pairs of vertebral ribs

F. Medical

Select the medical terms at the right that are described by the following statements. Consult your lecture text or medical dictionary.

1. Exaggerated lumbar curvature   herniation—1
2. Lateral curvature of the spine   kyphosis—2
3. "Slipped disk"   lordosis—3
4. Exaggerated thoracic curvature   scoliosis—4

Answers

Vertebral Differences	Fig. 13.3
1. _____	_____
2. _____	_____
3. _____	_____
4. _____	_____
5. _____	_____
Numbers	
1. _____	_____
2. _____	_____
3. _____	_____
4. _____	_____
5. _____	_____
6. _____	_____
7. _____	_____
8. _____	_____
9. _____	_____
10. _____	_____
Medical	
1. _____	
2. _____	Fig. 13.4
3. _____	_____
4. _____	_____
	_____

LABORATORY REPORT
14,15

Student: _____

Desk No.: _____ Section: _____

The Appendicular Skeleton and Articulations

A. Labels
Record the labels for all the illustrations of Exercises 14 and 15 in the appropriate columns.

B. Bone Names
Select the names of the bones described in the following statements:

1. Lateral forearm bone
2. Medial forearm bone
3. Bones of palm of hand
4. Upper arm bone
5. Fingers
6. Wrist bones
7. Components of shoulder girdle

carpals—1
clavicle—2
humerus—3
metacarpals—4
phalanges—5
radius—6
scapula—7
tarsals—8
ulna—9

C. Landmarks (Upper Extremities)
Identify the depressions and processes of the arm and shoulder described by the following statements:

acromion process—1 greater tubercle—8 olecranon process—14
capitulum—2 head—9 radial notch—15
coracoid process—3 lateral epicondyle—10 radial tuberosity—16
coronoid fossa—4 lesser tubercle—11 semilunar notch—17
coronoid process—5 medial epicondyle—12 styloid process—18
deltoid tuberosity—6 olecranon fossa—13 trochlea—19
glenoid cavity—7

1. Depression on scapula that articulates with humerus.
2. Depression on ulna that is in contact with head of radius.
3. Epicondyle of humerus adjacent to trochlea.
4. Distal medial condyle of humerus.
5. Distal lateral condyle of humerus.
6. Fossa on distal posterior surface of humerus.
7. Process on scapula that articulates with clavicle.
8. Process on anterior surface of ulna just below semilunar notch.
9. Process on scapula that lies anterior and superior to glenoid cavity.
10. Process on humerus to which deltoid muscle is attached.
11. Condyle of humerus that articulates with ulna.
12. Condyle of humerus that articulates with radius.
13. Part of humerus that articulates with scapula.
14. Epicondyle adjacent to capitulum.
15. Large process near head of humerus.
16. Fossa on distal anterior surface of humerus.
17. Small distal medial process of ulna.
18. Depression on proximal end of ulna that articulates with trochlea of humerus.
19. Proximal process of ulna, sometimes called the "funny bone."
20. Small process just above surgical neck of humerus.

Answers

Bone Names	Fig. 14.1
1. _____	_____
2. _____	_____
3. _____	_____
4. _____	_____
5. _____	_____
6. _____	_____
7. _____	_____

Fig. 14.2

Landmarks

1. _____
2. _____
3. _____
4. _____
5. _____
6. _____
7. _____
8. _____
9. _____
10. _____
11. _____
12. _____
13. _____
14. _____
15. _____
16. _____
17. _____
18. _____
19. _____
20. _____

D. Bones and Joints (Lower Extremities)

Identify the bones and joints of the lower extremities that are described by the following statements:

1. Upper bone of os coxa.
2. Anterior bone of os coxa.
3. Lower posterior bone of os coxa.
4. Heel bone.
5. Bones of toes.
6. Bones of fingers.
7. Bones surrounding the obturator foramen.
8. One half of pelvic girdle.
9. Pelvic girdle, sacrum, and coccyx.
10. Longest, heaviest bone of leg.
11. Strongest bone of lower leg.
12. Largest tarsal bone.
13. Tarsal bone that articulates with tibia.
14. Five elongated bones that form instep of foot.
15. Thin lateral bone of lower leg.
16. Bone adjacent to the sacroiliac joint.

calcaneous—1
femur—2
fibula—3
ilium—4
ischium—5
metatarsals—6
os coxa—7
patella—8
pelvis—9
phalanges—10
pubis—11
talus—12
tibia—13

E. Numbers

Supply the correct numbers for the following statements:

1. Number of phalanges in thumb.
2. Number of phalanges on each of the other four fingers.
3. Number of phalanges in the large toe.
4. Number of phalanges in each of the other toes.
5. Number designation for little finger metacarpal.
6. Number designation for thumb metacarpal.
7. Number designation for the small toe metatarsal.
8. Total number of tarsal bones in each foot.
9. Total number of metatarsals in each foot.
10. Total number of carpals in each wrist.

F. Comparisons

Compare the male and female pelves side-by-side to determine the validity (true or false) of these statements.

1. The greater pelvis (concavity of the expanded iliac bones above the pelvic brim) of the male is narrower than in the female.
2. The male sacrum and coccyx are more curved.
3. The muscular impressions on the female pelvis are less distinct (bone is smoother).
4. The sacrum of the female pelvis is shorter and wider.
5. The obturator foramina of the female pelvis are rounder and larger than in the male.
6. The iliac crests of the female pelvis protrude outward more than in the male.
7. The aperture of the female pelvis is larger and more oval than in the male pelvis.
8. The acetabula of the female face more anteriorly.
9. The aperture of the female pelvis is heart-shaped.
10. The obturator foramina of the female pelvis tend toward being triangular and smaller than in the male.

Answers

Bones and Joints	Fig. 14.3
1.	
2.	
3.	
4.	
5.	
6.	
7.	
8.	
9.	
10.	
11.	
12.	
13.	
14.	
15.	
16.	

Numbers	Fig. 14.4
1.	
2.	
3.	
4.	
5.	
6.	
7.	
8.	
9.	
10.	

Comparisons	
1.	
2.	
3.	
4.	
5.	
6.	
7.	
8.	
9.	
10.	

G. Landmarks (Lower Extremities)

Identify the following processes, fossae, and other structures of bones of the lower extremities:

acetabulum—1
greater trochanter—2
head of femur—3
head of fibula—4
iliac crest—5
intertrochanteric crest—6
intertrochanteric line—7
ischial spine—8
lateral condyle—9

lateral malleolus—10
lesser trochanter —11
medial condyle—12
medial malleolus—13
orbturator foramen—14
sacroiliac—15
symphysis pubis—16
tibial tuberosity—17
tuberosity of ischium—18

1. Joint between sacrum and ilium of os coxa.
2. Joint between pubic bones of ossa coxae.
3. Large prominent process of ischium.
4. Large process lateral to neck of femur.
5. Process just inferior to neck of femur on medial surface.
6. Proximal lateral process of tibia that articulates with femur.
7. Proximal medial process of tibia that articulates with femur.
8. Distal medial process of femur.
9. Distal lateral process of femur.
10. Fossa on os coxa that articulates with femur.
11. Uppermost process on posterior edge of ischium.
12. Uppermost ridge of os coxa.
13. Large opening in os coxa surrounded by pubis and ischium.
14. Part of femur that fits into depression of os coxa.
15. Oblique ridge between trochanters on posterior surface of femur.
16. Oblique ridge between trochanters on anterior surface of femur.
17. Distal lateral process of fibula that forms outer anklebone.
18. Distal medial process of tibia that forms inner prominence of ankle.
19. Proximal anterior process of tibia that protrudes below the kneecap.
20. End of fibula that articulates with upper end of tibia.

H. Joint Characteristics

Select the joints from the list on the right that have the following characteristics. More than one answer often applies.

1. Immovable.
2. Slightly movable.
3. Freely movable in one plane only.
4. Freely movable in two planes.
5. Movement in all directions.
6. Rotational movement only.
7. Fused joint of hyaline cartilage.
8. Contains fibrocartilaginous pad.
9. Condyle end in elliptical cavity.
10. Held together with interosseous ligaments.
11. Bone ends are flat or slightly convex.
12. Each bone end is concave in one direction, convex in another.
13. Bone ends joined by fibrous connective tissue.

ball and socket—1
condyloid—2
gliding—3
hinge—4
pivot—5
saddle—6
sutures—7
symphysis—8
synchondrosis—9
syndesmosis—10

Answers

Landmarks	Fig. 15.1
1. _____	_____
2. _____	_____
3. _____	_____
4. _____	_____
5. _____	_____
6. _____	_____
7. _____	_____
8. _____	_____
9. _____	_____
10. _____	_____
11. _____	
12. _____	Fig. 15.2
13. _____	_____
14. _____	_____
15. _____	_____
16. _____	_____
17. _____	_____
18. _____	_____
19. _____	_____
20. _____	_____

Characteristics	
1. _____	_____
2. _____	
3. _____	Fig. 15.3
4. _____	_____
5. _____	_____
6. _____	_____
7. _____	_____
8. _____	_____
9. _____	_____
10. _____	_____
11. _____	_____
12. _____	_____
13. _____	_____

I. Joint Classification

By examining and manipulating the bones of a skeleton, classify the following joints. Two terms apply to each joint.

1. Elbow joint
2. Knee joint
3. Phalanges to metacarpals
4. Phalanges to metatarsals
5. Hip joint
6. Shoulder joint
7. Intervertebral joint (between vertebral bodies)
8. Metaphysis of long bone of child
9. Interlocking joint between cranial bones
10. Symphysis pubis
11. Articulation between tibia and fibula at distal ends
12. Atlas to skull
13. Toe joints (between phalanges)
14. Finger joint (between phalanges)
15. Wrist (carpals to carpals)
16. Ankle (talus to tibia)
17. Ankle (tarsals to tarsals)
18. Axis-atlas rotation

Basic Types
slightly movable—1
freely movable—2
immovable—3

Subtypes
ball and socket—4
condyloid—5
gliding—6
hinge—7
pivot—8
saddle—9
suture—10
symphysis—11
synchondrosis—12
syndesmosis—13

J. Medical

Consult your lecture text, medical dictionary, or other source for the following:

ankylosis—1
bursitis—2
chronic fibrositis—3
dislocation—4
gouty arthritis—5
osteoarthritis—6
rheumatism—7
rheumatoid arthritis—8
sprain—9
strain—10
subluxation—11
tenosynovitis—12

1. Arthritis due to excess levels of uric acid in blood.
2. Arthritis resulting from degenerative changes in joint (wear and tear).
3. Arthritis of confusing etiology, characterized by deformity and immobility of joints; often treated with cortisone.
4. Inflammation of fibrous connective tissue causing tenderness and stiffness.
5. Inflammation of tendon and its sheath.
6. General term applied to soreness and stiffness in muscles and joints.
7. Inflammation of synovial bursa.
8. Partial or incomplete dislocation.
9. Displacement of bones from normal orientation in a joint.
10. Fixation or fusion of a joint; usually preceded by infection.
11. Injury to joint in which ligaments are stretched without swelling or discoloration.
12. Injury to joint by displacement, resulting in swelling and discoloration.

Answers

Fig. 15.4	Classification
_____	1. _____
_____	2. _____
_____	3. _____
_____	4. _____
_____	5. _____
_____	6. _____
_____	7. _____
_____	8. _____
_____	9. _____
_____	10. _____
_____	11. _____
_____	12. _____
_____	13. _____
_____	14. _____
_____	15. _____
_____	16. _____
_____	17. _____
_____	18. _____
_____	
_____	
_____	

Medical
1. _____
2. _____
3. _____
4. _____
5. _____
6. _____
7. _____
8. _____
9. _____
10. _____
11. _____
12. _____

LABORATORY REPORT
16

Student: _____

Desk No.: _____ Section: _____

Electronic Instrumentation

A. Transducers and Electrodes

1. What determines whether one uses an electrode or a transducer for monitoring a particular biological phenomenon? _____

   _____

   _____

2. Differentiate between:

   Input Transducer _____

   _____

   Output Transducer _____

   _____

3. Indicate what kind of pickup device (*electrode* or *transducer*) you would use for the following. If a transducer is to be used, indicate what specific type of transducer is necessary.

   Sound _____

   Light _____

   Muscle Pull _____

   Temperature _____

   Skin Conductivity _____

   Brain Waves _____

B. Unigraph Functions

Indicate after each of the following the specific control that you would use to produce the desired effect.

1. Turn on the unit. _____

2. Change the speed of the paper. _____

3. Increase the intensity of the tracing. _____

4. Record on the chart the exact time that a new event occurs. _____

5. Increase the sensitivity. _____

6. Balance the transducer. _____

7. Start and stop the movement of the paper. _____

C. Physiograph® Functions

Answer the following questions that pertain to the use of the Physiograph® Mark IIIS:

1. How do you increase or decrease the flow of ink to the pens? _____

   _____

2. Into what unit does one insert the cable from electrodes or transducers? _____

   _____

3. At what sensitivity setting *(low* or *high)* does one usually start an experiment? _____

4. At what speeds can the paper move on a Physiograph®? _____

5. What button is pressed to start the chart moving? _____

6. How is ink removed from the pens on the instrument at the end of the laboratory period? _____

   _____

D. Electronic Stimulator Functions

Indicate how you would set the controls on the Grass and Narco stimulators to get the following outputs:

Grass SD9	*Narco SM–1*

4.5 Volts:

Round Control Knob . . . . . . _____	Voltage Range Control . . . . . . _____
Decade Switch . . . . _____	Variable Control Knob . . . . . . _____

80 Impulses per Second:

Round Control Knob . . . . . . _____	Round Control Knob (Hz) . . . _____
Decade Switch . . . . _____	Rate Switch . . . . _____

Impulse Duration (Width) of 1.5 Milliseconds:

Round Control Knob . . . . . . _____	Round Control Knob . . . . _____
Decade Switch . . . . _____	

E. The Intelitool System

1. List several advantages that Intelitool systems have over conventional chart recorders.

   _____

   _____

   _____

   _____

2. What advantage does a Duograph or polygraph have over Intelitool systems?

   _____

   _____

LABORATORY REPORT
17,18

Student: _____

Desk No.: _____ Section: _____

Muscle Structure (Ex. 17)

A. Figure 17.1
Record the labels for figure 17.1 in the answer column.

B. Muscle Terminology
Match the terms of the right-hand column to the following statements:

1. Bundle of muscle fibers.
2. Sheath of fibrous connective tissue that surrounds each fasciculus.
3. Movable end of a muscle.
4. Immovable end of muscle.
5. Thin layer of connective tissue surrounding each muscle fiber.
6. Layer of connective tissue between the skin and deep fascia.
7. Flat sheet of connective tissue that attaches a muscle to the skeleton or to another muscle.
8. Layer of connective tissue that surrounds a group of fasciculi.
9. Outer connective tissue that covers an entire muscle.
10. Band of fibrous connective tissue that connects muscle to bone or to another muscle.

aponeurosis—1
deep fascia—2
endomysium—3
epimysium—4
fasciculus—5
insertion—6
ligament—7
origin—8
perimysium—9
superficial fascia—10
tendon—11

Body Movements (Ex. 18)

A. Figure 18.1
Record the letters for the various types of movement in this illustration in the answer column.

B. Muscle Types
Determine the types of muscles described by the following statements:

1. Prime movers.
2. Opposing muscles.
3. Muscles that assist prime movers.
4. Muscles that hold structures steady to allow agonists to act smoothly.

agonists—1
antagonists—2
fixation muscles—3
synergists—4

Answers

Terms

1. _____
2. _____
3. _____
4. _____
5. _____
6. _____
7. _____
8. _____
9. _____
10. _____

Types

1. _____
2. _____
3. _____
4. _____

Fig. 17.1

_____
_____
_____
_____
_____
_____
_____
_____
_____
_____
_____
_____
_____
_____

C. Movements

Identify the types of movements described by the following statements:

1. Movement of limb away from median line of body.
2. Movement of limb toward median line of body.
3. Movement around central axis.
4. Turning of sole of foot inward.
5. Turning of sole of foot outward.
6. Flexion of toes.
7. Backward movement of head.
8. Extension of foot at the ankle.
9. Angle between two bones is increased.
10. Angle between two bones is decreased.
11. Movement of hand from palm-down to palm-up position.
12. Conelike rotational movement.
13. Movement of hand from palm-up to palm-down position.

abduction—1
adduction—2
circumduction—3
dorsiflexion—4
eversion—5
extension—6
flexion—7
hyperextension—8
inversion—9
plantar flexion—10
pronation—11
rotation—12
supination—13

Answers	
Movements	Fig. 18.1
1. _____	_____
2. _____	_____
3. _____	_____
4. _____	_____
5. _____	_____
6. _____	_____
7. _____	_____
8. _____	_____
9. _____	_____
10. _____	_____
11. _____	
12. _____	
13. _____	

Student: _____

Desk No.: _____ Section: _____

Nerve and Muscle Tissues

A. Cell Drawings

On a separate sheet of paper, sketch representative cells of the various types of neurons and muscle cells from slides that are available. Staple the drawings to this report when it is handed in for grading.

B. Figure 19.1

Record the label numbers for this illustration in the answer column.

C. Nerve Cell Types

From the list of cells below, select those that are described by the following statements. More than one cell may apply in some cases.

1. Afferent neurons.
2. Efferent neurons.
3. Specialized cells of retina.
4. Pseudounipolar neurons.
5. Multipolar neurons.
6. Ganglion cells.
7. Golgi type II cells.
8. Neurons lacking in axons.
9. Predominant neurons in cerebellar cortex.
10. Linkage cells between Purkinje cells.
11. Linkage cells between pyramidal cells.
12. Linkage cells between sensory and motor neurons.
13. Predominant type of neuron in cerebral cortex.
14. Small cells covering the perikaryon of a sensory neuron.

amacrine cells—1
basket cells—2
interneurons—3
motor neurons—4
neurolemmocytes—5
Purkinje cells—6
pyramidal cells—7
satellite cells—8
sensory neurons—9
stellate cells—10

D. True–False

Indicate whether the following statements concerning nerve cells are true or false.

1. Neuroglial cells do not carry nerve impulses.
2. Perikarya of afferent neurons are located in the spinal ganglia.
3. Most neurons have many dendrites with only a single axon.
4. Motor neuron perikarya are located in the white matter of the spinal cord.
5. Myelinated fibers carry nerve impulses at a slower speed than unmyelinated ones.
6. Unmyelinated fibers lack neurolemmocytes.
7. Part of the axon of a sensory neuron functions like a dendrite.
8. Some neurons lack axons.
9. Both the myelin sheath and neurolemma are formed by Schwann cells.
10. Motor neurons carry nerve impulses away from the CNS.

Answers	
Nerve Cell Types	**Fig. 19.1**
1. _____	_____
2. _____	_____
3. _____	_____
4. _____	_____
5. _____	_____
6. _____	_____
7. _____	_____
8. _____	_____
9. _____	_____
10. _____	_____
11. _____	_____
12. _____	_____
13. _____	_____
14. _____	_____
True–False	
1. _____	_____
2. _____	_____
3. _____	
4. _____	
5. _____	
6. _____	
7. _____	
8. _____	
9. _____	
10. _____	

E. Neuron Structure

From the list of neuron structures, select those that are described by the following statements. More than one structure may apply in some cases.

1. Nerve cell body.
2. Side branch of an axon.
3. Long process on motor neuron.
4. Short process on motor neuron.
5. Lacking in unmyelinated nerve fibers.
6. Process that receives nerve impulses.
7. Slender filaments in perikaryon.
8. Formed by neurolemmocytes.
9. Synonym for neurolemmal sheath.
10. Interruptions in neurolemma.
11. Outer membrane covering myelin sheath.
12. Bodies on the ER that synthesize protein.
13. Promotes regeneration of damaged nerve fiber.
14. Facilitate the speed of nerve impulse transmission.
15. Bodies in perikaryon that stain well with basic aniline dyes.
16. Process that carries nerve impulses away from perikaryon.

axis cylinder—1
axon—2
collateral—3
dendrite—4
myelin sheath—5
neurolemma—6
neurolemmocytes—7
neurofibrils—8
Nissl bodies—9
nodes of Ranvier—10
perikaryon—11

F. Muscle Cell Characteristics

Identify the types of muscle fibers that possess the following characteristics:

1. Attached to bones.
2. Spindle-shaped cells.
3. Nonsyncytial.
4. Syncytial structure.
5. Voluntarily controlled (generally).
6. Lack cross-striae.
7. Long multinucleated cells.
8. In walls of the venae cavae where they enter the right atrium.
9. In walls of other veins and arteries.
10. Involuntarily controlled (generally).
11. Bifurcated cells.
12. Nuclei near surface of cell.

skeletal—1
smooth—2
cardiac—3

G. Muscle Cell Structures

Identify the muscle cell structures that are described by the following statements:

endomysium—1   intercalated disks—3   sarcolemma—5
fascicle—2   perimysium—4   sarcoplasm—6

1. Cell membrane of skeletal muscle fibers.
2. Cell membranes between individual cardiac muscle fibers.
3. Connective tissue between individual skeletal muscle fibers.
4. Cytoplasmic material of skeletal muscle fibers.
5. Bundle of skeletal muscle fibers.

Answers

Structure

1. _____
2. _____
3. _____
4. _____
5. _____
6. _____
7. _____
8. _____
9. _____
10. _____
11. _____
12. _____
13. _____
14. _____
15. _____
16. _____

Characteristics

1. _____
2. _____
3. _____
4. _____
5. _____
6. _____
7. _____
8. _____
9. _____
10. _____
11. _____
12. _____

Structures

1. _____
2. _____
3. _____
4. _____
5. _____

LABORATORY REPORT
20,21

The Neuromuscular Junction (Ex. 20)

A. Figure 20.1

Record the labels for this illustration in the answer column.

B. Summary

Supply the correct terms to complete the following statement concerning the neuromuscular junction:

The unmyelinated ending of a motor neuron consists of several knoblike structures called ___1___ . These knobs lie in depressions of the sarcolemma that are called ___2___ gutters. Between these knobs and the sarcolemma is a ___3___ cleft that is filled with a gelatinous substance. When a nerve impulse reaches the end of the neuron, a neurotransmitter, called ___4___ , bridges the gap to initiate an action potential in the muscle fiber.

Stimulation of the muscle fiber by the neurotransmitter is made possible by the presence of receptors in the ___5___ membrane, which are activated by the neurotransmitter. Although the synaptic vesicles contain large quantities of the neurotransmitter, the actual source of the neurotransmitter that bridges the gap is the ___6___ of the knoblike structures. Only after prolonged stimulation of the neuron do the ___7___ release the neurotransmitter.

Depolarization of the neurolemma results in ___8___ ions flowing into the neuron and ___9___ ions flowing outward. Depolarization in the muscle fiber progresses from the sarcolemma inward to the ___10___ system, and the release of ___11___ ions by the ___12___ .

Within 1/500 of a second after its passage across the gap, the neurotransmitter is inactivated by the enzyme ___13___ . Inactivation of the neurotransmitter is achieved by splitting it into its two components: ___14___ and ___15___ . Resynthesis of the neurotransmitter from these two components is accomplished by the enzyme ___16___ , which is produced by the ___17___ of the neuron and stored in the cytoplasm of the knoblike structures.

The Physicochemical Nature of Muscle Contraction (Ex. 21)

A. Length Changes

Record muscle fiber lengths before and after adding activants.

Original length _____ mm

Length after adding ATP, K^+, and Mg^{++} _____ mm

Answers

Summary

1. _____
2. _____
3. _____
4. _____
5. _____
6. _____
7. _____
8. _____
9. _____
10. _____
11. _____
12. _____
13. _____
14. _____
15. _____
16. _____
17. _____

Fig. 20.1

_____ | _____
_____ | _____
_____ | _____
_____ | _____
_____ | _____
_____ | _____
_____ | _____
_____ | _____

453

B. Drawings

Sketch the appearance of the muscle fibers as seen under high-dry or oil immersion before and during contraction.

Before Contraction   During Contraction

C. Structure

Select the muscle fiber components that are described by the following statements:

actin—1
F-actin protein—2
heavy meromyosin—3
light meromyosin—4
myosin—5
sarcomere—6
tropomyosin—7
troponin—8

1. Thick myofilament with cross-bridges.

2. Principal constituent of A band.

3. Two types of myofilaments that make up a myofibril.

4. Three components of actin.

5. Type of myofilament that makes up I band.

6. Protein of which cross-bridges are made.

7. Component of a myofibril that extends between two Z lines.

D. Physiology

Supply the terms that are missing in the following explanation of muscle contraction:

actin—1
ATP—2
calcium—3
cross-bridges—4

F-actin protein—5
magnesium—6
myosin—7
potassium—8

sarcoplasmic
  reticulum—9
troponin—10
tropomyosin—11

The interaction of actin and myosin during muscle contraction depends on the presence of ample amounts of ATP, calcium, and ___1___ ions. The calcium ions are released by the ___2___ during depolarization. When these ions are released, ___3___ myofilaments are pulled toward each other by the ratcheting action of ___4___ that are present on the ___5___ myofilaments. The function of the calcium ions is to combine with ___6___ , causing the ___7___ molecule to be drawn deeper, which, in turn, exposes active sites on the ___8___ filament. When active sites are exposed, the ___9___ pull the ___10___ filaments together, causing contraction. The energy for this ratcheting effect is derived from ___11___ .

Answers

Structure

1. _____
2. _____
3. _____
4. _____
5. _____
6. _____
7. _____

Physiology

1. _____
2. _____
3. _____
4. _____
5. _____
6. _____
7. _____
8. _____
9. _____
10. _____
11. _____

LABORATORY REPORT

22,24

Student: _____

Desk No.: _____ Section: _____

Muscle Contraction Experiments: Using Chart Recorders (Ex. 22)

A. Visible Twitches (Unrecorded)

Record here the **threshold stimuli** that cause visible twitches without using the recorder. Note that the muscle is stimulated directly and through the nerve.

A visible twitch with probe on **muscle** . . . . _____ volts

_____ msec

Extension of foot with probe on **muscle** . . . . _____ volts

_____ msec

A visible twitch with probe on **nerve** . . . . _____ volts

_____ msec

Extension of foot with probe on **nerve** . . . . _____ volts

_____ msec

Comparisons

Determine how much greater the stimulus requirement is for direct muscle stimulation by dividing one voltage into the other as follows:

$$\frac{\text{Voltage by Direct Muscle Stimulation}}{\text{Voltage by Nerve Stimulation}} =$$

for visible twitch . . . . _____

for foot extension . . . . _____

B. Recorded Twitches

Attach a segment of the chart in the space provided below. Use Scotch tape, glue, or staples.

C. Evaluation of Recorded Twitches

From the chart record above, determine the following:

Latent Period (A) . . . . _____ msec

Contraction Phase (B) . . . . _____ msec

Relaxation Phase (C) . . . . _____ msec

Entire Duration (A + B + C) . . . . _____ msec

D. Summation Results

Attach a segment of the chart that illustrates multiple motor unit summation. Be sure to indicate the voltages that were used for each stimulation.

Multiple Motor Unit Summation   **Wave Summation**

E. Definitions

Define the following terms:

*Motor Unit*_____

*Threshold Stimulus*_____

Recruitment _____

_____

*Tetany*_____

Electromyography (Ex. 24)

A. Recordings

Attach samples of myograms in the space below.

B. Applications

List several applications in which electromyography can be very useful: _____

_____

_____

LABORATORY REPORT

26

Student: _____

Desk No.: _____ Section: _____

Head and Neck Muscles

A. Illustrations

Record the labels for the illustrations of this exercise in the answer column.

B. Facial Expressions

Consult figure 26.1 and the text to identify the muscles described in the following statements:

1. Smiling	frontalis—1
2. Horror	orbicularis oculi—2
3. Irony	orbicularis oris—3
4. Sadness	platysma—4
5. Contempt or disdain	quadratus labii
6. Pouting lips	inferioris—5
7. Blinking or squinting	quadratus labii
8. Horizontal wrinkling of forehead	superioris—6
	triangularis—7

Origins   zygomaticus—8
9. On frontal bone
10. On zygomatic bone

Insertions
11. On lips
12. On eyebrows
13. On eyelid

C. Mastication

Consult figure 26.1 and the text to identify the muscles that perform the following movements in mastication of food:

1. Lowers the mandible.	buccinator—1
2. Raises the mandible.	external pterygoid—2
3. Retracts the mandible.	internal pterygoid—3
4. Protrudes the mandible.	masseter—4
5. Holds food in place.	temporalis—5

Origins
6. On zygomatic arch.
7. On sphenoid only.
8. On frontal, parietal, and temporal bones.
9. On maxilla, mandible, and pterygomandibular raphé.
10. On maxilla, sphenoid, and palatine bones.

Insertions
11. On orbicularis oris.
12. On coronoid process of mandible.
13. On medial surface of body of mandible.
14. On lateral surface of ramus and angle of mandible.
15. On neck of mandibular condyle and articular disk.

Answers

Facial	Fig. 26.1
1. _____	_____
2. _____	_____
3. _____	_____
4. _____	_____
5. _____	_____
6. _____	_____
7. _____	_____
8. _____	_____
9. _____	_____
10. _____	_____
11. _____	_____
12. _____	_____
13. _____	_____

Mastication	
1. _____	_____
2. _____	_____
3. _____	_____
4. _____	Fig. 26.2
5. _____	_____
6. _____	_____
7. _____	_____
8. _____	_____
9. _____	_____
10. _____	_____
11. _____	_____
12. _____	Fig. 26.3
13. _____	_____
14. _____	_____
15. _____	_____
	_____
	_____
	_____
	_____
	_____

D. Neck Muscles

Consult figures 26.2 and 26.3 to identify the muscles of the neck that are described below.

digastricus—1   splenius capitis—7
longissimus capitis—2   sternocleidomastoideus—8
mylohyoideus—3   sternohyoideus—9
omohyoideus—4   sternothyroideus—10
platysma—5   trapezius—11
semispinalis capitis—6

1. Broad sheetlike muscle on side of neck just under the skin.
2. A broad triangular muscle of the back that functions only partially as a neck muscle.
3. Provides support for the floor of the mouth.
4. A V-shaped muscle attached to hyoid and mandibular bones.
5. Posterior neck muscles that insert on the mastoid process.
6. Anterior neck muscle that is attached to the thyroid cartilage and part of the sternum.
7. Anterior neck muscle that extends from the hyoid bone to the clavicle and manubrium.
8. Long muscle of the neck that extends from the mastoid process to the clavicle and sternum.
9. Muscles that pull the head backward (hyperextension).
10. Muscles that pull the head forward.

Answers

Neck Muscles

1. _____
2. _____
3. _____
4. _____
5. _____
6. _____
7. _____
8. _____
9. _____
10. _____

Student: _____

Desk No.: _____ Section: _____

Trunk and Shoulder Muscles

A. Illustrations

Record the labels for the illustrations of this exercise in the answer column.

B. Anterior Trunk Muscles

By referring to figure 27.1 and the text, identify the muscles described by the following statements:

Origins
1. On upper eight or nine ribs.
2. On scapular spine and clavicle.
3. On third, fourth, and fifth ribs.
4. On clavicle, sternum, and costal cartilages.

deltoideus—1
intercostalis externi—2
intercostalis interni—3
pectoralis major—4
pectoralis minor—5
serratus anterior—6
subscapularis—7

Insertions
5. On coracoid process.
6. On deltoid tuberosity of humerus.
7. On anterior surface of scapula.
8. On upper anterior end of humerus.

Actions
9. Pulls scapula forward, downward, and inward.
10. Rotates arm medially.
11. Raises ribs (inhaling).
12. Lowers ribs (exhaling).
13. Adducts arm; also, rotates it medially.
14. Abducts arm.

C. Posterior Trunk Muscles

By referring to figure 27.2 and the text, identify the muscles described by the following statements:

infraspinatus—1   rhomboideus major—4   supraspinatus—7
latissimus dorsi—2   rhomboideus minor—5   teres major—8
levator scapulae—3   sacrospinalis—6   teres minor—9
                                               trapezius—10

Origins
1. Near inferior angle of scapula.
2. On lateral (axillary) margin of scapula.
3. On fossa of scapula above spine.
4. On infraspinous fossa of scapula.
5. On thoracic and lumbar vertebrae, sacrum, lower ribs, and iliac crest.
6. On lower part of ligamentum nuchae and first thoracic vertebra.
7. On transverse processes of first four cervical vertebrae.

Answers	
Anterior	**Fig. 27.1**
1. _____	_____
2. _____	_____
3. _____	_____
4. _____	_____
5. _____	_____
6. _____	_____
7. _____	_____
8. _____	**Fig. 27.2**
9. _____	
10. _____	_____
11. _____	_____
12. _____	_____
13. _____	_____
14. _____	_____
Posterior	_____
1. _____	_____
2. _____	_____
3. _____	_____
4. _____	_____
5. _____	_____
6. _____	_____
7. _____	

C. *Posterior Trunk Muscles (continued)*

infraspinatus—1 rhomboideus major—4 supraspinatus—7
latissimus dorsi—2 rhomboideus minor—5 teres major—8
levator scapulae—3 sacrospinalis—6   teres minor—9
                  trapezius—10

Insertions
 8. On lower third vertebral border of scapula.
 9. On vertebral border of scapula near origin of scapular spine.
10. On crest of lesser tubercle of humerus.
11. On intertubercular groove of humerus.
12. On middle facet of greater tubercle of humerus.
13. On ribs and vertebrae.

Actions
14. Abducts arm.
15. Raises scapula and draws it medially.
16. Extends spine to maintain erectness.
17. Adducts and rotates arm medially.
18. Adducts and rotates arm laterally.

Answers
Posterior
8. _____
9. _____
10. _____
11. _____
12. _____
13. _____
14. _____
15. _____
16. _____
17. _____
18. _____

LABORATORY REPORT

28

Student: _____

Desk No.: _____ Section: _____

Upper Extremity Muscles

A. Illustrations

Record the labels for figures 28.1, 28.2, and 28.3 in the answer column.

B. Arm Movements

Consult figure 28.1 to select the muscles that apply to the following statements. Verify your answers by referring to the text material.

1. Extends the forearm.
2. Adducts the arm.
3. Flexes the forearm.
4. Carries the arm forward in flexion.
5. Flexes forearm and rotates the radius outward to supinate the hand.

coracobrachialis—1
biceps brachii—2
brachialis—3
brachioradialis—4
triceps brachii—5

Origins

6. On lower half of humerus.
7. On coracoid process of scapula.
8. Above the lateral epicondyle of the humerus.
9. Three heads: One on scapula, another on posterior surface of humerus, and the third just below the radial groove of humerus.
10. Two heads: One on coracoid process; other in intertubercular groove of humerus.

Insertions

11. On the olecranon process.
12. On the radial tuberosity.
13. Near the middle, medial surface of humerus.
14. On front surface of coronoid process of ulna.
15. On the lateral surface of the radius just above the styloid process.

C. Hand Movements

Consult figures 28.2 and 28.3 to select the muscles that apply to the following statements. Verify as above.

1. Supinates the hand.
2. Pronates the hand.
3. Flexes and abducts hand.
4. Extends wrist and hand.
5. Flexes the thumb.
6. Extends all fingers except thumb.
7. Abducts the thumb.
8. Flexes all distal phalanges except the thumb.
9. Flexes all fingers except thumb.
10. Extends thumb.

abductor pollicis—1
extensor carpi radialis brevis—2
extensor carpi radialis longus—3
extensor carpi ulnaris—4
extensor digitorum communis—5
extensor pollicis longus—6
flexor carpi radialis—7
flexor carpi ulnaris—8
flexor digitorum profundus—9
flexor digitorum superficialis—10
flexor pollicis longus—11
pronator quadratus—12
pronator teres—13
supinator—14

Answers

Arm Movements	Fig. 28.1
1. _____	_____
2. _____	_____
3. _____	_____
4. _____	_____
5. _____	_____
6. _____	
7. _____	**Fig. 28.2**
8. _____	_____
9. _____	_____
10. _____	_____
11. _____	_____
12. _____	_____
13. _____	_____
14. _____	_____
15. _____	**Fig. 28.3**
	_____
Hand Movements	_____
1. _____	_____
2. _____	_____
3. _____	_____
4. _____	_____
5. _____	_____
6. _____	
7. _____	
8. _____	
9. _____	
10. _____	

C. *Hand Movements (continued)*

abductor pollicis—1
extensor carpi radialis brevis—2
extensor carpi radialis longus—3
extensor carpi ulnaris—4
extensor digitorum communis—5
extensor pollicis longus—6
flexor carpi radialis—7
flexor carpi ulnaris—8
flexor digitorum profundus—9
flexor digitorum superficialis—10
flexor pollicis longus—11
pronator quadratus—12
pronator teres—13
supinator—14

Origins

11. On interosseous membrane between radius and ulna.
12. On radius, ulna, and interosseous membrane.
13. On lateral epicondyle of humerus and part of ulna.
14. On humerus, ulna, and radius.
15. On medial epicondyle of humerus.

Insertions

16. On distal phalanges of second, third, fourth, and fifth fingers.
17. On distal phalanx of thumb.
18. On middle phalanges of second, third, fourth, and fifth fingers.
19. On radial tuberosity and oblique line of radius.
20. On upper lateral surface of radius.
21. On first metacarpal and trapezium.
22. On second metacarpal.
23. On proximal portion of second and third metacarpals.
24. On middle metacarpal.
25. On fifth metacarpal.

Answers
Hand Movements
11. _____
12. _____
13. _____
14. _____
15. _____
16. _____
17. _____
18. _____
19. _____
20. _____
21. _____
22. _____
23. _____
24. _____
25. _____

LABORATORY REPORT
29,30

Student: _____

Desk No.: _____ Section: _____

Abdominal, Pelvic, and Intercostal
Muscles (Ex. 29)

A. Illustrations
Record the labels for figures 29.1 and 29.3 in the answer column.

B. Abdomen
Consult figure 29.1 to identify the muscles that apply to the following statements. Verify your selections from the text.

external oblique—1   rectus abdominis—3
internal oblique—2   transversus abdominis—4

1. Flexion of spine in lumbar region.
2. Maintain intraabdominal pressure.
3. Antagonists of diaphragm.

Origins
4. On pubic bone.
5. On external surface of lower eight ribs.
6. On lateral half of inguinal ligament and anterior two-thirds of iliac crest.
7. On inguinal ligament, iliac crest, and costal cartilages of lower six ribs.

Insertions
8. On linea alba and crest of pubis.
9. On cartilages of fifth, sixth, and seventh ribs.
10. On the linea alba where fibers of the aponeurosis interlace.
11. On costal cartilages of lower three ribs, linea alba, and crest of pubis.

C. Pelvic Region
Consult figure 29.3 to identify the muscles that apply to the following statements. Verify your selections from the text.

iliacus—1
psoas major—2
quadratus lumborum—3

1. Flexes femur on trunk.
2. Flexes lumbar region of vertebral column.
3. Extension of spine at lumbar vertebrae.

Origins
4. On lumbar vertebrae.
5. On iliac crest, iliolumbar ligament, and transverse processes of lower four lumbar vertebrae.
6. On iliac fossa.

Insertions
7. On inferior margin of last rib and transverse processes of upper four lumbar vertebrae.
8. On lesser trochanter of femur.

Answers

Abdomen	Fig. 29.1
1. _____	_____
2. _____	_____
3. _____	_____
4. _____	_____
5. _____	_____
6. _____	_____
7. _____	_____
8. _____	
9. _____	Fig. 29.3
10. _____	_____
11. _____	_____

Pelvic	
1. _____	_____
2. _____	
3. _____	
4. _____	
5. _____	
6. _____	
7. _____	
8. _____	

Lower Extremity Muscles (Ex. 30)

A. Illustrations

Record the labels for this exercise in the answer column.

B. Thigh Movements

Consult figure 30.1 to select the muscles that apply to the following statements. More than one muscle may apply to a statement.

1. Gluteus muscle that extends femur.
2. Gluteus muscles that abduct femur.
3. Three muscles that pull femur toward median line.
4. Gluteus muscle that rotates femur outward.
5. Three muscles attached to linea aspera that flex femur.
6. Small muscle that abducts, extends, and rotates femur.

adductor brevis—1
adductor longus—2
adductor magnus—3
gluteus maximus—4
gluteus medius—5
gluteus minimus—6
piriformis—7

Origins

7. On anterior surface of sacrum.
8. On the pubis.
9. On external surface of ilium.
10. On ilium, sacrum, and coccyx.
11. On inferior surface of ischium and portion of pubis.

Insertions

12. On linea aspera of femur.
13. On anterior border of greater trochanter.
14. On upper border of greater trochanter of femur.
15. On lateral part of greater trochanter of femur.
16. On iliotibial tract and posterior part of femur.

C. Thigh Muscles

Identify the muscles of the thigh that are described by the following statements.

biceps femoris—1   sartorius—4   vastus intermedius—8
gracilis—2   semimembranosus—5   vastus lateralis—9
rectus femoris—3   semitendinosus—6   vastus medialis—10
                      tensor fasciae latae—7

1. The largest quadriceps muscle that is located on the side of the thigh.
2. That portion of the quadriceps that lies beneath the rectus femoris.
3. Portion of the quadriceps that is on the medial surface of the thigh.
4. Hamstring muscle that is on the lateral surface of the thigh.
5. The longest muscle of the thigh.
6. The most medial component of the hamstring muscles.
7. The smallest hamstring muscle.
8. A superficial muscle on the medial surface of the thigh that inserts with the sartorius on the tibia.
9. Four muscles of a group that extend the leg.
10. Flexes the thigh upon the pelvis.
11. Two muscles other than parts of quadriceps that rotate the thigh medially (inward).
12. Rotates the thigh laterally (outward).
13. Adducts the thigh.

Answers		
Thigh Movements	**Fig. 30.1**	**Fig. 30.10 Anterior**
1. _____	_____	_____
2. _____	_____	_____
3. _____	_____	_____
4. _____	_____	_____
5. _____	_____	_____
6. _____	_____	_____
7. _____	_____	_____
8. _____	_____	_____
9. _____	_____	_____
10. _____	**Fig. 30.2**	_____
11. _____	_____	_____
12. _____	_____	_____
13. _____	_____	_____
14. _____	_____	_____
15. _____	_____	_____
16. _____	_____	_____
Thigh Muscles	_____	_____
1. _____	_____	_____
2. _____	_____	**Fig. 30.10 Posterior**
3. _____	_____	
4. _____	**Fig. 30.3**	_____
5. _____	_____	_____
6. _____	_____	_____
7. _____	_____	_____
8. _____	_____	_____
9. _____	_____	_____
10. _____	_____	_____
11. _____	**Fig. 30.4**	_____
12. _____	_____	_____
13. _____	_____	_____
	_____	_____
	_____	_____
	_____	_____
	_____	_____
	_____	_____

D. Lower Leg and Foot Muscles

Identify the muscles of the lower leg and foot that are described by the following statements. In some instances several answers are applicable.

extensor digitorum longus—1
extensor hallucis longus—2
flexor digitorum longus—3
flexor hallucis longus—4
gastrocnemius—5
peroneus brevis—6

peroneus longus—7
peroneus tertius—8
soleus—9
tibialis anterior—10
tibialis posterior—11

1. Located on the front of the lower leg; causes dorsiflexion and inversion of the foot.
2. Deep portion of triceps surae.
3. Superficial portion of triceps surae.
4. Three muscles that originate on the fibula that cause various movements of the foot.
5. Two muscles that join to form the Achilles tendon.
6. Part of the triceps surae that originates on the femur.
7. Inserts on the proximal superior portion of the fifth metatarsal bone of the foot.
8. Inserts on the proximal inferior portion of the fifth metatarsal bone of the foot.
9. Flexes the great toe.
10. On the posterior surfaces of the tibia and fibula; causes plantar flexion and inversion of foot.
11. Originates on the tibia and fascia of the tibialis posterior; causes flexion of second, third, fourth, and fifth toes.
12. Flexes the calf of the thigh.
13. Originates on the tibia, fibula, and interosseous membrane; extends toes (dorsiflexion) and inverts foot.
14. Two muscles that insert on the calcaneus.
15. Peroneus muscle that causes dorsiflexion and eversion of foot.
16. Two peroneus muscles that cause plantar flexion.
17. Muscle on tibia that provides support for foot arches.

E. General Questions

Record the answers to the following questions in the answer column:

1. List the four muscles that are collectively referred to as the quadriceps femoris.
2. List two muscles associated with the iliotibial tract.
3. List three muscles that constitute the hamstrings.
4. What two muscles are associated with the Achilles tendon?
5. What two muscles constitute the triceps surae?

F. Figure 18.1

Refer back to figure 18.1 and identify the muscles that cause each type of movement. Record answers in column.

Answers

Lower Leg and Foot

1. _____   10. _____
2. _____   11. _____
3. _____   12. _____
4. _____   13. _____
5. _____   14. _____
6. _____   15. _____
7. _____   16. _____
8. _____   17. _____
9. _____

General Questions

1a. _____
b. _____
c. _____
d. _____
2a. _____
b. _____
3a. _____
b. _____
c. _____
4a. _____
b. _____
5a. _____
b. _____

Fig. 18.1

A. _____
B. _____
C. _____
D. _____
E. _____
F. _____
G. _____
H. _____
I. _____
J. _____
K. _____
L. _____
M. _____

LABORATORY REPORT

31

Student: _____

Desk No.: _____ Section: _____

The Spinal Cord, Spinal Nerves, and Reflex Arcs

A. Illustrations

Record the labels for figures 31.1, 31.2, and 31.4 in the answer columns.

B. Completion Questions

Supply the information that is necessary to complete the following statements:

1. The dural sac of the spinal cord is continuous with the _____ mater that surrounds the brain.
2. The caudal end of the spinal cord is called the _____ .
3. The caudal end of the spinal cord terminates at the lower border of the _____ lumbar vertebra.
4. The cluster of nerves that extends downward from the end of the spinal cord is called the _____ .
5. The coccygeal nerve emerges from the _____ plexus.
6. The femoral nerve is formed by the union of the following three lumbar nerves: _____, _____, and _____ .
7. The largest nerve emerging from the sacral plexus is the _____ nerve.
8. The iliohypogastric and ilioinguinal nerves are branches of the first _____ nerve.
9. The innermost meninx surrounding the spinal cord is the _____ mater.
10. The meninx that lies just within the dura mater is the _____ .
11. The spinal ganglion is an enlargement of the _____ root.
12. The central canal of the spinal cord is lined with _____ ependymal cells.
13. The cavity within the spinal cord that is continuous with the ventricles of the brain is called the _____ canal.
14. The inner portion of the spinal cord, which forms the pattern of a butterfly when seen in cross section, consists of _____ matter.
15. The space between the dura mater and vertebral bone is called the _____ space.
16. _____ (*vertebral* or *collateral*) ganglia are usually located close to the organ that is innervated.
17. A somatic reflex arc with one or more interneurons is said to be _____ .
18. Cell bodies of afferent neurons of visceral reflexes are located in _____ ganglia.
19. If a somatic reflex lacks an interneuron it is said to be _____ .
20. List the two divisions of the autonomic nervous system.

Answers

Completion

1. _____
2. _____
3. _____
4. _____
5. _____
6. _____
7. _____
8. _____
9. _____
10. _____
11. _____
12. _____
13. _____
14. _____
15. _____
16. _____
17. _____
18. _____
19. _____
20. _____

Fig. 31.1

_____	_____
_____	_____
_____	_____
_____	_____
_____	_____
_____	_____
_____	_____
_____	_____
_____	_____
_____	_____
_____	_____
_____	_____
_____	

C. True–False

Indicate the validity of the following statements with a T or F.

1. The spinal cord during fetal life fills the entire length of the spinal cavity.
2. The lumbar plexus is formed by the union of L_1, L_2, L_3, and most of the L_4 nerves.
3. The brachial plexus is formed by the union of the lower four cervical nerves and the first thoracic nerve.
4. The cervical plexus is formed by the union of the first three pairs of cervical nerves.
5. The filum terminale externa is a downward extension of the filum terminale interna.
6. The innermost fiber of the cauda equina (on the median line) is the filum terminale interna.
7. Cerebrospinal fluid fills the subarachnoid space.
8. The white matter of the spinal cord consists of myelinated fibers of nerve cells.
9. The posterior cutaneous nerve emerges from the lumbar plexus.
10. The sacral plexus is formed by the union of the femoral and sciatic nerves.
11. The effector in a somatic reflex is always skeletal muscle.
12. Most organs of the body are innervated by fibers of the sympathetic and parasympathetic divisions of the autonomic nervous system.
13. Collateral ganglia of the autonomic nervous system are united to form a chain along the vertebral column.
14. All visceral reflex arcs have two afferent neurons.
15. The central canal of the spinal cord is continuous with the ventricles of the brain.
16. The dura mater of the spinal cord extends peripherally to form a covering for spinal nerves.
17. The central canal of the spinal cord is lined with ciliated epithelium.
18. All somatic reflex arcs have at least one interneuron.
19. All visceral reflex arcs have two efferent neurons.
20. Spinal ganglia are located in the anterior roots of spinal nerves.

D. Numbers

Indicate the number of pairs of the following that are normally present:

1. Cervical nerves
2. Thoracic nerves
3. Lumbar nerves
4. Sacral nerves
5. Coccygeal nerves

Answers	
True–False	**Fig. 31.2**
1. _____	_____
2. _____	_____
3. _____	_____
4. _____	_____
5. _____	_____
6. _____	_____
7. _____	_____
8. _____	_____
9. _____	_____
10. _____	_____
11. _____	_____
12. _____	_____
13. _____	_____
14. _____	_____
15. _____	
16. _____	**Fig. 31.4**
17. _____	_____
18. _____	_____
19. _____	_____
20. _____	_____
Numbers	_____
1. _____	
2. _____	
3. _____	
4. _____	
5. _____	

LABORATORY REPORT

32

Student: _____

Desk No.: _____ Section: _____

Somatic Reflexes

A. Functional Nature

Unsuspended Frog: Does the spinal frog attempt to control the position of its hind legs? _____1_____ . Do leg muscles of the spinal frog seem *flaccid* or *firm?* _____2_____. Does tonus exist in these muscles? _____3_____ . Does the animal jump when prodded? _____4_____ . Can the animal swim? _____5_____ . Does the animal float or sink? _____6_____ .

Suspended Frog: Does the frog attempt to remove the acid? _____7_____ . Do you think that the frog experiences a burning sensation? _____8_____ .

Explain: _____

_____

Does the addition of acid to the abdomen or other leg cause a response in the animal? _____9_____ . If you try to restrain the animal's attempt to remove the acid, does it attempt some other maneuver? _____10_____ .

Make a statement as to the apparent functional nature of reflexes. _____

_____

_____

_____

Answers
Functional Nature
1. _____
2. _____
3. _____
4. _____
5. _____
6. _____
7. _____
8. _____
9. _____
10. _____
Inhibition
1. _____
2. _____

B. Reaction Time

Record the reaction time (seconds) for each concentration in the table below. Plot these times in the graph.

HCl (Percentage)	Time (Seconds)
0.05	
0.1	
0.2	
0.3	
0.4	
0.5	
1.0	

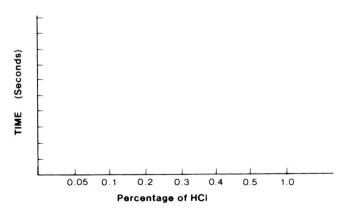

Conclusion: _____

_____

C. Reflex Radiation

Describe the response of the spinal frog to increased electrical stimulation of the foot. _____

_____

_____

_____

_____

D. Reflex Inhibition

Was it possible to inhibit right foot withdrawal from the acid by electrically stimulating the left foot?

_____1_____ . If so, how much voltage was required? _____2_____ . At what pH? _____. At what

acid concentration was reflex action inhibited? _____ At what voltage? _____

E. Synaptic Fatigue

Where do you believe fatigue occurred in this experiment? _____

_____

F. Diagnostic Reflex Tests

Record the responses for the first five reflexes with a 0 for *no response,* a + for *diminished response,* and ++ for *good response.* For Hoffmann's reflex place a check in the proper space. List the nerves in the last column that are affected in abnormal responses.

REFLEX	RESPONSES		Nerves Affected When Response Is Abnormal
	Right Side	Left Side	
1. Biceps			
2. Triceps			
3. Brachioradialis			
4. Patellar			
5. Achilles			
6. Hoffmann's	Absent (Normal)	Present (Abnormal)	
7. Plantar Flexion	Present (Normal)	Babinski's Sign (Abnormal)	

34

Brain Anatomy: External

A. Illustrations

Record the labels for all the illustrations of this exercise in the answer columns.

B. Brain Dissections

Answer the following questions that pertain to your observations of the sheep brain dissections.

1. Does the arachnoid mater appear to be attached to the dura

   mater? _____ to the brain? _____

2. Describe the appearance of the dura mater.

   _____

   _____

   _____

3. Within what structure is the sagittal sinus enclosed?

   _____

4. Differentiate:

   Gyrus: _____

   Sulcus: _____

5. Of what value are the sulci and gyri of the cerebrum?

   _____

   _____

6. List five structures that you were able to identify on the dorsal

   surface of the midbrain: _____

   _____

   _____

7. Is the cerebellum of the sheep brain divided on the median line

   as is the case with the human brain? _____

   _____

8. What significance is there to the size differences of the olfactory
   bulbs between the sheep and human brains?

   _____

   _____

   _____

9. Which cranial nerve is the largest in diameter?

   _____

Answers	
Fig. 34.1	**Fig. 34.4**
_____	_____
_____	_____
_____	_____
_____	_____
_____	_____
_____	_____
_____	_____
_____	_____
_____	_____
_____	_____
Fig. 34.2	_____
_____	_____
_____	_____
_____	_____
_____	_____
_____	_____
_____	_____
_____	**Fig. 34.5**
_____	_____
_____	_____
_____	_____
_____	_____
_____	_____
_____	_____
	_____

C. Meninges

Select the meninx or meningeal space described by the following statements:

1. Fibrous outer meninx.
2. The middle meninx.
3. Meninx that surrounds the sagittal sinus.
4. Meninx that forms the falx cerebri.
5. Meninx attached to the brain surface.
6. Space that contains cerebrospinal fluid.

arachnoid mater—1
dura mater—2
epidural space—3
pia mater—4
subarachnoid space—5
subdural space—6

D. Fissures and Sulci

Select the fissure or sulcus that is described by the following statements:

1. Between the frontal and parietal lobes.
2. On the median line between the two cerebral hemispheres.
3. Between the occipital and parietal lobes.
4. Between the temporal and parietal lobes.

anterior central sulcus—1
central sulcus—2
lateral cerebral fissure—3
longitudinal cerebral fissure—4
parieto-occipital fissure—5

E. Brain Functions

Select the part of the brain that is responsible for the following activities:

1. Maintenance of posture.
2. Cardiac control.
3. Respiratory control.
4. Vasomotor control.
5. Consciousness.
6. Voluntary muscular movements.
7. Brightness and sound discrimination (animals only).
8. Contains fibers that connect the two cerebral hemispheres.
9. Contains nuclei of fifth, sixth, seventh, and eighth cranial nerves.
10. Coordination of complex muscular movements.
11. Function in humans is unknown.

cerebellum—1
cerebrum—2
corpora quadrigemina—3
corpus callosum—4
medulla oblongata—5
midbrain—6
pineal body—7
pons Varolii—8

F. Cerebral Functional Localization

Select those areas of the cerebrum that are described by the following statements:

Location
1. On postcentral gyrus.
2. On parietal lobe.
3. On superior temporal gyrus.
4. On temporal lobe.
5. On frontal lobe.
6. On occipital lobe.
7. On precentral gyrus.
8. On angular gyrus.
9. In area anterior to precentral gyrus.

association area—1
auditory area—2
common integrative area—3
motor speech area—4
olfactory area—5
somatomotor area—6
somatosensory area—7
premotor area—8
visual area—9

Answers

Meninges	Fissures and Sulci
1. _____	1. _____
2. _____	2. _____
3. _____	3. _____
4. _____	4. _____
5. _____	
6. _____	

Brain Functions

Fig. 34.6	
_____	1. _____
_____	2. _____
_____	3. _____
_____	4. _____
_____	5. _____
_____	6. _____
_____	7. _____
_____	8. _____
_____	9. _____
_____	10. _____
_____	11. _____

Localization

_____	1. _____
_____	2. _____
_____	3. _____
_____	4. _____
_____	5. _____
_____	6. _____
_____	7. _____
_____	8. _____
_____	9. _____

F. Cerebral Functional Localization (continued)

Function

10. Speech production.
11. Speech understanding.
12. Cutaneous sensibility.
13. Sight.
14. Sense of smell.
15. Visual interpretation.
16. Voluntary muscular movement.
17. Integration of sensory association areas.
18. Influences motor area function.

association area—1
auditory area—2
common integrative area—3
motor speech area—4
olfactory area—5
somatomotor area—6
somatosensory area—7
premotor area—8
visual area—9

G. Cranial Nerves

Record the number of the cranial nerve that applies to the following statements. More than one nerve may apply to some statements.

1. Sensory nerves.
2. Mixed nerves.
3. Emerges from midbrain.
4. Emerges from pons Varolii.
5. Emerges from medulla.

Structures Innervated

6. Cochlea of ear.
7. Heart.
8. Salivary glands.
9. Abdominal viscera.
10. Retina of eye.
11. Receptors in nasal membranes.
12. Taste buds at back of tongue.
13. Lateral rectus muscle of eye.
14. Taste buds of anterior two-thirds of tongue.
15. Semicircular canals of ear.
16. Thoracic viscera.
17. Three extrinsic eye muscles (superior rectus, medial rectus, inferior oblique) and levator palpebrae.
18. Superior oblique muscle of eye.
19. Pharynx, upper larynx, uvula, and palate.

1. Olfactory (I)
2. Optic (II)
3. Oculomotor (III)
4. Trochlear (IV)
5. Trigeminal (V)
6. Abducens (VI)
7. Facial (VII)
8. Vestibulocochlear (VIII)
9. Glossopharyngeal (IX)
10. Vagus (X)
11. Accessory (XI)
12. Hypoglossal (XII)

H. Trigeminal Nerve

Select the branch of the trigeminal nerve that innervates the following structures:

1. All lower teeth
2. All upper teeth
3. Tongue
4. Lacrimal gland
5. Outer surface of nose
6. Buccal gum tissues of mandible
7. Lower teeth, tongue, muscles of mastication, and gum surfaces

incisive—1
infraorbital—2
inferior alveolar—3
lingual—4
long buccal—5
mandibular—6
maxillary—7
ophthalmic—8

Answers

Localization

10. _____
11. _____
12. _____
13. _____
14. _____
15. _____
16. _____
17. _____
18. _____

Cranial Nerves

1. _____
2. _____
3. _____
4. _____
5. _____
6. _____
7. _____
8. _____
9. _____
10. _____
11. _____
12. _____
13. _____
14. _____
15. _____
16. _____
17. _____
18. _____
19. _____

Trigeminal Nerve

1. _____
2. _____
3. _____
4. _____
5. _____
6. _____
7. _____

I. *General Questions*

Select the correct answers that complete the following statements:

1. Shallow furrows on the surface of the cerebrum are called
   (1) sulci.   (2) fissures.   (3) gyri.

2. Deep furrows on the surface of the cerebrum are called
   (1) sulci.   (2) fissures.   (3) gyri.

3. The infundibulum supports the
   (1) mammillary body.   (2) hypophysis.   (3) hypothalamus.

4. The mammillary bodies are a part of the
   (1) medulla.   (2) midbrain.   (3) hypothalamus.

5. The corpora quadrigemina are located on the
   (1) cerebrum.   (2) medulla.   (3) midbrain.

6. The major ganglion of the trigeminal nerve is the
   (1) gasserian.   (2) sphenopalatine.   (3) ciliary.

7. Convolutions on the surface of the cerebrum are called
   (1) sulci.   (2) fissures.   (3) gyri.

8. The hypophysis is located on the inferior surface of the
   (1) medulla.   (2) midbrain.   (3) hypothalamus.

9. The spinal bulb is the
   (1) medulla oblongata.   (2) pons Varolii.   (3) midbrain.

10. The pineal gland is located on the
    (1) medulla.   (2) pons Varolii.   (3) midbrain.

Answers
General Questions
1. _____
2. _____
3. _____
4. _____
5. _____
6. _____
7. _____
8. _____
9. _____
10. _____

LABORATORY REPORT

35

Student: _____

Desk No.: _____ Section: _____

Brain Anatomy: Internal

A. Illustrations

Record the labels for all the illustrations of this exercise in the answer columns.

B. Location

Identify that part of the brain in which the following structures are located.

cerebellum—1   diencephalon—3   midbrain—5
cerebrum—2   medulla oblongata—4   pons Varolii—6

1. Caudate nuclei	8. Lateral ventricles
2. Cerebral aqueduct	9. Mammillary bodies
3. Cerebral peduncles	10. Putamen
4. Corpus callosum	11. Rhinencephalon
5. Fornix	12. Thalamus
6. Globus pallidus	13. Third ventricle
7. Hypothalamus	

C. Functions

Identify the structures that perform the following functions:

arachnoid granulations—1   infundibulum—9
caudate nuclei—2   intermediate mass—10
cerebral aqueduct—3   interventricular foramen—11
cerebral peduncles—4   lateral aperture—12
choroid plexus—5   lentiform nucleus—13
corpus callosum—6   median aperture—14
fornix—7   rhinencephalon—15
hypothalamus—8   thalamus—16

1. Relay station for all messages to cerebrum.
2. Temperature regulation.
3. Entire olfactory mechanism of cerebrum.
4. Assists in muscular coordination.
5. Fiber tracts of olfactory mechanism.
6. Connects the two halves of the thalamus.
7. Exerts steadying effect on voluntary movements.
8. Allows cerebrospinal fluid to pass from third to fourth ventricle.
9. Consists of ascending and descending tracts.
10. Supporting stalk of hypophysis.
11. Allows cerebrospinal fluid to return to blood.
12. A commissure that unites the cerebral hemispheres.
13. Secretes cerebrospinal fluid into ventricle.
14. Coordinates autonomic nervous system.

D. General Questions

Select the best answer that completes the following statements:

1. The brain stem consists of the
   (1) cerebrum, pons, midbrain, and medulla.
   (2) cerebellum, medulla, and pons.
   (3) pons, medulla, and midbrain.
2. The lateral ventricles are separated by the
   (1) thalamus.   (2) fornix.   (3) septum pellucidum.

Answers

Location	Fig. 35.2
1. _____	_____
2. _____	_____
3. _____	_____
4. _____	_____
5. _____	_____
6. _____	_____
7. _____	_____
8. _____	_____
9. _____	_____
10. _____	_____
11. _____	_____
12. _____	_____
13. _____	_____

Functions	
1. _____	_____
2. _____	_____
3. _____	_____
4. _____	_____
5. _____	_____
6. _____	
7. _____	Fig. 35.3
8. _____	_____
9. _____	_____
10. _____	_____
11. _____	_____
12. _____	_____
13. _____	_____
14. _____	_____

General	
1. _____	_____
2. _____	_____
	_____
	_____
	_____
	_____

3. Pain is perceived in the
   (1) somatomotor area.   (2) thalamus.   (3) pons.
4. The hypothalamus regulates
   (1) the hypophysis, appetite, and wakefulness.
   (2) body temperature, hypophysis, and vision.
   (3) reproductive functions, body temperature, and voluntary movements.
5. Feelings of pleasantness and unpleasantness appear to be associated with the
   (1) somatomotor area.   (2) hypothalamus.   (3) thalamus.
   (4) medulla oblongata.
6. The reticular formation consists of
   (1) gray matter.   (2) white matter.
   (3) an interlacement of white and gray matter.
7. Alert consciousness is partially regulated by
   (1) caudate nuclei.   (2) lentiform nuclei.
   (3) nuclei of the reticular formation.
8. The intermediate mass passes through the
   (1) midbrain.   (2) third ventricle.   (3) hypothalamus.
9. The reticular formation is seen in the
   (1) spinal cord.   (2) cerebrum.   (3) brain stem.
   (4) spinal cord, brain stem, and diencephalon.
10. Centers for vomiting, coughing, swallowing, and sneezing are located in the
    (1) pons.   (2) medulla.   (3) midbrain.   (4) thalamus.

E. Cerebrospinal Fluid

You should be able to trace the path of the cerebrospinal fluid from its point of origin to where it is reabsorbed into the blood. If you can complete the following paragraph without referring to the text, you know the sequence fairly well. If you can state the sequence from memory, better yet.

Ventricles
lateral—1
third—2
fourth—3
Passageways
acoustic meatus—4
aqueduct of Sylvius—5
foramen magnum—6
intermediate foramen—7
lateral aperture—8
median aperture—9

Structures
arachnoid granulations—10
cerebellum—11
cerebrum—12
choroid plexus—13
cisterna cerebellomedullaris—14
cisterna superior—15
dura mater—16
sagittal sinus—17
septum pellucidum—18
subarachnoid space—19

Cerebrospinal fluid is secreted into each ventricle by a ___1___ . From the ___2___ ventricles, which are located in the cerebral hemispheres, the fluid passes to the ___3___ ventricle through an opening called the ___4___ . A canal, called the ___5___ , allows the fluid to pass from the latter ventricle to the ___6___ ventricle. From this last ventricle the cerebrospinal fluid passes into a subarachnoid space called the ___7___ through three foramina: one ___8___ and two ___9___ . From this cavity the fluid passes over the cerebellum into another subarachnoid space called the ___10___ . It also passes ___11___ *(up, down)* the posterior side of the spinal cord and ___12___ *(up, down)* the anterior side of the spinal cord. From the cisterna superior the cerebrospinal fluid passes to the subarachnoid space around the ___13___ . This fluid is reabsorbed back into the blood through delicate structures called the ___14___ . The blood vessel that receives the cerebrospinal fluid is the ___15___ .

Answers	
General	**Fig. 35.4**
3._____	_____
4._____	_____
5._____	_____
6._____	_____
7._____	_____
8._____	_____
9._____	_____
10._____	_____
Cerebrospinal Fluid	_____
1._____	_____
2._____	_____
3._____	_____
4._____	_____
5._____	
6._____	
7._____	
8._____	
9._____	
10._____	
11._____	
12._____	
13._____	
14._____	
15._____	

LABORATORY REPORT

36,37

Student: _____

Desk No.: _____ Section: _____

Anatomy of the Eye (Ex. 36)

A. Illustrations

Record the labels from figures 36.1, 36.2, and 36.3 in the answer columns.

B. Structures

Identify the structures described by the following statements:

1. Small nonphotosensitive area on retina.
2. Small pit in retina of eye.
3. Outer layer of wall of eye.
4. Fluid between lens and retina.
5. Fluid between lens and cornea.
6. Inner light-sensitive layer.
7. Round yellow spot on retina.
8. Delicate membrane that lines eyelids.
9. Middle vascular layer of wall of eyeball.
10. Drainage tubes for tears in eyelids.
11. Clear transparent portion of front of eyeball.
12. Tube that drains tears from lacrimal sac.
13. Chamber between iris and cornea.
14. Conical body in medial corner of the eye.
15. Chamber between the iris and lens.
16. Circular color band between lens and cornea.
17. Circular band of smooth muscle tissue surrounding lens.
18. Cartilaginous loop through which superior oblique muscle acts.
19. Connective tissue between lens perimeter and surrounding muscle.
20. A semicircular fold of conjunctiva in the medial canthus of the eye.

anterior chamber—1
aqueous humor—2
blind spot—3
caruncula—4
choroid coat—5
ciliary body—6
cornea—7
conjunctiva—8
fovea centralis—9
iris—10
lacrimal ducts—11
lacrimal sac—12
macula lutea—13
medial canthus—14
nasolacrimal duct—15
plica semilunaris—16
posterior chamber—17
pupil—18
retina—19
scleroid coat—20
suspensory ligament—21
trochlea—22
vitreous body—23

C. Extrinsic Muscles

Select the muscles of the eye that are described by the following statements:

1. Inserted on top of eyeball.
2. Inserted on side of eyeball.
3. Inserted on medial surface.
4. Inserted on bottom of eyeball.
5. Inserted on eyelid.
6. Raises the eyelid.
7. Rotates eyeball inward.
8. Rotates eyeball downward.
9. Rotates eyeball outward.
10. Rotates eyeball upward.

inferior oblique—1
inferior rectus—2
lateral rectus—3
medial rectus—4
superior levator palpebrae—5
superior oblique—6
superior rectus—7

Answers

Structures	Fig. 36.1
1. _____	_____
2. _____	_____
3. _____	_____
4. _____	_____
5. _____	_____
6. _____	_____
7. _____	_____
8. _____	_____
9. _____	_____
10. _____	_____
11. _____	_____
12. _____	Fig. 36.2
13. _____	_____
14. _____	_____
15. _____	_____
16. _____	_____
17. _____	_____
18. _____	_____
19. _____	_____
20. _____	_____

Muscles	
1. _____	_____
2. _____	_____
3. _____	_____
4. _____	_____
5. _____	_____
6. _____	_____
7. _____	_____
8. _____	_____
9. _____	_____
10. _____	_____
	_____
	_____

D. Functions

Select the part of the eye that performs the following functions. More than one answer may apply in some cases.

aqueous humor—1
blind spot—2
choroid coat—3
ciliary body—4
conjunctiva—5
iris—6
lacrimal ducts—7
lacrimal puncta—8
lacrimal sac—9
macula lutea—10

nasolacrimal duct—11
pupil—12
retina—13
scleral venous sinus—14
scleroid coat—15
suspensory ligament—16
trabeculae—17
trochlea—18
vitreous body—19

1. Place where nerve fibers of retina leave eyeball.
2. Furnishes blood supply to retina and sclera.
3. Large vessel in wall of the eye that collects aqueous humor from trabeculae.
4. Exerts force on the lens, changing its contour.
5. Controls the amount of light that enters the eye.
6. Maintains firmness and roundness of eyeball.
7. Part of retina where critical vision occurs.
8. Provides most of the strength to the wall of the eyeball.

Answers

Functions	Fig. 36.3
1. _____	_____
2. _____	_____
3. _____	_____
4. _____	_____
5. _____	_____
6. _____	
7. _____	
8. _____	

E. Beef Eye Dissection

Answer the following questions that pertain to the dissection of the beef eye:

1. What is the shape of the pupil?_____

2. Why do you suppose it is so difficult to penetrate the sclera with a sharp scalpel? _____

3. What is the function of the black pigment in the eye? _____

4. Compare the consistency of the two fluids in the eye.

   aqueous humor: _____

   vitreous humor: _____

5. When you hold the lens up and look through it, what is unusual about the image? _____

6. Does the lens magnify printed matter when placed directly on it? _____

7. Compare the consistencies of the following portions of the lens:

   center: _____

   edge: _____

8. What is the reflective portion of the choroid coat called? _____

9. Is there a macula lutea on the retina of the beef eye? _____

F. Ophthalmoscopy

Record your ophthalmoscope measurements here, and answer the questions.

1. **Diopter Measurements**

   The diopter value of a lens is the reciprocal of its focal length (f) in meters, or $D = \dfrac{1}{f}$.

   A lens of one diopter (1D) has a focal length of 1 meter ($D = \dfrac{1}{f} = \dfrac{1}{1} = 1$); a 2D lens has a focal length of 0.5 meter, or $\dfrac{1}{f} = \dfrac{1}{0.5} = 2$; etc.

   Record here your measurements for: 5D_____; 10D_____; 20D_____; and 40D_____.

   Calculated diopter distances: 5D_____; 10D_____; 20D_____; and 40D_____.

   Do your calculations match the measured distances? _____

2. If you were able to examine your laboratory partner's eye with "0" in the diopter window of the ophthalmoscope, what would this indicate about the curvature of the lens of your eye? _____ about your laboratory partner's eye? _____

Visual Tests (Ex. 37)

A. The Purkinje Tree

Describe in a few sentences the image you observed. _____

_____

_____

_____

_____

_____

B. Tabulations

Record the results that were obtained in the following five eye tests:

TEST	RIGHT EYE	LEFT EYE
Blind Spot (inches)		
Near Point (inches)		
Scheiner's Experiment (inches)		
Visual Acuity (X/20)		
Astigmatism (present or absent)		

Visual Tests

C. Color Blindness Test

Have your laboratory partner record your responses to each test plate. The mark *x* indicates inability to respond correctly.

PLATE NUMBER	SUBJECT'S RESPONSE	NORMAL RESPONSE	RESPONSE IF RED-GREEN DEFICIENCY				RESPONSE IF TOTALLY COLOR-BLIND
1		12	12				12
2		8	3				x
3		5	2				x
4		29	70				x
5		74	21				x
6		7	x				x
7		45	x				x
8		2	x				x
9		x	2				x
10		16	x				x
11		traceable	x				x
			Protan		Deuteran		
			Strong	Mild	Strong	Mild	
12		35	5	(3)5	3	3(5)	
13		96	6	(9)6	9	9(6)	
14	can trace two lines	can trace two lines	purple	purple (red)	red	red purple	x

Conclusion: _____

D. Pupillary Reflexes

Answer the following questions related to your observations in the two experiments on pupillary reflexes:

1. **Light Intensity and Pupil Size**
   a. What pupillary size change occurred in this experiment? _____

   b. Trace the pathways of the nerve impulses in effecting the response. _____

   c. Can you suggest a possible benefit that might result from this phenomenon? _____

2. **Accommodation to Distance**
   a. Did the pupil of the unexposed eye become smaller when the right eye was exposed to light?

   b. Can you suggest a benefit that might result from this happening? _____

480

LABORATORY REPORT

38

Student: _____

Desk No.: _____ Section: _____

The Ear:
Its Role in Hearing

A. Figure 38.2
Record the labels for this figure in the answer column.

B. Structure
Identify the structures described by the following statements:

basilar membrane—1   malleus—6   scala tympani—11
endolymph—2   perilymph—7   scala vestibuli—12
hair cells—3   reticular lamina—8 stapes—13
helicotrema—4   rods of Corti—9   tectorial membrane—14
incus—5   scala media—10   tympanic membrane—15
                                        vestibular membrane—16

1. Ossicle that fits in oval window.
2. Ossicle activated by tympanic membrane.
3. Chamber of cochlea into which round window opens.
4. Chamber of cochlea into which oval window opens.
5. Membrane set in vibration by sound waves in the air.
6. Membrane that contains 20,000 fibers of varying lengths.
7. Membrane in which tips of hair cells are embedded.
8. Two fluids found in cochlea.
9. Fluid within scala tympani.
10. Fluid within cochlear duct.
11. Fluid within scala vestibuli.
12. Parts of spiral organ.
13. Transfers vibrations from basilar membrane to reticular lamina.
14. Receptors that initiate action potentials in cochlear nerve.
15. Opening between scala vestibuli and scala tympani.

C. Physiology of Hearing (True–False)
Indicate the validity of each of the following statements:
1. The loudness of sound is directly proportional to the square of the amplitude.
2. The pitch of a sound is determined by the amplitude of the sound wave.
3. Quality and pitch are two different characteristics of sound.
4. Some hearing takes place through the bones of the skull rather than through the auditory ossicles.
5. While there is no cure for nerve deafness, conduction deafness can often be corrected.
6. Endolymph in the scala media flows into the scala vestibuli through the helicotrema.
7. Nerve deafness can be detected by placing a tuning fork on the mastoid process.
8. The frequency range used in speech is from 30 to 20,000 cps.
9. The bending of hairs of hair cells causes an alternating electrical charge known as the endocochlear potential.
10. Before using an audiometer, one must calibrate it.

Answers

Structure	Fig. 38.2
1. _____	_____
2. _____	_____
3. _____	_____
4. _____	_____
5. _____	_____
6. _____	_____
7. _____	_____
8. _____	_____
9. _____	_____
10. _____	_____
11. _____	_____
12. _____	_____
13. _____	_____
14. _____	_____
15. _____	_____

Physiology	
1. _____	_____
2. _____	_____
3. _____	_____
4. _____	_____
5. _____	_____
6. _____	_____
7. _____	_____
8. _____	_____
9. _____	_____
10. _____	

D. Screening Test

Record the distances at which you could hear the watch tick. Record the same data for three other persons.

Subject	INCHES FROM EAR			
	Right Ear		Left Ear	
	Approaching	Receding	Approaching	Receding

E. Tuning Fork Methods

Record the results of the two tuning fork methods.

Rinne Test

Right Ear _____

Left Ear _____

Weber Test

Right Ear _____

Left Ear _____

F. Audiometry

Record the threshold levels in the right ear with a red "O" and in the left ear with a blue "X." Connect the points with appropriately colored lines.

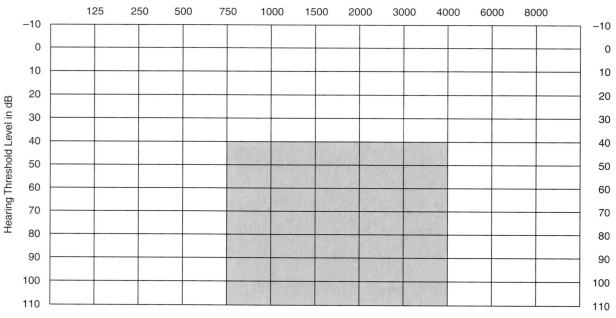

Frequency (Cycles per Second)

(Shaded area indicates critical area of speech interpretation)

LABORATORY REPORT
39

Student: _____

Desk No.: _____ Section: _____

The Ear:
Its Role in Equilibrium

A. Anatomy

Answer the following questions that pertain to the anatomy of the vestibular apparatus:

1. What are the sensory receptors for static equilibrium called?
2. Where are the static equilibrium sensory receptors located (two places)?
3. What are the small calcium carbonate crystals of the macula called?
4. Where are the sensory receptors of dynamic equilibrium located?
5. What nerve branch of the eighth cranial nerve supplies the vestibular apparatus?
6. What part of the vestibular apparatus connects directly to the cochlear duct?
7. What is the name of the fluid within the vestibular apparatus?
8. What structures in joints prevent a sense of malequilibrium when one tilts the head to one side?
9. What fluid lies between the semicircular ducts and the surrounding bone of the semicircular canals?
10. Within how many planes do the semicircular ducts lie?

Answers

Anatomy

1. _____
2. _____
   _____
3. _____
4. _____
5. _____
6. _____
7. _____
8. _____
9. _____
10. _____

B. Nystagmus

After having observed nystagmus, answer these questions:

1. What direction do the eyes move in relation to the direction of rotation? _____

2. What seems to be the function of slow movement of the eye? _____

   _____

   _____

3. What seems to be the function of fast movement of the eye? _____

   _____

   _____

4. What part of the vestibular apparatus triggers the reflexes that control the eye muscles? _____

   _____

C. *Proprioceptive Influences*

Record your observations here.

1. At-Rest Reactions

a. Was the subject able to place the heel of one foot on the toes of the other foot with eyes closed?

_____

b. Was the subject able to touch nose with eyes closed? _____

c. Explain what mechanisms in the body make this possible:_____

_____

d. Describe the subject's ability to touch the pencil eraser with eyes closed after practicing with eyes open. _____

_____

2. Effects of Rotation

a. Was the subject, with eyes open, able to point directly to the pencil eraser with finger after rotation? _____

_____

b. When blindfolded subject tried to locate pencil eraser after rotation. What was the result?

_____

_____

LABORATORY REPORT

40-45

Student: _____

Desk No.: _____ Section: _____

Hematological Tests

Test Results

Except for blood typing, record all test results in table 2 on page 488. Calculations for blood cell counts should be performed as stated below.

A. Differential White Blood Cell Count

As you move the slide in the pattern indicated in figure 40.4, record all the different types of cells in table 1. Refer to figures 40.1 and 40.2 for cell identification. Use this method of tabulation: ᚶᚶᚶ ᚶᚶᚶ 1 1. Identify and tabulate 100 leukocytes. Divide the total of each kind of cell by 100 to determine percentages.

Table I Leukocyte tabulation.

NEUTROPHILS	LYMPHOCYTES	MONOCYTES	EOSINOPHILS	BASOPHILS
Totals				
Percent				

B. Calculations for Blood Cell Counts

1. **Total White Blood Cell Count**
   Total WBCs counted in 4 W areas $\times$ 50 = **WBCs per cu mm**

2. **Total Red Blood Cell Count**
   Total RBCs counted in 5 R areas $\times$ 10,000 = **RBCs per cu mm**

C. Blood Typing

Record your blood type here: _____

1. If you needed a blood transfusion, what types of blood could you be given?

   _____

2. If your blood were to be given to someone in need, what type should that person have?

   _____

Questions

Although most of the answers to the following questions can be derived from this manual, it will be necessary to consult your lecture text or medical dictionary for some of the answers.

A. Blood Components

Identify the blood components described by the following statements. More than one answer may apply.

antibodies—1	fibrin—5	neutrophils—10
basophils—2	fibrinogen—6	platelet factor—11
eosinophils—3	lymphocytes—7	platelets—12
erythrocytes—4	lysozyme—8	prothrombin—13
	monocytes—9	thromboplastin—14

1. Cells that transport O_2 and CO_2.
2. Phagocytic cell concerned, primarily, with local infections.
3. Phagocytic cell concerned, primarily, with generalized infections.
4. Protein substances in blood that inactivate foreign protein.
5. Noncellular formed elements essential for blood clotting.
6. Substances necessary for blood clotting.
7. Gelatinous material formed in blood clotting.
8. Substance released by injured cells that initiates blood clotting.
9. Enzyme in blood plasma that destroys some kinds of bacteria.
10. Leukocyte distinguished by having distinctive red-stained granules.
11. Leukocyte that probably produces heparin.
12. Most numerous type of leukocyte.
13. Contains hemoglobin.
14. Most numerous type of blood cell.

B. Blood Diseases

Identify the pathological conditions characterized by the following statements. More than one condition may apply to some statements.

anemia—1	leukemia—5	neutropenia—9
eosinophilia—2	leukocytosis—6	neutrophilia—10
erythroblastosis fetalis—3	leukopenia—7	pernicious anemia—11
hemophilia—4	lymphocytosis—8	polycythemia—12
		sickle-cell anemia—13

1. Too few red blood cells.
2. Too many red blood cells.
3. Too few leukocytes.
4. Too many neutrophils.
5. Too many lymphocytes.
6. Too little hemoglobin.
7. Defective RBCs due to stomach enzyme deficiency.
8. Rh factor incompatibility present at birth.
9. Hereditary blood disease of blacks; RBCs of abnormal shape.
10. Heritable bleeder's disease.
11. Cancerlike condition in which there are too many leukocytes.

Answers

Components

1. _____
2. _____
3. _____
4. _____
5. _____
6. _____
7. _____
8. _____
9. _____
10. _____
11. _____
12. _____
13. _____
14. _____

Diseases

1. _____
2. _____
3. _____
4. _____
5. _____
6. _____
7. _____
8. _____
9. _____
10. _____
11. _____

C. Terminology

Differentiate between the following pairs of related terms:

Plasma _____

Lymph _____

Coagulation _____

Agglutination _____

Antibody _____

Antigen _____

D. True–False

Indicate in the answer column whether the following statements are true or false.

1. The most common types of blood are types O and A.
2. A universal donor has O-Rh negative blood.
3. Diapedesis is the ability of leukocytes to move between the cells of the capillary walls.
4. Anemia is usually due to an iodine deficiency.
5. Hemoglobin combines with both oxygen and carbon dioxide.
6. The maximum life span of an erythrocyte is about 30 days.
7. White blood cells live much longer than red blood cells.
8. The vitamin that is essential to blood clotting is vitamin D.
9. An infant born with erythroblastosis would best be treated with a transfusion of Rh negative instead of Rh positive blood.
10. Worm infestations may cause eosinophilia.
11. Viral infections may cause an increase in the number of lymphocytes or monocytes.
12. Blood clotting can be enhanced with dicumarol.
13. Megakaryocytes of bone marrow produce blood platelets.
14. Bloods that are matched for ABO and Rh factor can be mixed with complete assurance of no incompatibility.
15. The rarest type of blood is AB negative.
16. Heparin is essential to fibrin formation.
17. Blood typing may be used to prove that an individual could not be the father in a paternity suit.
18. Blood typing may be used to prove that an individual is the father in a paternity suit.
19. Approximately 87% of the population is Rh negative.
20. A deficiency of calcium will result in the failure of blood to clot.
21. Fibrin is formed when prothrombin reacts with fibrinogen.

Answers

True–False

1. _____
2. _____
3. _____
4. _____
5. _____
6. _____
7. _____
8. _____
9. _____
10. _____
11. _____
12. _____
13. _____
14. _____
15. _____
16. _____
17. _____
18. _____
19. _____
20. _____
21. _____

E. Summarization of Results

Record blood cell counts and other test results in the following table:

Table 2

TEST	NORMAL VALUES	TEST RESULTS	EVALUATION (over, under, normal)
Differential WBC Count	Neutrophils: 50%–70%		
	Lymphocytes: 20%–30%		
	Monocytes: 2%–6%		
	Eosinophils: 1%–5%		
	Basophils: 0.5%–1%		
Total WBC Count	5000–9000 per cu mm		
RBC Count	Males: 4.8–6.0 million cu mm		
	Females: 4.1–5.1 million cu mm		
Hemoglobin Percentage	Males: 13.4–16.4 gm/100 ml		
	Females: 12.2–15.2 gm/100 ml		
Hematocrit (VPRC)	Males: 40%–54% (Av. 47%)		
	Females: 37%–47% (Av. 42%)		
Coagulation Time	2–6 minutes		

F. Materials

Identify the various blood tests in which the following supplies are used:

1. Hemacytometer
2. Hemoglobinometer
3. Centrifuge
4. Capillary tubes
5. Wright's stain
6. Microscope slides
7. Hemolysis applicators
8. Seal-Ease
9. Diluting fluid
10. Cover glass

hematocrit (VPRC)—1
clotting time—2
RBC and WBC counts—3
differential WBC count—4
hemoglobin determination—5

Materials

1. _____
2. _____
3. _____
4. _____
5. _____
6. _____
7. _____
8. _____
9. _____
10. _____

Student: _____

Desk No.: _____ Section: _____

Anatomy of the Heart

A. Illustrations

Record the labels for figures 46.1 and 46.2 in the answer columns.

B. Structures

Identify the structures of the heart described by the following statements:

aorta—1	myocardium—13
aortic semilunar valve—2	papillary muscles—14
bicuspid valve—3	parietal pericardium—15
chordae tendineae—4	pulmonary trunk—16
endocardium—5	pulmonic semilunar valve—17
epicardium—6	right atrium—18
inferior vena cava—7	right pulmonary artery—19
left atrium—8	right ventricle—20
left pulmonary artery—9	superior vena cava—21
left ventricle—10	tricuspid valve—22
ligamentum arteriosum—11	ventricular septum—23
mitral valve—12	visceral pericardium—24

1. Lining of the heart.
2. Partition between right and left ventricles.
3. Fibroserous saclike structure surrounding the heart.
4. Two chambers of the heart that contain deoxygenated blood.
5. Large artery that carries blood from the right ventricle.
6. Blood vessel that returns blood to heart from head and arms.
7. Remnant of a functional prenatal vessel between the pulmonary trunk and the aorta.
8. Structure formed from ductus arteriosus.
9. Valve at the base of the pulmonary trunk.
10. Two arteries that are branches of the pulmonary trunk.
11. Synonym for bicuspid valve.
12. Structures on the cardiac wall to which the chordae tendineae are attached.
13. Muscular portion of cardiac wall.
14. Thin covering on surface of heart (2 names).
15. Large vein that empties blood into top of right atrium.
16. Two chambers of the heart that contain oxygenated blood.
17. Chamber of the heart that receives blood from the lungs.
18. Large artery that carries blood out from left ventricle.
19. Large vein that empties blood into the lower part of the right atrium.
20. Atrioventricular valves.
21. Valve at the base of the aorta.
22. Blood vessel in which openings to the coronary arteries are located.
23. Atrium into which the coronary sinus empties.
24. Valvular restraints that prevent the atrioventricular valve cusps from being forced back into the atria.

Answers

Structures	Fig. 46.1
1. _____	_____
2. _____	_____
3. _____	_____
4. _____	_____
5. _____	_____
6. _____	_____
7. _____	_____
8. _____	_____
9. _____	_____
10. _____	_____
11. _____	
12. _____	
13. _____	
14. _____	
15. _____	
16. _____	
17. _____	
18. _____	
19. _____	
20. _____	
21. _____	
22. _____	
23. _____	
24. _____	

C. Coronary Circulation

Identify the arteries and veins of the coronary circulatory system that are described by the following statements:

anterior interventricular
   artery—1
circumflex artery—2
coronary sinus—3
great cardiac vein—4
left coronary artery—5

middle cardiac vein—6
posterior descending right
   coronary artery—7
posterior interventricular vein—8
right coronary artery—9
small cardiac vein—10

1. Two principal branches of the left coronary artery.
2. Coronary vein that lies in the anterior interventricular sulcus.
3. Coronary artery that lies in the right atrioventricular sulcus.
4. Vein that receives blood from the great cardiac vein.
5. Vein that lies in the posterior interventricular sulcus.
6. Two major coronary arteries that originate in the wall of the aorta.
7. Large coronary vessel that empties into the right atrium.
8. Coronary vein that parallels the right coronary artery in the right atrioventricular sulcus.
9. Branch of the right coronary artery on the posterior surface of the heart.
10. Vein that lies alongside of the posterior descending right coronary artery.

D. Sheep Heart Dissection

After completing the sheep heart dissection, answer the following questions. Some of these questions were encountered during the dissection; others pertain to structures not shown in figures 46.1 and 46.2.

1. Identify the following structures:

   Pectinate muscle: _____

   _____

   Moderator band: _____

   _____

   Ligamentum arteriosum: _____

   _____

2. How many papillary muscles did you find in the right

   ventricle? _____ left ventricle? _____

3. How many pouches are present in each of the following?

   pulmonary semilunar valve: _____

   aortic semilunar valve: _____

4. Where does blood enter the myocardium?

   _____

5. Where does blood leave the myocardium and return to the circulatory system?

Answers	
Coronary Circulation	Fig. 46.2
1. _____	_____
2. _____	_____
3. _____	_____
4. _____	_____
5. _____	_____
6. _____	_____
7. _____	_____
8. _____	_____
9. _____	_____
10. _____	_____
	_____
	_____
	_____
	_____
	_____
	_____
	_____
	_____
	_____
	_____
	_____
	_____
	_____
	_____
	_____
	_____

47,48

Cardiovascular Sounds (Ex. 47)

A. Phonocardiogram

Attach below a segment of a phonocardiogram that you made on the recorder. Identify on the chart those segments that represent first and second sounds.

B. Questions

1. What causes the first sound? _____
   _____

2. What causes the second sound? _____
   _____

3. What causes the third sound? _____
   _____

4. What causes a murmur? _____
   _____

5. Why does the aortic valve close sooner than the pulmonic valve during inspiration? _____
   _____
   _____

6. What effect would complete right bundle branch block have on the splitting of the second sound?
   _____
   _____
   _____

7. Explain why pulmonary stenosis is associated with greater splitting of the second heart sound.
   _____
   _____
   _____

Electrocardiogram Monitoring:
Using Chart Recorders (Ex. 48)

A. Tracings

Attach three tracings of your experiment in the spaces provided below. It will be necessary to trim excess paper from the tracings to fit them into the allotted spaces. Attach with Scotch tape.

TRACINGS	EVALUATION
Calibration	
	*Heart Rate: _____ per min QR Potential: _____ millivolts **Duration of cycle: _____ msec
Subject at Rest	
	*Heart Rate: _____ per min QR Potential: _____ millivolts **Duration of cycle: _____ msec
After Exercise	

*Space between two margin lines is 3 seconds.
**Time from beginning of P wave to end of T wave.

B. Questions

1. During what part of the ECG wave pattern do atrial depolarization and contraction occur? _____

   _____

2. During what part of the ECG wave pattern do ventricular depolarization and contraction occur?

   _____

3. Would a heart murmur necessarily show up on an ECG? _____

   Explain. _____

   _____

LABORATORY REPORT

50

Student: _____

Desk No.: _____ Section: _____

Pulse Monitoring

A. *Tracings*

Attach samples of tracings made in this experiment in the spaces provided below and on the next page. Use the right-hand column for relevant comments.

TRACINGS	EVALUATION
Transducer at Heart Level	Pulse Rate: _____
Arm Slowly Raised above Heart Level	Pulse Rate: _____
Arm Lowered to Floor	Pulse Rate: _____
Holding Breath for 15 Seconds	Pulse Rate: _____

TRACINGS	EVALUATION
	Pulse Rate: _____
Brachial Artery Occluded for 15–20 Seconds	
	Pulse Rate: _____
Chart Speed Increased to 25 mm per Second	
	Pulse Rate: _____
Subject Startled	

B. Pulse Tabulations

Record your pulse (while at rest) on the chalkboard. Once the pulse rates of all students are recorded, record the highest, lowest, and median pulses:

Highest _____ Lowest _____ Median _____

Significance _____

_____

C. Questions

1. What conceivably might be an explanation for changes occurring in the dicrotic notch with aging?

   _____

   _____

2. What might be an explanation for an alteration in the pulse amplitude during smoking? _____

   _____

3. The finger pulse is a manifestation of blood volume, velocity, direction, and pressure. Which one of these factors is directly responsible for the dicrotic notch and why? _____

   _____

Laboratory Report 51,52

Student: ______________________

Desk No.: __________ Section: __________

Blood Pressure Monitoring (Ex. 51)

A. Stethoscope Method

Record here the blood pressures of the test subject.

	BEFORE EXERCISE	AFTER EXERCISE
Systolic		
Diastolic		
*Pulse Pressure		
**Mean Arterial Pressure		

* Pulse pressure = systolic pressure minus diastolic pressure.
**Mean arterial pressure ≃ diastolic pressure + ⅓ pulse pressure.

B. Recording Method

In the spaces provided below and on the next page tape or staple chart recordings for the various portions of the experiments.

Calibration of Recorder
Subject in Chair

Subject Standing Erect	Subject Reclining

After Exercise

C. Questions

1. Give the systolic and diastolic pressures for normal blood pressure: _____

2. What systolic pressure indicates hypertension? _____

3. Why is a low diastolic pressure considered harmful? _____

4. Why are diuretics of value in treating hypertension? _____

   _____

5. Why is reduced salt intake of value in regulating blood pressure? _____

   _____

   _____

Peripheral Circulation Control (Ex. 52)

A. Results

1. What observable change occurred in the frog's foot when histamine was applied to the foot?

   _____

   _____

2. What was the effect of histamine on the arterioles to produce this effect (*vasodilation* or *vasocon-striction*)? _____

3. What observable change occurred in the frog's foot when epinephrine was applied to the foot?

   _____

   _____

LABORATORY REPORT

53

Student: _____

Desk No.: _____ Section: _____

The Arteries and Veins

A. Illustrations

Record the labels for figures 53.2, 53.3, and 53.4 in the answer columns.

B. Arteries

Refer to the illustration in figure 53.2 to identify the arteries described by the following statements:

anterior tibial—1
aorta—2
aortic arch—3
axillary—4
brachial—5
brachiocephalic—6
celiac—7
common iliac—8
deep femoral—9
external iliac—10
femoral—11
inferior mesenteric—12

internal iliac—13
left common carotid—14
left subclavian—15
popliteal—16
posterior tibial—17
radial—18
renal—19
right common carotid—20
subclavian—21
superior mesenteric—22
ulnar—23

1. Artery of the shoulder.
2. Artery of upper arm.
3. Artery of the armpit.
4. Gives rise to right common carotid and right subclavian.
5. Lateral artery of the forearm.
6. Medial artery of the forearm.
7. Three branches of the aortic arch.
8. Gives rise to femoral artery.
9. Major artery of the thigh.
10. Artery of knee region.
11. Artery of calf region.
12. Large branch of common iliac.
13. Small branch of common iliac.
14. Supplies the stomach, spleen, and liver.
15. Major artery of chest and abdomen.
16. Supplies the kidney.
17. Also known as the hypogastric artery.
18. Supplies blood to most of small intestines and part of colon.
19. Artery in interior portion of lower leg.
20. Branch of femoral that parallels medial surface of femur.
21. Curved vessel that receives blood from left ventricle.
22. Supplies the large intestine and rectum.

Answers

Arteries	Fig. 53.2
1. _____	_____
2. _____	_____
3. _____	_____
4. _____	_____
5. _____	_____
6. _____	_____
7. _____	_____
8. _____	_____
9. _____	_____
10. _____	_____
11. _____	_____
12. _____	_____
13. _____	_____
14. _____	_____
15. _____	_____
16. _____	_____
17. _____	_____
18. _____	_____
19. _____	_____
20. _____	_____
21. _____	_____
22. _____	_____
	_____
	_____
	_____
	_____
	_____

C. Veins

Refer to figures 53.3 and 53.4 to identify the veins described by the following statements:

accessory cephalic—1
axillary—2
basilic—3
brachial—4
brachiocephalic—5
cephalic—6
common iliac—7
coronary—8
dorsal venous arch—9
external iliac—10
external jugular—11
femoral—12
great saphenous—13
hepatic—14
inferior vena cava—15
inferior mesenteric—16
internal iliac—17
internal jugular—18
median cubital—19
popliteal—20
portal—21
posterior tibial—22
pyloric—23
subclavian—24
superior mesenteric—25
superior vena cava—26

1. Largest vein in neck.
2. Small vein in neck.
3. Vein of armpit.
4. Empty into brachiocephalic veins.
5. Collects blood from veins of head and arms.
6. Short vein between basilic and cephalic veins.
7. Collects blood from veins of chest, abdomen, and legs.
8. On posterior surface of humerus.
9. Large veins that unite to form the inferior vena cava.
10. On medial surface of upper arm.
11. On lateral portion of forearm.
12. On lateral portion of upper arm.
13. Vein from liver to inferior vena cava.
14. Receives blood from descending colon and rectum.
15. Empties into popliteal.
16. Collects blood from two brachiocephalic veins.
17. Collects blood from top of foot.
18. Superficial vein on medial surface of leg.
19. Vein of knee region.
20. Collects blood from intestines, stomach, and colon.
21. Vein that receives blood from posterior tibial.
22. Receives blood from ascending colon and part of ileum.
23. Empties into great saphenous.
24. Small vein that empties into common iliac at juncture of external iliac.
25. Vein that great saphenous empties into.
26. Two veins that drain blood from stomach into portal vein.

Answers

Veins	Fig. 53.3
1. _____	_____
2. _____	_____
3. _____	_____
4. _____	_____
5. _____	_____
6. _____	_____
7. _____	_____
8. _____	_____
9. _____	_____
10. _____	_____
11. _____	_____
12. _____	_____
13. _____	_____
14. _____	_____
15. _____	_____
16. _____	_____
17. _____	_____
18. _____	_____
19. _____	_____
20. _____	_____
21. _____	_____
22. _____	_____
23. _____	_____
24. _____	_____
25. _____	Fig. 53.4
26. _____	_____

LABORATORY REPORT
54,55

Student: _____

Desk No.: _____ Section: _____

Fetal Circulation (Ex. 54)

A. Figure 54.1

Record the labels for this illustration in the answer column.

B. Questions

Record the answers for the following questions in the answer column:

1. What blood vessel in the umbilical cord supplies the fetus with nutrients?
2. What blood vessels in the umbilical cord return blood to the placenta from the fetus?
3. What blood vessel shunts blood from the pulmonary artery to the aorta?
4. What vessel in the liver carries blood from the umbilical vein to the inferior vena cava?
5. What structure in the liver is formed from the ductus venosus?
6. What structure forms from the ductus arteriosus?
7. What structure forms from the umbilical vein?
8. What structure enables blood to flow from the right atrium to the left atrium before birth?
9. Give two occurrences at childbirth that prevent excessive blood loss by the infant through the cut umbilical cord.
10. What structures near the bladder form from the hypogastric arteries?

The Lymphatic System and the Immune Response (Ex. 55)

A. Figure 55.1

Record the labels for this illustration in the answer column.

B. Components

By referring to figures 55.1 and 55.2, identify the following structures in the lymphatic system:

1. Large vessel in thorax and abdomen that collects lymph from lower extremities.
2. Short vessel that collects lymph from right arm and right side of head.
3. Microscopic lymphatic vessels situated among cells of tissues.
4. Saclike structure that receives chyle from intestine.
5. Blood vessel that receives fluid from thoracic duct.
6. Filters of the lymphatic system.
7. Vessels of arms and legs that convey lymph to collecting ducts.
8. Depression in lymph node through which blood vessels enter node.
9. Outer covering of lymph node.
10. Structural units of lymph node that are packed with lymphocytes.
11. Cells involved in organ transplant rejections.
12. Cells that produce antibodies.

capsule—1
cisterna chyli—2
germinal centers—3
hilum—4
left subclavian vein—5
lymphatics—6
lymph capillaries—7
lymphoblasts—8
lymph nodes—9
plasma cells—10
right lymphatic duct—11
thoracic duct—12
none of these—13

Answers

Questions

1. _____
2. _____
3. _____
4. _____
5. _____
6. _____
7. _____
8. _____
9a. _____
b. _____
10. _____

Components	Fig. 54.1
1. _____	_____
2. _____	_____
3. _____	_____
4. _____	_____
5. _____	_____
6. _____	_____
7. _____	_____
8. _____	_____
9. _____	_____
10. _____	_____
11. _____	_____
12. _____	_____

C. Questions

1. List three forces that move lymph through the lymphatic vessels:

    a. _____

    b. _____   c. _____

2. How does lymph in the lymphatics of the legs differ in composition from the lymph in the thoracic duct?

    _____

    _____

3. Where do stem cells for all lymphocytes originate? _____

    _____

4. Give the cell type that accounts for each type of immunity:

    Humoral immunity: _____

    Cellular immunity: _____

5. What tissues program stem cells for the following?

    Humoral immunity: _____

    Cellular immunity: _____

6. What type of tissue is common to the thymus gland and bursal tissues?

    _____

Answers
Fig. 55.1
_____
_____
_____
_____
_____
_____
_____
_____
_____
_____
_____
_____
_____
_____
_____
_____
_____

LABORATORY REPORT

56

The Respiratory Organs

Student: _____

Desk No.: _____ Section: _____

A. Illustrations

Record the labels for the illustrations of this exercise in the answer column.

B. Cat Dissection

After completing the dissection of the cat, answer the following questions. Both of these questions are asked in the course of the dissection.

1. Are the cartilaginous rings of the trachea continuous all the way around the organ? _____

2. Which lung has four lobes? _____

   Which has three lobes? _____

C. Sheep Pluck Dissection

After completing the sheep pluck dissection, answer the following questions:

1. Are the cartilaginous rings of the trachea continuous all the way around the organ? _____

2. Describe the texture of the surface of the lung. _____

   _____

   _____

3. How many lobes exist on the right lung? _____
   On the left lung? _____

4. Why does lung tissue collapse so readily when you quit blowing into it with a straw? _____

   _____

   _____

5. What does the "pulmonary membrane" consist of? _____

   _____

   _____

   _____

D. Histological Study

On a separate sheet of plain paper make drawings, as required by your instructor, of nasal, tracheal, and lung tissues.

Answers

Fig. 56.1	Fig. 56.2
_____	_____
_____	_____
_____	_____
_____	_____
_____	_____
_____	_____
_____	_____
_____	_____
_____	_____
_____	_____
_____	**Fig. 56.3**
_____	_____
_____	_____
_____	_____
_____	_____
_____	_____
_____	_____
_____	_____
_____	_____
_____	

501

E. Organ Identification

Identify the respiratory structures according to the following statements:

alveoli—1
bronchioles—2
bronchi—3
cricoid cartilage—4
epiglottis—5
hard palate—6
larynx—7
lingual tonsils—8
nasal cavity—9
nasal conchae—10

nasopharynx—11
oral cavity proper—12
oral vestibule—13
oropharynx—14
palatine tonsils—15
pharyngeal tonsils—16
pleural cavity—17
soft palate—18
thyroid cartilage—19
trachea—20

1. Partition between nasal and oral cavities.
2. Cartilage of Adam's apple.
3. Small air sacs of lung tissue.
4. Cavity above soft palate.
5. Cavity above hard palate.
6. Cavity that contains the tongue.
7. Cavity near palatine tonsils.
8. Voice box.
9. Small tubes leading into alveoli.
10. Another name for adenoids.
11. Tube between larynx and bronchi.
12. Fleshy lobes in nasal cavity.
13. Cavity between lips and teeth.
14. Tonsils located on sides of pharynx.
15. Tonsils attached to base of tongue.
16. Flexible flaplike cartilage over larynx.
17. Most inferior cartilaginous ring of larynx.
18. Tubes formed by bifurcation of trachea.
19. Potential cavity between lung and thoracic wall.

F. Organ Function

Select the structures in the above list that perform the following functions:

1. Initiates swallowing.
2. Prevents food from entering the larynx when swallowing.
3. Provides supporting walls for vocal folds.
4. Assists in destruction of harmful bacteria in the oral region.
5. Warms the air as it passes through the nasal cavity.
6. Prevents food from entering nasal cavity during chewing and swallowing.
7. Provides surface for gas exchange in the lungs.
8. Essential for speech.

Answers

Organ Identification

1. _____
2. _____
3. _____
4. _____
5. _____
6. _____
7. _____
8. _____
9. _____
10. _____
11. _____
12. _____
13. _____
14. _____
15. _____
16. _____
17. _____
18. _____
19. _____

Organ Function

1. _____
2. _____
3. _____
4. _____
5. _____
6. _____
7. _____
8. _____

Student: _____

Desk No.: _____ Section: _____

Hyperventilation and Rebreathing

A. Questions for Nonrecording Portion of Experiment

1. Why does breathing into a bag after hyperventilating restore normal breathing more rapidly than breathing into the air? _____

_____

_____

_____

_____

2. How long were you able to hold your breath with only one deep inspiration? _____

3. How long were you able to hold your breath after hyperventilating? _____

_____

4. After exercising, how long were you able to hold your breath? _____

_____

Generalizations on above: _____

_____

_____

_____

B. Tracings

In the spaces provided below and on the next page, tape or staple chart recordings for the various portions of this experiment.

Normal Respiration (Slow Speed)　　　　　　Speed: _____ cm/sec

B. *Tracings* (continued)

Normal Respiration (Fast Speed)   Speed: _____ cm/sec

Rebreathing and Recovery

Hyperventilation

C. *Generalizations*

What generalizations can you make from the above tracings?

_____

_____

_____

_____

_____

_____

_____

LABORATORY REPORT

58,59

The Diving Reflex (Ex. 58)

A. *Tracings*

Tape or staple the chart recordings for the various portions of this experiment in the spaces provided below and on the next page.

Tidal Breathing before Immersion

Holding Breath after Tidal Inspiration

Immersion in Water at 70° Fahrenheit

Immersion in Water at 60° Fahrenheit

A. *Tracings* (continued)

Hyperventilation Followed by Holding Breath (No Immersion)

Hyperventilation Followed by Immersion   Total Time Immersion: _____

B. *Generalizations*

_____

_____

_____

_____

_____

_____

The Valsalva Maneuver (Ex. 59)

A. *Tracing*

In the space below, tape or staple a chart recording made during a Valsalva maneuver.

LABORATORY REPORT
60,61

Student: _____

Desk No.: _____ Section: _____

Spirometry:
Lung Capacities (Ex. 60)

A. Tabulation

Record in the following table the results of your spirometer readings and calculations.

LUNG CAPACITIES	NORMAL (ml)	YOUR CAPACITIES
Tidal Volume (TV)	500	
Minute Respiratory Volume (MRV) (MRV = TV × Resp. Rate)	6000	
Expiratory Reserve Volume (ERV)	1100	
Vital Capacity (VC) (VC = TV + ERV + IRV)	See appendix A (Tables II and III)	
Inspiratory Capacity (IC) (IC = VC − ERV)	3000	
Inspiratory Reserve Volume (IRV) (IRV = IC − TV)	2500	

B. Evaluation

List below any of your lung capacities that are significantly low (22% or more). _____

_____

_____

_____

507

Spirometry:
The FEV$_T$ Test (Ex. 61)

A. Expirogram

Trim the tracing of your expirogram to as small a piece as you can without removing volume values and tape it in space shown. Do not tape sides and bottom.

— — — — — — — — — —

    Tape Here

B. Calculations

Record your computations for determining the various percentages here.

1. **Vital Capacity** From the expirogram determine the total volume of air expired (Vital Capacity) .... _____

2. **Corrected Vital Capacity** By consulting table 61.1, determine the corrected vital capacity (step 4 on page 341).
    VC × Conversion Factor = ... _____

    Math:

3. **One-Second Timed Vital Capacity (FEV$_1$)** After determining the FEV$_1$ (steps 1, 2, and 3 on page 341) record your results here .... _____

4. **Corrected FEV$_1$** Correct the FEV$_1$ for the temperature of the spirometer (step 5 on page 341).

    FEV$_1$ × Conversion Factor = ... _____

5. **FEV$_1$ Percentage** Divide the corrected FEV$_1$ by the corrected vital capacity to determine the percentage expired in the first second (step 6 on page 341).

    $$\frac{\text{Corrected FEV}_1}{\text{Corrected Vital Capacity}} =$$ .... _____

    Math:

6. **Percent of Predicted VC** Determine what percentage your vital capacity is of the predicted vital capacity (step 7 on page 341).

    $$\frac{\text{Your Corrected VC}}{\text{Predicted VC}} =$$ .... _____

    Math:

C. Questions

1. Why isn't a measure of one's vital capacity as significant as the FEV$_T$? _____

    _____

    _____

    _____

2. What respiratory diseases are most readily detected with this test? _____

    _____

    _____

LABORATORY REPORT

63

Student: _____

Desk No.: _____ Section: _____

Anatomy of the Digestive System

A. Illustrations

Record the labels for the illustrations of this exercise in the answer columns.

B. Microscopic Studies

Record here the drawings of the microscopic examinations that are required for this exercise. If there is insufficient space for all required drawings, utilize a separate sheet of paper for all of them.

Answers

Fig. 63.1	Fig. 63.2
_____	_____
_____	_____
_____	_____
_____	_____
_____	_____
_____	_____
_____	_____
_____	_____
_____	Fig. 63.3
_____	_____
_____	_____
_____	_____
_____	_____
_____	_____
_____	_____
_____	_____
_____	_____
_____	_____
_____	_____
_____	
_____	

509

C. *Alimentary Canal*

Identify the parts of the digestive system described by the following statements:

anus—1	esophagus—10
cecum—2	ileum—11
colon—3	pharynx—12
colon, sigmoid—4	rectum—13
duct, common bile—5	stomach, fundus—14
duct, cystic—6	stomach, pyloric portion—15
duct, hepatic—7	valve, cardiac—16
duct, pancreatic—8	valve, ileocecal—17
duodenum—9	valve, pyloric—18

1. Place where swallowing (peristalsis) begins.
2. Entrance opening of stomach.
3. Exit opening of stomach.
4. Tube between mouth and stomach.
5. Distal coiled portion of small intestine.
6. First twelve inches of small intestine.
7. Proximal pouch or compartment of large intestine.
8. Section of large intestine between descending colon and rectum.
9. Most active portion of stomach.
10. Duct that drains the liver.
11. Valve between small and large intestines.
12. Duct that drains the gallbladder.
13. Part of small intestine where most digestion occurs.
14. Part of small intestine where most absorption occurs.
15. Duct that joins the common bile duct before entering the intestine.
16. Structure on which appendix is located.
17. Duct that conveys bile to intestine.
18. Exit of alimentary canal.
19. Part of tract where most water absorption (conservation) occurs.
20. Last six inches of alimentary canal.

Answers

Alimentary Canal	Fig. 63.4
1. _____	_____
2. _____	_____
3. _____	_____
4. _____	_____
5. _____	_____
6. _____	
7. _____	_____
8. _____	
9. _____	
10. _____	_____
11. _____	
12. _____	_____
13. _____	
14. _____	
15. _____	
16. _____	
17. _____	
18. _____	
19. _____	
20. _____	

D. Oral Cavity

Identify the structures of the mouth that are described by the following statements:

arch, glossopalatine—1
arch, pharyngopalatine—2
buccae—3
duct, parotid—4
duct, submandibular—5
ducts, lesser sublingual—6
frenulum, labial—7
frenulum, lingual—8
gingiva—9
glands, parotid—10
glands, sublingual—11

glands, submandibular—12
mucosa—13
papillae, filiform—14
papillae, foliate—15
papillae, fungiform—16
papillae, vallate—17
tonsil, lingual—18
tonsil, palatine—19
tonsil, pharyngeal—20
uvula—21

1. Lining of the mouth.
2. Name for the cheeks.
3. Duct that drains the parotid gland.
4. Fold of skin between the lip and gums.
5. Duct that drains the submandibular gland.
6. Fold of skin between the tongue and floor of mouth.
7. Portion of mucosa around the teeth.
8. Tonsils seen at back of mouth.
9. Tonsils located at root of tongue.
10. Fingerlike projection at end of soft palate.
11. Vertical ridges on side of tongue.
12. Rounded papillae on dorsum of tongue.
13. Small tactile papillae on surface of tongue.
14. Large papillae at back of tongue.
15. Ducts that drain sublingual gland.
16. Place where taste buds are located.
17. Salivary glands located inside and below the mandible.
18. Membrane located in front of palatine tonsil.
19. Membrane located in back of palatine tonsil.
20. Salivary glands located under the tongue.
21. Salivary glands located in cheeks.

E. The Teeth

Select the correct answer that completes each of the following statements:

1. All deciduous teeth usually erupt by the time a child is
   (1) one year old. (2) two years old. (3) four years old.
2. A complete set of primary teeth consists of
   (1) ten teeth. (2) twenty teeth. (3) thirty-two teeth.
3. The permanent teeth with the longest roots are the
   (1) incisors. (2) cuspids. (3) molars.
4. The smallest permanent molars are the
   (1) first molars. (2) second molars. (3) third molars.
5. A complete set of permanent teeth consists of
   (1) twenty-four teeth. (2) twenty-eight teeth.
   (3) thirty-two teeth.
6. The primary dentition lacks
   (1) molars. (2) incisors. (3) bicuspids.

Answers

Oral Cavity	Fig. 63.5
1.	
2.	
3.	
4.	
5.	
6.	
7.	
8.	
9.	
10.	
11.	Fig. 63.7
12.	
13.	
14.	
15.	
16.	
17.	
18.	
19.	
20.	
21.	

Teeth	
1.	
2.	
3.	
4.	
5.	
6.	

7. A child's first permanent tooth usually erupts during the
   (1) second year.   (2) third year.
   (3) fifth year.   (4) sixth year.
8. Trifurcated roots exist on
   (1) upper cuspids.   (2) upper molars.   (3) lower molars.
9. Bifurcated roots exist on
   (1) upper first bicuspids and lower molars.
   (2) cuspids and lower bicuspids.   (3) all molars.
10. A tooth is considered "dead" or devitalized if
    (1) the enamel is destroyed.   (2) the pulp is destroyed.
    (3) the periodontal membrane is infected.
11. Dentin is produced by cells called
    (1) odontoblasts.   (2) ameloblasts.   (3) cementocytes.
12. Cementum is secreted by cells called
    (1) odontoblasts.   (2) ameloblasts.   (3) cementocytes.
13. The wisdom tooth is
    (1) a supernumerary tooth.   (2) a succedaneous tooth.
    (3) the third molar.
14. Caries (cavities) are caused primarily by
    (1) using the wrong toothpaste.   (2) acid production by bacteria.
    (3) fluorides in water.
15. Most teeth that have to be extracted have become useless because of
    (1) caries.   (2) gingivitis.   (3) periodontitis (pyorrhea).
16. Enamel covers the
    (1) entire tooth.   (2) clinical crown only.
    (3) anatomical crown only.
17. The tooth receives nourishment through
    (1) the apical foramen.   (2) the periodontal membrane.
    (3) both the apical foramen and the periodontal membrane.
18. The tooth is held in the alveolus by
    (1) bone.   (2) dentin.   (3) periodontal membrane.

Answers

Teeth

7. _____
8. _____
9. _____
10. _____
11. _____
12. _____
13. _____
14. _____
15. _____
16. _____
17. _____
18. _____

LABORATORY REPORT 64,65

Student: ______________________

Desk No.: __________ Section: __________

The Chemistry of Hydrolysis (Ex. 64)

A. Carbohydrate Digestion

1. **IKI Test** Record here the presence (+) or absence (−) of starch as revealed by IKI test on spot plates.

1A	2A	3A	4A	5A

a. Did the pancreatic extract hydrolyze starch? ______________________

b. Which tube is your evidence for this conclusion? ______________________

c. How do tubes 1A and 5A compare in color? ______________________

d. What do you conclude from question *c?* ______________________

e. What is the value of tube 4A? ______________________

____

f. What is the value of tube 2A? ______________________

____

2. **Barfoed's and Benedict's Tests** After performing Barfoed's and Benedict's tests on tubes 1A and 5A, what are your conclusions about the degree of digestion of starch by amylase? ______________________

____

____

B. Protein Digestion

Record the color and optical densities (OD) of each tube in the following chart.

1P	2P	3P	4P	5P
Color				
OD				

1. Did pancreatic juice hydrolyze the BAPNA? ______________________
2. Which tube is your evidence for this conclusion? ______________________

3. What other tube shows protein hydrolysis? _____

4. What can you conclude from the observation in question 3? _____

_____

5. Why is there no hydrolysis in tube 4P? _____

6. Why is there no hydrolysis in tube 3P? _____

7. Why didn't tube 2P show hydrolysis? _____

_____

C. Fat Digestion

Determine the pH of each tube by comparing with a bromthymol blue color standard. Record these values in the table below.

1L	2L	3L	4L	5L	6L

1. Did the pancreatic extract hydrolyze fat in the presence of bile? _____

2. Did the pancreatic extract hydrolyze fat in the absence of bile? _____

3. How do tubes 1L and 6L compare as to degree of hydrolysis? _____

_____

4. What can you conclude from the observation in question 3? _____

_____

5. What is the value of tube 2L? _____

_____

D. Summarization

Make an itemized statement of all the facts you learned in this experiment.

Enzyme Review

Consult your text and lecture notes to answer the following questions concerning digestive enzymes:

A. Digestive Juices

Select the enzymes that are present in the following digestive juices. Use the enzymes listed in the right column.

1. Saliva
2. Gastric juice
3. Pancreatic fluid
4. Intestinal juice (succus entericus)

Carbohydrases
   amylase, pancreatic—1
   amylase, salivary—2
   lactase—3
   maltase—4
   sucrase—5

Proteases
   carboxypeptidase—6
   chymotrypsin—7
   erepsin (peptidase)—8
   pepsin—9
   rennin—10
   trypsin—11

Lipases
   lipase, pancreatic—12

B. Substrates

Select the enzymes that act on the following substrates:

1. Casein
2. Fats
3. Lactose
4. Maltose
5. Peptides
6. Proteins
7. Proteoses and peptones
8. Starch
9. Sucrose

C. End Products

Select the enzymes from the above list that produce the following end products in digestion.

1. Amino acids
2. Fatty acids
3. Fructose
4. Galactose
5. Glucose
6. Glycerol
7. Maltose
8. Polypeptides

Answers

Juices

1. _____
2. _____
3. _____
4. _____

Substrates

1. _____
2. _____
3. _____
4. _____
5. _____
6. _____
7. _____
8. _____
9. _____

End Products

1. _____
2. _____
3. _____
4. _____
5. _____
6. _____
7. _____
8. _____

Factors Affecting Hydrolysis (Ex. 65)

A. Tabulations

If the class has been subdivided into separate groups to perform different parts of this experiment, these tabulations should be recorded on the chalkboard so that all students can copy the results onto their own Laboratory Report sheets.

1. **Temperature** After IKI solution has been added to all ten tubes, record the color and degree of amylase activity in each tube in the following tables. Refer to the carbohydrate differentiation separation outline in appendix B pertaining to the IKI test.

Table I   Amylase action at 20° C.

TUBE NO.	1	2	3	4	5	6	7	8	9	10
Color										
Starch Hydrolysis										

515

Factors Affecting Hydrolysis

Table II Amylase action at 37° C.

TUBE NO.	1	2	3	4	5	6	7	8	9	10
Color										
Starch Hydrolysis										

2. **Hydrogen Ion Concentration** After IKI solution has been added to all nine tubes, record, as above, the color and degree of amylase action for each tube in the following table.

Table III pH and amylase action at 37° C.

TIME	2 MINUTES			4 MINUTES			6 MINUTES		
Tube No.	1	2	3	4	5	6	7	8	9
pH	5	7	9	5	7	9	5	7	9
Color									
Starch Hydrolysis									

B. Conclusions

Answer the following questions, which are related to the results of this experiment.

1. How long did it take for complete starch hydrolysis at 20° C? _____ At 37° C? _____

2. Why would you expect the above results to turn out as they have? _____

   _____

   _____

3. What effect does boiling have on amylase? _____

4. At which pH was starch hydrolysis most rapid? _____

   How does this pH compare with the pH of normal saliva? _____

   _____

5. What is the pH of gastric juice? _____

   What do you suppose happens to the action of amylase in the stomach? _____

   _____

Anatomy of the Urinary System

A. Illustrations

Record the labels for the three illustrations of this exercise in the answer columns. Note that figure 66.2 is on the reverse side of this sheet.

B. Microscopy

If the space provided below is adequate for your histology drawings, use it. Use a separate sheet of paper if several sketches are to be made.

C. Anatomy

Identify the structures of the urinary system that are described by the following statements:

calyces—1   medulla—6   renal pelvis—11
collecting tubule—2   nephron—7   renal pyramids—12
cortex—3   renal capsule—8   ureters—13
glomerular capsule—4 renal column—9   urethra—14
glomerulus—5   renal papilla—10

1. Tubes that drain the kidneys.
2. Tube that drains the urinary bladder.
3. Portion of kidney that contains renal corpuscles.
4. Cone-shaped areas of medulla.
5. Portion of kidney that consists primarily of collecting tubules.
6. Basic functioning unit of the kidney.
7. Two portions of renal corpuscle.
8. Distal tip of renal pyramid.
9. Short tubes that receive urine from renal papillae.
10. Funnel-like structure that collects urine from calyces of each kidney.
11. Structure that receives urine from several nephrons.
12. Cup-shaped membranous structure that surrounds glomerulus.
13. Thin fibrous outer covering of kidney.
14. Cortical tissue between renal pyramids.
15. Tuft of capillaries that produces dilute urine.

Answers

Anatomy	Fig. 66.1
1. _____	_____
2. _____	_____
3. _____	_____
4. _____	_____
5. _____	_____
6. _____	_____
7. _____	_____
8. _____	_____
9. _____	_____
10. _____	_____
11. _____	
12. _____	Fig. 66.3
13. _____	
14. _____	
15. _____	

Anatomy of the Urinary System

D. *Physiology*

Select the answer that best completes the following statements about the physiology of urine production. For definitions of medical terms consult a medical dictionary or your lecture text.

1. Blood enters the glomerulus through the
   (1) efferent arteriole.   (2) arcuate artery.
   (3) afferent arteriole.   (4) None of these is correct.

2. Surgical removal of a kidney is called
   (1) nephrectomy.   (2) nephrotomy.   (3) nephrolithotomy.

3. Water reabsorption from the glomerular filtrate into the peritubular blood is facilitated by
   (1) antidiuretic hormone.   (2) renin.
   (3) aldosterone.   (4) Both 1 and 3 are correct.

4. The following substances are reabsorbed through the walls of the nephron into the peritubular blood:
   (1) glucose and water.   (2) urea and water.
   (3) glucose, amino acids, salts, and water.
   (4) glucose, amino acids, urea, salts, and water.

5. The amount of urine produced is affected by
   (1) blood pressure.   (2) environmental temperature.
   (3) amount of solute in glomerular filtrate.
   (4) 1, 2, 3 and additional factors are correct.

6. The amount of urine normally produced in 24 hours is about
   (1) 100 ml.   (2) 500 ml.   (3) 1.5 liters.   (4) 4.5 liters.

7. The reabsorption of sodium ions from the glomerular filtrate into the peritubular blood draws the following back into the blood:
   (1) potassium ions.   (2) chloride ions.
   (3) water.   (4) chloride ions and water.

8. Cells of the walls of collecting tubules secrete the following substances into urine:
   (1) amino acids.   (2) uric acid.
   (3) ammonia.   (4) Both 2 and 3 are correct.

9. Renal diabetes is due to
   (1) a lack of ADH.   (2) a lack of insulin.
   (3) faulty reabsorption of glucose in the nephron.
   (4) Both 1 and 2 are correct.

10. The presence of glucose in the urine and a low level of insulin in the blood would be diagnosed as
    (1) diabetes insipidus.   (2) diabetes mellitus.
    (3) renal diabetes.   (4) None of these is correct.

11. Approximately 80% of water in the glomerular filtrate is reabsorbed in the
    (1) proximal convoluted tubule.   (2) Henle's loop.
    (3) distal convoluted tubule.   (4) None of these is correct.

12. The desire to micturate normally occurs when the following amount of urine is present in the bladder:
    (1) 100 ml.   (2) 200 ml.   (3) 300 ml.   (4) 400 ml.

13. Removal of calculus (kidney stones) is called
    (1) nephrectomy.   (2) nephrotomy.
    (3) nephrolithotomy.   (4) Both 2 and 3 are correct.

Answers

Physiology

1. _____
2. _____
3. _____
4. _____
5. _____
6. _____
7. _____
8. _____
9. _____
10. _____
11. _____
12. _____
13. _____

Fig. 66.2

518

LABORATORY REPORT

67

Student: _____

Desk No.: _____ Section: _____

Urine: Composition and Tests

A. Test Results

Record on the chart below the results of all urine tests performed. If the tests are done as a demonstration, the results will be tabulated in columns A and B. If students perform the tests on their own urine, the last column will be used for their results.

TEST	NORMAL VALUES	ABNORMAL VALUES	A POSITIVE TEST CONTROL	B UNKNOWN SAMPLE	C STUDENT'S URINE
COLOR	Colorless Pale straw Straw Amber	Milky Reddish-amber Brownish-yellow Green Smoky brown			
CLOUDINESS	Clear	+ slight ++ moderate +++ cloudy ++++ very cloudy			
SP. GRAVITY	1.001–1.060	Above 1.060			
pH	4.8–7.5	Below 4.8 Above 7.5			
ALBUMIN (Protein)	None	+ barely visible ++ granular +++ flocculent ++++ large flocculent			
MUCIN	None	Visible amounts			
GLUCOSE	None	+ yellow green ++ greenish yellow +++ yellow ++++ orange			
KETONES	None	+ pink-purple ring			
HEMOGLOBIN	None	See color chart			
BILIRUBIN					

B. *Microscopy*

Record here by sketch and word any structures seen on microscopic examination.

C. *Interpretation*

Indicate the probable significance of each of the following in urine. More than one condition may apply in some instances.

bladder infection—1 diabetes mellitus—4 glomerulonephritis—7
cirrhosis of liver—2 exercise (extreme)—5 gonorrhea—8
diabetes insipidus—3 kidney tumor—6 hepatitis—9
              normally present—10

1. Albumin (constantly)
2. Albumin (periodic)
3. Bacteria
4. Bile pigments
5. Blood
6. Creatinine
7. Glucose
8. Mucin
9. Porphyrin
10. Pus cells (neutrophils)
11. Sodium chloride
12. Specific gravity (high)
13. Specific gravity (low)
14. Urea
15. Uric acid

D. *Terminology*

Select the terms described by the following statements:

albuminuria—1 cystitis—4 glycosuria—7
anuria—2 diuretic—5 nephritis—8
catheter—3 enuresis—6 pyelitis—9
            uremia—10

1. Absence of urine production.
2. High urea level in blood.
3. Sugar in urine.
4. Albumin in urine.
5. Inflammation of nephrons.
6. Inflammation of pelvis of kidney.
7. Bladder infection.
8. Device for draining bladder.
9. Involuntary bed-wetting during sleep.
10. Chemical that stimulates urine production.

Answers

Interpretation

1. _____
2. _____
3. _____
4. _____
5. _____
6. _____
7. _____
8. _____
9. _____
10. _____
11. _____
12. _____
13. _____
14. _____
15. _____

Terminology

1. _____
2. _____
3. _____
4. _____
5. _____
6. _____
7. _____
8. _____
9. _____
10. _____

LABORATORY REPORT

68

The Endocrine Glands

A. Illustrations
Record the labels for figures 68.12 and 68.13 in the answer columns.

B. Microscopy
Make drawings of the various microscopic studies on separate drawing paper. Include these drawings with the Laboratory Report.

C. Sources
Select the glandular tissues that produce the following hormones:

1. ACTH
2. ADH
3. Aldosterone
4. Calcitonin
5. Chorionic gonadotropin
6. Cortisol
7. Epinephrine
8. Estrogens
9. FSH
10. Glucagon
11. ICSH
12. Insulin
13. LH
14. Melatonin
15. Norepinephrine
16. Oxytocin
17. Parathyroid hormone
18. PIF
19. Prolactin
20. Progesterone
21. Somatostatin
22. Testosterone
23. Thymosin
24. Thyrotropin
25. Thyroxine
26. Triiodothyronine
27. Vasopressin

adrenal cortex
   zona fasciculata—1
   zona glomerulosa—2
   zona reticularis—3
adrenal medulla—4
hypothalamus—5
ovary
   corpus luteum—6
   Graafian follicle—7
pancreas
   alpha cells—8
   beta cells—9
   delta cells—10
   acini cells—11
parathyroid gland—12
pineal gland—13
pituitary gland
   adenohypophysis—14
   neurohypophysis—15
placenta—16
testis—17
thymus gland—18
thyroid gland
   follicular cells—19
   parafollicular cells—20

D. Physiology
Select the hormones that produce the following physiological effects. Note that a separate list of hormones is provided for each group.

GROUP I

1. Increases metabolism.
2. Increases blood pressure.
3. Increases strength of heartbeat.
4. Promotes sodium absorption in nephron.
5. Promotes gluconeogenesis.
6. Causes vasoconstriction in all organs.
7. Causes vasoconstriction in all organs except muscles and liver.
8. Causes vasodilation in skeletal muscles.
9. Inhibits the estrus cycle in lower animals.

aldosterone—1
epinephrine—2
glucocorticoids—3
melatonin—4
norepinephrine—5
none of these—6

Answers

Sources	Fig. 68.12
1. _____	_____
2. _____	_____
3. _____	_____
4. _____	_____
5. _____	_____
6. _____	_____
7. _____	_____
8. _____	_____
9. _____	_____
10. _____	_____
11. _____	_____
12. _____	_____
13. _____	_____
14. _____	_____
15. _____	_____
16. _____	_____
17. _____	_____
18. _____	_____
19. _____	_____
20. _____	_____
21. _____	_____
22. _____	_____
23. _____	_____
24. _____	_____
25. _____	_____
26. _____	_____
27. _____	_____

Physiology Group 1	
1. _____	_____
2. _____	_____
3. _____	_____
4. _____	_____
5. _____	_____
6. _____	_____
7. _____	_____
8. _____	_____
9. _____	_____

D. Physiology (continued)

Select the hormones that produce the following physiological effects:

GROUP II

1. Inhibits insulin production.
2. Inhibits production of SH.
3. Stimulates osteoclasts.
4. Mobilizes fatty acids.
5. Inhibits glucagon production.
6. Raises blood calcium level.
7. Increases cardiac output and respiratory rate.
8. Impedes phosphorus reabsorption in renal tubules.
9. Promotes storage of glucose, fatty acids, and amino acids.
10. Mobilizes glucose, fatty acids, and amino acids.
11. Stimulates metabolism of all cells in body.
12. Lowers blood phosphorus levels.
13. Decreases blood calcium when injected intravenously.

calcitonin—1
glucagon—2
insulin—3
parathyroid hormone—4
somatostatin—5
thyroid hormone—6
none of these—7

GROUP III

1. Promotes breast enlargement.
2. Promotes rapid bone growth.
3. Stimulates testicular descent.
4. Causes milk ejection from breasts.
5. Promotes maturation of spermatozoa.
6. Causes constriction of arterioles.
7. Stimulates production of thyroxine.
8. Stimulates glucocorticoid production.
9. Inhibits prolactin production.
10. Causes thickening and vascularization of endometrium.
11. Causes interstitial cells to produce testosterone.
12. Causes myometrium contraction during childbirth.
13. Promotes milk production after childbirth.
14. Promotes stratification of vaginal epithelium.
15. Suppresses development of female genitalia in male.
16. Prepares the endometrium for egg cell implantation.
17. Promotes growth of all cells in the body.

ACTH—1
ADH—2
chorionic gonadotropin—3
estrogens—4
FSH—5
LH—6
oxytocin—7
PIF—8
progesterone—9
somatotropin—10
testosterone—11
thyrotropin—12
none of these—13

E. Hormonal Imbalance

Select the hormones that are responsible for the following disorders. After each hormone number indicate with a (+) or (−) whether the condition is due to *excess* (+) or *deficiency* (−). More than one hormone may apply in some cases.

1. Acromegaly
2. Addison's disease
3. Cretinism
4. Cushing's syndrome
5. Diabetes insipidus
6. Diabetes mellitus
7. Dwarfism
8. Exophthalmos
9. Gigantism
10. Hyperglycemia
11. Hypothyroidism
12. Myxedema
13. Tetany

ADH—1
glucagon—2
glucocorticoids—3
insulin—4
mineralocorticoids—5
parathyroid hormone—6
somatotropin—7
thyroid hormone—8
none of these—9

Answers

Physiology Group II	Fig. 68.13
1. _____	_____
2. _____	_____
3. _____	_____
4. _____	
5. _____	
6. _____	
7. _____	
8. _____	
9. _____	
10. _____	
11. _____	
12. _____	
13. _____	

Group III	
1. _____	
2. _____	
3. _____	
4. _____	
5. _____	Imbalance
6. _____	1. _____
7. _____	2. _____
8. _____	3. _____
9. _____	4. _____
10. _____	5. _____
11. _____	6. _____
12. _____	7. _____
13. _____	8. _____
14. _____	9. _____
15. _____	10. _____
16. _____	11. _____
17. _____	12. _____
	13. _____

Student: _____

Desk No.: _____ Section: _____

The Reproductive System

A. Labels
Record the labels for the illustrations of this exercise in the answer columns.

B. Microscopy
Draw the various stages of spermatogenesis on a separate piece of paper to be included with this Laboratory Report.

C. Male Organs
Identify the structures described by the following statements:

common ejaculatory duct—1
corpora cavernosa—2
corpus spongiosum—3
bulbourethral gland—4

epididymis—5
glans penis—6
penis—7
prepuce—8
prostate gland—9

prostatic urethra—10
seminal vesicle—11
seminiferous tubule—12
testes—13
vas deferens—14

1. Copulatory organ of male.
2. Source of spermatozoa.
3. Erectile tissue of penis (3 answers).
4. Fold of skin over end of penis.
5. Structure that stores spermatozoa.
6. Contribute fluids to semen (3 answers).
7. Duct that passes from urinary bladder through the prostate gland.
8. Coiled-up structures in testes where spermatozoa originate.
9. Tube that carries sperm from epididymis to ejaculatory duct.
10. Distal end of penis.

D. Female Organs
Identify the structures described by the following statements:

cervix—1
clitoris—2
external os—3
fimbriae—4
greater vestibular glands—5

hymen—6
infundibulum—7
internal os—8
labia majora—9
labia minora—10

mons pubis—11
myometrium—12
paraurethral glands—13
ovaries—14
uterine tubes—15
vagina—16

1. Copulatory organ of female.
2. Source of ova in female.
3. Muscular wall of uterus.
4. Neck of uterus.
5. Provides vaginal lubrication during coitus.
6. Fingerlike projections around edge of infundibulum.
7. Protuberance of erectile tissue sensitive to sexual excitation.
8. Entrance to uterus at vaginal end.
9. Homologous to bulbourethral glands of male.
10. Homologous to penis of male.
11. Homologous to scrotum of male.
12. Homologous to prostate gland of male.
13. Ducts that convey ovum to uterus.
14. Membranous fold of tissue surrounding entrance to vagina.
15. Open funnel-like end of uterine tube.

Answers

Male Organs	Fig. 69.1
1. _____	_____
2. _____	_____
3. _____	_____
4. _____	_____
5. _____	_____
6. _____	_____
7. _____	_____
8. _____	_____
9. _____	_____
10. _____	_____

Female Organs	
1. _____	_____
2. _____	_____
3. _____	_____
4. _____	_____
5. _____	_____
6. _____	_____
7. _____	_____
8. _____	_____
9. _____	_____
10. _____	_____
11. _____	_____
12. _____	_____
13. _____	_____
14. _____	_____
15. _____	_____

E. *Ligaments*

Identify the following supporting structures that hold the ovaries, uterus, and uterine tubes in place:

broad ligament—1   mesosalpinx—3   sacrouterine ligament—6
infundibulopelvic   ovarian ligament—4   suspensory ligament—7
    ligament—2   round ligament—5

1. A ligament that extends from the uterus to the ovary.
2. Ligament between the cervix and sacral part of the pelvic wall.
3. A ligament that extends from the infundibulum to the wall of the pelvic cavity.
4. A ligament that extends from the corpus of the uterus to the body wall.
5. A large flat ligament of peritoneal tissue that extends from the uterine tube to the cervical part of the uterus.
6. A mesentery between the ovary and uterine tubes that contains blood vessels leading to the uterine tube.
7. A fold of peritoneum that attaches the ovary to the uterine tube.

F. *Germ Cells*

Identify the types of cells of oogenesis and spermatogenesis described by the following statements. More than one answer may apply.

Spermatogenesis
primary spermatocyte—1
secondary spermatocyte—2
spermatids—3
spermatogonia—4
spermatozoa—5

Oogenesis
oogonia—6
polar bodies—7
primary oocyte—8
secondary oocyte—9

1. Cells that are haploid.
2. Cells that contain tetrads.
3. Cells that contain monads.
4. Cells that contain dyads.
5. Cells that are diploid.
6. Cells at periphery of ovary that produce all ova.
7. Cells that undergo mitosis.
8. Cells in which first meiotic division occurs.
9. Cells at periphery of seminiferous tubule that give rise to all spermatozoa.
10. Cells that develop directly into mature spermatozoa.
11. Cells in which second meiotic division occurs.

G. *Physiology of Reproduction and Development*

Select the *best answer* for each of the following statements:

1. The ovum gets from the ovary to the uterus by
   (1) amoeboid movement.   (2) ciliary action.
   (3) peristalsis and ciliary action of uterine tube.
2. The vaginal lining is normally
   (1) alkaline.   (2) acid.   (3) neutral.
3. The seminal vesicle secretion is
   (1) alkaline.   (2) acid.   (3) neutral.
4. The birth canal consists of the
   (1) vagina.   (2) uterus.   (3) the vagina and uterus.
5. Spermatozoan viability is enhanced by a temperature that is
   (1) 98.6° F.   (2) above 98.6° F.   (3) below 98.6° F.
6. The prostatic secretion is
   (1) alkaline.   (2) acid.   (3) neutral.

Answers

Ligaments	Fig. 69.2
1. _____	_____
2. _____	_____
3. _____	_____
4. _____	_____
5. _____	_____
6. _____	_____
7. _____	_____

Germ Cells	
1. _____	_____
2. _____	_____
3. _____	_____
4. _____	_____
5. _____	**Fig. 69.3**
6. _____	_____
7. _____	_____
8. _____	_____
9. _____	_____
10. _____	_____
11. _____	_____

Physiology	
1. _____	_____
2. _____	_____
3. _____	_____
4. _____	_____
5. _____	_____
6. _____	_____

7. The fetal stage of the human is
   (1) the first 8 weeks.　　(2) from the ninth week until birth.
   (3) the entire prenatal term.
8. Circumcision involves excisement of the
   (1) prepuce.　　(2) perineum.　　(3) glans penis.
9. Fertilization of the human ovum usually occurs in the
   (1) uterus.　　(2) vagina.　　(3) uterine tube.
10. Implantation of the fertilized ovum occurs usually
    (1) within 2 days after fertilization.
    (2) within 6 to 8 days after fertilization.
    (3) after 10 days.
11. The innermost fetal membrane surrounding the embryo is the
    (1) amnion.　　(2) chorion.　　(3) decidua.
12. The life span of an ovum without fertilization is
    (1) 6 hours.　　(2) 24–28 hours.　　(3) 7 days.
13. Uterine muscle consists of
    (1) smooth muscle.　　(2) striated muscle.
    (3) both smooth and striated muscle.
14. The outermost fetal membrane surrounding the embryo is the
    (1) amnion.　　(2) chorion.　　(3) decidua.
15. The lining of the uterus is the
    (1) myometrium.　　(2) endometrium.　　(3) epimetrium.
16. Milk production by the breasts usually occurs
    (1) immediately after birth.　　(2) on the second day.
    (3) on the third or fourth day.
17. The afterbirth refers to the
    (1) placenta.　　(2) damaged uterus.
    (3) the placenta, umbilical vessels, and fetal membranes.
18. Parturition is the
    (1) process of giving birth.　　(2) period of pregnancy.
    (3) first few days of postnatal life.
19. The following substances pass from the mother to the fetus
    through the placenta:
    (1) nutrients, gases, and blood cells.
    (2) nutrients and all blood cells except red blood cells.
    (3) nutrients, gases, hormones, and antibodies.
20. Abortion is correctly known as
    (1) criminal emptying of the uterus.
    (2) interruption of pregnancy during fetal life.
    (3) interruption of pregnancy during embryonic life.

H. Menstrual Cycle

Select the correct answer to the following statements:

1. The cessation of menstrual flow in a woman in her late forties is
   known as
   (1) menarche.　　(2) menopause.　　(3) amenorrhea.
2. The first menstrual flow is known as
   (1) menarche.　　(2) climacteric.　　(3) menopause.
3. A distinct rise in the basal body temperature usually occurs
   (1) at ovulation.　　(2) during the proliferative phase.
   (3) during the quiescent period.

Answers	
Physiology	**Fig. 69.4**
7.＿＿＿	＿＿＿
8.＿＿＿	＿＿＿
9.＿＿＿	＿＿＿
10.＿＿＿	＿＿＿
11.＿＿＿	＿＿＿
12.＿＿＿	＿＿＿
13.＿＿＿	＿＿＿
14.＿＿＿	＿＿＿
15.＿＿＿	＿＿＿
16＿＿＿	＿＿＿
17.＿＿＿	＿＿＿
18.＿＿＿	＿＿＿
19.＿＿＿	＿＿＿
20.＿＿＿	＿＿＿
Menstrual Cycle	＿＿＿
1.＿＿＿	＿＿＿
2.＿＿＿	＿＿＿
3.＿＿＿	＿＿＿

4. Menstrual flow is the result of
   (1) a deficiency of progesterone and estrogen.
   (2) a deficiency of estrogen.
   (3) an excess of progesterone and estrogen.
5. Ovulation usually (not always) occurs the following number of days before the next menstrual period:
   (1) three.　(2) seven.　(3) fourteen.　(4) eighteen.
6. Pain during ovulation, as experienced by some women, is known as
   (1) dysmenorrhea.　(2) oligomenorrhea.　(3) mittelschmerz.
7. The proliferative phase in the menstrual cycle begins at about the
   (1) second day.　(2) fifth day.　(3) seventh day of menstrual cycle.
8. Repair of the endometrium after menstruation is due to
   (1) estrogen.　(2) progesterone.　(3) luteinizing hormone.
9. Progesterone secretion ceases entirely within the following number of days after the onset of menses:
   (1) ten.　(2) fourteen.　(3) twenty-six.　(4) twenty-eight.
10. Excessive blood flow during menstruation is known as
    (1) oligomenorrhea.　(2) dysmenorrhea.　(3) menorrhagia.
11. Excessive discomfort and pain during menstruation is known as
    (1) dysmenorrhea.　(2) oligomenorrhea.　(3) menorrhagia.
12. Occasional or irregular menses is known as
    (1) amenorrhea.　(2) oligomenorrhea.　(3) dysmenorrhea.
13. Absence of menstruation is called
    (1) dysmenorrhea.　(2) oligomenorrhea.　(3) amenorrhea.

I. Medical

Select the type of surgery or condition that is described by the following statements:

anteflexion—1　　hysterectomy—6　　retroflexion—11
cryptorchidism—2　mastectomy—7　　salpingitis—12
endometritis—3　　oophorectomy—8　　salpingectomy—13
episiotomy—4　　oophorhysterectomy—9　syphilis—14
gonorrhea—5　　oophoroma—10　　vasectomy—15

1. Surgical removal of breast.
2. Failure of testes to descend into scrotum.
3. Surgical removal of uterus.
4. Spirochaetal sexually transmitted disease.
5. Sterilization procedure in males.
6. Type of incision made in perineum at childbirth to prevent excessive damage to anal sphincter.
7. Surgical removal of one or both ovaries.
8. Ovarian malignancy.
9. Sexually transmitted disease that affects the mucous membranes rather than the blood.
10. Surgical removal or sectioning of uterine tubes.
11. Most common sexually transmitted disease.
12. Inflammation of uterine wall.
13. Malpositioned uterus (2 types).
14. Sexually transmitted disease caused by a coccoidal (spherical) organism.
15. Surgical removal of ovaries and uterus.

Answers

Menstrual Cycle

4. _____
5. _____
6. _____
7. _____
8. _____
9. _____
10. _____
11. _____
12. _____
13. _____

Medical

1. _____
2. _____
3. _____
4. _____
5. _____
6. _____
7. _____
8. _____
9. _____
10. _____
11. _____
12. _____
13. _____
14. _____
15. _____

Table I International Atomic Weights

ELEMENT	SYMBOL	ATOMIC NUMBER	ATOMIC WEIGHT
Aluminum	Al	13	26.97
Antimony	Sb	51	121.76
Arsenic	As	33	74.91
Barium	Ba	56	137.36
Beryllium	Be	4	9.013
Bismuth	Bi	83	209.00
Boron	B	5	10.82
Bromine	Br	35	79.916
Cadmium	Cd	48	112.41
Calcium	Ca	20	40.08
Carbon	C	6	12.010
Chlorine	Cl	17	35.457
Chromium	Cr	24	52.01
Cobalt	Co	27	58.94
Copper	Cu	29	63.54
Fluorine	F	9	19.00
Gold	Au	79	197.2
Hydrogen	H	1	1.0080
Iodine	I	53	126.92
Iron	Fe	26	55.85
Lead	Pb	82	207.21
Magnesium	Mg	12	24.32
Manganese	Mn	25	54.93
Mercury	Hg	80	200.61
Nickel	Ni	28	58.69
Nitrogen	N	7	14.008
Oxygen	O	8	16.0000
Palladium	Pd	46	106.7
Phosphorus	P	15	30.98
Platinum	Pt	78	195.23
Potassium	K	19	39.096
Radium	Ra	88	226.05
Selenium	Se	34	78.96
Silicon	Si	14	28.06
Silver	Ag	47	107.880
Sodium	Na	11	22.997
Strontium	Sr	38	87.63
Sulfur	S	16	32.066
Tin	Sn	50	118.70
Titanium	Ti	22	47.90
Tungsten	W	74	183.92
Uranium	U	92	238.07
Vanadium	V	23	50.95
Zinc	Zn	30	65.38
Zirconium	Zr	40	91.22

Table II   Predicted Vital Capacities for Females

HEIGHT IN CENTIMETERS AND INCHES

AGE	CM 152 / IN 59.8	154 60.6	156 61.4	158 62.2	160 63.0	162 63.7	164 64.6	166 65.4	168 66.1	170 66.9	172 67.7	174 68.5	176 69.3	178 70.1	180 70.9	182 71.7	184 72.4	186 73.2	188 74.0
16	3,070	3,110	3,150	3,190	3,230	3,270	3,310	3,350	3,390	3,430	3,470	3,510	3,550	3,590	3,630	3,670	3,715	3,755	3,800
17	3,055	3,095	3,135	3,175	3,215	3,255	3,295	3,335	3,375	3,415	3,455	3,495	3,535	3,575	3,615	3,655	3,695	3,740	3,780
18	3,040	3,080	3,120	3,160	3,200	3,240	3,280	3,320	3,360	3,400	3,440	3,480	3,520	3,560	3,600	3,640	3,680	3,720	3,760
20	3,010	3,050	3,090	3,130	3,170	3,210	3,250	3,290	3,330	3,370	3,410	3,450	3,490	3,525	3,565	3,605	3,645	3,695	3,720
22	2,980	3,020	3,060	3,095	3,135	3,175	3,215	3,255	3,290	3,330	3,370	3,410	3,450	3,490	3,530	3,570	3,610	3,650	3,685
24	2,950	2,985	3,025	3,065	3,100	3,140	3,180	3,220	3,260	3,300	3,335	3,375	3,415	3,455	3,490	3,530	3,570	3,610	3,650
26	2,920	2,960	3,000	3,035	3,070	3,110	3,150	3,190	3,230	3,265	3,300	3,340	3,380	3,420	3,455	3,495	3,530	3,570	3,610
28	2,890	2,930	2,965	3,000	3,040	3,070	3,115	3,155	3,190	3,230	3,270	3,305	3,345	3,380	3,420	3,460	3,495	3,535	3,570
30	2,860	2,895	2,935	2,970	3,010	3,045	3,085	3,120	3,160	3,195	3,235	3,270	3,310	3,345	3,385	3,420	3,460	3,495	3,535
32	2,825	2,865	2,900	2,940	2,975	3,015	3,050	3,090	3,125	3,160	3,200	3,235	3,275	3,310	3,350	3,385	3,425	3,460	3,495
34	2,795	2,835	2,870	2,910	2,945	2,980	3,020	3,055	3,090	3,130	3,165	3,200	3,240	3,275	3,310	3,350	3,385	3,425	3,460
36	2,765	2,805	2,840	2,875	2,910	2,950	2,985	3,020	3,060	3,095	3,130	3,165	3,205	3,240	3,275	3,310	3,350	3,385	3,420
38	2,735	2,770	2,810	2,845	2,880	2,915	2,950	2,990	3,025	3,060	3,095	3,130	3,170	3,205	3,240	3,275	3,310	3,350	3,385
40	2,705	2,740	2,775	2,810	2,850	2,885	2,920	2,955	2,990	3,025	3,060	3,095	3,135	3,170	3,205	3,240	3,275	3,310	3,345
42	2,675	2,710	2,745	2,780	2,815	2,850	2,885	2,920	2,955	2,990	3,025	3,060	3,100	3,135	3,170	3,205	3,240	3,275	3,310
44	2,645	2,680	2,715	2,750	2,785	2,820	2,855	2,890	2,925	2,960	2,995	3,030	3,060	3,095	3,130	3,165	3,200	3,235	3,270
46	2,615	2,650	2,685	2,715	2,750	2,785	2,820	2,855	2,890	2,925	2,960	2,995	3,030	3,060	3,095	3,130	3,165	3,200	3,235
48	2,585	2,620	2,650	2,685	2,715	2,750	2,785	2,820	2,855	2,890	2,925	2,960	2,995	3,030	3,060	3,095	3,130	3,160	3,195
50	2,555	2,590	2,625	2,655	2,690	2,720	2,755	2,785	2,820	2,855	2,890	2,925	2,955	2,990	3,025	3,060	3,090	3,125	3,155
52	2,525	2,555	2,590	2,625	2,655	2,690	2,720	2,755	2,790	2,820	2,855	2,890	2,925	2,955	2,990	3,020	3,055	3,090	3,125
54	2,495	2,530	2,560	2,590	2,625	2,655	2,690	2,720	2,755	2,790	2,820	2,855	2,885	2,920	2,950	2,985	3,020	3,050	3,085
56	2,460	2,495	2,525	2,560	2,590	2,625	2,655	2,690	2,720	2,755	2,790	2,820	2,855	2,885	2,920	2,950	2,980	3,015	3,045
58	2,430	2,460	2,495	2,525	2,560	2,590	2,625	2,655	2,690	2,720	2,750	2,785	2,815	2,850	2,880	2,920	2,945	2,975	3,010
60	2,400	2,430	2,460	2,495	2,525	2,560	2,590	2,625	2,655	2,685	2,720	2,750	2,780	2,810	2,845	2,875	2,915	2,940	2,970
62	2,370	2,405	2,435	2,465	2,495	2,525	2,560	2,590	2,620	2,655	2,685	2,715	2,745	2,775	2,810	2,840	2,870	2,900	2,935
64	2,340	2,370	2,400	2,430	2,465	2,495	2,525	2,555	2,585	2,620	2,650	2,680	2,710	2,740	2,770	2,805	2,835	2,865	2,895
66	2,310	2,340	2,370	2,400	2,430	2,460	2,495	2,525	2,555	2,585	2,615	2,645	2,675	2,705	2,735	2,765	2,800	2,825	2,860
68	2,280	2,310	2,340	2,370	2,400	2,430	2,460	2,490	2,520	2,550	2,580	2,610	2,640	2,670	2,700	2,730	2,760	2,795	2,820
70	2,250	2,280	2,310	2,340	2,370	2,400	2,425	2,455	2,485	2,515	2,545	2,575	2,605	2,635	2,665	2,695	2,725	2,755	2,780
72	2,220	2,250	2,280	2,310	2,335	2,365	2,395	2,425	2,455	2,480	2,510	2,540	2,570	2,600	2,630	2,660	2,685	2,715	2,745
74	2,190	2,220	2,245	2,275	2,305	2,335	2,360	2,390	2,420	2,450	2,475	2,505	2,535	2,565	2,590	2,620	2,650	2,680	2,710

From: Archives of Environmental Health
February 1966, Vol. 12, pp. 146–189
E. A. Gaensler, MD and G. W. Wright, MD

Table III   Predicted Vital Capacities for Males

HEIGHT IN CENTIMETERS AND INCHES

AGE	CM 152 / IN 59.8	154 / 60.6	156 / 61.4	158 / 62.2	160 / 63.0	162 / 63.7	164 / 64.6	166 / 65.4	168 / 66.1	170 / 66.9	172 / 67.7	174 / 68.5	176 / 69.3	178 / 70.1	180 / 70.9	182 / 71.7	184 / 72.4	186 / 73.2	188 / 74.0
16	3,920	3,975	4,025	4,075	4,130	4,180	4,230	4,285	4,335	4,385	4,440	4,490	4,540	4,590	4,645	4,695	4,745	4,800	4,850
18	3,890	3,940	3,995	4,045	4,095	4,145	4,200	4,250	4,300	4,350	4,405	4,455	4,505	4,555	4,610	4,660	4,710	4,760	4,815
20	3,860	3,910	3,960	4,015	4,065	4,115	4,165	4,215	4,265	4,320	4,370	4,420	4,470	4,520	4,570	4,625	4,675	4,725	4,775
22	3,830	3,880	3,930	3,980	4,030	4,080	4,135	4,185	4,235	4,285	4,335	4,385	4,435	4,485	4,535	4,585	4,635	4,685	4,735
24	3,785	3,835	3,885	3,935	3,985	4,035	4,085	4,135	4,185	4,235	4,285	4,330	4,380	4,430	4,480	4,530	4,580	4,630	4,680
26	3,755	3,805	3,855	3,905	3,955	4,000	4,050	4,100	4,150	4,200	4,250	4,300	4,350	4,395	4,445	4,495	4,545	4,595	4,645
28	3,725	3,775	3,820	3,870	3,920	3,970	4,020	4,070	4,115	4,165	4,215	4,265	4,310	4,360	4,410	4,460	4,510	4,555	4,605
30	3,695	3,740	3,790	3,840	3,890	3,935	3,985	4,035	4,080	4,130	4,180	4,230	4,275	4,325	4,375	4,425	4,470	4,520	4,570
32	3,665	3,710	3,760	3,810	3,855	3,905	3,950	4,000	4,050	4,095	4,145	4,195	4,240	4,290	4,340	4,385	4,435	4,485	4,530
34	3,620	3,665	3,715	3,760	3,810	3,855	3,905	3,950	4,000	4,045	4,095	4,140	4,190	4,225	4,285	4,330	4,380	4,425	4,475
36	3,585	3,635	3,680	3,730	3,775	3,825	3,870	3,920	3,965	4,010	4,060	4,105	4,155	4,200	4,250	4,295	4,340	4,390	4,435
38	3,555	3,605	3,650	3,695	3,745	3,790	3,840	3,885	3,930	3,980	4,025	4,070	4,120	4,165	4,210	4,260	4,305	4,350	4,400
40	3,525	3,575	3,620	3,665	3,710	3,760	3,805	3,850	3,900	3,945	3,990	4,035	4,085	4,130	4,175	4,220	4,270	4,315	4,360
42	3,495	3,540	3,590	3,635	3,680	3,725	3,770	3,820	3,865	3,910	3,955	4,000	4,050	4,095	4,140	4,185	4,230	4,280	4,325
44	3,450	3,495	3,540	3,585	3,630	3,675	3,725	3,770	3,815	3,860	3,905	3,950	3,995	4,040	4,085	4,130	4,175	4,220	4,270
46	3,420	3,465	3,510	3,555	3,600	3,645	3,690	3,735	3,780	3,825	3,870	3,915	3,960	4,005	4,050	4,095	4,140	4,185	4,230
48	3,390	3,435	3,480	3,525	3,570	3,615	3,655	3,700	3,745	3,790	3,835	3,880	3,925	3,970	4,015	4,060	4,105	4,150	4,190
50	3,345	3,390	3,430	3,475	3,520	3,565	3,610	3,650	3,695	3,740	3,785	3,830	3,870	3,915	3,960	4,005	4,050	4,090	4,135
52	3,315	3,353	3,400	3,445	3,490	3,530	3,575	3,620	3,660	3,705	3,750	3,795	3,835	3,880	3,925	3,970	4,010	4,055	4,100
54	3,285	3,325	3,370	3,415	3,455	3,500	3,540	3,585	3,630	3,670	3,715	3,760	3,800	3,845	3,890	3,930	3,975	4,020	4,060
56	3,255	3,295	3,340	3,380	3,425	3,465	3,510	3,550	3,595	3,640	3,680	3,725	3,765	3,810	3,850	3,895	3,940	3,980	4,025
58	3,210	3,250	3,290	3,335	3,375	3,420	3,460	3,500	3,545	3,585	3,630	3,670	3,715	3,755	3,800	3,840	3,880	3,925	3,965
60	3,175	3,220	3,260	3,300	3,345	3,385	3,430	3,470	3,500	3,555	3,595	3,635	3,680	3,720	3,760	3,805	3,845	3,885	3,930
62	3,150	3,190	3,230	3,270	3,310	3,350	3,390	3,440	3,480	3,520	3,560	3,600	3,640	3,680	3,730	3,770	3,810	3,850	3,890
64	3,120	3,160	3,200	3,240	3,280	3,320	3,360	3,400	3,440	3,490	3,530	3,570	3,610	3,650	3,690	3,730	3,770	3,810	3,850
66	3,070	3,110	3,150	3,190	3,230	3,270	3,310	3,350	3,390	3,430	3,470	3,510	3,550	3,600	3,640	3,680	3,720	3,760	3,800
68	3,040	3,080	3,120	3,160	3,200	3,240	3,280	3,320	3,360	3,400	3,440	3,480	3,520	3,560	3,600	3,640	3,680	3,720	3,760
70	3,010	3,050	3,090	3,130	3,170	3,210	3,250	3,290	3,330	3,370	3,410	3,450	3,480	3,520	3,560	3,600	3,640	3,680	3,720
72	2,980	3,020	3,060	3,100	3,140	3,180	3,210	3,250	3,290	3,330	3,370	3,410	3,450	3,490	3,530	3,570	3,610	3,650	3,680
74	2,930	2,970	3,010	3,050	3,090	3,130	3,170	3,200	3,240	3,280	3,320	3,360	3,400	3,440	3,470	3,510	3,550	3,590	3,630

From: Archives of Environmental Health
February 1966, Vol. 12, pp. 146–189
E. A. Gaensler, MD and G. W. Wright, MD

B Appendix
Tests and Methods

Carbohydrate Differentiation

Several experiments in this book require that sugars and starches be differentiated. An understanding of the application of these tests can be derived from the separation outline at the bottom of this page.

Barfoed's Test   To 5 ml of reagent in test tube, add 1 ml of unknown. Place in boiling water bath for 5 minutes. Interpretation is based on separation outline.

Benedict's Test   This is a semiquantitative test for amounts of reducing sugar present. To 5 ml of reagent in test tube, add 8 drops of unknown. Place in boiling water bath for 5 minutes. A green, yellow, or orange-red precipitate determines the amount of reducing sugar present. See Ex. 65.

Electronic Equipment Adjustments

Lengthy instructions pertaining to recorder and transducer manipulations that are used in several experiments are outlined here for reference.

Balancing Transducer on Unigraph
(Ex. 22)

The Unigraph is designed to function in a sensitivity range of 0.1 MV/CM to 2 MV/CM. When an experiment is begun, the sensitivity control is set at 2 MV/CM, its least sensitive position. As the experiment progresses it may become necessary to shift to a more sensitive setting, such as 1 MV/CM or 0.5 MV/CM. If the Wheatstone bridge in the transducer is not balanced, the shifting from one sensitivity position to another will cause the stylus to change position. This produces an unsatisfactory record. Thus, it is essential that one go through the following steps to balance the transducer:

1. Connect the transducer to the transducer jack in the end of the Unigraph. The small upper socket is the one to use.
2. Before plugging in the Unigraph, make sure that the power switch is off, the chart control switch is on STBY, the gain selector knob is on 2 MV/CM, the sensitivity knob is turned completely counterclockwise, the mode selector control is on TRANS, and the Hi Filter and Mean switches are positioned toward Normal. Now plug in the Unigraph.

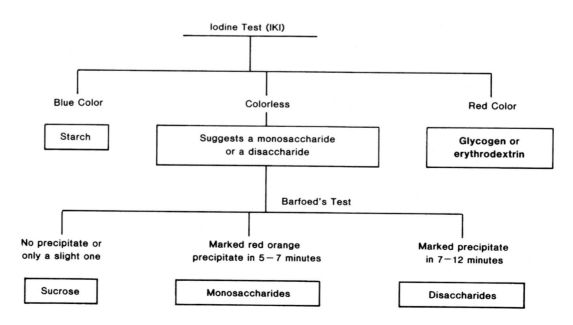

3. Turn on the Unigraph power switch and set the heat control knob at the two o'clock position.
4. Place the speed control lever at the slow position and the chart control (c.c.) lever at Chart On.
5. As the chart moves along, bring the stylus to the center of the paper by turning the centering knob.
6. Turn the sensitivity control completely clockwise to maximum sensitivity. If the bridge is unbalanced, the stylus will move away from the center. To return the stylus to the center, unlock the TRANS-BAL control by pushing the small lever on the control and turning the knob in either direction to return the stylus to the center of the chart.
7. Set the gain control to 1. If the bridge is unbalanced, the stylus will not be centered. Center with the TRANS-BAL control.
8. Set the gain control to 0.5 and recenter the stylus with the TRANS-BAL control. Repeat this same procedure for 0.2 and 0.1 settings of the gain control. The bridge is now completely balanced at its highest gain and sensitivity. Lock the TRANS-BAL control with the lock lever.

Note: Although we have gone through this lengthy process to achieve a balanced transducer, we may find that we will still get baseline shifts when going from one setting to another. However, they will be minor compared to one that has not been through this procedure. Any minor deviations should be corrected with the centering control.

9. Return the gain control to 2 and the chart control switch to STBY.

Calibration of Unigraph with Strain Gauge (Ex. 22)

When the transducer has been balanced, it is necessary to calibrate the tracing on the Unigraph to a known force on the transducer. Calibration establishes a linear relationship between the magnitude of deflection and the degree of strain (extent of bending) of the transducer leaf before the muscle is attached to the transducer. Calibrations must be made for each gain at maximum sensitivity that one expects to use in the experiment (your muscle preparations will probably require gain settings of 2, 1, or 0.5). After calibration, it is possible to directly determine the force of each contraction by simply reading the maximum height of the tracing.

Materials:
small paper clip weighing around 0.5 gram

1. Weigh a convenient object, such as a small paper clip, to the nearest 0.1 mg.
2. With the chart control switch on STBY, suspend the paper clip on the two smallest leaves of the transducer.
3. Put the c.c. switch to Chart On and note the extent of deflection on the chart. Record about 1 centimeter on the chart and place the c.c. switch on STBY. Label this deflection in mm/mg.
4. Set the gain control at 1, c.c. to Chart On, and record for another centimeter. Place the c.c. switch on STBY. Label the deflection.
5. Repeat the above procedure for the other two gain settings (0.5, 0.2, and 0.1). Note how the stylus vibrates as you increase the sensitivity of the Unigraph. This is normal.
6. Return the c.c. switch to STBY, gain to 2, and remove the paper clip. The Unigraph is now calibrated and the muscle can now be attached to it.

Calibration for EMG on Unigraph (Ex. 24)

Calibration of the Unigraph for EMG measurements is necessary to establish that 1 centimeter deflection of the stylus equals 0.1 millivolt. When calibration is accomplished, the recording is made with the EEG mode since there is no EMG mode. An EMG can be made very well in this mode at lower sensitivities than would be used when making an EEG. Proceed as follows to calibrate:

1. Turn the mode selector control to CC/Cal (Capacitor Coupled Calibrate).
2. Turn the small main switch to ON.
3. Turn the stylus heat control to the two o'clock position.
4. Check the speed selector lever to see that it is in the slow position. The free end of the lever should be pointed toward the styluses.
5. Move the chart-stylus switch to Chart On and observe the width of the tracing that appears on the paper. Adjust the stylus heat control to produce the desired width of tracing.
6. Set the gain knob to 0.1 MV/CM.
7. Push down the 0.1 MV button to determine the amount of deflection. Hold the button down for 2 seconds before releasing. If it is released too quickly the tracing will not be perfect. Adjust the deflection so that it deflects exactly 1 centimeter in each direction by turning the sensitivity knob. The unit is now calibrated so that you get 1 cm deflection for 0.1 millivolt.

8. Reset the gain knob to 0.5 MV/CM. This reduces the sensitivity of the instrument.
9. Return the mode selector control to EEG.
10. Return the chart-stylus switch to STBY. Place the speed selector in the fast position. The instrument is now ready for use.

Balancing the Myograph to Physiograph® *Transducer Coupler* (Ex. 22)

Balancing (matching) of the transducer signal with the amplifier is necessary to get the recording pen to stay on a preset baseline when the RECORD button is in either the OFF or the ON position. Proceed as follows:

1. Before starting, make sure that the RECORD switch is OFF and that the myograph jack is inserted into the transducer coupler.
2. Set the sensitivity control (outer knob) to its lowest numbered setting (highest sensitivity).

3. Set the paper speed at 0.5 cm/sec and lower the pen lifter.
4. Adjust the pen position so that the pen is writing *exactly* on the center line.
5. Place the RECORD switch in the ON position. The pen will probably be moving a large distance either up or down.
6. With the BALANCE control, adjust the pen so that it is again writing *exactly* on the center line.
7. Check your match, or balance, by placing the RECORD button in the OFF position. **Your system is balanced if the pen remains on the center line.**
8. If the pen does not remain on the center line, relocate the pen to the center line again with the position control, place the RECORD button in the ON position, and repeat the balancing procedure. You have a balanced system *only* when the pen remains on a preset baseline with the RECORD button in both OFF and ON positions.

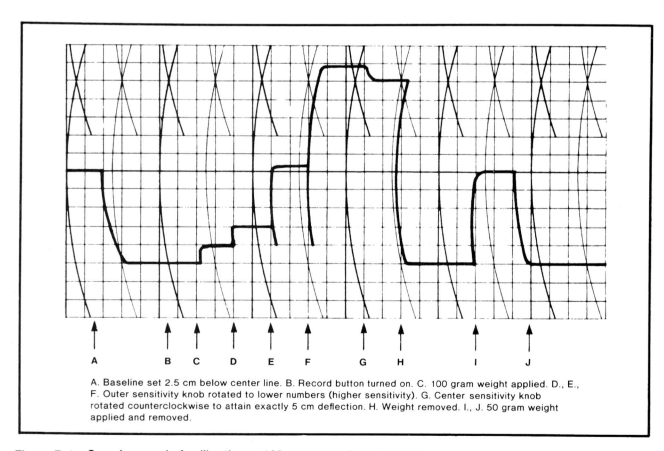

A. Baseline set 2.5 cm below center line. B. Record button turned on. C. 100 gram weight applied. D., E., F. Outer sensitivity knob rotated to lower numbers (higher sensitivity). G. Center sensitivity knob rotated counterclockwise to attain exactly 5 cm deflection. H. Weight removed. I., J. 50 gram weight applied and removed.

Figure B-1   Sample record of calibration at 100 grams per 5 centimeters.

Calibration of Myograph Transducer on Physiograph® (Ex. 22)

Calibration of the transducer is necessary so that we know exactly how much tension, in grams, is being exerted by the muscle during contraction. The channel amplifier must be balanced first. Our calibration here will be to get 5 cm pen deflection with 100 grams. *These instructions apply to couplers that lack built-in calibration buttons.* Figure B-1 on the previous page illustrates the various steps.

Materials:
  100 gram weight
  50 gram weight

1. Start the paper drive at 0.1 cm/sec and lower the pen lifter so that we are recording on the desired channel.
2. Check to see that the channel amplifier is balanced.
3. With the pen position control, set the baseline exactly 2.5 cm (5 blocks) below the channel center line.
4. Rotate the outer knob of the sensitivity control fully counterclockwise to the lowest sensitivity level (i.e., 1000 setting). Be sure that the inner knob of the sensitivity control is in its fully clockwise "clicked" position.
5. Place the RECORD button in the ON position.
6. Suspend a 100 gram weight to the actuator of the myograph. Note that the addition of the weight has caused the baseline to move upward approximately one block (0.5 cm). See step C, figure B-1.
7. Rotate the *outer* knob of the sensitivity control clockwise until the pen exceeds 5 cm (10 blocks) of deflection from the original baseline. See step F, figure B-1.
8. Rotate the *inner* knob of the sensitivity control counterclockwise to bring the pen back downward until it is *exactly* 5 cm (10 blocks) of deflection from the original baseline.
9. Remove the weight. The pen should return to the original baseline. If it does not return exactly to the baseline, reset the baseline with the position control, reapply the weight, and rotate the inner knob until the pen is writing exactly 5 cm above the original baseline.
10. Attach a 50 gram weight to the myograph to see if you get 5 blocks of pen deflection, as shown in the last deflection in figure B-1. The unit is now properly calibrated.

Calibration of Myograph Transducer with Calibration Button (Ex. 22)

When the coupler has a calibration button, the use of a 100 gram weight is not necessary. To calibrate, all that is necessary is to press the 100 gram button, and follow the steps above to produce the 5 cm pen deflection.

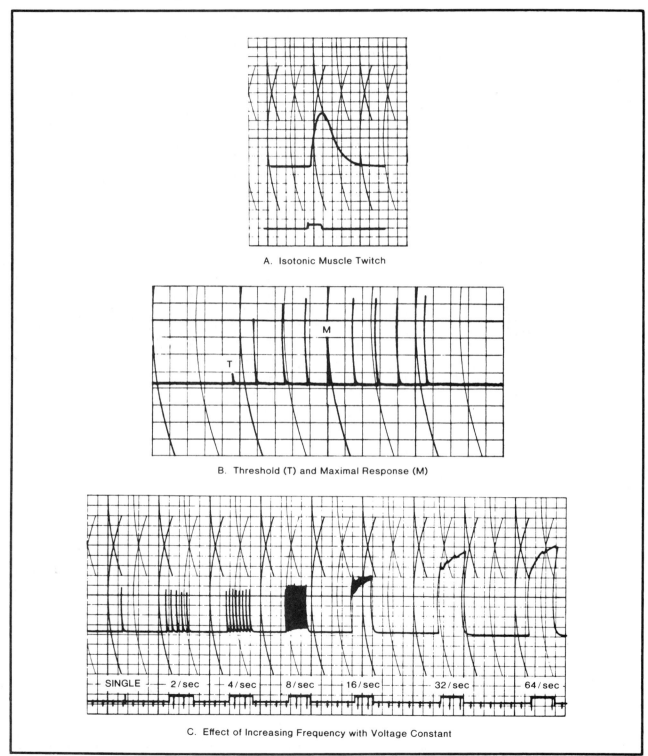

A. Isotonic Muscle Twitch

B. Threshold (T) and Maximal Response (M)

SINGLE — 2/sec — 4/sec — 8/sec — 16/sec — 32/sec — 64/sec

C. Effect of Increasing Frequency with Voltage Constant

Figure SR-1 Physiograph® sample records of frog muscle contraction (Ex. 22).

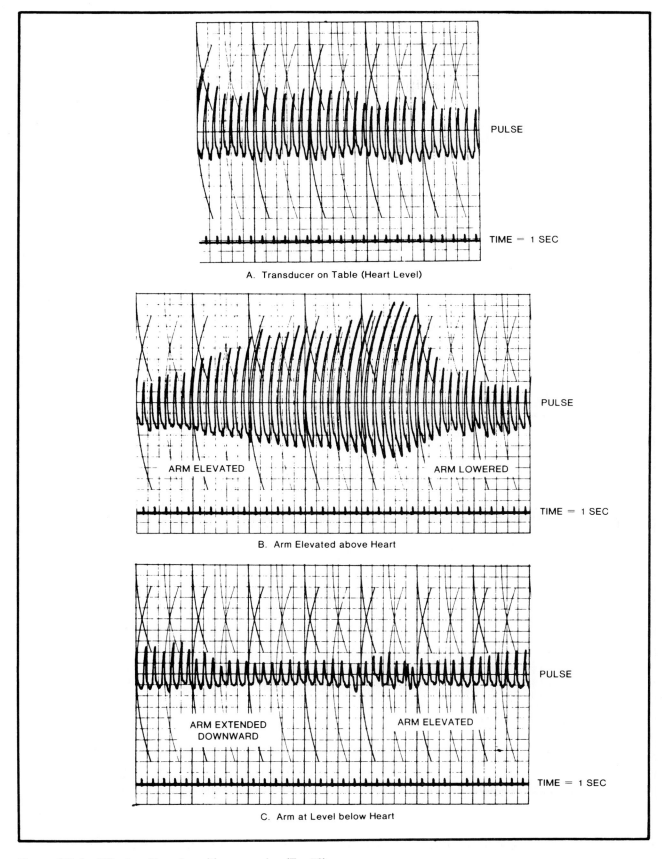

A. Transducer on Table (Heart Level)

B. Arm Elevated above Heart

C. Arm at Level below Heart

Figure SR-2 Effects of hand position on pulse (Ex. 50).

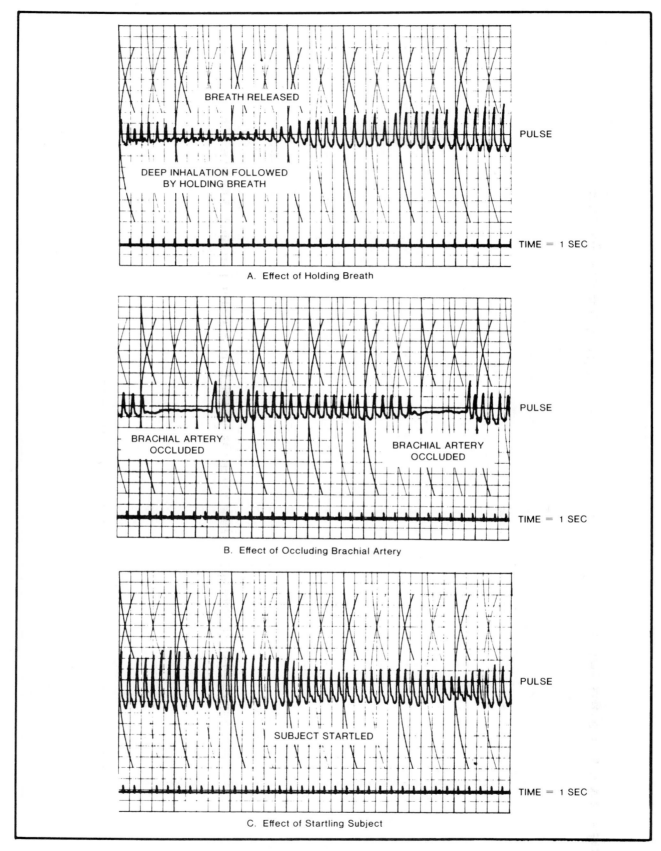

BREATH RELEASED

PULSE

DEEP INHALATION FOLLOWED
BY HOLDING BREATH

TIME = 1 SEC

A. Effect of Holding Breath

PULSE

BRACHIAL ARTERY
OCCLUDED

BRACHIAL ARTERY
OCCLUDED

TIME = 1 SEC

B. Effect of Occluding Brachial Artery

PULSE

SUBJECT STARTLED

TIME = 1 SEC

C. Effect of Startling Subject

Figure SR-3   Pulse variations due to various factors (Ex. 50).

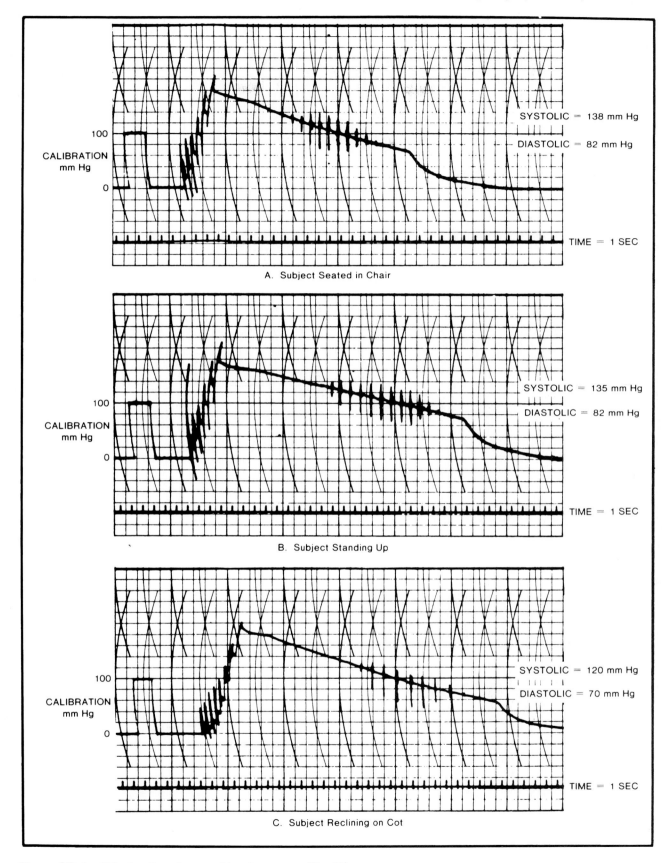

A. Subject Seated in Chair

B. Subject Standing Up

C. Subject Reclining on Cot

Figure SR-4 Effects of posture on blood pressure (Ex. 51).

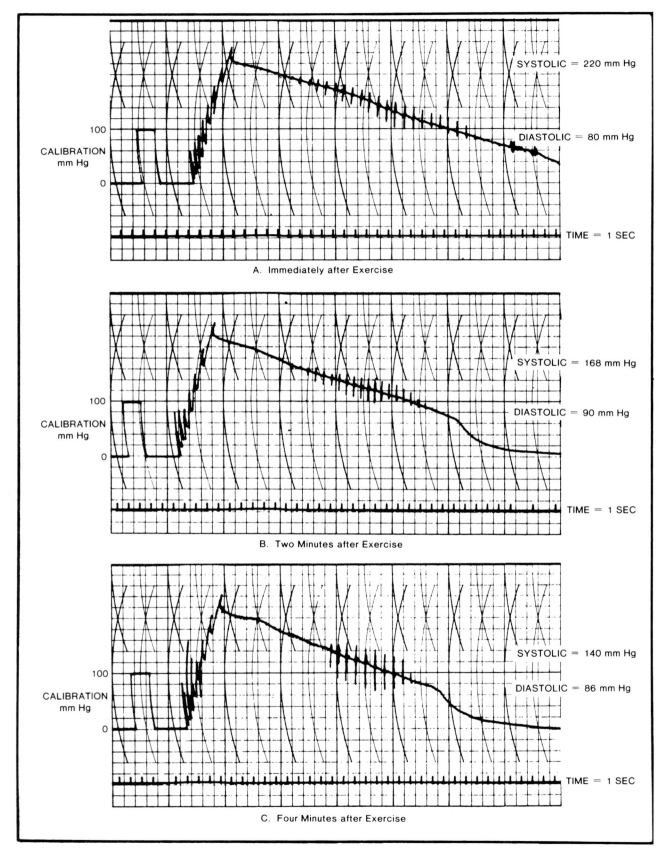

A. Immediately after Exercise

B. Two Minutes after Exercise

C. Four Minutes after Exercise

Figure SR-5   Sample records of blood pressure monitoring after exercise (Ex. 51).

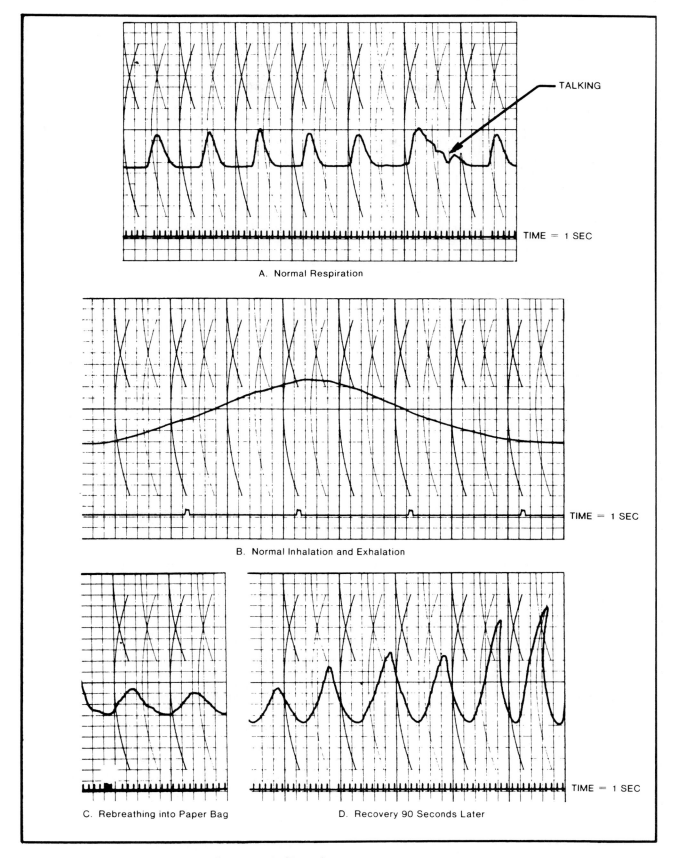

TALKING

TIME = 1 SEC

A. Normal Respiration

TIME = 1 SEC

B. Normal Inhalation and Exhalation

TIME = 1 SEC

C. Rebreathing into Paper Bag

D. Recovery 90 Seconds Later

Figure SR-6 Pneumograph recording records (Ex. 57).

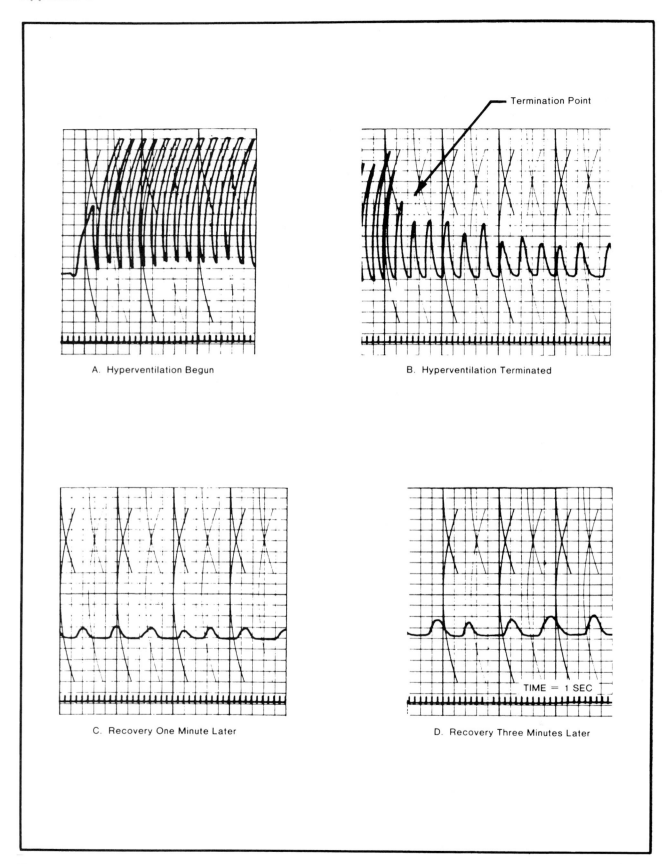

A. Hyperventilation Begun

B. Hyperventilation Terminated

Termination Point

C. Recovery One Minute Later

D. Recovery Three Minutes Later

TIME = 1 SEC

Figure SR-7   Hyperventilation sample record (Ex. 57).

Reading References

Anatomy and Physiology

Hole, John W., Jr., 1994. *Human anatomy and physiology*. 5th ed. Dubuque, IA: WCB/McGraw-Hill.

Jacob, Stanley W., and Ashworth, Clarice A. 1989. *Elements of anatomy and physiology*. 2d ed. Philadelphia: W. B. Saunders Co.

Tortora, Gerald J., and Garbrowski, Sandra R., 1996. *Principles of anatomy and physiology*. 7th ed. Reading, MA: Addison Wesley.

Van De Graaff, Kent M., and Fox, Stuart Ira, 1994. *Concepts of human anatomy and physiology*. 4th ed. Dubuque, IA: WCB/McGraw-Hill.

Anatomy

Clemente, Carmine D., et al., ed., 1987. *Anatomy: a regional atlas of the human body*. 3d ed. Baltimore: Williams & Wilkins.

Crouch, James E., 1983. *Introduction to human anatomy: a laboratory manual*. 6th ed. Baltimore: Williams & Wilkins.

Crouch, James E., and Carr, Micheline, 1989. *Anatomy and physiology: a laboratory manual*. Baltimore: Williams & Wilkins.

Greenblatt, Gordon M., 1981. *Cat musculature: a photographic atlas*. 2d ed. Chicago: Univ. of Chicago Press.

Netter, Frank H., 1983–1990. *The CIBA collection of medical illustrations*. 13 vols. Summit, NJ: CIBA Pharmaceutical Products, Inc.

Schlossberg, Leon, and Zuidema, George D., 1986. *The Johns Hopkins atlas of human functional anatomy*. 3d ed. Baltimore: Johns Hopkins.

Van De Graaff, Kent M., 1997. *Human anatomy*. 5th ed. Dubuque, IA: WCB/McGraw-Hill.

Physiology

Ganong, William, 1995. *Review of medical physiology*. 17th ed. Los Angeles: Appleton & Lang.

Guyton, Arthur C., and Hall, John E., 1995. *Textbook of medical physiology*. 9th ed. Philadelphia: W. B. Saunders Co.

Histology

Bergman, Ronald A., and Afifi, Adel K., 1974. *Atlas of microscopic anatomy*. Philadelphia: W. B. Saunders Co.

Cormack, David H., 1993. *Essential histology*. Philadelphia: Lippincott-Raven.

Cormack, David H., 1997. *Clinically integrated histology*. Philadelphia: Lippincott-Raven.

Chemistry

Campbell, P. N., and Marshall, R. D., 1985. *Essays in biochemistry*. Vol. 20. Orlando, FL: Academic Press.

Gilman, Alfred G., et al., eds., 1990. *The pharmacological basis of therapeutics*. 8th ed. New York: McGraw-Hill.

Richterich, Roland, and Columbo, J. P., 1981. *Clinical chemistry: theory, practice, and interpretation*. New York: Wiley Interscience.

Tietz, Norbert W., ed., 1997. *Clinical guide to laboratory tests*. 3d ed. Philadelphia: W. B. Saunders Co.

Instrumentation

Cobbold, Richard S., 1974. *Transducers for biochemical measurements: principles and applications*. New York: Wiley Interscience.

Dewhurst, D. J., 1975. *An introduction to biomedical instrumentation*. 2d ed. Elmsford, NY: Pergamon Press.

DuBorg, Joseph L., 1978. *Introduction to biomedical electronics*. New York: McGraw-Hill.

Geddes, L. A., 1991. *Handbook of blood pressure measurement*. Totowa, NJ.: Humana.

Geddes, L. A., and Baker, L. E., 1989. *Principles of applied medical instrumentation*. 3d ed. New York: Wiley Interscience.

Miller, H., and Harrison, D. C., 1974. *Biomedical electrode technology*. Orlando, FL: Academic Press.

Tischler, Morris, 1981. *Experiments in amplifiers, filters and oscillators*. New York: McGraw-Hill.

Tischler, Morris, 1981. *Experiments in general and biomedical instrumentation*. New York: McGraw-Hill.

Tischler, Morris, 1985. *Optoelectronics: a text-lab manual*. New York: McGraw-Hill.

Tischler, W. A., and Geddes, L. A. 1980. *Electrical defibrillation*. Reprint by CRC.

Muscle Physiology

Basmajian, John V., and Wolf, Steven, 1990. *Therapeutic exercise*. 5th ed. Baltimore: Williams & Wilkins.

Basmajian, John V., and MacConaill, M. A., 1977. *Muscles and movements: a basis for human kinesiology.* Rev. ed. Baltimore: Williams & Wilkins.

Jabre, Joe F., and Hackett, Earl R., 1983. *EMG manual.* Springfield, IL: C. C. Thomas.

Murray, J. M., and Weber, A., 1974. *The cooperative action of muscle proteins.* Scientific Offprint #1290. San Francisco: W. H. Freeman Co.

Murray, Jim, and Karpovich, Peter, 1982. *Weight training in athletics.* Englewood Cliffs, NJ: Prentice-Hall.

Shepard, Roy J. l, 1984. *Biochemistry of physical activity.* Springfield, IL: C. C. Thomas.

Shepard, Roy J., *Physiology and biochemistry of exercise.* New York: Praeger.

Eye and Ear

Browning, G. G. 1986. *Clinical otology and audiology.* Stoneham, MA: Butterworth.

Corboy, John M., 1995. *The retinoscopy book: an introductory manual for eye care professionals.* 4th ed. Thorofare, NJ: SLACK, Inc.

Masland, Richard H., December, 1986, Scientific American. *The functional architecture of the retina.* New York: Scientific American, Inc.

Schnapf, Julie L., and Baylor, Denis A., April, 1987, Scientific American. *How photoreceptor cells respond to light.* New York: Scientific American, Inc.

Miscellaneous

Adamovich, David R., 1984. *The heart: fundamentals of electrocardiography, exercise physiology, and exercise stress testing.* Hempstead, NY: Sports Medicine Books.

Basmajian, John V., 1989. *Biofeedback: principles and practice for clinicians.* 3d ed. Oradell, NJ: Medical Economics.

Diggs, L. W., et al. *The morphology of human blood cells,* 4th ed. Chicago: Abbott Laboratories.

Duke, Richard C., Ojcius, D. M., and Young, John, Ding-E, December, 1996, Scientific American. *Cell suicide in health and disease.* New York: Scientific American, Inc.

Hughes, John R., 1994. *EEG in clinical practice.* Stoneham, MA: Butterworth.

Kalashnikov, V., 1986. *Beat the box: the insider's guide to outwitting the lie detector.* Staten Island, NY: Gordon Press.

Keeley, Eloise, 1984. *Lie detector manual.* Marshfield, MA: TelShare Pub. Co.

Matte, James, 1980. *The art and science of the polygraph technique.* Springfield, IL: C. C. Thomas.

Rothman, James E., and Orci, Lelio, March, 1996, Scientific American. *Budding vesicles in living cells.* New York: Scientific American, Inc.

Short, Charles E., 1993. *Alpha-2 agents in animals sedation, analgesia and anesthesia.* Vet Practice.

Wintrobe, Maxwell M., et al., 1997. *Clinical hematology.* 10th ed. Baltimore: Williams & Wilkins.

Index